AF505922

High Calorie Diet and the Human Brain

Akhlaq A. Farooqui

High Calorie Diet and the Human Brain

Metabolic Consequences of Long-Term Consumption

 Springer

Akhlaq A. Farooqui
Columbus, OH, USA

ISBN 978-3-319-15253-0   ISBN 978-3-319-15254-7   (eBook)
DOI 10.1007/978-3-319-15254-7

Library of Congress Control Number: 2015932526

Springer Cham Heidelberg New York Dordrecht London

Printed on acid-free paper

Springer International Publishing AG Switzerland is part of Springer Science+Business Media
(www.springer.com)

This monograph is dedicated to my parents, teachers, and colleagues, whose guidance and influence continues to inspire me.

Akhlaq A. Farooqui

Preface

It is becoming increasingly evident that diet patterns not only produce healthy effects, but also modulate the onset and pathogenesis of metabolic diseases (obesity, type II diabetes, and metabolic syndrome) and age-related acute and chronic neurological disorders (stroke, Alzheimer disease, and depression). Thus, long-term consumption of high calorie diet, which is enriched in saturated fat, cholesterol, n-6 fatty acids, simple sugars, and salt, produces chronic oxidative stress and inflammation processes. These processes are accompanied by overproduction of reactive oxygen species, advanced glycation products, eicosanoids, and proinflammatory cytokines (tumor necrosis factor-α, interleukine-1β, and interleukin-6) in the visceral tissues and brain. Generation of above-mentioned mediators contributes to cognitive decline and neurodegeneration. In contrast, Mediterranean diet which is low in saturated fats and simple sugars, but high in olive oil, n-3 fatty acids, whole grains, fruits, and vegetables produces neuroprotective effects supporting the view that Mediterranean diet may be able to influence the onset and development of metabolic and neurological disorders. Like Mediterranean diet, the consumption of ketogenic diet also decreases or retards obesity, metabolic syndrome, diabetes, epilepsy amyotrophic lateral sclerosis, Alzheimer disease, Parkinson disease, and some mitochondriopathies. Metabolic and neurological diseases are accompanied by different biochemical, neurochemical, and neuropathological features. However, oxidative stress and inflammation are shared by both types of diseases.

Information on the effects of various diet patterns is scattered throughout the literature in the form of original papers, reviews, and some books. These books mainly describe effects of diet patterns on visceral organs. At present, there are no books on effects of long-term consumption of high calorie diet on the brain. The overarching objective of this monograph is to provide readers with a comprehensive and cutting edge information on the effects of long-term consumption of high calorie diet on the brain in a manner that is not only useful to students and teachers, but also to researchers, dietitians, nutritionists, exercise physiologists, and physicians. This monograph has ten chapters. Chapter 1 describes the effects of long-term consumption of high calorie diet and calorie restriction on health. Chapter 2 provides information on the neurochemical effects of long-term consumption of high fat diet on health. Chapter 3 deals with neurochemical effects of long-term consumption of simple carbohydrates containing diet on health. Chapter 4 focuses on cutting edge information on long-term consumption of animal protein-enriched diet on health. Chapter 5 describes the effect of long-term soft drink consumption on human health. Chapter 6 narrates the effect of long-term consumption of high salt in inducing biochemical changes in the brain. Chapter 7 describes the cutting edge information on the importance and roles of fiber in the diet. Chapter 8 provides readers with information on the effects of long-term consumption of high calorie diet on the development of chronic visceral disease. Chapter 9 deals with cutting edge information on effects of long-term consumption of high calorie diet on neurological disorders. Finally, Chap. 10 deals with perspective and direction for future research on modifications of high calorie diet needed for optimal health of human brain.

My presentation and demonstrated ability to present complicated information on signal transduction processes associated with the effects of various types of diet patterns and neurological disorders will make this book particularly accessible to neuroscience and food science graduate students, teachers, and fellow researchers. It can be used as a supplemental text for a range of neuroscience, nutrition, and biochemistry courses. Clinicians, neuroscientists, neurologists, dietitians, and nutritionists will find this book useful for understanding the molecular aspects of various types of diet patterns on human health. To the best of my knowledge, this monograph will be the first to provide a comprehensive description of signal transduction processes associated with long-term consumption of various types of diet patterns on the brain.

The choices of topics presented in this monograph are personal. They are not only based on my interest on the effects of diet on the brain, but also in areas where major progress has been made. The key objective of this monograph is to critically evaluate the effect of long-term consumption of each component of high calorie diet on metabolic processes in the brain. Each chapter of this monograph contains an extensive list of references, which are arranged alphabetically to works that are cited in the text. I have tried to ensure uniformity and mode of presentation as well as a logical progression of subjects from one topic to another and have provided an extensive bibliography. For the sake of simplicity and uniformity, a large number of figures with chemical structures of dietary components along with line diagrams of colored signal transduction pathways are also included. I hope that my attempt to integrate and consolidate the knowledge on the effect of long-term consumption of high calorie diet will initiate more studies on molecular mechanisms associated with beneficial effects of various diet patterns on human health in general and brain in particular. This knowledge will be useful for the optimal health of young, boomer, and pre-boomer American generations.

Columbus, OH   Akhlaq A. Farooqui

Acknowledgments

I thank my wife, Tahira, for critical reading of this monograph, offering valuable advice, useful discussion, and evaluation of subject matter. Without her help and participation, this monograph neither could nor would have been completed. I would also like to express my gratitude to Gregory Baer and Melissa Higgs of Springer, New York, for their quick responses to my queries and professional manuscript handling.

Akhlaq A. Farooqui

Contents

Abbreviations

AD	Alzheimer disease
AGE	Advanced glycation endproducts
ALS	Amyotrophic lateral sclerosis
APP	Amyloid precursor protein
ARA	Arachidonic acid
BBB	Blood—brain barrier
CAT	Catalase
COX	Cyclooxygenase
CR	Calorie restriction
DHA	Docosahexaenoic acid
EPA	Eicosapentaenoic acid
EPOX	Epoxygenase
GPx	Glutathione peroxidase
HD	Huntington disease
IGF	Insulin growth factor
IL	Interleukin
Ins-1,4,5-P_3	Inositol-1,4,5-trisphosphate
LOX	Lipoxygenase
LTP	Long-term potentiation
MetS	Metabolic syndrome
NO	Nitric oxide
PD	Parkinson disease
PKC	Protein kinase C
PLA$_2$	Phospholipase A$_2$
PtdCho	Phosphatidylcholine
PtdEtn	Phosphatidylethanolamine
PtdIns	Phosphatidylinositol
PtdIns(4,5)P_2	Phosphatidylinositol 4,5-bisphosphate
PtdIns4P	Phosphatidylinositol 4-phosphate

PUFA	Polyunsaturated fatty acid
RAGE	Receptor for advanced glycation endproducts
SOD	Superoxide dismutase
TLR	Toll-like receptor
TNF-α	Tumor necrosis factor-alpha

About the Author

Akhlaq A. Farooqui is a leader in the field of signal transduction, brain phospholipases A_2, bioactive ether lipid metabolism, polyunsaturated fatty acid metabolism, glycerophospholipid-, sphingolipid-, and cholesterol-derived lipid mediators, glutamate-induced neurotoxicity, and modulation of signal transduction by phytochemicals. Dr. Farooqui has discovered the stimulation of plasmalogen-selective phospholipase A_2 (PlsEtn-PLA$_2$) and diacyl- and monoacylglycerol lipases in brains from patients with Alzheimer disease. Stimulation of PlsEtn-PLA$_2$ produces plasmalogen deficiency and increases levels of eicosanoids that may be related to the loss of synapses in brains of patients with Alzheimer disease. Dr. Farooqui has published cutting edge research on the generation and identification of glycerophospholipid-, sphingolipid-, and cholesterol-derived lipid mediators in kainic acid-mediated neurotoxicity by lipidomics. Dr. Farooqui has authored ten monographs: Glycerophospholipids in Brain: Phospholipase A_2 in Neurological Disorders (2007), Neurochemical Aspects of Excitotoxicity (2008), Metabolism and Functions of Bioactive Ether Lipids in Brain (2008), and Hot Topics in Neural Membrane Lipidology (2009), Beneficial Effects of Fish Oil in Human Brain (2009), Neurochemical Aspects of Neurotraumatic and Neurodegenerative Diseases (2010), Lipid Mediators and their Metabolism in the Brain (2011), Phytochemicals, Signal Transduction, and Neurological Disorders; Metabolic Syndrome: An Important Risk Factor for Stroke, Alzheimer Disease, and Depression and Inflammation and Oxidative Stress in Neurological Disorders. All monographs are published by Springer, New York.

In addition, Dr. Akhlaq A. Farooqui has edited seven books (Biogenic Amines: Pharmacological, Neurochemical and Molecular Aspects in the CNS (2010) Nova Science Publisher, Hauppauge, N.Y.; Molecular Aspects of Neurodegeneration and Neuroprotection, Bentham Science Publishers Ltd. (2011); Phytochemicals and Human Health: Molecular and pharmacological Aspects (2011), Nova Science Publisher, Hauppauge, N.Y.; Molecular Aspects of Oxidative Stress on Cell Signaling in Vertebrates and Invertebrates (2012), Wiley

Blackwell Publishing Company, New York; Beneficial effects of propolis on Human Health in Chronic Diseases (2012) Vol 1, Nova Science Publishers, Hauppaauge, New York; Beneficial effects of propolis on Human Health in Chronic Diseases (2012) Vol 2, Nova Science Publishers, Hauppaauge, New York; and Metabolic Syndrome and Neurological Disorders (2013) Wiley Blackwell Publishing Company, New York.

Chapter 1
Effect of Long Term Consumption of High Calorie Diet and Calorie Restriction on Human Health

1.1 Introduction

Diet in sufficient amount and composition is essential for good health in humans. Total dietary energy intake is modulated by food portion size (Rolls et al. 2006), food energy density (Rolls 2010), the number of eating occasions (Bellisle et al. 1997; Popkin and Duffey 2010), and the energy intake from energy-rich beverages (sugary and alcoholic drinks) (Pan and Hu 2011; Yeomans 2010). Diet and gene interactions play a central role in maintaining health throughout life. Major functions of diet are not only to provide energy and building material for the growth, and development, but also to maintain cellular functions for survival. The effects of diet on gene expression are exerted throughout the life-cycle, with prenatal and early postnatal life being especially critical periods for optimal development. Changes in gene expression may be dynamic and short-term, stable and long-term, and even heritable between cell divisions and across generations. Palatable foods activate the same brain regions of reward and pleasure that are active in drug addiction (Volkow et al. 2012), supporting the view that neuronal mechanism of food addiction may lead to overeating and obesity (Davis 2014; Dileone et al. 2012; Volkow et al. 2012; Ziauddeen and Fletcher 2013; Potenza 2014). Dopamine, a neurotransmitter, which directly activates reward and pleasure centers influences both mood and food intake (Black et al. 2002; Cawley et al. 2013). These observations further support the link between psychology and eating behaviors. It is well known that rodents and humans repeatedly eat a particular food to experience this positive feeling of gratification. This type of repetitive behavior of food consumption induces the activation of brain reward pathways that eventually overrides other signals of satiety and hunger (Singh 2014). Thus, a gratification habit through a favorable food may cause overeating and morbid obesity. Overeating and obesity stems from many biological factors engaging both central and peripheral systems in a bi-directional manner involving mood and emotions. Emotional eating and altered mood can also lead to altered food choice and intake leading to overeating and obesity (Singh 2014). Major components of diet

© Springer International Publishing Switzerland 2015
A.A. Farooqui, *High Calorie Diet and the Human Brain*,
DOI 10.1007/978-3-319-15254-7_1

(carbohydrate, lipids, and proteins) have ability to modulate genes associated with protection against acute and chronic diseases associated with aging (Farooqui 2013). If diet intake does not meet cellular nutritional needs, cellular metabolism and function slows down or even stop resulting into death (Farooqui 2013).

Effects of diet on brain and cognitive function are multifactorial and very complex. These effects not only involve the impact of a single dietary component or chemical on the brain but with numerous interactions between multiple nutrients, metabolites, foods, and other environmental and genetic factors (Dauncey 2012; Gomez-Pinilla and Tyagi 2013). In addition to major macronutrients, diet also contains micronutrients (vitamins, minerals, and trace elements), which play important roles in maintaining optimal health. Among body tissues, brain is responsible for a large amount of energy consumption. Although, brain accounts for only 2 % body weight, it consumes 20 % of the oxygen and 25 % of glucose. Such large energy consumption is required for the maintenance of ionic balance in neurons, production of action potential and initiation and maintenance of post-synaptic currents, and recycling of neurotransmitters (Rolfe and Brown 1997). Because of high oxygen consumption, brain is highly susceptible to oxidative and nitrosative damage. Large portions and long term consumption of high fat and high carbohydrate diet (high calorie diet) increases the chances of brain damage not only through oxidative processes, but also due to alterations in hormones (insulin, leptin and ghrelin) along with changes in neural cell homeostasis leading to metabolic abnormalities, behavioral disturbances, and motor and cognitive impairments.

High calorie diet (a diet consumed by Western societies) not only contains high amounts of processed macronutrients (fats, cholesterol, proteins and sugars), but also has high salt (sodium chloride). High calorie diet is low in fiber. High calorie diet is characterized by higher consumption of refined grains, potato, tea, whole grains, hydrogenated fats, legumes, and casserole. It can also be defined as a diet rich in high-fat sandwich spreads, red meat, potatoes, butter and lard, eggs, sandwich meat, sauces, pizza, soda, and sweets and desserts (Villegas et al. 2004, 2010; Lau et al. 2009; Esmaillzadeh et al. 2007). Collective evidence indicates that high calorie diet provides about 50 % of total daily calories from refined carbohydrates, 35 % calories from fat and refined oils, and 15 % from proteins of animal origin. In addition, in this diet the ratio between n-6 (arachidonic acid, ARA) to n-3 fatty acid (docosahexaenoic acid, DHA) is about 20:1 (Farooqui 2009). Present day high calorie diet also contains some vegetables and fruits, which are not only sprayed with herbicides and pesticides, but are also genetically modified. In contrast, Paleolithic diet on which our forefathers lived and survived thousands of years contained carbohydrates (40 %), fats (21 %), and proteins (39 %). Furthermore, fruits and vegetables that were consumed in Paleolithic era contained no herbicides and were not genetically modified. In Paleolithic era, the ratio between n-6 and n-3 fatty acids in Paleolithic diet was 1:1 (Simopoulos 2009, 2013; Bengmark 2013) (Table 1.1). Collective evidence suggests that human diet has changed dramatically over millennia. With technological and economic advancements in western countries particularly over the past 60 years, diet-related pathological conditions are the leading causes of premature death (Dean et al. 2011). With globalization, the consumption of high calorie diets

Table 1.1 Composition of Paleolithic and high calorie diets

Food components	Paleolithic diet	High calorie
Proteins	39 % (animal and plant origin)	15 % (animal origin)
Carbohydrates	40 % (whole grains)	50 % (refined sugars)
Fats	21 % (animal and plant origin i.e., olive oil)	35 % (partially hydrogenated oil)
n-6 to n-3 fatty acid ratio	1:1	20:1
Sodium levels	Low	High
Fiber	High	Low
Fruits and vegetables	High	Low
Preservatives, herbicides and pesticides	Absent	Present
Presence of genetically modified foods	Absent	Present

coupled with inactivity has contributed to the pandemic of lifestyle-related conditions (Carrera-Bastos et al. 2012).

Long term consumption of high calorie diet produces obesity, a pathological condition, mainly caused by a positive energy balance for several months to years. Such a balance/imbalance is regulated by a complex metabolic network of signals and signal transduction processes that connect the endocrine system with the central nervous system. The consumption of high calorie diet also promotes oxidative stress through not only induction of mitochondrial dysfunction and peroxisomal oxidation of fatty acids, but also diminished activities of antioxidant enzymes such as super-oxide dismutase (SOD), catalase (CAT), and glutathione peroxidase (GPx) (Fernández-Sánchez et al. 2011). Finally, high ROS production and the decrease in antioxidant capacity may also cause endothelial dysfunction, which is characterized by a reduction in the bioavailability of vasodilators, particularly nitric oxide (NO), and an increase in endothelium-derived contractile factors (Fernández-Sánchez et al. 2011). Obesity is also associated with impaired muscle microcirculation. These processes can impair whole body insulin sensitivity not only by hindering the entry of insulin and glucose into skeletal muscle and decreasing their availability to muscle cells, but also by inducing abnormal information processing, which leads to network dysfunction through the production of high levels of arachidonic acid-derived lipid mediators (eicosanoids and platelet activating factor) (Farooqui et al. 2012; Farooqui 2013). Collective evidence suggests that high calorie diet increases the risk of visceral diseases (obesity, insulin resistance, type 2 diabetes, metabolic syndrome, and cardiovascular disease), and neurological disorders (stroke, AD, PD, and depression) in experimental animals and humans (Fig. 1.1) (Buettner et al. 2007; Pinheiro et al. 2007; Brown and Goldstein 2008; Oliveira et al. 2009; Oliveira Junior et al. 2010; Farooqui et al. 2012; Farooqui 2013), supporting the view that the consumption of high calorie diet is associated with the endocrine disturbances, abnormalities in cardiac function, interstitial fibrosis, and altered molecular expression of contractile proteins, including the synthesis of β-myosin heavy-chain (MHC) isoforms (Buettner et al. 2007; Pinheiro et al. 2007; Oliveira Junior et al. 2010;

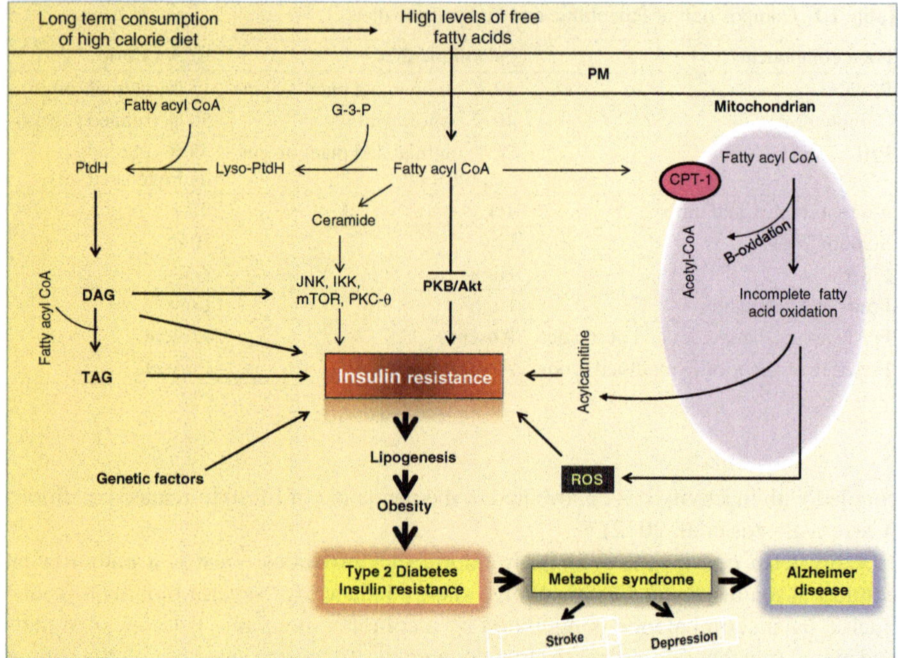

Fig. 1.1 Induction of insulin resistance by lipid mediators. *PM* plasma membrane, *G-3-P* glycerol-3-phosphate, *lyso-PtdH* lysophosphatidic acid, *PtdH* phosphatic acid, *DAG* diacylglycerol, *TAG* triacylglycerol, *JNK* c-Jun N-terminal kinase, IKK IκB kinase, *PKCθ* PKC theta, *Akt* cytosolic protein kinase, *PKB* protein kinase B, *CPT-I* Carnitine palmitoyltransferase-I

Poudyal et al. 2012). Collective evidence suggests that long term consumption of high calorie diet impairs systemic metabolic homeostasis, which is a metabolic stressor associated with oxidative and endoplasmic reticulum stress.

1.2 Effect of Long-Term Consumption of High Calorie Diet on Insulin Resistance, Inflammation, and Oxidative Stress in Visceral Tissues

Long term consumption of high calorie diet induces insulin resistance, a process that is defined as an inadequate response by insulin-sensitive tissues (liver, skeletal muscle, and adipose tissue) to normal circulating levels of insulin (Schenk et al. 2008). Insulin regulates glucose uptake and free fatty acid (FFA) levels in the blood. In adipose tissue, insulin decreases lipolysis thereby reducing FFA efflux from adipocytes. In liver, insulin inhibits gluconeogenesis by reducing key enzyme activities, and in skeletal muscle insulin predominantly induces glucose uptake by stimulating the translocation of the GLUT4 glucose transporter to the plasma membrane.

Insulin resistance is regulated by both genetic and acquired factors. Although, very little is known about the genetic causes or predispositions of insulin resistance in pre-diabetic populations, but it is proposed that defects in oxidative metabolism and inherited defects in the basic insulin signaling cascade are closely associated with genetic causes of insulin resistance (Fig. 1.1) (Morino et al. 2006, 2008; Thaler and Schwartz 2010). It is also stated that genetic component may interact with environmental factors to promote a pronounced pathophysiological abnormality during insulin resistance (Thaler and Schwartz 2010). The most common acquired factors that produce insulin resistance are obesity, sedentary lifestyle, and aging. All these processes are interrelated and connected with each other with signal transduction processes (Mokdad et al. 2003; Hamilton et al. 2007; Thaler and Schwartz 2010). The molecular mechanisms that contribute to insulin resistance are not fully understood. However, accumulation of lipids (free fatty acids, triacylglycerol, diacylglycerol, acylcarnitines, and ceramide) in the liver is considered to be one of the primary mechanisms involved in insulin resistance (Fig. 1.1) (Itani et al. 2002; Adams et al. 2004, 2009; Holland et al. 2007a). Recent studies have also indicated that activity of stearoyl-CoA desaturase (SCD) positively correlates with insulin resistance, abdominal adiposity and hyperlipidemia (Attie et al. 2002; Warensjo et al. 2007, Mar-Heyming et al. 2008, Paillard et al. 2008). In addition, adipose tissue-derived palmitoleate has been proposed to act as a "lipokine," a lipid-derived circulating factor that controls energy homeostasis and insulin resistance in mice (Cao et al. 2008), supporting the view that insulin resistance is modulated by many lipid mediators that are generated in the tissues by our diet. In high calorie diet, saturated fatty acids are key contributors of insulin resistance. They regulate activities of enzymes associated with lipid and carbohydrate metabolism by modulating proinflammatory gene expression and transcription factors that modulate glucose and lipid metabolism (Staiger et al. 2005; Rioux and Legrand 2007). Since, polyunsaturated fatty acids also modulate insulin resistance, therefore in rodents, supplementation with n-3 PUFA from fish oil reduces and reverses insulin resistance not only through their anti-inflammatory effect, but also through the regulation of the expression of genes associated with carbohydrate and lipid metabolism (Farooqui 2009). n-6 fatty acids do not produce insulin resistance reversal effect in rodents. However, in humans, n-6 PUFA may improve insulin resistance and also decrease diabetes risk (Riserus et al. 2009). Based on available information, it is suggested that desaturation indexes may also be linked with adiposity and insulin resistance. For example, the ratios of palmitic and stearic acid (180) to palmitoleic (161) and oleic acid (181) respectively, have been correlated with adiposity and insulin resistance, while the ratio of dihomo-gammalinoleic acid (203n-6) to arachidonic acid (204n-6) is negatively associated with obesity and insulin resistance (Jumpertz et al. 2012).

Among insulin resistance inducing lipids, DAG modulates isoforms of protein kinase C (PKCθ and PKCε), which can regulate insulin-mediated signal transduction via serine phosphorylation of insulin receptor substrate (IRS)-1 (Yu et al. 2002). Ceramides promote insulin resistance by preventing insulin-stimulated Akt serine phosphorylation, activation, and translocation of Akt to its substrate (Summers et al. 1998). In addition, ceramide initiates inflammatory signaling pathways, leading to

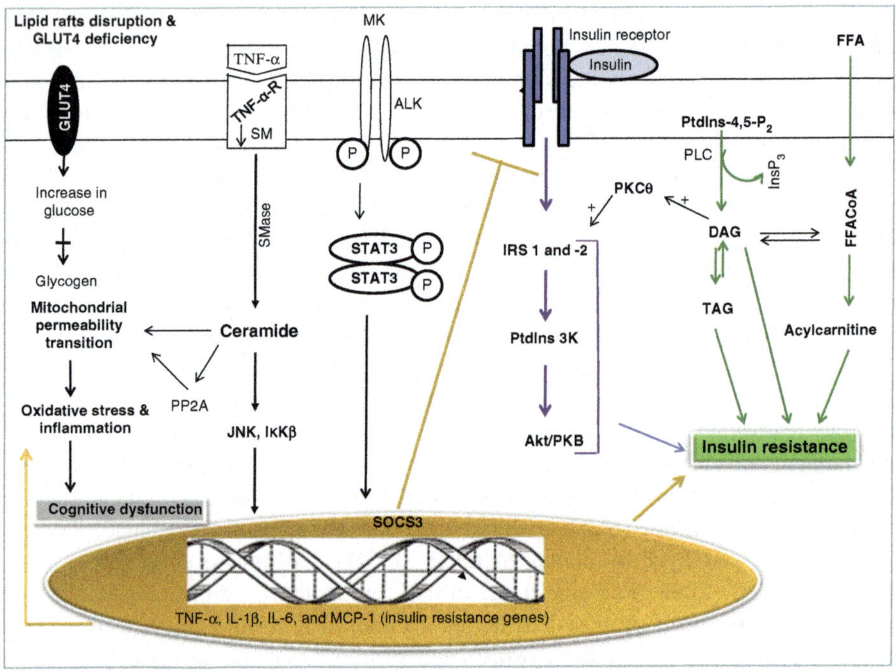

Fig. 1.2 Metabolic processes contributing to insulin resistance. *SM* Sphingomyelin, *SMase* sphingomyelinase, *PtdIns 4,5-bisP2* phosphatidylinositol 4,5-bis phosphate, *InsP3* inositol 1,4,5-trisphosphate, *DAG* diacylglycerol, *TAG* triacylglycerol, *PLC* phospholipase C, *PKCθ* protein kinase θ, *TNF-α* tumor necrosis factor-alpha, *TNF-α-R* tumor necrosis factor-alpha receptor, *MK* midkine, *ALK* anaplastic lymphoma kinase, *STAT3* signal transducer and activator of transcription 3, *SOCS3* suppressor of cytokine signaling3, *PP2A* protein phosphatase 2A, *JNK* c-Jun N-terminal kinase, *IRS* insulin receptor substrate, *PtdIns 3 K* phosphatidylinositol 3 kinase, *Akt/PKB* protein kinase B, *ROS* reactive oxygen species

the activation of both c-jun NH_2-terminal kinase (JNK) and nuclear factor κB/inducer of κ kinase (Fig. 1.2) (Ruvolo 2003). In-efficient tissue long-chain fatty acid beta-oxidation, due in part to a relatively low tricarboxylic acid cycle capacity, increases tissue accumulation of acetyl-CoA and generates chain-shortened acylcarnitine molecules that activate proinflammatory pathways implicated in insulin resistance (Adams et al. 2009; Schooneman et al. 2013). Collective evidence suggests that the consumption of saturated fatty acids (palmitate) leads to tissue accumulation of lipotoxic fatty acid derivatives such as diacylglycerol (DAG) and ceramide that promote activation of (i) DAG-sensitive PKCs, (ii) atypical PKCs and (iii) PP2A that, in turn, impair IRS- and Akt-directed insulin signaling by mechanisms involving IRS serine phosphorylation or repression of Akt activation/phosphorylation (Copps and White 2012; Galbo et al. 2011). Strikingly, dietary substitution of SFAs or co-supplementation with unsaturated fatty acids antagonize the insulin desensitising effects of SFAs, enhance energy expenditure and improve muscle lipid composition and serum acylcarnitine profiles in humans (Kien et al. 2011, 2013;

Fedor and Kelley 2009). Such observations are in line with cell-based studies demonstrating that monounsaturated fatty acids (MUFAs), such as palmitoleate and oleate, not only attenuate the SFA-induced loss in mitochondrial integrity and function but also repress their proinflammatory and ER stress-inducing potential (Xue et al. 2012; Yuzefovych et al. 2010; Macrae et al. 2013). These findings support the view that all above mentioned lipids exert their insulin resistance inducing effects not only because of their role in energy homeostasis, but also through the inhibition of insulin signaling (Morino et al. 2006; Holland et al. 2007b; Cai et al. 2005; Chung et al. 2008; Ikonen and Vainio 2005).

Adipocytes in the visceral fat synthesize and release a number of cytokines and adipokines, such as leptin, adiponectins, tumor necrosis factor-alpha (TNF-α), interleukin-1beta (IL-1β), interleukin-6 (IL-6), MCP-1, and midkine (MK), which modify insulin signaling and initiate the development of insulin resistance leading to type II diabetes and metabolic syndrome (Fonseca-Alaniz et al. 2007; Ouchi et al. 2011; Fernández-Sánchez et al. 2011; Fan et al. 2014; Farooqui 2013). MK is a 13-kDa heparin-binding growth factor, which promotes cell proliferation, differentiation, survival and migration, inflammation also contributes to insulin resistance (Kadomatsu et al. 2013; Weckbach et al. 2011, 2012). The molecular mechanism involved in midkine-mediated insulin resistance is not fully understood. However, In vitro studies have indicated that MK impairs insulin signaling in 3 T3-L1 adipocytes, as indicated by reduction in phosphorylation of Akt and IRS-1 and decrease in translocation of glucose transporter 4 (GLUT4) to the plasma membrane in response to insulin stimulation. Moreover, MK also activates the STAT3-suppressor of cytokine signaling 3 (SOCS3) pathways in adipocytes (Fan et al. 2014).

Under diabetic conditions, prolonged hyperglycemia and the accompanying production of excess amounts of advanced glycation end products (AGEs) can also activate NF-κB. JNK promotes insulin resistance through the phosphorylation of serine residues in insulin receptor signaling (IRS)-1 that negatively regulates normal signaling through the insulin receptor/IRS-1 axis. As stated above, NF-κB induces insulin resistance by promoting the expression of numerous target genes as those for TNF-α, IL-6, IL-8, MCP-1, MIP-1α, MIP-2, ICAM-1, and VCAM-1 (Shoelson et al. 2006). Proinflammatory cytokines and adipokines also entail the functions of regulating food intake, therefore exerting a direct effect on weight control. For example, leptin acts on the limbic system by stimulating dopamine uptake, creating a feeling of fullness. Hyperglycemia may also induce glycation of various structural and functional proteins including plasma proteins and collagen (Negre-Salvayre et al. 2009). In addition, glycation reactions lead to the production of more reactive oxygen species (ROS) leading to oxidative stress. The non-enzymatic glycation of plasma proteins such as albumin, fibrinogen and globulins may produce various deleterious effects including platelet activation, generation of oxygen free radicals, and impairment in immune system regulation (Negre-Salvayre et al. 2009).

Insulin resistance may contribute to essential hypertension through the formation of methylglyoxal (MG) and uric acid. These metabolites are derived from high levels of glucose and fructose, which are major components of high calorie diet. MG reacts with proteins to form advanced glycation end products (AGEs) and contribute to oxidative stress. The generation of MG may not only cause alterations in functioning of the renin-angiotensin system, but trigger the production of proinflammatory cytokines which result in the release of more ROS. Generation of uric acid inhibits nitric oxide synthase and reduces the bioavailability of NO resulting in impairment of endothelium-dependent vasodilation (Cheng et al. 2005). Collective evidence suggests that insulin resistance contributes to defects in insulin receptor function, abnormalities in insulin signaling, alterations in glucose metabolism, induction of hyperinsulinemia, hyperglycemia and increases blood pressure (Wang and Jin 2009) leading to metabolic syndrome, a pathological condition, which is an important risk factor for chronic visceral diseases (obesity, cardiovascular diseases, diabetes, metabolic syndrome, arthritis, and cancer) (Farooqui 2013). Acute hyperglycemia decreases endothelial nitric oxide production leading to impairment in normal vascular function (Beckman et al. 2002), adversely impacting coronary microcirculation (Fujimoto et al. 2006), and markedly attenuating signal transduction pathways critical to endogenous cardioprotective and cerebroprotective responses (Kersten et al. 2001).

Long term consumption of high calorie diet also induces endoplasmic reticulum (ER) stress, which also promotes the development of insulin resistance (Fig. 1.3). ER is a well-organized protein-folding machine composed of protein chaperones,

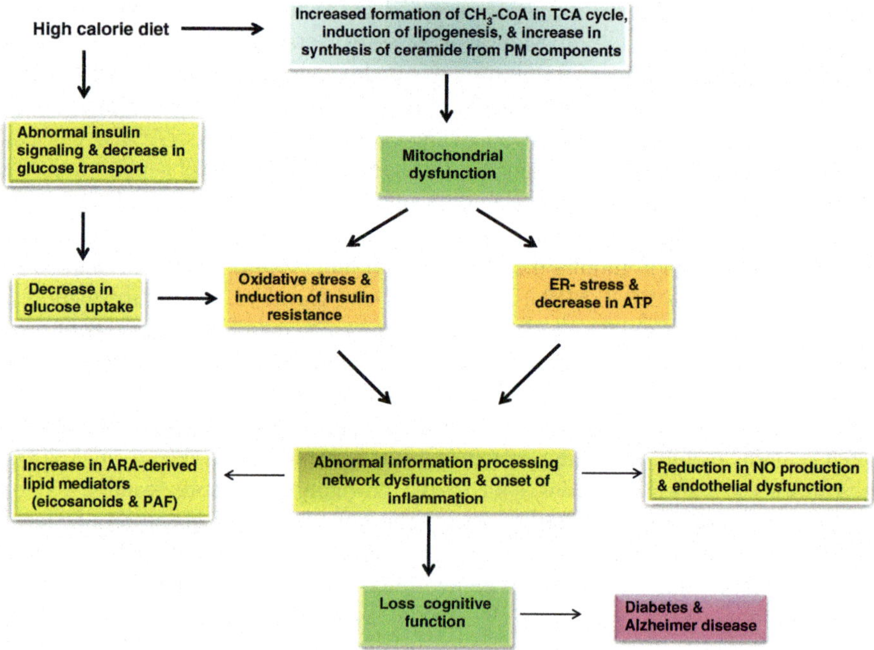

Fig. 1.3 Processes associated with the pathogenesis of type Ii diabetes and neurological disorders. *TCA* Tricarboxylic acid cycle, *PAF* platelet activating factor, *ARA* arachidonic acid, *NO* nitric oxide

proteins that catalyze protein folding, and sensors that detect the presence of misfolded or unfolded proteins (Malhotra and Kaufman 2007). The efficiency of protein-folding reactions not only depends on appropriate environmental and genetic factors, but also on the consumption of macronutrients. Conditions that disrupt protein folding threaten cells with decrease in viability and longevity. Accumulation of unfolded proteins in ER lumen initiates activation of an adaptive signaling cascade known as the unfolded protein response (UPR) (Malhotra and Kaufman 2007). An increase in mitochondrial oxidative phosphorylation due to increase in influx of macronutrients increases mitochondrial superoxide ($O_2^{\cdot-}$) production may also contribute to the pathogenesis of insulin resistance in animal and cellular models (Hoehn et al. 2009). Based on these results, it is suggested that mitochondrial $O_2^{\cdot-}$ production may represent an important link between mitochondrial function and insulin resistance (Hoehn et al. 2009). Collective evidence suggests that mitochondrial oxidative stress, ER stress, intracellular ceramide accumulation, and the induction of JNK, IKK, or PKCθ may contribute to the development of insulin resistance and diabetic symptoms in rodents consuming high calorie diet (Savage et al. 2007). Patients with type II diabetes show late diabetic complications—atherosclerosis, hypertension, and dyslipidemia (Roglic and Unwin 2010; Fowler 2011). Chronic hyperglycemia is accompanied by oxidative stress, dyslipoproteinemia, glycation of proteins, and endothelial dysfunction (Wu et al. 2008; Gradinaru et al. 2013). Among elderly subjects (65 years and above) the late diabetic complications are more common than young subjects. This view is supported by studies on levels of markers of free radical-induced lipid peroxidation and antioxidant status in diabetic patients (Maritim et al. 2003; de Rekeneire et al. 2006).

1.3   Effect of Long-Term Consumption of High Calorie Diet on Insulin Resistance, Inflammation, and Oxidative Stress in the Brain

Brain monitors peripheral energy balance via hormones that are actively transported across the blood brain barrier (El Messari et al. 2002). Insulin, a hormone produced and secreted by the pancreas in response to increased blood glucose levels, has receptors that are widely distributed throughout the brain (Banks 2004). Administration of insulin into the brain produces many effects, including effects on food intake (Stockhorst et al. 2004), cognitive performance (Benedict et al. 2011), autonomic outflow (Landsberg 2001), and peripheral infusions of insulin increase the ability of the arterial baroreflex to alter peripheral sympathetic activity in response to changes in blood pressure (Young et al. 2010). As stated above, insulin resistance is a pathological condition in which the effects of insulin on peripheral tissues are reduced, primarily as a result of overeating, obesity and lack of physical activity. As insulin resistance develops, the transport of insulin across the blood brain barrier is reduced, thereby altering its ability to affect the central nervous system (Banks 2004; Woods et al. 2003). As stated above, long-term consumption

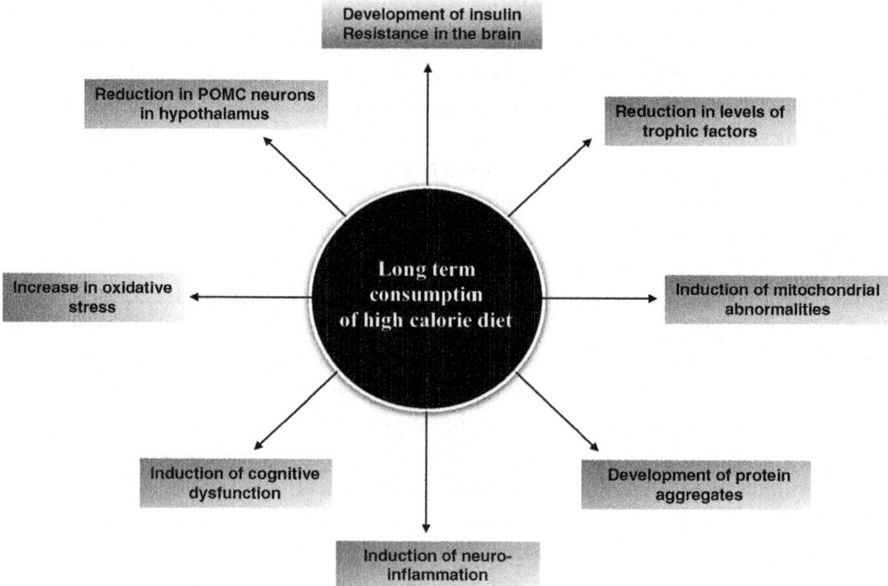

Fig. 1.4 Effect of long-term consumption of high calorie diet activities in the brain. *POMC* Proopiomelanocortin, *BDNF* brain-derived neurotrophic factor

of high calorie diet induces many changes in brain metabolism and functions (Fig. 1.4). Long-term consumption of high calorie diet not only produces mitochondrial abnormalities, facilitates the formation of protein aggregates (Aβ, neurofibrillary tangles, Lewy Bodies, and lipofuscin ceroid), but also reduces levels trophic factors, and promotes the induction of hyperglycemia, neuroinflammation and oxidative stress (Farooqui 2013). Within the brain, hyperglycemia causes structural abnormalities that resemble the progressive, widespread atrophy often associated with biological aging (Biessels et al. 2006). These changes may cause functional impairments in the brain. The severity of brain impairments is highly dependent on the areas (brain region) and extent the damage to the brain (Thaler et al. 2012). The mediobasal hypothalamus regulates energy balance and prevents obesity by adjusting appetite and food intake in response to signals of metabolic status including insulin and leptin. It is shown that high calorie diet-mediated increase in the expression of inhibitor of κB kinaseβ (IKKβ) in the hypothalamus directs the activation of NF-κB (Zhang et al. 2008). Forced expression of IKKβ in the CNS interrupts leptin and insulin signaling, resulting in increased in intake of high calorie food and weight gain compared to wild type mice. The targeted disruption of IKKβ in the hypothalamus lowers food intake, and protect mice from obesity as well as diet-induced insulin resistance and glucose intolerance (Zhang et al. 2008). The energy circuitry mechanism of the hypothalamus is critical for body weight homeostasis (Schwartz et al. 2000). Multiple regions of the hypothalamus interact to modulate energy homeostasis by integrating cues that signal metabolic state. The neurons in the arcuate nucleus not only produce orexigenic (neuropeptide Y, NPY) and anorexigenic (proopiomelanocortin, POMC) neuropeptides, but also express receptors for sensing and integration of peripheral signals, such as insulin receptor and the long form of the leptin receptor (Cone 2005). The consumption of high calorie diet causes alterations in the expression of *Npy, Pomc, Insr, and Lepr* genes, suggesting that alterations at the level of the hypothalamic appetite regulatory mechanism may contribute to the development of insulin resistance and obesity in these rats (Srinivasan et al. 2008).

Unlike inflammation in visceral tissues, which develops as a consequence of obesity and insulin resistance after weeks to months, the onset of hypothalamic inflammation occurs both in rats and mice within 1–3 days after the start of high calorie diet and prior to substantial weight gain. Hypothalamic inflammation is accompanied by reactive gliosis involving both microglial and astroglial cell populations along with increase in markers of neuronal injury (TNF-α, IL1-β, and IL-6) within a week. The presence of high levels of n-6 fatty acids in high calorie diet and their metabolism may result in the generation of high levels of proinflammatory eicosanoids (prostaglandins, leukotrienes, and thromboxanes), which along with proinflammatory cytokines may support and contribute to inflammation in the hypothalamus. Although these responses temporarily subside due to the onset of neuroprotective mechanisms, which may initially limit the damage, but with continuation of high calorie diet uptake, inflammation and gliosis return permanently to the mediobasal hypothalamic region (Thaler and Schwartz 2010; Thaler et al. 2012). These observations on rodents are supported by MRI studies in humans, which indicate that there is an increase in gliosis in the mediobasal hypothalamus of obese humans. Collective evidence suggests that in both human and rodent consumption of high calorie diet is associated with oxidative stress, inflammation, and neuronal injury in the hypothalamus, an area of the brain involved in body weight control (Thaler et al. 2012). Saturated fatty acids, a major component of high calorie diet promote hypothalamic inflammation by activating signal transduction though TLR4, which leads to endoplasmic reticulum stress, in situ expression of inflammatory cytokines and eventually, apoptosis of neurons, all contributing to the development of adipostatic hormone resistance and anomalous expression of the neurotransmitters involved in the regulation of energy homeostasis (Milanski et al. 2009; Moraes et al. 2009). Like visceral tissues, consumption of high calorie diet modulates inflammation in hypothalamus through several mechanisms. These mechanisms include activation of TLR4, induction of endoplasmic reticulum stress, IKKβ/NF-κB signaling, and induction of SOCS3 along with other intracellular inflammatory signals associated with high levels of circulating saturated fatty acids (Thaler et al. 2012; Fessler et al. 2009; Zhang et al. 2008) that exacerbate the inflammatory response and facilitate insulin resistance. The relative contribution of these mechanisms in the onset and maintenance remains uncertain. However, much earlier onset of inflammation in hypothalamus relative to peripheral tissues suggests that different processes may be associated with the inflammation in peripheral tissues (Thaler and Schwartz 2010). In addition, the consumption of high calorie diet is also associated with the alterations in hippocampal morphology/plasticity and impairment of cognitive function in normal rats (Granholm et al. 2008; Stranahan et al. 2008). High calorie diet-mediated changes in hippocampal morphology/plasticity are of considerable interest because this region is involved in learning and memory formation. At the molecular level these impairments might involve changes in glutamate-receptor subtypes, in second-messenger systems and in protein kinases (Farooqui 2013).

In normal C57BL/6 mice the consumption of high calorie diet results in the loss of working memory correlated with striking increase in neuroinflammatory and marked increase in APP processing (Li et al. 2007; Thirumangalakudi et al. 2008). Many studies have indicated that high calorie diet-mediated abnormalities in insulin signaling and function (insulin resistance) and alterations in glucose metabolism are linked with the pathogenesis of type II diabetes and Alzheimer Disease (AD) (de la Monte and Wands 2005; Talbot et al. 2012; Farooqui 2013) and the term "type III diabetes" has been used to describe AD (Figs. 1.4 and 1.5) (de la Monte and Wands 2008; de la Monte and Tong 2013). Several potential mechanisms have been proposed to link type II diabetes with AD. First, hyperglycemia may cause increased oxidative stress and accumulation of advanced glycation end-products, leading to

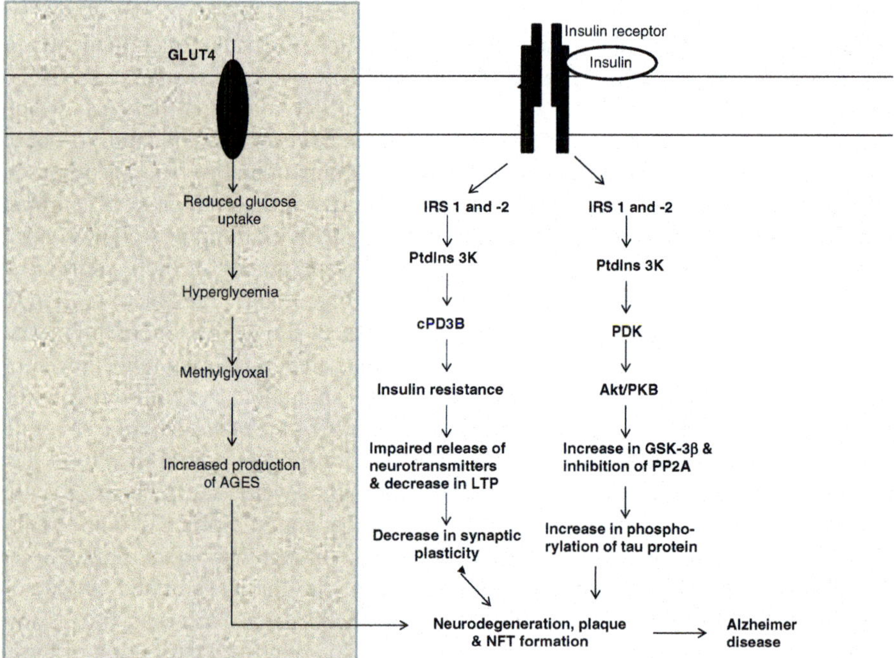

Fig. 1.5 Generation of advanced glycation products and alterations in insulin signaling contribute to the pathogenesis of Alzheimer disease. *AGEs* advanced glycation products, *IRS1 and 2* insulin receptor substrate 1 and 2, *PtdIns 3 K* phosphatidylinositol 3 kinase, *Akt/PKB* protein kinase B, *LTP* long term potentiation, *GLUT4* glucose transporter4, *PDK* phosphoinositide-dependent kinase, *GSK-3β* glycogen synthase kinase-3beta, *PP2A* protein phosphatase 2A, *NFTs* neurofibrillary tangles

progressive functional and structural abnormalities in the brain (Farooqui 2013; de la Monte 2009; de la Monte and Tong 2013). Second, although the cause and progression of AD remains undetermined, β-amyloid peptides deposits are considered as the fundamental cause of the disease. Glycation of proteins produces changes in neurochemical activity of proteins and their degradation processes. In brain, protein cross-linking by AGE results in the formation of detergent-insoluble and protease-resistant aggregates. Such aggregates may interfere with both axonal transport and intracellular protein traffic in neurons (Farooqui 2013). As stated above, Type II diabetes is associated with insulin resistance and hyperinsulinaemia, which may interfere with β-amyloid peptides metabolism (Craft 2007) through the reduction of insulin-degrading enzyme activity. In addition, mitochondrial dysfunction, ER stress, abnormal protein processing, abnormalities in insulin signaling, dysregulated glucose metabolism, hypercholesterolemia, and the activation of inflammatory pathways are features common to type II diabetes and AD (Sims-Robinson et al. 2010; Farooqui 2013). Endothelial cells, which cover the entire inner surface of all blood vessels and help to form the blood brain barrier (BBB) is known to prevent the entry of circulating molecules into the brain parenchyma. Anomalies in apoptosis of cells which make up the BBB, especially that of brain microvascular endothelial cells may represent a key step in hyperglycemia-induced BBB disruption. Indeed, the apoptotic effect of hyperglycemia on endothelial cells derived from large arteries and venous system e.g., aorta and umbilical vein is well-documented (Nakagami et al. 2001). Although the mechanisms involved remain unclear, hyperglycaemia is known to elicit cellular injury by increasing the quantities of pro-apoptotic proteins (Bad, Bax and Bid) and possibly by decreasing those of anti-apoptotic bcl-2 proteins (Bcl-2 and bcl-XL) (Yang et al. 2008). Sufficient elevations in Bax to Bcl-2 ratio in turn trigger the conversion of inactive procaspase-3 to caspase-3, the major executor of apoptotic process (Nakagami et al. 2001). Hyperglycemia can also produce apoptosis through stimulation of oxidative stress characterized by the exaggerated synthesis or release of free radicals, in particular superoxide anion ($O_2{}^{\bullet-}$) (Cifarelli et al. 2011). Higher concentrations of $O_2{}^{\bullet-}$ may mediate the rates of endothelial cell apoptosis through activation of a variety of redox-sensitive mechanisms including mitochondrial dysfunction, Bax protein over-expression and p38 mitogen-activated protein kinase (p38MAPK) (Yang et al. 2008; van den Oever et al. 2010). As the main source of $O_2{}^{\bullet-}$ in vasculature, NADPH oxidase is anticipated to be intimately involved in BMEC survival/death (Allen and Bayraktutan 2009). Based on these studies, it is suggested that high calorie diet-mediated neurochemical and cellular changes are associated with a higher risk for developing type II diabetes and metabolic syndrome, pathological conditions, which are risk factors for stroke, AD, and depression as well as other neurodegenerative diseases (Fig. 1.1) (Seneff et al. 2011; Farooqui 2013).

High calorie diet enhances the production of ROS and reactive nitrogen species (RNS) in the brain, which utilizes about 20 % of respired oxygen for normal function, even though it represents only 5 % of the body weight. Under physiological conditions the antioxidant defense systems within neural cells can easily neutralize the levels of ROS and RNS produced by low oxidative stress through cellular

Table 1.2 Effect of high levels of ROS on NF-κB-mediated gene expression

Genes	Expression	Reference
Phospholipase A_2	Increased	Farooqui and Horrocks (2007)
Cyclooxygenase-2	Increased	Frank-Cannon et al. (2009), Glass et al. (2010), Farooqui (2013)
Nitric oxide synthase	Increased	Frank-Cannon et al. (2009), Glass et al. (2010), Farooqui (2013)
Matrix metalloproteinase	Increased	Frank-Cannon et al. (2009), Glass et al. (2010), Farooqui (2013)
Tumor necrosis factor-α	Increased	Frank-Cannon et al. (2009), Glass et al. (2010), Farooqui (2013)
Interleukin-1β	Increased	Frank-Cannon et al. (2009), Glass et al. (2010), Farooqui (2013)
Interleukin-6	Increased	Frank-Cannon et al. (2009), Glass et al. (2010), Farooqui (2013)
Vascular cell adhesion molecule-1	Increased	Frank-Cannon et al. (2009), Glass et al. (2010), Farooqui (2013)
Intercellular adhesion molecule 1	Increased	Frank-Cannon et al. (2009), Glass et al. (2010), Farooqui (2013)
Growth factors	Increased	Frank-Cannon et al. (2009), Glass et al. (2010), Farooqui (2013)

antioxidative defense systems. However, generation of high levels of ROS results in the severe oxidative stress, which promotes neural cell injury and death (Farooqui 2010). Thus, the generation of high ROS in neural cells facilitates the translocation of transcription factor (NF-κB) from cytoplasm to the nucleus where it promotes the expression of proinflammatory enzymes, proinflammatory cytokines and chemokines, growth factors, cell cycle regulatory molecules, and adhesion molecules (Table 1.2). Collective evidence suggests that high calorie diet produces inflammatory and neurodestructive effects in the brain (Zilberter 2011). In addition, high calorie diet also damages the immune system. The consumption of high calorie diet is also typified by reduced exposure to microorganisms, increased exposure to pollutions, heightened levels of stress, and many other variables that may contribute to immune dysfunction (Rook 2010). Long term consumption of high calorie diet also results in fewer white blood cells (RBC) to fight infection. Low numbers of RBC decreases phagocytosis capability (Nieman et al. 1999; Ghanim et al. 2004). While a complex interplay among hormonal, metabolic, and immunological processes is involved in biologic responses in the obese subjects, the resultant immune dysfunction increases the risk of infections of the gums, respiratory system, and of surgical sites after an operation (Falagas and Kompoti 2006; Ylostalo et al. 2008). The molecular mechanisms involved in the interplay of hormonal, metabolic, and immunological processes are not fully understood. However, it is becoming increasingly evident that the consumption of high calorie diet results in increased levels of leptin in the blood. All mononuclear immune cells have leptin receptors and their activation leads to an increase in proinflammatory cytokines (IL-1, IL-6, and TNF-α)

(Frederich et al. 1995). Leptin stimulates Natural killer cells, activates the transcription factor STAT3, and reduces the anti-inflammatory T-regulatory (Treg) cells (Carbone et al. 2012). In general, the release of adiponectin produces opposing effects on immunity and interestingly the ratio of the two can predict the development of coronary artery disease in diabetic individuals (Finucane et al. 2009; Farooqui 2013).

1.4 Reasons for Increased Life Expectancy in Developed Societies

In spite of the consumption of high calorie diet, there has been a dramatic increase in average life expectancy during the twentieth century throughout the world. In 2012, life expectancy at birth for both sexes globally was 70 years, ranging from 62 years in low-income countries to 79 years in high-income countries, giving a ratio of 1.3 between the two income groups (Leeson 2014). Women live longer than men all around the world. The gap in life expectancy between the sexes was 5 years in 1990 and had remained the same by 2012. The gap is much larger in high-income countries (more than 6 years) than in low-income countries (around 3 years). This due to availability of clean water, immunization against infectious and parasitic diseases, advances in medicine, and survival of children at the birth along with better living standards and availability of nutritious food at low prices. These advances have prolonged the life of many people with age-related chronic diseases. These diseases may not kill but they consume a lot of health care resources and threaten the quality of life of the sufferers (Lam and Lauder 2000; Bredow et al. 2008). It should be noted that longevity is a very complex process. In addition to diet, longevity is influenced by environmental, behavioral, and socio-demographic factors (Chrysohoou and Stefanadis 2013). These factors not only influence the life-expectancy, but also effect physiological pathways of aging. Increase in life expectancy and quality of life are two different issues. If the quality of life is poor in spite of increase life expectancy, the individual becomes a burden on caregiver and healthcare system. In twenty-first century developed countries, which have reduced burden of infectious diseases, constitute an environment where metabolic, cardiovascular, and autoimmune diseases thrive (Manzel et al. 2014). The consumption of high calorie diet in developed countries has resulted in obesity, metabolic syndrome, and cardiovascular disease. These diseases are in turn risk factor for neurological and autoimmune diseases (Farooqui 2013). The projected cost to Medicare for treating these neurological and autoimmune diseases is estimated to be about several trillion dollars by 2050. This number does not include other visceral, neurological diseases, and cancer. Such an amount will not only burst NIH budget, but will seriously affect U.S. economy. Although, available drugs may not reverse neurological and autoimmune disorders, but healthy diet, regular exercise and 6–7 h of sleep may not only retard neuroinflammation and oxidative stress caused by high calorie diet but also produce beneficial effects on motor and cognitive functions and memory deficits that occur to some extent during normal aging and to a large extent in neurological disorders (Farooqui 2012). I propose that delaying the onset of neuroinflammation and oxidative stress in neurological disorders by a few years by consuming a healthy diet may lead to significant reduction in the prevalence of neurological disorders and, consequently relieving the burden on health care system.

1.4.1 Beneficial Effects of Calorie Restriction on Brain Health

Calorie restriction (CR) refers to a reduced calorie intake (a reduction of food intake by 20–40 %) without malnutrition. CR with adequate micronutrient supplementation promotes metabolic fitness, longevity, and disease protection in rodent models of aging. Longevity is a highly conserved nutrient sensing process linking nutrient signaling with aging and aging-related diseases. Mediators of caloric restriction, including growth hormone, insulin, insulin-like growth factor (IGF-1), the circulating longevity hormone, Klotho, which binds to the IGF-1 receptor, as well as the Sirtuin family of longevity genes, have been shown to modulate both organismal lifespan and healthspan across species (Narasimhan et al. 2009). Neurochemical changes during the caloric restriction also activate stress pathways that increase animal resistance to subsequent stress or nutritional limitation. This process is known as hormesis. Sirtuin1 (SIRT1) is a NAD^+-dependent histone deacetylase, which catalyzes a reaction that couples lysine deacetylation to NAD^+ hydrolysis. During this reaction, NAD^+ is hydrolyzed to nicotinamide (NAM) and O-acetyl-ADP ribose. NAM is a strong inhibitor of SIRT1 deacetylase activity (Sauve and Schramm 2003). SIRT1 activity is closely associated with insulin signaling in insulin-sensitive organs (adipose tissue, liver and skeletal muscle) as well as in the brain and plays an important role in inducing beneficial effects of CR (Haigis and Guarente 2006). In the liver, SIRT1 deacetylates and activates the transcriptional co-activator PGC1-alpha and the transcription factor FOXO1 to promote gluconeogenesis (Frescas et al. 2005). Under fasting or CR, SIRT1 senses and responses to metabolic status such as elevation in intracellular levels of NAD, then activates and deaceyltes several proteins associated with stress resistance, mitochondrial function, and aging, whereas the consumption of high fat diet (high calorie diet) decreases SIRT1 in the liver (Deng et al. 2007). In the adipose tissue, SIRT1 improves pancreatic β-cell function in part by inhibiting uncoupling protein 2 (UCP2) that is a negative player of glucose-induced insulin secretion (Haigis and Guarente 2006). Thus, by negatively regulating PPAR-γ, SIRT1 plays an important role in this adaptive response to nutrient scarcity (Picard et al. 2004). Positive regulation of PPAR-γ coactivator 1 (PGC1)-α by SIRT1 increases hepatic glucose production and utilization in skeletal muscle during starvation (Deng et al. 2007). Collectively, these findings support the view that SIRT1 links the status of energy content with cellular function in visceral and brain tissues triggering stress response and signaling to the transcriptional level (Fulco and Sartorelli 2008). SIRT1 also deacetylates many regulatory proteins, such as PGC-1α, p53, FOXO, HSF and HIF-2α to trigger resistance to metabolic,

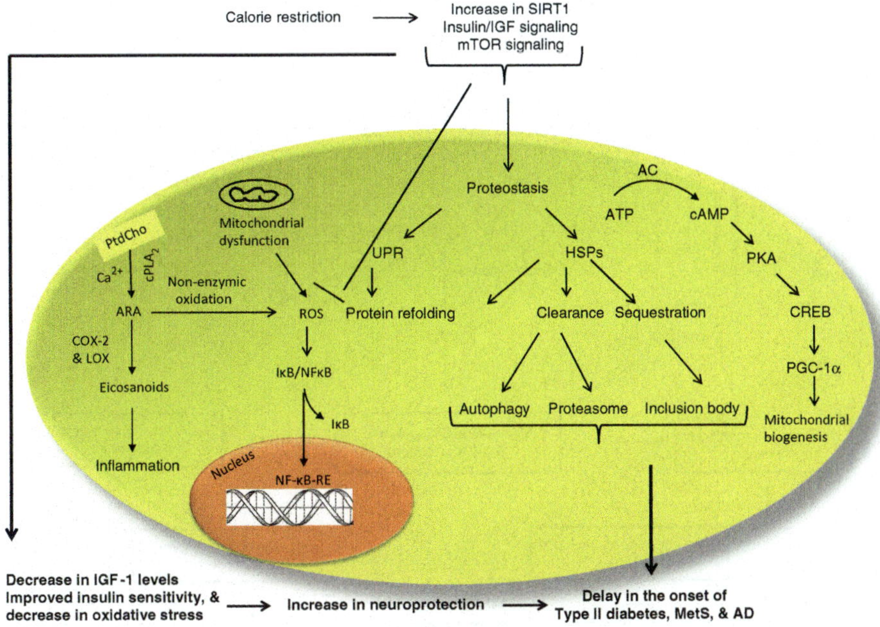

Fig. 1.6 Calorie restriction-mediated modulation of longevity and delay on the onset of type II diabetes II, metabolic syndrome and Alzheimer disease. *PtdCho* Phosphatidylcholine, *ARA* arachidonic acid, *cPLA₂* cytosolic phospholipase A₂, *COX-2* cyclooxygenase-2, *LOX* lipoxygenase, *ROS* reactive oxygen species, *NF-κB* nuclear factor kappaB, *NF-κB-RE* nuclear factor κB-response element, *IκB* inhibitory subunit of NF-κB, *UPR* unfolded protein response, *HSPs* heat shock proteins, *MetS* metabolic syndrome, *AD* Alzheimer disease, *AC* adenylylate cyclase, *PKA* protein kinase A, *(PGC)-1α* peroxisome proliferator-activated receptor-γ coactivator

oxidative, heat and hypoxic stress (Figs. 1.6 and 1.7) (Guarente 2009). During CR, SIRT1 activity is closely associated the longevity. The original hypothesis that CR mediates its effects passively by suppressing metabolic rate or reducing damage caused by ROS has been replaced with a fundamentally different hypothesis. In this hypothesis in CR produces an active defense response, which promotes cell survival during harsh conditions. This defense response involves longevity regulatory pathways, which include insulin/insulin-like growth factor 1 (IGF-1), the mammalian target of rapamycin (mTOR), AMP-activated kinase (AMPK), and sirtuins (Figs. 1.6 and 1.7) (North and Sinclair 2012). CR also decreases age-related learning and memory impairments in animals and human (Witte et al. 2009), probably through higher expression of an NMDA-receptor subunit in the hippocampus. In addition, CR also attenuates age-related brain atrophy in monkeys (Colman et al. 2009) and stabilizes the expression of synaptic protein expression to avoid aging-related changes (Mladenovic Djordjevic et al. 2010). Collective evidence suggests that CR lowers plasma insulin levels and mediates greater sensitivity to insulin by decreasing leptin and increasing adiponectin, lowing body temperatures; reducing cholesterol, triglycerides and blood pressure, and increasing levels of human growth hormone.

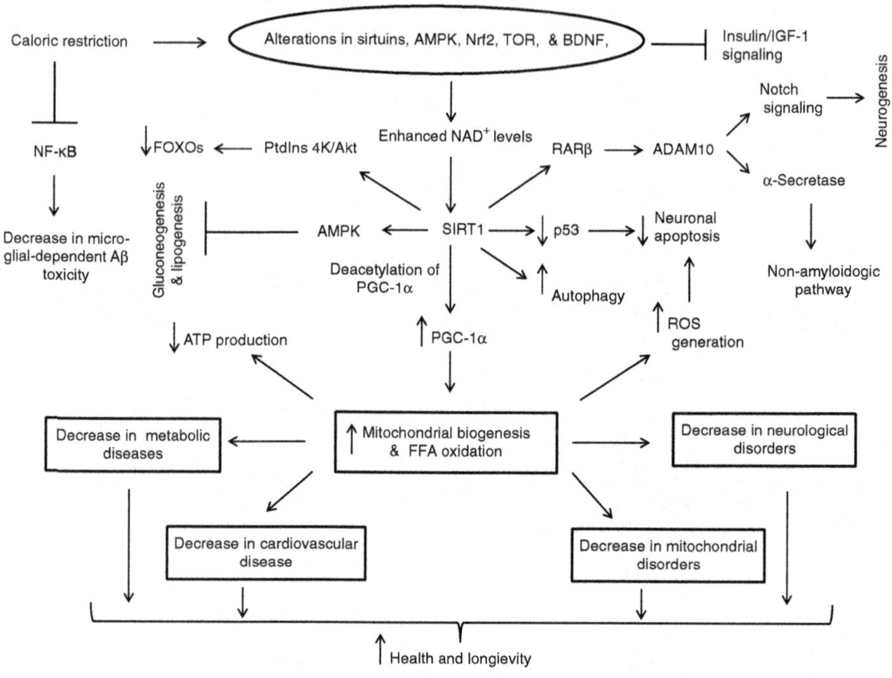

Fig. 1.7 Effect of increase in SIRT1on health and longevity during calorie restriction. *NF-κB* nuclear factor kappaB; *FOXO* forkhead box protein O, *Nrf2* NF-E2-related factor 2, *AMPK* AMP-activated protein kinase, *BDNF* brain-derived neurotrophic factor, *SIRT1* sirtuin1, *(PGC)-1α* PPARγ coactivator-1α, *ROS* reactive oxygen species. SIRT1-mediated modulation of insulin/IGF and TOR signaling modulates mitochondrial function, genomic stability, and proteostasis resulting in longevity

CR not only regulates intracellular redox status through its antioxidative actions and improves the resilience of synapses to metabolic and oxidative damage, but also modulates the total number, structure, and functional status of synapses (Mattson 2012). These observations support the view that CR induces neuroprotective responses by regulating electrical and synaptic activity throughout neuronal circuits through the modulation of NMDA and AMPA types of glutamate receptor and secretion of neurotrophic factors (BDNF) (Stranahan and Mattson 2011; Adams et al. 2008; Rothman et al. 2012). This trophic factor has been reported to promote neuroprotection, neurite expression, axonal and dendritic growth, remodeling, neuronal differentiation, and synaptic plasticity such as synaptogenesis in arborizing axon terminals, and synaptic transmission efficacy (Mattson 2012; Stranahan and Mattson 2011; Rothman et al. 2012).

A better knowledge of hypothalamic SIRT1 activities is very importance since the hypothalamus is the primary brain region that interprets adiposity or nutrient related inputs to regulate energy homeostasis. Specifically, the anorexigenic proopiomelanocortin (POMC) neurons and the orexigenic NPY/Agouti-related protein (AgRP) neurons in the arcuate nucleus (ARC) of the hypothalamus are the major regulators of feeding and energy expenditure (Morton et al. 2006; Sasaki and Kitamura 2010). In contrast to visceral tissues, levels of SIRT1 protein in the hypothalamus are decreased with fasting and increased with feeding (Sasaki et al. 2010). The hypothalamic control of food intake and body weight involves the action of metabolic sensors including the mammalian target of rapamycin (mTOR) (Cota et al. 2006) and AMP activated kinase (AMPK) (Minokoshi et al. 2004). Because of its dependence on NAD^+, SIRT1 also functions as a metabolic sensor (Fulco et al. 2008). Based on the above information, it is proposed that SIRT1 activities in hypothalamus may be another metabolic factor controlling food intake. Inhibition of SIRT1 activity through inhibitors or by knocking-down SIRT1 gene expression in the hypothalamic region is associated with the decrease in food intake and body weight gain. CR or intermittent fasting increases hypothalamic NAD^+ levels and SIRT1 protein content. Inhibition of hypothalamic SIRT1 activity reverses the fasting induced decrease of FOXO1 acetylation leading to increase in expression of SIRT1 in POMC and decrease in AgRP neurons.

The effects of CR on the aging brain are not only regionally specific, but also dependent on the neuronal and synaptic substrates of that specific area and its neuronal circuits (Mora et al. 2007). For example, it is reported that the gray matter volume in the caudate nucleus area of the brain decreases with age in control animals, but is preserved in calorie-restricted monkeys (Colman et al. 2009). In contrast, other areas of the monkey brain, including the frontal and temporal cortex display a significant reduction in gray matter volume, which is not decreased by a reduction in food intake (Colman et al. 2009). In addition, it is also shown that caloric restriction increases the levels of BDNF in several areas of the brain, particularly the hippocampus (Lee et al. 2002). Thus, recent studies have indicated that 40 % CR in food intake induces a marked increase in neurotransmitters and BDNF levels in rats throughout their entire lifespan (Del Arco et al. 2011). In addition certain nutrients modulate the activities of specific molecular substrates important for learning, memory, and other cognitive functions (Gomez-Pinilla 2008; Farooqui 2009). An example of one of those nutrients is the omega-3 fatty acids (docosahexaenoic acid), which are considered essential for maintaining synaptic function and plasticity (Gomez-Pinilla 2008; Farooqui 2009). In fact, the omega-3 fatty acids,, are an important component of neuronal membranes and it has been found that dietary supplementation with this fatty acid elevates the levels of BDNF in the hippocampus and counteracts rat learning disabilities after traumatic brain injury (Wu et al. 2008; Farooqui 2009). Moreover, different micronutrients, including flavonoids, contained in fruits and vegetables may counteract the aging process by improving cognitive functions (2012).

Life-long prospective food-restriction studies in humans have not been performed. Two human studies in healthy adults have clearly shown that reducing the intake of calories by about 20 % in nutritionally adequate diets over periods of 2 years (Walford et al. 2002) and 3–15 years (Fontana et al. 2004) lowers body weight, blood pressure, body fat, blood cholesterol, blood triglycerides, blood glucose, and blood insulin. These studies have indicated that significant reduction in the risk factors occurs for cardiovascular disease and diabetes. Furthermore, carotid artery intima-media thickness, a measure of atherosclerotic development, is markedly reduced by calorie restriction (Fontana et al. 2004). Similar results have been obtained on rhesus monkey, which has been maintained on 30 % CR over 12 years in studies conducted at the National Institute of Aging in Baltimore (Lane et al. 1999). Collective evidence suggests that long-term reductions in food intake lower body weight and reduce the risk of cardiovascular disease, diabetes, and metabolic syndrome, by delaying their onset.

1.5   Conclusion

Long term consumption of high calorie diet initiates oxidative stress and neuroinflammation causing early onset of visceral and brain diseases, where as food restriction delays the onset of these pathological conditions. High calorie diet is enriched in saturated and trans fats, n-6 fatty acids, cholesterol, low in fiber, and high in refined sugars and salt. These components elevate levels of insulin and increase the risk of diabetes, metabolic syndrome and cardiovascular disease, as well as stroke, AD, dementia, and depression. CR in animals not only reduces metabolic rate and oxidative stress, and alters neuroendocrine and sympathetic nervous system function, by but improves insulin sensitivity. CR also regulates intracellular redox status through its antioxidative actions. Therefore, it is safe to suggest that much of the longevity phenomenon can be attributed to improved insulin signaling and reduction in oxidative stress and chronic inflammation.

References

Adams JM II, Pratipanawatr T, Berria R, Wang E, DeFronzo RA, Sullards MC, Mandarino LJ (2004) Ceramide content is increased in skeletal muscle from obese insulin-resistant humans. Diabetes 53:25–31

Adams MM, Shi L, Linville MC et al (2008) Caloric restriction and age affect synaptic proteins in hippocampal CA3 and spatial learning ability. Exp Neurol 211:141–149

Adams SH, Hoppel CL, Lok KH, Zhao L, Wong SW, Minkler PE, Hwang DH, Newman JW, Garvey WT (2009) Plasma acylcarnitine profiles suggest incomplete long-chain fatty acid beta-oxidation and altered tricarboxylic acid cycle activity in type 2 diabetic African-American women. J Nutr 139:1073–1081

Allen CL, Bayraktutan U (2009) Antioxidants attenuate hyperglycaemia-mediated brain endothelial cell dysfunction and blood–brain barrier hyperpermeability. Diabetes Obes Metab 11:480–490

Attie AD, Krauss RM, Gray-Keller MP, Brownlie A, Miyazaki M, Kastelein JJ, Lusis AJ, Stalenhoef AF, Stoehr JP, Hayden MR, Ntambi JM (2002) Relationship between stearoyl-CoA desaturase activity and plasma triglycerides in human and mouse hypertriglyceridemia. J Lipid Res 43:1899–1907

Banks WA (2004) The source of cerebral insulin. Eur J Pharmacol 490:5–12

Beckman JA, Goldfine AB, Gordon MB, Garrett LA, Creager MA (2002) Inhibition of protein kinase Cbeta prevents impaired endothelium-dependent vasodilation caused by hyperglycemia in humans. Circ Res 90:107–111

Bellisle F, McDevitt R, Prentice A (1997) Meal frequency and energy balance. Br J Nutr 77(S1): S57–S70

Benedict C, Frey WH 2nd, Schiöth HB, Schultes B, Born J, Hallschmid M (2011) Intranasal insulin as a therapeutic option in the treatment of cognitive impairments. Exp Gerontol 46:112–115

Bengmark S (2013) Processed foods, dysbiosis, systemic inflammation, and poor health. Curr Nutri Food Sci 9:113–143

Biessels GJ, Staekenborg S, Brunner E, Brayne C, Scheltens P (2006) Risk of dementia in diabetes mellitus: a systematic review. Lancet Neurol 5:64–74

Black KJ, Hershey T, Koller JM, Videen TO, Mintun MA, Price JL, Perlmutter JS (2002) A possible substrate for dopamine-related changes in mood and behavior: prefrontal and limbic effects of a D3-preferring dopamine agonist. Proc Natl Acad Sci U S A 99:17113–17118

Bredow T, Peterson S, Sandau K (2008) Health-related quality of life. In: Peterson S (ed) Middle-range theory: application to nursing research. Lippincott Williams & Wilkins, Philadelphia, pp 273–289

Brown MS, Goldstein JL (2008) Selective versus total insulin resistance: a pathogenic paradox. Cell Metab 7:95–96

Buettner R, Schölmerich J, Bollheimer LC (2007) High-fat diets: modeling the metabolic disorders of human obesity in rodents. Obesity 12:798–808

Cai D, Yuan M, Frantz DF, Melendez PA, Hansen L, Lee J, Shoelson SE (2005) Local and systemic insulin resistance resulting from hepatic activation of IKK-beta and NF-kappaB. Nat Med 11:183–190

Cao H, Gerhold K, Mayers JR, Wiest MM, Watkins SM, Hotamisligil GS (2008) Identification of a lipokine, a lipid hormone linking adipose tissue to systemic metabolism. Cell 134:933–944

Carbone F, La Rocca C, Matarese G (2012) Immunological functions of leptin and adiponectin. Biochimie 94:2082–2088

Carrera-Bastos P, Fontes-Villalba M, Keefe JHO, Lindeberg S, Cordain L (2012) The western diet and lifestyle and diseases of civilization. Res Rep Clin Cardiol 2:15–35

Cawley EI, Park S, Aan Het Rot M, Sancton K, Benkelfat C, Young SN, Boivin DB, Leyton M (2013) Dopamine and light: dissecting effects on mood and motivational states in women with subsyndromal seasonal affective disorder. J Psychiatry Neurosci 38:388–397

Cheng ZJ, Vapaatalo H, Mervaala E (2005) Angiotensin II and vascular inflammation. Med Sci Monit 11:RA194–RA205

Chrysohoou C, Stefanadis C (2013) Longevity and diet. Myth or pragmatism? Maturitas 76:303–307

Chung J, Nguyen AK, Henstridge DC, Holmes AG, Chan MH, Mesa JL, Lancaster GI, Southgate RJ, Bruce CR, Duffy SJ, Horvath I, Mestril R, Watt MJ, Hooper PL, Kingwell BA, Vigh L, Hevener A, Febbraio MA (2008) HSP72 protects against obesity-induced insulin resistance. Proc Natl Acad Sci U S A 105:1739–1744

Cifarelli V, Geng X, Styche A, Lakomy R, Trucco M, Luppi P (2011) C-peptide reduces high-glucose-induced apoptosis of endothelial cells and decreases NAD(P)H-oxidase reactive oxygen species generation in human aortic endothelial cells. Diabetologia 54:2702–2712

Colman RJ, Anderson RM, Johnson SC, Kastman EK, Kosmatka KJ, Beasley TM, Allison DB, Cruzen C, Simmons HA, Kemnitz JW, Weindruch R (2009) Caloric restriction delays disease onset and mortality in rhesus monkeys. Science 325:201–204

Cone RD (2005) Anatomy and regulation of the central melanocortin system. Nat Neurosci 8:571–578

Copps KD, White MF (2012) Regulation of insulin sensitivity by serine/threonine phosphorylation of insulin receptor substrate proteins IRS1 and IRS2. Diabetologia 55:2565–2582

Cota D, Proulx K, Smith KA, Kozma SC, Thomas G, Woods SC, Seeley RJ (2006) Hypothalamic mTOR signaling regulates food intake. Science 312:927–930

Craft S (2007) Insulin resistance and Alzheimer's disease pathogenesis: potential mechanisms and implications for treatment. Curr Alzheimer Res 4:147–152

Dauncey MJ (2012) Recent advances in nutrition, genes and brain health. Proc Nutr Soc 71: 581–591

Davis C (2014) Evolutionary and neuropsychological perspectives on addictive behaviors and addictive substances: relevance to the "food addiction" construct. Subst Abuse Rehabil 5:129–137

de la Monte SM (2009) Insulin resistance and Alzheimer's disease. BMB Rep 42:475–481

de la Monte SM, Tong M (2013) Insulin resistance and metabolic failure underlie Alzheimer disease. In: Farooqui T, Farooqui AA (eds) Metabolic syndrome and neurological disorders. Wiley, Oxford, pp 1–30

de la Monte SM, Wands JR (2005) Review of insulin and insulin-like growth factor expression, signaling, and malfunction in the central nervous system: relevance to Alzheimer's disease. J Alzheimers Dis 7:45–61

de la Monte SM, Wands JR (2008) Alzheimer's disease is type 3 diabetes-evidence reviewed. J Diabetes Sci Technol 2:1101–1113

Dean E, Al-Obaidi S, de Andrade AD, Gosselink R, Umerah G, Al-Abdelwahab S, Anthony J, Bhise AR, Bruno S, Butcher S, Fagevik-Olsén M, Frownfelter D, Gappmaier E, Gylfadóttir S, Habibi M, Hanekom S, Hasson S, Jones A, LaPier T, Lomi C, Mackay L, Mathur S, O'Donoghue G, Playford K, Ravindra S, Sangroula K, Scherer S, Skinner M, Wong WP (2011) The first physical therapy summit on global health: implications and recommendations for the 21st century. Physiother Theory Pract 27:531–547

Del Arco A, Segovia G, De Blas M, Garrido P, Acuña-Castroviejo D, Pamplona R, Mora F (2011) Prefrontal cortex, caloric restriction and stress during aging: studies on dopamine acetylcholine release, BDNF and working memory. Behav Brain Res 216:136–145

Deng XQ, Chen LL, Li NX (2007) The expression of SIRT1 in nonalcoholic fatty liver disease induced by high-fat diet in rats. Liver Int 27:708–715

de Rekeneire N, Peila R, Ding J, Colbert LH, Visser M, Shorr RI, Kritchevsky SB, Kuller LH, Strotmeyer ES, Schwartz AV, Vellas B, Harris TB (2006) Diabetes, hyperglycemia, and inflammation in older individuals: the health, aging and body composition study. Diabetes Care 29:1902–1908

Dileone RJ, Taylor JR, Picciotto MR (2012) The drive to eat: comparisons and distinctions between mechanisms of food reward and drug addiction. Nat Neurosci 15:1330–1335

El Messari S, Aït-Ikhlef A, Ambroise DH, Penicaud L, Arluison M (2002) Expression of insulin-responsive glucose transporter GLUT4 mRNA in the rat brain and spinal cord: an in situ hybridization study. J Chem Neuroanat 24:225–242

Esmaillzadeh A, Kimiagar M, Mehrabi Y, Azadbakht L, Hu FB, Willett WC (2007) Dietary patterns, insulin resistance, and prevalence of the metabolic syndrome in women. Am J Clin Nutr 85:910–918

Falagas ME, Kompoti M (2006) Obesity and infection. Lancet Infect Dis 6:438–446

Fan N, Sun H, Wang Y, Zhang L, Xia Z, Peng L, Hou Y, Shen W, Liu R, Peng Y (2014) Midkine, a potential link between obesity and insulin resistance. PLoS One 9:e88299

Farooqui AA (2009) Beneficial effects of fish oil on human brain. Springer, New York

Farooqui AA (2010) Neurochemical aspects of neurotraumatic and neurodegenerative diseases. Springer, New York

Farooqui AA (2012) Phytochemical, signal transduction, and neurological disorders. Springer, New York

Farooqui AA (2013) Metabolic syndrome: an important risk factor for stroke, Alzheimer disease and depression. Springer, New York

Farooqui AA (2014) Inflammation and Oxidative Stress in Neurological Disorders. Springer International Publishing Switzerland

Farooqui AA, Horrocks LA (2007) Glycerophospholipids in brain. Springer, New York

Farooqui AA, Farooqui T, Panza F, Frisardi V (2012) Metabolic syndrome as a risk factor for neurological disorders. Cell Mol Life Sci 69:741–762

Fedor D, Kelley DS (2009) Prevention of insulin resistance by n-3 polyunsaturated fatty acids. Curr Opin Clin Nutr Metab Care 12:138–146

Fernández-Sánchez A, Madrigal-Santillán E, Bautista M, Esquivel-Soto J, Morales-González A, Esquivel-Chirino C, Durante-Montiel I, Sánchez-Rivera G, Valadez-Vega C, Morales-González JA (2011) Inflammation oxidative stress, and obesity. Int J Mol Sci 12:3117–3132

Fessler MB, Rudel LL, Brown JM (2009) Toll-like receptor signaling links dietary fatty acids to the metabolic syndrome. Curr Opin Lipidol 20:379–385

Finucane FM, Luan J, Wareham NJ, Sharp SJ, O'Rahilly S, Balkau B, Flyvbjerb A, Walker M, Hojlund K, Nolan JJ, Savage DB (2009) Correlation of the leptin: adiponectin ratio with measures of insulin resistance in non-diabetic individuals. Diabetologia 52:2345–2349

Fonseca-Alaniz MH, Takada J, Alonso-Vale MI, Lima FB (2007) Adipose tissue as an endocrine organ: from theory to practice. J Pediatr 83(Suppl 5):S192–S203

Fontana L, Meyer TE, Klein S, Holloszy JO (2004) Long-term caloric restriction is highly effective in reducing the risk for atherosclerosis in humans. Proc Natl Acad Sci U S A 101:6659–6663

Fowler MJ (2011) Microvascular and macrovascular complications of diabetes. Clin Diabet 29:116–122

Frank-Cannon TC, Alto LT, McAlpine FE, Tansey MG (2009) Does neuroinflammation fan the flame in neurodegenerative diseases? Mol Neurodegener 4:47

Frederich RC, Hamann A, Anderson S, Lollmann B, Lowell BB, Flier JS (1995) Leptin levels reflect body lipid content in mice: evidence for diet-induced resistance to leptin action. Nat Med 1:1311–1314

Frescas D, Valenti L, Accili D (2005) Nuclear trapping of the forkhead transcription factor FoxO1 via Sirt-dependent deacetylation promotes expression of glucogenetic genes. J Biol Chem 280:20589–20595

Fujimoto K, Hozumi T, Watanabe H, Tokai K, Shimada K, Yoshiyama M, Homma S, Yoshikawa J (2006) Acute hyperglycemia induced by oral glucose loading suppresses coronary microcirculation on transthoracic Doppler echocardiography in healthy young adults. Echocardiography 23:829–834

Fulco M, Sartorelli V (2008) Comparing and contrasting the roles of AMPK and SIRT1 in metabolic tissues. Cell Cycle 7:3669–3679

Fulco M, Cen Y, Zhao P, Hoffman EP, McBurney MW, Sauve AA, Sartorelli V (2008) Glucose restriction inhibits skeletal myoblast differentiation by activating SIRT1 through AMPK-mediated regulation of Nampt. Dev Cell 14:661–673

Galbo T, Olsen GS, Quistorff B, Nishimura E (2011) Free fatty acid-induced PP2A hyperactivity selectively impairs hepatic insulin action on glucose metabolism. PLoS One 6:e27424

Ghanim H, Aljada A, Hofmeyer D, Syed T, Mohanty P, Dandona P (2004) Circulating mononuclear cells in the obese are in a proinflammatory state. Circulation 110:1564–1571

Glass CK, Saijo K, Winner B, Marchetto MC, Gage FH (2010) Mechanisms underlying inflammation in neurodegeneration. Cell 140:918–934

Gomez-Pinilla F (2008) Brain foods: the effects of nutrients on brain function. Nat Rev Neurosci 9:568–578

Gomez-Pinilla F, Tyagi E (2013) Diet and cognition: Interplay between cell metabolism and neuronal plasticity. Curr Opin Clin Nutr Metab Care 16:726–733

Gradinaru D, Borsa C, Ionescu C, Margina D (2013) Advanced oxidative and glycoxidative protein damage markers in the elderly with type 2 diabetes. J Proteomics 92:313–322

Granholm AC, Bimonte-Nelson HA, Moore AB, Nelson ME, Freeman LR, Sambamurti K (2008) Effects of a saturated fat and high cholesterol diet on memory and hippocampal morphology in the middle-aged rat. J Alzheimers Dis 14:133–145

Guarente L (2009) Hypoxic hookup. Science 324:1281–1282

Haigis MC, Guarente LP (2006) Mammalian sirtuins—emerging roles in physiology, aging, and calorie restriction. Genes Dev 20:2913–2921

Hamilton MT, Hamilton DG, Zderic TW (2007) The role of low energy expenditure and sitting on obesity, metabolic syndrome, type 2 diabetes, and cardiovascular disease. Diabetes 56:2655–2667

Hoehn KL, Salmon AB, Hohnen-Behrens C, Turner N, Hoy AJ, Maghzal GJ, Stocker R, Van Remmen H, Kraegen EW, Cooney GJ, Richardson AR, James DE (2009) Insulin resistance is a cellular antioxidant defense mechanism. Proc Natl Acad Sci U S A 106:17787–17792

Holland WL, Brozinick JT, Wang LP, Hawkins ED, Sargent KM, Liu Y, Narra K, Hoehn KL, Knotts TA, Siesky A, Nelson DH, Karathanasis SK, Fontenot GK, Birnbaum MJ, Summers SA (2007a) Inhibition of ceramide synthesis ameliorates glucocorticoid-, saturated-fat- and obesity-induced insulin resistance. Cell Metab 5:167–179

Holland WL, Knotts TA, Chavez JA, Wang LP, Hoehn KL, Summers SA (2007b) Lipid mediators of insulin resistance. Nutr Rev 65:S39–S46

Ikonen E, Vainio S (2005) Lipid microdomains and insulin resistance: is there a connection? Sci STKE 2005(268):e3

Itani SI, Ruderman NB, Schmieder F, Boden G (2002) Lipid-induced insulin resistance in human muscle is associated with changes in diacylglycerol, protein kinase C, and IkappaB-alpha. Diabetes 51:2005–2011

Jumpertz R, Guijarro A, Pratley RE, Mason CC, Piomelli D, Krakoff J (2012) Associations of fatty acids in cerebrospinal fluid with peripheral glucose concentrations and energy metabolism. PLoS One 7:e41503

Kadomatsu K, Kishida S, Tsubota S (2013) The heparin-binding growth factor midkine: the biological activities and candidate receptors. J Biochem 153:511–521

Kersten JR, Toller WG, Tessmer JP, Pagel PS, Warltier DC (2001) Hyperglycemia reduces coronary collateral blood flow through a nitric oxide-mediated mechanism. Am J Physiol Heart Circ Physiol 281:H2097–H2104

Kien CL, Everingham KI, Stevens D, Fukagawa NK, Muoio DM (2011) Short-term effects of dietary fatty acids on muscle lipid composition and serum acylcarnitine profile in human subjects. Obesity (Silver Spring) 19:305–311

Kien CL, Bunn JY, Tompkins CL, Dumas JA, Crain KI, Ebenstein DB, Koves TR, Muoio DM (2013) Substituting dietary monounsaturated fat for saturated fat is associated with increased daily physical activity and resting energy expenditure and with changes in mood. Am J Clin Nutr 97:689–697

Lam CL, Lauder IJ (2000) The impact of chronic diseases on the health-related quality of life (HRQOL) of Chinese patients in primary care. Fam Pract 17:159–166

Landsberg L (2001) Insulin-mediated sympathetic stimulation: role in the pathogenesis of obesity-related hypertension (or, how insulin affects blood pressure, and why). J Hypertens 19:523–528

Lane MA, Ingram DK, Roth GS (1999) Caloric restriction in nonhuman primates: effects on diabetes and cardiovascular risk. Toxicol Sci 52(Suppl2):41–48

Lau C, Toft U, Tetens I, Carstensen B, Jørgensen T, Pedersen O, Borch-Johnsen K (2009) Dietary patterns predict changes in two-hour post-oral glucose tolerance test plasma glucose concentrations in middle-aged adults. J Nutr 139:588–593

Lee J, Duan W, Mattson MP (2002) Evidence that brain derived neurotrophic factor is required for basal neurogenesis and mediates, in part, the enhancement of neurogenesis by dietary restriction in the hippocampus of adult mice. J Neurochem 82:1367–1375

Leeson GW (2014) Future prospects for longevity. Post Reprod Health 20:11–15

Li ZG, Zhang W, Sima AA (2007) Alzheimer-like changes in rat models of spontaneous diabetes. Diabetes 56:1817–1824

Macrae K, Stretton C, Lipina C, Blachnio-Zabielska A, Baranowski M, Gorski J, Marley A, Hundal HS (2013) Defining the role of DAG, mitochondrial function and lipid deposition in palmitate-induced proinflammatory signaling and its counter-modulation by palmitoleate. J Lipid Res 54:2366–2378

Malhotra JD, Kaufman RJ (2007) The endoplasmic reticulum and the unfolded protein response. Semin Cell Dev Biol 18:716–731

Manzel A, Muller DN, Hafler DA, Erdman SE, Linker RA, Kleinewietfeld M (2014) Role of "Western diet" in inflammatory autoimmune diseases. Curr Allergy Asthma Rep 14:404

Mar-Heyming R, Miyazaki M, Weissglas-Volkov D, Kolaitis NA, Sadaat N, Plaisier C, Pajukanta P, Cantor RM, de Bruin TW, Ntambi JM, Lusis AJ (2008) Association of stearoyl-CoA desaturase 1 activity with familial combined hyperlipidemia. Arterioscler Thromb Vasc Biol 28:1193–1199

Maritim AC, Sanders RA, Watkins JB III (2003) Diabetes, oxidative stress, and antioxidants: a review. J Biochem Mol Toxicol 17:24–38

Mattson MP (2012) Energy intake and exercise as determinants of brain health and vulnerability to injury and disease. Cell Metab 16:706–722

Milanski M, Degasperi G, Coope A, Morari J, Denis R, Cintra DE, Tsukumo DM, Anhe G, Amaral ME, Takahashi HK, Curi R, Oliveira HC, Carvalheira JB, Bordin S, Saad MJ, Velloso LA (2009) Saturated fatty acids produce an inflammatory response predominantly through the activation of TLR4 signaling in hypothalamus: implications for the pathogenesis of obesity. J Neurosci 29:359–370

Minokoshi Y, Alquier T, Furukawa N, Kim YB, Lee A, Xue B, Mu J, Foufelle F, Ferré P, Birnbaum MJ, Stuck BJ, Kahn BB (2004) AMP-kinase regulates food intake by responding to hormonal and nutrient signals in the hypothalamus. Nature 428:569–574

Mladenovic Djordjevic A, Perovic M, Tesic V, Tanic N, Rakic L, Ruzdijic S, Kanazir S (2010) Long-term dietary restriction modulates the level of presynaptic proteins in the cortex and hippocampus of the aging rat. Neurochem Int 56:250–255

Mokdad AH, Ford ES, Bowman LA, Dietz WH, Vinicor E, Bales VS, Mark JS (2003) Prevalence of obesity, diabetes, and obesity-related health risk factors. JAMA 289:76–79

Mora F, Segovia G, del Arco A (2007) Aging, plasticity and environmental enrichment: structural changes and neurotransmitter dynamics in several areas of the brain. Brain Res Rev 55:78–88

Moraes JC, Coope A, Morari J, Cintra DE, Roman EA, Pauli JR, Romanatto T, Carvalheira JB, Alexandre LR, Oliveira M, Saad J, Velloso LA (2009) High-fat diet induces apoptosis of hypothalamic neurons. PLoS One 4:e5045

Morino K, Petersen KF, Shulman GI (2006) Molecular mechanisms of insulin resistance in humans and their potential links with mitochondrial dysfunction. Diabetes 55(Suppl 2):S9–S15

Morino K, Neschen S, Bilz S, Sono S, Tsirigotis D, Reznick RM, Moore I, Nagai Y, Samuel V, Sebastian D, White M, Philbrick W, Shulman GI (2008) Muscle-specific IRS-1 Ser→Ala transgenic mice are protected from fat-induced insulin resistance in skeletal muscle. Diabetes 57:2644–2651

Morton GJ, Cummings DE, Baskin DG, Barsh GS, Schwartz MW (2006) Central nervous system control of food intake and body weight. Nature 443:289–295

Nakagami H, Morishita R, Yamamoto K (2001) Phosphorylation of p38 mitogen-activated protein kinase downstream of bax-caspase-3 pathway leads to cell death induced by high D-glucose in human endothelial cells. Diabetes 50:1472–1481

Narasimhan SD, Mukhopadhyay A, Tissenbaum HA (2009) InAKTivation of insulin/IGF-1 signaling by dephosphorylation. Cell Cycle 8:3878–3884

Negre-Salvayre A, Salvayre R, Augé N, Pamplona R, Portero-Otín M (2009) Hyperglycemia and glycation in diabetic complications. Antioxid Redox Signal 11:3071–3109

Nieman DC, Henson DA, Nehlsen-Cannarella SL, Ekkens M, Utter AC, Butterworth DE, Fagoaga OR (1999) Influence of obesity on immune function. J Am Diet Assoc 99:294–299

North BJ, Sinclair DA (2012) The intersection between aging and cardiovascular disease. Circ Res 110:1097–1108

Oliveira Junior SA, Dal Pai-Silva M, Martinez PF, Campos DH, Lima-Leopoldo AP, Leopoldo AS, Nascimento AF, Okoshi MP, Okoshi K, Padovani CR, Cicogna AC (2010) Differential nutritional, endocrine, and cardiovascular effects in obesity-prone and obesity-resistant rats fed standard and hypercaloric diets. Med Sci Monit 12:BR208–BR217

Oliveira SA Jr, Okoshi K, Lima-Leopoldo AP, Leopoldo AS, Campos DH, Martinez PF, Okoshi MP, Padovani CR, Pai-Silva MD, Cicogna AC (2009) Nutritional and cardiovascular profiles of normotensive and hypertensive rats kept on a high fat diet. Arq Bras Cardiol 12:526–533

Ouchi N, Parker JL, Lugus JJ, Walsh K (2011) Adipokines in inflammation and metabolic disease. Nat Rev Immunol 11:85–97

Paillard F, Catheline D, Duff FL, Bouriel M, Deugnier Y, Pouchard M, Daubert JC, Legrand P (2008) Plasma palmitoleic acid, a product of stearoyl-coA desaturase activity, is an independent marker of triglyceridemia and abdominal adiposity. Nutr Metab Cardiovasc Dis 18:436–440

Pan A, Hu FB (2011) Effects of carbohydrates on satiety: differences between liquid and solid food. Curr Opin Clin Nutr Metab Care 14:385–390

Picard F, Kurtev M, Chung N, Topark-Ngarm A, Senawong T, Machado De Oliveira R, Leid M, McBurney MW, Guarente L (2004) Sirt1 promotes fat mobilization in white adipocytes by repressing PPAR-gamma. Nature 429:771–776

Pinheiro AR, Cunha AR, Aguila MB, Mandarim-de-Lacerda CA (2007) Beneficial effects of physical exercise on hypertension and cardiovascular adverse remodeling of diet-induced obese rats. Nutr Metab Cardiovasc Dis 12:365–375

Popkin BM, Duffey K (2010) Does hunger and satiety drive eating anymore? Increasing eating occasions and decreasing time between eating occasions in the United States. Am J Clin Nutr 91:1342–1347

Potenza MN (2014) Obesity, food, and addiction: emerging neuroscience and clinical and public health implications. Neuropsychopharmacology 39:249–250

Poudyal H, Panchal SK, Ward LC, Waanders J, Brown L (2012) Chronic high-carbohydrate, high-fat feeding in rats induces reversible metabolic, cardiovascular, and liver changes. Am J Physiol Endocrinol Metab 12:E1472–E1482

Rioux V, Legrand P (2007) Saturated fatty acids: simple molecular structures with complex cellular functions. Curr Opin Clin Nutr Metab Care 10:752–758

Riserus U, Willett WC, Hu FB (2009) Dietary fats and prevention of type 2 diabetes. Prog Lipid Res 48:44–51

Roglic G, Unwin N (2010) Mortality attributable to diabetes: estimates for the year 2010. Diabetes Res Clin Pract. 87:15–19

Rolfe DFS, Brown GC (1997) Cellular energy utilization and molecular origin of standard metabolic rate in mammals. Physiol Rev 77:731–758

Rolls BJ (2010) Plenary Lecture 1: Dietary strategies for the prevention and treatment of obesity. Proc Nutr Soc 69:70–79

Rolls BJ, Roe LS, Meengs JS (2006) Larger portion sizes lead to a sustained increase in energy intake over 2 days. J Am Diet Assoc 106:543–549

Rook GA (2010) 99th Dahlem conference on infection, inflammation and chronic inflammatory disorders: Darwinian medicine and the 'hygiene' or 'old friends' hypothesis. Clin Exp Immunol 160:70–79

Rothman SM, Griffioen KJ, Wan R, Mattson MP (2012) Brain-derived neurotrophic factor as a regulator of systemic and brain energy metabolism and cardiovascular health. Ann N Y Acad Sci 1264:49–63

Ruvolo PP (2003) Intracellular signal transduction pathways activated by ceramide and its metabolites. Pharmacol Res 47:383–392

Sasaki T, Kitamura T (2010) Roles of FoxO1 and Sirt1 in the central regulation of food intake. Endocr J 57:939–946

Sasaki T, Kim HJ, Kobayashi M, Kitamura YI, Yokota-Hashimoto H, Shiuchi T, Minokoshi Y, Kitamura T (2010) Induction of hypothalamic Sirt1 leads to cessation of feeding via agouti-related peptide. Endocrinology 151:2556–2566

Sauve AA, Schramm VL (2003) Sir2 regulation by nicotinamide results from switching between base exchange and deacetylation chemistry. Biochemistry 42:9249–9256

Savage DB, Petersen KF, Shulman GI (2007) Disordered lipid metabolism and the pathogenesis of insulin resistance. Physiol Rev 87:507–520

Schenk S, Saberi M, Olefsky JM (2008) Insulin sensitivity: modulation by nutrients and inflammation. J Clin Invest 118:2992–3002

Schooneman MG, Vaz FM, Houten SM, Soeters MR (2013) Acylcarnitines: reflecting or inflicting insulin resistance? Diabetes 62:1–8

Schwartz MW, Woods SC, Porte D Jr, Seeley RJ, Baskin DG (2000) Central nervous system control of food intake. Nature 404:661–671

Seneff S, Wainwright G, Mascitelli L (2011) Nutrition and Alzheimer's disease: the detrimental role of a high carbohydrate diet. Eur J Intern Med 22:134–140

Shoelson SE, Lee J, Goldfine AB (2006) Inflammation and insulin resistance. J Clin Invest 116:1793–1801

Simopoulos AP (2009) Evolutionary aspects of the dietary omega-6:omega-3 fatty acid ratio: medical implications. World Rev Nutr Diet 100:1–21

Simopoulos AP (2013) Dietary omega-3 fatty acid deficiency and high fructose intake in the development of metabolic syndrome, brain metabolic abnormalities, and non-alcoholic fatty liver disease. Nutrients 5:2901–2923

Sims-Robinson C, Kim B, Rosko A, Feldman EL (2010) How does diabetes accelerate Alzheimer disease pathology? Nat Rev Neurol 6:551–559

Singh M (2014) Mood, food, and obesity. Front Psychol 5:925

Srinivasan M, Mitrani P, Sadhanandan G, Dodds C, Shbeir-ElDika S, Thamotharan S, Ghanim H, Dandona P, Devaskar SU, Patel MS (2008) A high-carbohydrate diet in the immediate postnatal life of rats induces adaptations predisposing to adult-onset obesity. J Endocrinol 197:565–574

Staiger H, Staiger K, Haas C, Weisser M, Machicao F, Haring HU (2005) Fatty acid-induced differential regulation of the genes encoding peroxisome proliferator-activated receptor-gamma coactivator-1alpha and -1beta in human skeletal muscle cells that have been differentiated in vitro. Diabetologia 48:2115–2118

Stockhorst U, de Fries D, Steingrueber H-J, Scherbaum WA (2004) Insulin and the CNS: effects on food intake, memory, and endocrine parameters and the role of intranasal insulin administration in humans. Physiol Behav 83:47–54

Stranahan AM, Mattson MP (2011) Bidirectional metabolic regulation of neurocognitive function. Neurobiol Learn Mem 96:507–516

Stranahan AM, Norman ED, Lee K, Cutler RG, Telljohann RS, Egan JM, Mattson MP (2008) Diet-induced insulin resistance impairs hippocampal synaptic plasticity and cognition in middle-aged rats. Hippocampus 18:1085–1088

Summers SA, Garza L, Zhou H, Birnbaum MJ (1998) Regulation of insulin-stimulated glucose transporter GLUT4 translocation and Akt kinase activity by ceramide. Mol Cell Biol 18:5457–5464

Talbot K, Wang HY, Kazi H, Han LY, Bakshi KP, Stucky A, Fuino RL, Kawaguchi KR, Samoyedny AJ, Wilson RS, Arvanitakis Z, Schneider JA, Wolf BA, Bennett DA, Trojanowski JQ, Arnold SE (2012) Demonstrated brain insulin resistance in Alzheimer's disease patients is associated with IGF-1 resistance, IRS-1 dysregulation, and cognitive decline. J Clin Invest 122:1316–1338

Thaler JP, Schwartz MW (2010) Minireview: Inflammation and obesity pathogenesis: the hypothalamus heats up. Endocrinology 151:4109–4115

Thaler JP, Yi CX, Schur EA, Guyenet SJ, Hwang BH, Dietrich MO, Zhao X, Sarruf DA, Izgur V, Maravilla KR, Nguyen HT, Fischer JD, Matsen ME, Wisse BE, Morton GJ, Horvath TL, Baskin DG, Tschöp MH, Schwartz MW (2012) Obesity is associated with hypothalamic injury in rodents and humans. J Clin Invest 122:153–162

Thirumangalakudi L, Prakasam A, Zhang R, Bimonte-Nelson H, Sambamurti K, Kindy MS, Bhat NR (2008) High cholesterol-induced neuroinflammation and amyloid precursor protein processing correlate with loss of working memory in mice. J Neurochem 106:475–485

van den Oever IA, Raterman HG, Nurmohamed MT, Simsek S (2010) Endothelial dysfunction, inflammation, and apoptosis in diabetes mellitus. Mediators Inflamm 2010:792393

Villegas R, Salim A, Flynn A, Perry IJ (2004) Prudent diet and the risk of insulin resistance. Nutr Metab Cardiovasc Dis 14:334–343

Villegas R, Yang G, Gao YT, Cai H, Li H, Zheng W, Shu XO (2010) Dietary patterns are associated with lower incidence of type 2 diabetes in middle-aged women: the Shanghai Women's Health Study. Int J Epidemiol 39:889–899

Volkow ND, Wang GJ, Fowler JS, Tomasi D, Baler R (2012) Food and drug reward: overlapping circuits in human obesity and addiction. Curr Top Behav Neurosci 11:1–24

Walford RL, Mock D, Verdery R, Mac Callum T (2002) Calorie restriction in Biosphere 2: alterations in physiologic, hematologic, hormonal, and biochemical parameters in humans restricted for a 2 year period. J Gerontol A Biol Sci Med Sci 57A:B211–B224

Wang Q, Jin T (2009) The role of insulin signaling in the development of beta-cell dysfunction and diabetes. Islets 1:95–101

Warensjo E, Ingelsson E, Lundmark P, Lannfelt L, Syvanen AC, Vessby B, Riserus U (2007) Polymorphisms in the SCD1 gene: associations with body fat distribution and insulin sensitivity. Obesity (Silver Spring) 15:1732–1740

Weckbach LT, Muramatsu T, Walzog B (2011) Midkine in inflammation. ScientificWorldJournal 11:2491–2505

Weckbach LT, Groesser L, Borgolte J, Pagel JI, Pogoda F et al (2012) Midkine acts as proangiogenic cytokine in hypoxia-induced angiogenesis. Am J Physiol Heart Circ Physiol 303:H429–H438

Witte AV, Fobker M, Gellner R, Knecht S, Floel A (2009) Caloric restriction improves memory in elderly humans. Proc Natl Acad Sci U S A 106:1255–1260

Woods SC, Seeley RJ, Baskin DG, Schwartz MW (2003) Insulin and the blood-brain barrier. Curr Pharm Des 9:795–800

Wu A, Ying Z, Gomez-Pinilla F (2008) Docosahexaenoic acid dietary supplementation enhances the effects of exercise on synaptic plasticity and cognition. Neuroscience 26:751–759

Xue B, Yang Z, Wang X, Shi H (2012) Omega-3 polyunsaturated fatty acids antagonize macrophage inflammation via activation of AMPK/SIRT1 pathway. PLoS One 7:e45990

Yang Z, Mo X, Gong Q (2008) Critical effect of VEGF in the process of endothelial cell apoptosis induced by high glucose. Apoptosis 13:1331–1343

Yeomans MR (2010) Alcohol, appetite and energy balance: is alcohol intake a risk factor for obesity? Physiol Behav 100:82–89

Ylostalo P, Suominen-Taipale L, Reunanen A, Knuuttila M (2008) Association between body weight and periodontal infection. J Clin Periodontol 35:297–304

Young CN, Deo SH, Chaudhary K, Thyfault JP, Fadel PJ (2010) Insulin enhances the gain of arterial baroreflex control of muscle sympathetic nerve activity in humans. J Physiol 588:3593–3603

Yu C, Chen Y, Cline GW, Zhang D, Zong H, Wang Y, Bergeron R, Kim JK, Cushman SW, Cooney GJ, Atcheson B, White MF, Kraegen EW, Shulman GI (2002) Mechanism by which fatty acids inhibit insulin activation of insulin receptor substrate-1 (IRS-1)-associated phosphatidylinositol 3-kinase activity in muscle. J Biol Chem 277:50230–50236

Yuzefovych L, Wilson G, Rachek L (2010) Different effects of oleate vs. palmitate on mitochondrial function, apoptosis, and insulin signaling in L6 skeletal muscle cells: role of oxidative stress. Am J Physiol Endocrinol Metab 299:E1096–E1105

Zhang X, Zhang G, Zhang H, Karin M, Bai H, Cai D (2008) Hypothalamic IKKbeta/NF-kappaB and ER stress link overnutrition to energy imbalance and obesity. Cell 135:61–73

Ziauddeen H, Fletcher PC (2013) Is food addiction a valid and useful concept? Obes Rev 14:19–28

Zilberter T (2011) Carbohydrate-biased control of energy metabolism: the darker side of the selfish brain. Front Neuroenergetics 3:8

Chapter 2
Neurochemical Effects of Long Term Consumption of High Fat Diet

2.1 Introduction

Our diet consists of various types of fats (lipids), a group of naturally occurring organic compounds, which are soluble in nonpolar organic solvents and insoluble in water. The term lipid is ambiguous because the above definition is somewhat misleading. There are many substances that are regarded as lipids which are soluble in both water and nonpolar solvents. Lipids are not only a good source of high energy and essential fatty acids, but are required for the absorption of fat-soluble vitamins (Salvati et al. 2000). Lipids are an important constituent of the brain, not only as components of myelin, but also because of the large surface-to volume ratio of neurons-neurons contain a higher proportion of lipid than other cells because lipid is the main constituent of the neuronal cell membrane (Greenwood and Young 2001). Dietary lipids include fatty acids (saturated, monounsaturated, and polyunsaturated), glycerides, (neutral glycerides and phosphoglycerides), non-glyceride lipids, and sterols. Phosphoglycerides include phospholipids, whereas nonglyceride lipids include sphingolipids (Farooqui and Horrocks 2007). Dietary fatty acids have been reported to play an essential role in brain function (Kaplan and Greenwood 1998). Although, the mechanisms of action of fatty acids are not clear, the fatty acid composition of neural membranes is at least partly controlled by the diet (Fisher and Levine 1991; Farooqui and Horrocks 2007).

Dietary lipids are digested and absorbed by the digestive tract. Among these lipids, fatty acids behave not only as signaling molecules, but also act as precursors to families of lipid mediators, and components of both simple and compound lipids (Bourre et al. 1992; Rapoport 2005; Farooqui 2011). Following digestion, fatty acids are incorporated in noticeable amounts in chylomicron phospholipids during the process of gastrointestinal absorption, and are also packaged into VLDL triacylglycerols by the liver. Human brain lacks the ability to synthesize essential fatty acids from their precursor α-linolenic and linoleic acids. Fatty acids are transported in the blood either bound to albumin or in the form of triacylglycerol esterified with

© Springer International Publishing Switzerland 2015

A.A. Farooqui, *High Calorie Diet and the Human Brain*,

DOI 10.1007/978-3-319-15254-7_2

circulating very low density lipoproteins along with ApoE (a glycoprotein of 34 kDa). Only a small proportion of total plasma fatty acids is found in the free fatty acid pool. Through the action of lipoprotein lipase bound to the luminal surface of endothelial cells, fatty acids are cleaved from circulating triacylglycerol where they can act as ligands or taken up by peripheral tissues as well as brain (Polozova and Salem 2007). Understanding mechanisms by which fatty acids cross the blood brain barrier (BBB) for their utilization by neurons and glial cells is not fully understood. However, it is proposed that passive diffusion and intracellular fatty acid binding proteins, which are expressed on the luminal side of the endothelial cells that line the neurovascular unit, play important roles in fatty acid transport related processes (Fig. 2.1) (Kalant and Cianflone 2004). These fatty acids are then transferred to astrocytes that provide support to endothelial and other CNS cells. From astrocytes, fatty acids are transferred into cellular membranes and neurons. Brain has capability to synthesize phospholipids, sphingolipids, and cholesterol. Brain biomembranes (neural membranes) are enriched in phospholipids, sphingolipids, and cholesterol. Phospholipids and sphingolpids contain fatty acids as components, whereas cholesterol forms ester with fatty acids. In the nervous system, where fatty acids are found in huge amounts, they participate in its development and maintenance throughout life. Among fatty acids, long-chain polyunsaturated fatty acids (PUFAs), which have more than 16 carbon atoms and more than one *cis* double bond, are of critical nutritional importance due to their role in optimal brain development and function

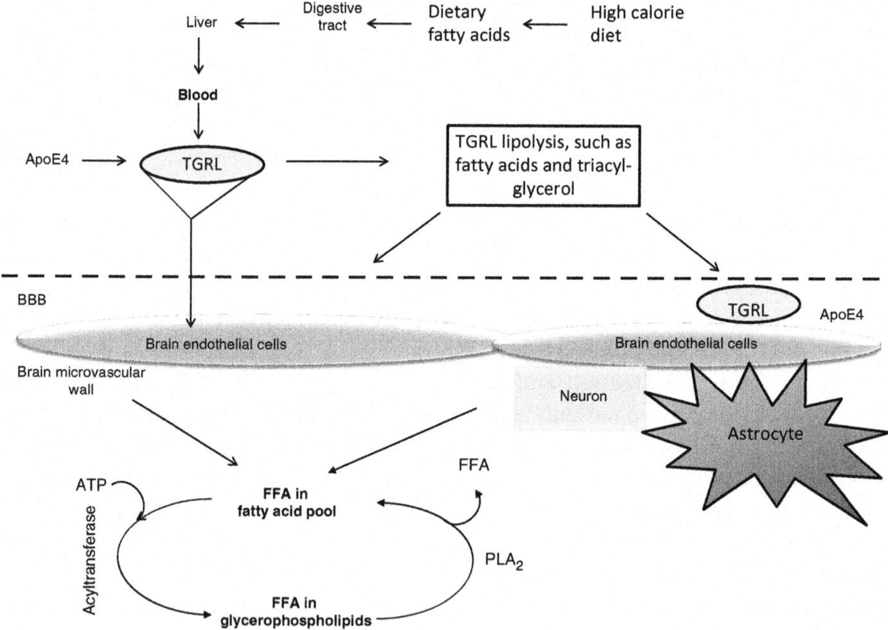

Fig. 2.1 Transport of fatty acids containing TAG and phospholipids from blood to the brain. *FFA* Free fatty acid, *PLA₂* phospholipase A₂, *ApoE4* apolipoprotein E4, *TAG* triacylglycerol

(Farooqui 2009). Because physiological properties of PUFAs largely depend on the position of the first double bond from the terminal methyl group of the carbon chain, therefore PUFAs are categorized into n-3, n-6, and n-9 PUFAs by its position. n-3 PUFAs are found in plants and fish and are the main constituents of fish oil used as a dietary supplement; n-6 PUFAs are found in vegetable oils; and n-9 fatty acids are found in olive oil. PUFAs are incorporated into cell membranes and are metabolized into lipid signaling molecules and other compounds (Farooqui 2009). Mammalian tissues are capable of synthesizing phospholipids, sphingolipids, and cholesterol through *de novo* synthesis (Farooqui 2011). Phospholipids, sphingolipids, and cholesterol not only perform many key biological functions, such as acting as structural components of neural membranes, energy storage sources, and intermediates in signaling pathways, but also provide neural membranes with a suitable environment, fluidity and ion permeability (Farooqui and Horrocks 2007). These lipids are under tight homeostatic control (Oresic et al. 2008), and exhibit spatial and dynamic complexity at multiple levels. Among these lipids, phospholipids are most abundant (Farooqui et al. 2000a). They are asymmetrically distributed between the two bilayers of neural membranes. These bilayers are often conceptualized as undulating fields of identical molecules. In fact, they are complex matrices of several hundred molecularly distinct species (Farooqui et al. 2000b). Their composition is in constant flux. Carbon chains and defining polar head groups are dynamically exchanged by activated phospholipases and lysophospholipid transferases in response to environmental stimuli. The binding between phospholipids and proteins is necessary for vertical positioning and tight integration of proteins into the membrane. Among above mentioned lipids, phospholipids are derivatives of *sn*-glycero-3-phosphoric acid. They contain an O-acyl or O-alkyl or O-alk-1′-enyl residue at the *sn*-1 position and an O-acyl residue at the *sn*-2 position of the glycerol moiety. They are defined on the basis of the substituents on the phosphoric acid at the *sn*-3 position. In neural membranes degradation of phospholipids, sphingolipids, and cholesterol is linked with various receptors located in plasma, nuclear, mitochondrial, and microsomal membranes (Farooqui and Horrocks 2007).

2.2 Consumption of High Fat Diet and Generation of Lipid Mediators in the Brain

High calorie diet, which contains saturated fat and cholesterol, produces damaging affects in the brain. Detrimental changes in the brain include behavioral changes, reduction in hippocampal neurogenesis (Hwang et al. 2008), increase in oxidative stress (Morrison et al. 2010; Freeman et al. 2013), and alteration in microvascular pathology (Franciosi et al. 2009). Furthermore, high fat diet induces obesity, which correlates with reduction in focal gray matter volume and enlargement in white matter in the frontal lobe (Pannacciulli et al. 2006), altering learning, memory, and executive function in humans (Elias et al. 2003) and cognitive deficits in a rodent model (Winocur and Greenwood 2005). These changes are linked to changes in stress and reward pathways in the CNS, supporting the view that the rewarding properties of such foods may override energy balance signals (Franken and Muris 2005). High fat diet activates brain reward centers in a manner similar to drugs of abuse (Cagniard et al. 2006). Thus, it is likely that behaviors and motivation for overeating and drug abuse share common underlying mechanisms. As stated above, dietary lipids are digested and components of lipids are absorbed by the digestive tract, transported through circulation to brain and visceral tissues, where they are incorporated in biomembranes as phospholipids and sphingolipids. Brain is capable of synthesizing its own cholesterol pool and dietary cholesterol becomes incorporated in biomembranes of visceral tissues, arteries and veins. Interactions of agonists with their receptors result in phospholipases A_2 (PLA_2) and sphingomyelinase-mediated release of free fatty acids and ceramide from phospholipids and sphingolipids respectively. PLA_2 liberates fatty acids from the sn-2 position of neural membrane phospholipids with the generation of lysophospholipid (Farooqui and Horrocks 2007), whereas sphingomyelinase generates phosphorylcholine and ceramide (Farooqui 2011). Lysophospholipid is converted into platelet activating factor (PAF, 1-O-alkyl-2-acetyl-sn-glycero-3-phosphocholine), a bioactive metabolite, which plays important roles in microcirculatory disturbances and neuroinflammation (Farooqui et al. 2008; Farooqui 2011), whereas ceramide is metabolized to ceramide 1-phosphate, sphingosine and sphingosine 1-phosphate (Farooqui 2011).

It well known that n-6 and n-3 PUFA are highly concentrated in neural phospholipids, including subcellular membranes. Although the circulation contains at least ten times more n-6 arachidonic acid (ARA) than n-3 PUFA, docosahexaenoic acid (DHA) predominates in the retina and brain. Furthermore, DHA is particularly concentrated at neural synapses (Farooqui 2009). Importantly, DHA levels in neural membranes vary according to dietary consumption of this fatty acid and there is evidence that low PUFA and high cholesterol levels reduce phospholipid fluidity (Yehuda et al. 2002). Fatty acids are released by the action of PLA_2 include ARA, and DHA. The enzymic oxidation of ARA through cyclooxygenases (COXs); lipoxygenases (LOXs); and epoxygenases (EPOX) results in the synthesis of eicosanoids, a family of lipid mediators, which include prostaglandins (PGs), leukotriene (LTs), lipoxins (LXs), and thromboxanes (TXs), as well as hydroxyeicosatetraenoic acid (HETE) and epoxyeicosatetraenoic acids (EETs), and dihydroxyeicosatrienoic acids (DHETs) (Fig. 2.2) (Phillis et al. 2006). Eicosanoids produce a wide range of biological actions including potent effects on inflammation, vasodialation, vasoconstriction, apoptosis and immune responses (Farooqui 2011) through interactions with eicosanoid receptors. Non-enzymic peroxidation of ARA is a multistep process involving the abstraction of electrons from ARA by metals or free radicals (hydroxyl radicals, $OH^\bullet$) leading to the formation of alkoxyl and peroxyl radicals. These radicals in turn oxidize additional substrates, thereby perpetuating an autocatalytic cycle, which is ultimately terminated by radical-radical interactions. The overall process leads to the formation of many non-radical intermediates and end products including 4-hydroxynonenal (4-HNE), isoprostanes, isofuran, isoketal, acrolein, and malondialdehyde (Fig. 2.3). Of these, the 4-HNE is considered as one of the most abundant and bioactive specie (Esterbauer et al. 1991; Farooqui 2011). 4-HNE has been studied

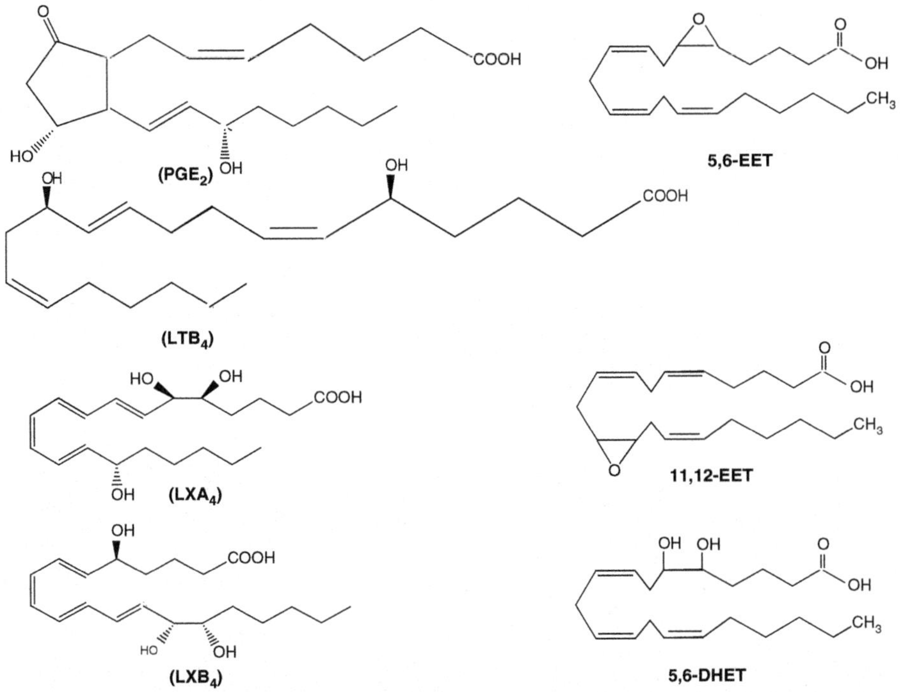

Fig. 2.2 Chemical structures of ARA-derived eicosanoids

extensively both in the context of physiological and pathological vascular events as well as in neurological disorders (Farooqui 2011). It reacts avidly with nucleophilic groups found in glutathione, proteins, phospholipids, and nucleic acids. Chemical reactions of 4-HNE with nucleophiles incite a range of effects. At low concentrations, oxidized lipids have the capacity to promote a protected cellular phenotype, whereas at high concentrations they disrupt cell structure and function leading to either necrotic or apoptotic cell death (Farooqui 2011). Increased generation of reactive oxygen species (ROS) following consumption of high fat diet also results in activation and translocation of NF-κB to the nucleus, where it induces the transcription of proinflammatory cytokines and chemokines (TNF-α, IL-1β, IL-6 and MCP1). These cytokines stimulate activities of PLA$_2$ (Farooqui 2011). In addition, ROS also contribute to the dysfunction of the blood–brain barrier (BBB) and damage to the brain parenchymal cells.

Insulin is an important hormone that regulates food intake. Central administration of insulin decreases food intake in rodents and primates, and hypothalamus is implicated in insulin's action in reducing food intake. Under physiological conditions, insulin inhibits hepatic glucose production, promotes skeletal muscle glucose uptake, and retards lipolysis. Insulin resistance causes impairments in insulin-mediated suppression of hepatic glucose production, skeletal muscle glucose disposal, and inhibition of lipolysis, leading to relative hyperglycemia and increased plasma levels

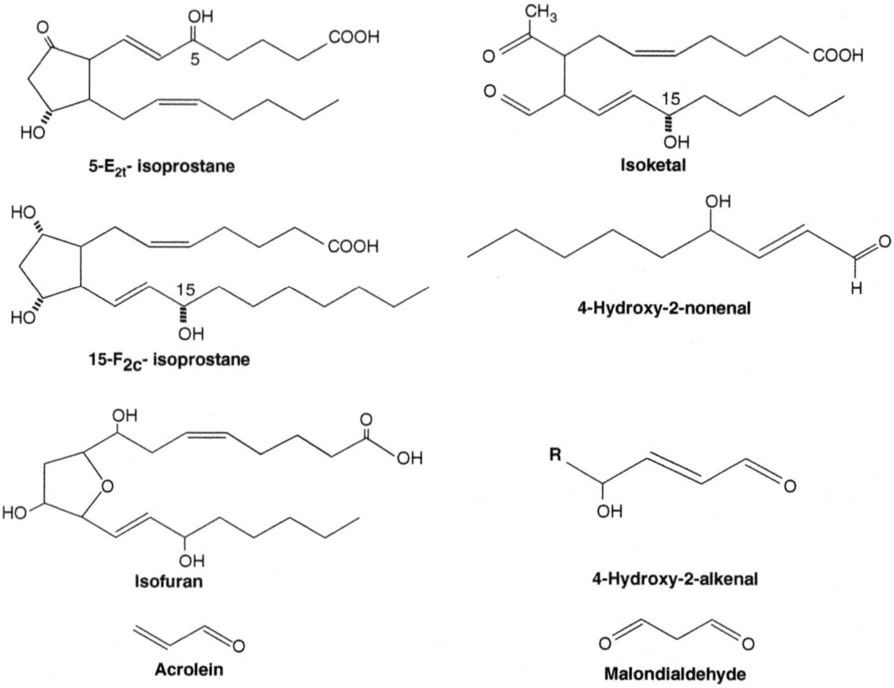

Fig. 2.3 Chemical structures of ARA-derived non-enzymic lipid mediators

of FFA (Olefsky and Glass 2010). Consumption of high fat diet, which is enriched in saturated and n-6 fatty acids not only increases body weight, but also impairs peripheral insulin sensitivity as demonstrated by reduction in insulin signaling (Savage et al. 2007), decrease in adiponectin (an adipocyte-derived protein with insulin sensitizing, and anti-inflammatory properties), increase in proinflammatory adipokines, as well as blunts skeletal muscle response. These processes can be ameliorated by reducing dietary fat content (Muurling et al. 2002). Circulating insulin crosses the BBB and acts on insulin receptor in the hypothalamus to lower food intake and body weight. It is not known how does and to what extent systemic insulin resistance is manifested in the brain. However, it is known that chronic or acute intracerebroventricular administration of insulin decreases food consumption and obesity in baboons, mice, and rats (Woods 1996). Administration of insulin antibodies and selective knocking out of neuronal insulin receptors by insulin receptor antisense oligonucleotides reduce insulin activity and insulin signaling cascade (Plum et al. 2006), resulting in hyperphagia and obesity (McGowan et al. 1992; Brüning et al., 2000; Obici et al. 2002; Plum et al. 2006).

The hypothalamus plays an important role in monitoring fatty acid metabolism as part of its energy-sensing function (Kim et al. 2002; Obici et al. 2003). Fatty acids can be taken up into the cell either from the blood circulation through transporters or synthesized de novo inside the cell. Within the cell, fatty acids either undergo oxidation in the mitochondria for energy production or may be directed towards glycerolipid synthesis (triglycerides or phospholipids) for later energy production or membrane function. Entry of free fatty acids in mitochondria via palmitoyltransferase (CPT) 1-dependent pathway plays an important role in the appetite regulation (Lam et al. 2005). Activation of central AMPK in response to elevated cellular AMP/ATP ratio results in the phosphorylation and inactivation of acyl CoA carboxylase that, in turn, reduces production of malonyl CoA, the allosteric inhibitor of CPT1 (Zammit and Arduini 2008; Andrews et al. 2008). The depletion of malonyl CoA removes its allosteric inhibition on CPT1 and facilitates mitochondrial fatty acid entry, oxidation and increased ATP production required for neuronal activation. In fact, recent studies indicate that the availability of neuron-intrinsic free fatty acids and their metabolism may provide the bioenergetic requirements for hypothalamic neuronal activation and firing (Andrews et al. 2008). Neuron-intrinsic free fatty acids are associated with the initiation of food-seeking behavior. The mechanisms responsible for regulating neuronal availability of free fatty acids are unclear and more studies are required on this aspect of appetite control (Lam et al. 2005). In the brain fatty acids not only play an important role in the regulation of glucose metabolism, but are required for general energy homeostasis (Seeley and Woods 2003; Lam et al. 2005). Thus, hypothalamic malonyl-CoA responds to the level of circulating glucose and leptin, which are associated with maintenance and modulation of energy homeostasis (Wolfgang et al. 2007). Leptin, a 16 kDa protein mainly produced by adipocytes, is known to decrease hypothalamic 5′-adenosine monophosphate-activated kinase (AMPK), an energy-sensing kinase, which responds to changes in the energy levels of the cell and the whole body in order to maintain adequate ATP levels in the cell (Minokoshi et al. 2004). During energy sensing, AMPK phosphorylates and inactivates acetyl-CoA carboxylase, thereby inhibiting fatty acid synthesis by decreasing the availability of malonyl-CoA (Frederich and Balschi 2002). Ghrelin, a hormone, which is released from the stomach, duodenum, and ileum and cannabinoids stimulate hypothalamic AMPK leading to an increase in appetite while inhibiting AMPK activity in the liver and adipose tissue, thereby leading to lipogenic effects (van Thuijl et al. 2008).

Unlike saturated and n-6 FFA, consumption of n-3 fatty acids regulates neural membrane fluidity, permeability and receptor function along with suppression of neuroinflammation (Farooqui 2009) and reduction in insulin resistance. In visceral tissues, n-3 PUFA have been reported to regulate the expression of a number of genes involved in carbohydrate and lipid metabolism by modulating the activity or expression of a number of transcription factors including peroxisome proliferator activated receptors (PPAR), sterol regulatory element binding protein-1c (SREBP-1c), hepatic nuclear factors (HNF), retinoid X receptors (RXR) and liver X receptor (LXR) (Jump et al. 2013) (Table 2.1). n-3 Fatty acids induce anti-inflammatory and insulin-sensitizing effects by increasing the adiponectin. It is not known whether or not n-3 FFA can directly stimulate adiponectin secretion from human adipocytes. Recent studies have indicated that n-3 FFA increase adiponectin secretion from human adipocytes via a peroxisome proliferator-activated receptor γ-dependent mechanism (Tishinsky 2013). The effects of n-3 PUFA on adiponectin

Table 2.1 Modulation of transcription factors by n-3 fatty acids

Transcription factor	Effect	Reference
NF-κB	Inhibition	Jump et al. (2013)
LXR	Inhibition	Jump et al. (2013)
SREBP	Inhibition	Jump et al. (2013)
PPAR	Stimulation	Jump et al. (2013)

secretion are additive when combined with the thiazolidinedione and rosiglitazone. In addition, incorporation of n-3 FFA into a high saturated FFA diet inhibits the impairment in adiponectin response and restores impairment in insulin response in rodent skeletal muscle. The molecular mechanism associated with these processes is not fully understood. However, it is proposed that n-3 FFA may act by preventing saturated FFA-induced increases in expression of toll-like receptor 4 (Tishinsky 2013). These effects of n-3 FFA may also contribute to reduction in plasma triglyceride (TAG) level, arrhythmias, oxidative stress, inflammation, and improve endothelial dysfunction. n-3 FFA may also produce antithrombotic effects including increased platelet response to clopidogrel and decreased thrombin formation (Farooqui 2009). Collective evidence suggests that the long term consumption of high-fat diets may modulate the central mechanisms of feeding regulation as well as peripheral metabolism by changing the plasma lipid profile and gene expression (Jump 2008). Animal studies have found that saturated and n-6 and n-9 fatty acids are more involved in the genesis of insulin resistance, while n-3 PUFAs have been associated with reduction in insulin resistance (Magdeldin et al. 2009; Nuernberg et al. 2011). Furthermore, diet rich in n-6 FFA has been reported to promote abdominal obesity, possibly by increasing sterol regulatory element binding protein 1c (SREBP1c) in visceral adipose tissues of rats, thereby stimulating lipogenic pathways (Muhlhausler et al. 2010). Conversely, diet rich in n-3 FFA produces anti-obesogenic effects and leads to a decrease in fat deposition and fasting serum triglyceride concentrations concomitant with an increase in energy expenditure and fatty acid oxidation in human and rodent studies (Buckley and Howe 2009).

Like ARA, DHA is oxidized by 15-LOX generating D series resolvins (RvDs), protectins/neuroprotectins (PDs/NPDs), and maresins (MaRs) (Fig. 2.4). These DHA-derived metabolites have potent anti-inflammatory and proresolution properties. They retard excessive inflammatory responses and promote resolution by enhancing clearance of apoptotic cells and debris from inflamed brain tissue (Serhan et al. 2008, 2011; Bazan 2009). Detailed investigations have been performed on NPD_1 in retinal pigment epithelial (RPE) cells, and brain (Marcheselli et al. 2003; Lukiw et al. 2005; Bazan 2009). NPD_1 up-regulates antiapoptotic proteins (Bcl-2 and Bcl-xL) and down-regulates proapoptotic proteins (Bax and Bad) in response to cellular oxidative stress and cytokine activation leading to an overall prosurvival transcriptome (Marcheselli et al. 2003; Lukiw et al. 2005; Bazan 2009). Generation of NPD_1 provides a specific endogenous mechanism to explain DHA-mediated modulation of neuroinflammation and neuroprotection (Fig. 2.5). In addition, NPD_1 also regulates adiponectin (González-Périz et al. 2009), a 244 amino acid

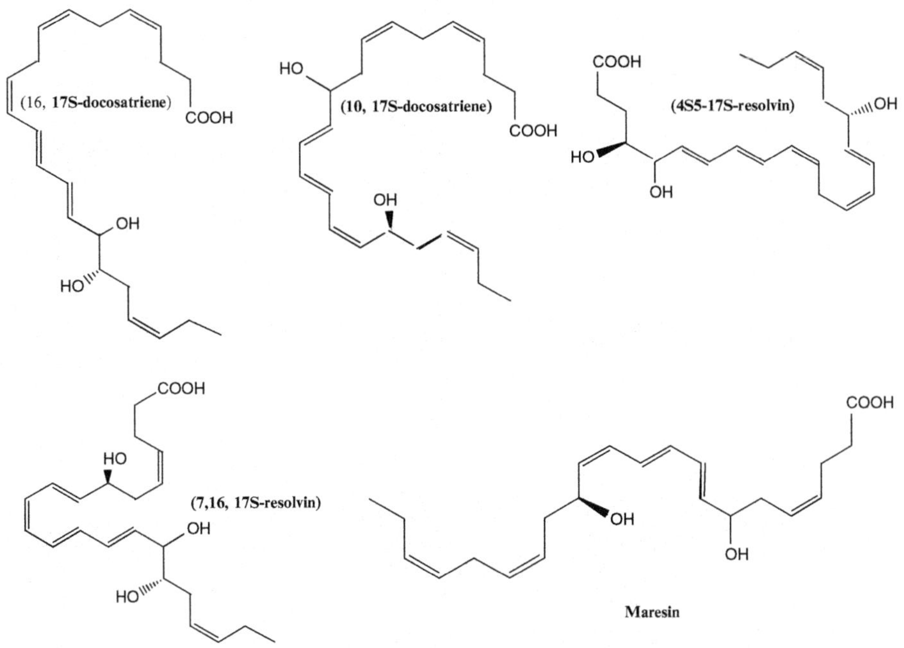

Fig. 2.4 Chemical structures of DHA-derived lipid mediators

multimeric protein (30 kDa) solely produced and secreted by adipose tissue, acts through its seven transmembrane receptors distinct from G-protein coupled receptors (AdipoR1 and AdipoR2), which are expressed throughout the brain (Ahima 2005; Rondinone 2006). Adiponectin modulates hypothalamic and brainstem neuronal activity, and acts centrally to control peripheral metabolism (Hoyda et al. 2007; Kubota et al. 2007). Adiponectin not only activates cAMP-dependent protein kinase A, but also inhibits endothelial nuclear factor $\kappa\beta$ (NF-$\kappa\beta$) signaling. Adiponectin is a potent anti-inflammatory protein which suppresses TNF-α-induced NF-κB activation and block TNF-α release from endothelial cells and macrophages. Thus, activities of adiponectin are inversely proportional to obesity, diabetes, and other insulin-resistant states. Collective evidence suggests that at the molecular level, adiponectin exerts insulin-sensitizing effects by increasing glucose uptake, NO production, and free fatty acid oxidation (Chen et al. 2003; Soodini and Hamdy 2004; Dyck 2009) and shows an antiinflammatory activity mainly through a cAMP-mediated interference with NF-κB signaling (Ouchi et al. 2000). Based on above mentioned studies, it can be suggested that n-3 FFA-derived lipid mediators (NPD$_1$ and resolvins) are generally anti-inflammatory, anti-thrombotic and vasodilatory, balancing and counteracting pro-inflammatory, vasoconstricting actions of eicosanoids derived from n-6 FFA (Farooqui 2009). Like ARA, DHA is highly prone to non-enzymic oxidation due to their large number of double bonds and their position within the fatty acid chain (Shahidi and Zhong 2010). The non-enzymic

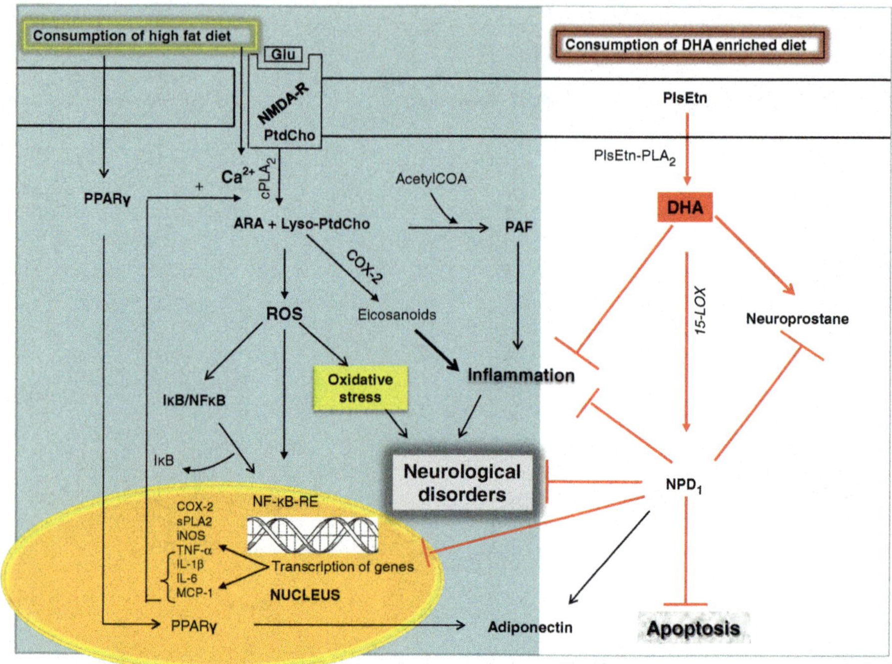

Fig. 2.5 Interactions between ARA and DHA-derived lipid mediators. *PM* plasma membrane, *NMDA-R* N-methyl-D-aspartate receptor, *Glu* glutamate, *PtdCho* phosphatidylcholine, *PlsEtn* ethanolamine plasmalogen, *PlsEtn-PLA₂* plasmalogen-selective phospholipase A₂, *cPLA₂* cytosolic phospholipase A₂, *COX-2* cyclooxygenase, *15-LOX* 15-lipoxygenase, *ARA* arachidonic acid, *DHA* docosahexaenoic acid, *ROS* reactive oxygen species, *NF-κB* nuclear factor kappaB, *NF-κB-RE* nuclear factor kappaB response element, *IκB* inhibitory subunit of NF-κB, *TNF-α* tumor necrosis factor-α, *IL-1β* Interleukin-1 beta, *IL-6* Interleukin-6, *PGs* prostaglandins, *LTs* leukotrienes, *TXs* thromboxanes

oxidation of DHA results in the generation of 4-hydroxyhexanal, neuroprostane, and neurofuran. Non-enzymic lipid mediators of ARA and DHA metabolism, contributing to oxidative stress (Farooqui 2011).

2.2.1 Harmful Effects of Saturated Fatty Acids Present in the High Fat Diet

Consumption of saturated fat has been linked to an increased risk of cardiovascular disease, and this effect is mediated primarily by increased concentrations of LDL cholesterol. Major dietary sources of saturated fatty acids in the United States are full-fat dairy products and red meat (US Department of Health and Human Services 2005). Saturated fats modulate vascular function by increasing the selective uptake of cholesterol in the arterial wall, resulting in increased atherogenesis in mouse models (Seo et al. 2005). The high fat diet is not only enriched in saturated fatty acids and n-6 fatty acid (ARA), which are found in red meat and vegetable oils (such as sunflower, safflower, and corn oils), but has significant levels of trans fatty acids. It produces inflammation through several mechanisms. Saturated fatty acids (lauric, palmitic and, stearic acids) in high fat diet exert proinflammatory effects through the activation of toll-like receptor (TLR-2 and TLR4) signaling (Suganami et al. 2007; Ajuwon and Spurlock 2005; Lee et al. 2003). Mice with either targeted disruption or naturally occurring mutations of TLR4 do not develop obesity-associated insulin resistance (Poggi et al. 2007; Tsukumo et al. 2007). However, genetic models interfering with inflammatory signaling such as conventional JNK1- and TLR4-deficient mice develop obesity-associated insulin resistance (Hirosumi et al. 2002). More recently, activation of inflammatory signaling has also been detected in the hypothalamus of obese rats (De Souza et al. 2005; Wisse and Schwartz 2009). It is shown that inhibiting IKK2 action in the brain (Zhang et al. 2008) or ameliorating ER-stress activation in the CNS (Ozcan et al. 2009) prevents neuronal leptin resistance. At the molecular level, the activation of TLR4 by saturated fatty acids triggers NFκB-mediated pro-inflammatory gene expression and subsequent cytokine secretion from macrophages and microglial cells. These cells play a complex role in the brain. Although a set number of quiescent microglial cells are always present, and needed for normal function, activation of these inflammatory cells is typically correlated with the occurrence of an inflammatory event (Kettenmann et al. 2011). Microglial cells function as macrophages in the brain, with the job of surveying the area and controlling any disturbance/foreign invader via phagocytosis (Nakajima et al. 2003). The stimulation of microglial cells releases both pro- or anti-inflammatory cytokines and chemokines (Kettenmann et al. 2011). Although negative feedback factors like suppressor of cytokine signaling 3 (SOCS3) (Narazaki et al. 1998) and monocyte chemoattractant 1-induced protein (MCP1-IP) act to suppress pro-inflammatory cytokine signaling, these feedback factors have been reported to be dysfunctional in obese humans with type II diabetes (Scheele et al. 2012). High fat diet also increases levels of triacylglycerols, which have been reported to impair the transport of leptin across the BBB (Banks et al. 2004). This process may account in part for the peripheral leptin resistance seen in individuals consuming high fat diet. Thus, triacylglycerols can impair cognition by preventing leptin from reaching the brain regions important for learning and memory. Triglycerides may also affect cognition through their ability to modify release of feeding peptides (Farr et al. 2008), many of which affect cognition through nitric oxide-dependent pathways (Diano et al. 2006; Gaskin et al. 2003).

In addition, high-fat diet reduces the expression of insulin receptors, inhibits the oxidation of fatty acids in skeletal muscles, diminishes the mRNA expression and intracellular protein content of GLUT4, and reduces the translocation of GLUT4 to the cell membrane of visceral tissues (Lee et al. 2006; Eller et al. 2013). These factors may be responsible for the observed hyperglycemia and hyperinsulinemia in animals (Sundaresan et al. 2011). In hepatocytes, saturated fatty acids contribute to the induction of endoplasmic reticulum stress (ER stress). Induction of ER stress by saturated FFA upregulates the proapoptotic activities of the borine (BH3)-contain proteins, such as Bim and PUMA (Cazanave et al. 2010). Both Bim and PUMA contribute to lipoapoptosis (Cazanave et al. 2010). These proteins also initiate and participate in processes triggering the mitochondrial pathway of apoptosis (Cazanave et al. 2010). It is also suggested that the upregulation of PUMA and Bim is transcriptionally modulated by saturated FFA-induced ER stress (Cazanave et al. 2010).

Consumption of high calorie diet with fats of animal origin is mainly responsible for the dramatic rise in metabolic diseases through mechanisms, which involve low grade inflammation (Hotamisligil 2006; Shoelson et al. 2006; Farooqui 2013). At the molecular level, TNF-α is continuously released from adipose tissues during obesity activates protein kinase C (PKC), which phosphorylates insulin receptor substrates on serine residues such as Ser-307. This phosphorylation leads to the inactivation of insulin signaling and hence to insulin resistance (Tanti et al. 2004). However, the origin of the antigens responsible for the inflammatory process is not fully understood. However, it is proposed that intestinal microbiota (*Bacteroidetes* species) in animals consuming a fatty diet may contribute to obesity, insulin resistance, and low grade inflammation (Ley et al. 2005; Turnbaugh et al. 2006). Collectively, these studies indicate that the intestinal microbiota and the interactions between the host and the microbiota are involved in the control of energy metabolism and low grade inflammation. This process requires bacterial antigen receptors such as toll-like receptors (TLR) (Keita et al. 2008) and nod-like receptors (NLR) (Kufer et al. 2008), through which viable and dead bacteria and their components (lipopolysaccharide) initiate the activation of innate immune cells and provoke a non-resolving low grade inflammation. A recent study has indicated that high doses of LPS activate the NF-κB and the PtdIns 3K pathways leading to a strong and transient expression of pro- and anti-inflammatory mediators. In contrast, low doses of LPS, which are associated with obesity activate transcription factor (ATF)2 through IL-1R-associated kinases (IRAK)-1 and Toll-interesting protein (Tollip)-mediated generation of mitochondrial ROS as well as suppress PtdIns 3K activity, leading to induction and maintenance of low grade inflammation (Maitra et al. 2012). Furthermore, cells from the innate immune system infiltrate the adipose depots during metabolic diseases (Weisberg et al. 2006). Based on these observations, it is suggested that during high-fat diet-induced diabetes, commensal intestinal bacteria translocate in a pathological manner from the intestine towards the tissues where they trigger a local inflammation.

2.2.2 Harmful Effects of High Levels of Polyunsaturated Fatty Acids in the High Fat Diet

As stated above, current high fat diet contains high levels of ARA. At present, the ratio of ARA to DHA in high fat diet is about 20:1. The Paleolithic diet on which human beings have evolved, and lived for most of their existence had a ratio of 1:1. In addition, Paleolithic diet was high in fiber, rich in fruits, vegetables, lean meat, and fish

(Simopoulos 2008, 2011; Cordain et al. 2005). Long term consumption of saturated fatty acids and ARA in high calorie diet not only elevates triacylglycerol levels, which are an important risk factor for cardiovascular and cerebrovascular diseases. High fat diet also increases levels of PGs, LTs, and TXs and upregulates the expression of proinflammatory genes including genes for cytokines (TNF-α, and IL-1β) and enzymes (secretory phospholipase A_2, cyclooxygenase-2, and nitric oxide synthase) in brain and peripheral tissues. These cytokines and enzymes initiate and maintain neuroinflammation. When they are present in high quantities, above mentioned metabolites, cytokines, and enzymes influence various metabolic activities besides inflammation such as platelet aggregation, haemorrhage, vasoconstriction, and vasodilation (Farooqui 2011). In addition, n-6 FFA enriched diet not only induces an increase in hypothalamic neuronal activation, but also inhibits insulin-induced hypophagia and Akt serine phosphorylation (Carvalheira et al. 2003).

In contrast, the long term consumption of Paleolithic diet (DHA-enriched diet) lowers triacylglycerol in the blood and produces anti-inflammatory effects that partly cause repression of genes that code for pro-inflammatory cytokines in the brain and peripheral tissues. Reduction of insulin resistance by DHA is based on its anti-inflammatory properties. This process is probably mediated through TLRs (Thomas and Pfeiffer 2011). Contrary to the proinflammatory profile of saturated fatty acids, DHA inhibits TLR-2 and TLR-4 (Lee et al. 2004). Additional effects of DHA on insulin action may be related to beneficial alterations in membrane fluidity, increased binding affinity of the insulin receptor, and improved glucose transport into cells via glucose transporters (Lovejoy 2002; Farooqui 2009), as well as effects on circulating triglycerides and low-density lipoprotein particles (Fedor and Kelley 2009). Finally, effects on the regulation of various genes that are involved in lipid and carbohydrate metabolism have been shown which include peroxisome proliferator-activated receptors, Sterol Regulatory Element-Binding Protein (SREBP-1c), hepatic nuclear factors, retinoid X receptors, and liver X receptors (Thomas and Pfeiffer 2011). Studies on diet enriched in fish oil as a source of n-3 FFA have indicated that this diet abolishes serotonin-induced hypophagia and impairs hypothalamic serotonin turnover along with changes in levels of 5-HT 2C receptor (Watanabe et al. 2010). Low levels of DHA and high levels of ARA in high calorie diet are linked with acute and chronic neurodegenerative and neuropsychiatric disorders, such as stroke, Alzheimer disease, and depression (Farooqui 2009, 2010, 2011, 2013). Collective evidence suggests that a higher ratio of n-6:n-3 fatty acids is closely associated with neuroinflammatory changes in the brain (Shelton and Miller 2010).

2.2.3 Harmful Effects of Trans Fatty Acids in High Fat Diet

Trans fatty acids are unsaturated fatty acids with at least one double bond in the trans configuration. These fatty acids are produced when liquid oils are processed into solid fats like shortening and hard margarine through hydrogenation.

Hydrogenation not only increases the shelf life, but also elevates flavor and stability of foods having shortening or margarine. Trans fatty acids are also generated during heating and frying of oils at high temperatures. The major sources of trans fatty acids are partially hydrogenated fats commonly used for manufacturing processed food and food cooked in fast food restaurants. Trans fatty acids produce many harmful effects in humans. Consumption of trans fatty acids increases the risk of cardiovascular disease along with unfavorable lipid profile, accentuation of systemic inflammation, endothelial dysfunction, and disruption of glucose homeostasis (Mozaffarian 2006; Micha and Mozaffarian 2009). The average consumption of trans fatty acids in the United States is estimated to be ~4 g/day. This intake corresponds to ~2 % of total energy intake (Harnack et al. 2003). However, on an individual basis, a person may consume ≤50 g trans fatty acids from a single high-fat meal or snack in the United States (Stender et al. 2008). Importantly, prolonged consumption of trans fats activates TLR4/NFκB pathway-mediated induction of inflammatory cytokines in the hypothalamus and cytokine-induced impairment of central insulin hypophagia (Pimentel et al. 2012). The hypothalamus is a key regulator of energy homeostasis, through the production of orexigenic and anorexigenic neuropeptides targeted by the peripheral hormones leptin and insulin, which exert pivotal control of food intake and energy expenditure (see below). Although the molecular mechanism(s) involved in these processes are not fully understood, but several possibilities for trans fatty acid-mediated pathologic effects have been proposed. They include (a) activation of TLR4/NFκB pathway and induction of inflammatory cytokines (b) changes in cholesteryl ester transfer protein metabolism, which may play a prominent role in transferring cholesterol from HDL to LDL and VLDL, (c) modulation of endothelial cell function by trans fatty acids, and (d) trans fatty acid-mediated ligand-dependent effects on peroxisome proliferator-activated receptor (PPAR) or retinoid X receptor (RXR) pathways (Katan 1998; Harvey et al. 2008; Mozaffarian and Willett 2007; Harnack et al. 2003; Micha and Mozaffarian 2009). It is proposed that the effects of trans fatty acids induce insulin resistance, which may be mediated through an alterations in gene expression (Katan 1998; Hunter 2006; Mozaffarian and Willett 2007; Stender et al. 2008). The harmful effects of trans fatty acids on human heart have also been linked to cancer, type II diabetes, maternal, and macular degeneration, but the relationship of trans fatty acids with above mentioned conditions has been controversial and more studies are required on harmful effects of trans fatty acids on human health (Hunter 2006). It is also shown that the maternal intake of hydrogenated vegetable fats that are rich in trans fatty acids during pregnancy and lactation triggers changes in the lipid metabolism and decreases serum levels of adiponectin in 21-day-old pups (Mennitti et al. 2014). These findings were accompanied by increases in TNF-α gene expression and the protein expression of TRAF-6 (TNF receptor-associated factor 6) in the adipose tissue (Pisani et al. 2008a; de Oliveira et al. 2011). The consumption of trans fatty acids during lactation may not only contribute to insulin resistance, but also to increase in gene expression of plasminogen activator inhibitor type-1 (PAI-1) in the adipose tissues of adult offspring (Pisani et al. 2008b; Osso et al. 2008). PAI-1, a pro-inflammatory adipokine, which is mainly produced and secreted by the visceral adipose tissue and vascular endothelium is modulated by TNF-α, insulin, free fatty acids and glucocorticoids (Skurk and Hauner 2004; Hajer et al. 2008). Increase in PAI-1 levels in serum is also closely associated with pro-thrombotic effects and obesity, which increase the risk for cardiovascular disease (Hajer et al. 2008; Giordano et al. 2011).

High intake of trans fatty acid is associated with a higher risk of developing AD. The consumption of trans fatty acids increases levels of trans fatty acids in the brain in a dose-related manner. Very high trans fatty acid consumption produces substantial modification in the brain fatty acid profile by increasing mono-unsaturated fatty acids and decreasing polyunsaturated fatty acids (PUFA) (Phivilay et al. 2009). The consumption of very high trans fatty acids by 3xTg-AD mice pro-duces a shift from DHA toward n-6 docosapentaenoic acid (DPA, 22:5n-6) without altering the n-3:n-6 PUFA ratio in the cortex. This observation supports the view that trans fatty acids consumption may modulate brain fatty acid profiles but cause no significant effect on major brain neuropathological hallmarks of AD in this animal model of AD (Phivilay et al. 2009).

2.3 Studies on iPLA2γ Deficient Mice Consuming High Fat Diet

The pathogenic mechanisms associated with high fat-mediated visceral and brain diseases are not fully understood. However, it is becoming increasingly evident that in visceral tissues consumption of high fat diet produces accelerated body weight gain (obesity), deposition of visceral fat, and insulin resistance. While the exact mechanisms regarding how obesity-mediates its detrimental effects on health is not fully understood, but increased chronic inflammation and oxidative stress have been reported to be two key physiological features of obesity (Hotamisligil 2006). Obesity mediated chronic inflammation is characterized by abnormal cytokine pro-duction, increase in levels of proinflammatory lipid mediators, and activation of a network of inflammatory signaling pathways (Chandalia and Abate 2007). Indeed, levels of inflammatory mediators not only correlate tightly with the degree of insulin resistance (Farooqui 2013), but also suggestive of risk of development of vascular disease (Rader 2000). Inflammatory and innate immune responses are also activated by increased levels of serum lipids, such as cholesterol and saturated long-chain fatty acids (Kennedy et al. 2009; Averill and Bornfeldt 2009).

The biochemical mechanisms associated with high fat diet-mediated obesity is not fully understood. However, recent studies indicate the involvement of calcium-independent PLA_2 isoform, namely $iPLA_2\gamma$. Studies on $iPLA_2\gamma^{-/-}$ mice indicate that these mice are completely resistant to high fat diet-induced obesity and the subse-quent development of insulin resistance (Mancuso et al. 2010). Mice null for $iPLA_2\gamma$ are characterized by multiple bioenergetic dysfunctions, including growth retardation, cold intolerance, reduction in exercise endurance, marked increase in mortality from cardiac stress after transverse aortic constriction, abnormal mitochondrial function, and a marked reduction in myocardial cardiolipin content with alteration in cardiolipin molecular species composition (Mancuso et al. 2007, 2009). These changes cannot be explained by either decrease in consumption of high fat diet or malabsorption of ingested dietary fat. Instead, iPLA$_2\gamma^{-/-}$ mice exhibits multiple cell type-specific alterations in mitochondrial function, including uncoupling of mitochondrial respiration from ATP generation due to abnormalities in cardiolipin metabolism resulting from the absence of iPLA$_2\gamma$ in skeletal muscles. iPLA$_2\gamma$ deficient mice have an unanticipated elevation in epididymal adipocyte oxidative capacity not only due to an increase in adipocyte mitochondrial mass and in levels of uncoupling proteins3 (*UCP3*), but also due to increase in electron transport chain uncoupling (Mancuso et al. 2010). The iPLA$_2\gamma^{-/-}$ mice also show considerable defect in glucose-stimulated insulin secretion, which may be caused by mitochondrial dysfunction in pancreatic β cells. Collective evidence suggests that these biochemical alterations may protect the iPLA$_2\gamma^{-/-}$ mice from the downstream sequelae of high fat feeding that represent the biochemical progenitors of the metabolic syndrome, a complex entity consisting of a constellation of metabolic risk factors including central (or vascular) obesity, insulin resistance/impaired glucose tolerance, dyslipidemia (hypertriglyceridemia and low HDL-C) and hypertension associated with an atherogenic, procoagulant and inflammatory state (Farooqui et al. 2012). It is speculated that iPLA$_2\gamma$ plays an important role in animal bioenergetics through modulation of adipocyte and skeletal muscle metabolism, insulin release and insulin sensitivity, and cellular mitochondrial function (Mancuso et al. 2010).

2.4   High Fat Diet and Onset of Inflammation in Hypothalamus and Hippocampus

At the molecular level long term consumption of high fat diet results in elevating the generation of ROS, proinflammatory lipid mediators (eicosanoids, cytokines, and chemokines), activation of the Toll-like receptor 4, inhibition of NF-κB kinase subunit β/NK-κB and activation of c-Jun amino-terminal kinase 1 (Williams 2012; Meng and Cai 2011; Milanski et al. 2009) in hypothalamus and hippocampus. Hypothalamic sensing of consumption of high fat diet and increase in circulating lipids involves two bona fide mechanisms that modulate energy homeostasis at the whole body level. Key enzymes, such as AMP-activated protein kinase (AMPK) and fatty acid synthase (FAS), as well as intermediate metabolites, such as malonyl-CoA and long-chain fatty acids-CoA have been shown to play a major role in this neuronal network, integrating peripheral signals with classical neuropeptide-based mechanisms (Martinez de Morentin et al. 2010). Most of the hypothalamic changes induced by a high-fat diet seem to be causally linked not only with the induction of neuroinflammation, mitochondrial dysfunction, and oxidative stress, but also with onset of endoplasmic reticulum (ER) stress, and defect in autophagy, a process by which cells engulf and breakdown intracellular proteins and organelles (Fig. 2.6). As stated above, consumption of high fat diet, which is enriched in ARA results in

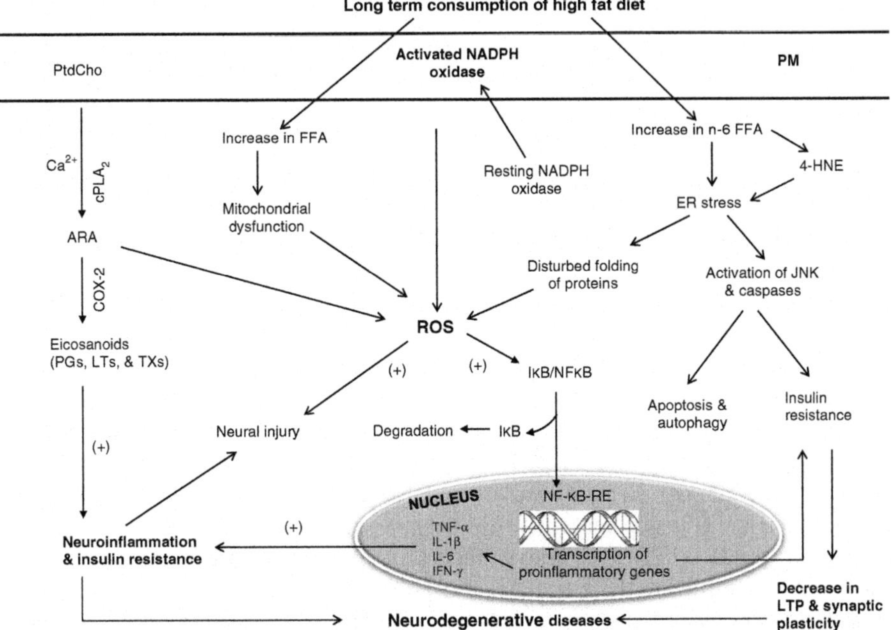

Fig. 2.6 Long term consumption of high fat diet and mechanisms responsible for the generation of oxidative stress and neuroinflammation. *PM* plasma membrane, *PtdCho* phosphatidylcholine, *cPLA₂* cytosolic phospholipase A₂, *COX-2* cyclooxygenase, *15-LOX* 15-lipoxygenase, *ARA* arachidonic acid, *ROS* reactive oxygen species, *NF-κB* nuclear factor kappaB, *NF-κB-RE* nuclear factor kappaB response element, *IκB* inhibitory subunit of NF-κB, *TNF-α* tumor necrosis factor-α, *IL-1β* Interleukin-1 beta, *IL-6* Interleukin-6, *PGs* prostaglandins, *LTs* leukotrienes, *TXs* thromboxanes

increase in levels of 4-HNE. This metabolite stimulates autophagy. The molecular mechanism through which 4-HNE stimulates autophagy is not fully understood. However, recent studies indicate that 4-HNE forms protein adducts that accumulates in the ER. 4-HNE triggers components of the unfolded protein response (UPR) to stimulate autophagy either by a JNK-dependent mechanism or through the activation of several stress-sensitive transcription factors, such as ATF3, ATF4, and CHOP (Vladykovskaya et al. 2012; Haberzettl and Hill 2013). The ER is the major signal transducing organelle that senses and responds to the changes in cellular homeostasis. ER stress has been reported to contribute to the development of insulin resistance in over-nourished or obese rodents (Schroder and Kaufman 2005; Malhotra and Kaufman 2007). ER is a well-organized protein-folding machine composed of protein chaperones and proteins that catalyze protein folding, and sensors that detect the presence of misfolded or unfolded proteins (Malhotra and Kaufman 2007). Furthermore, ER also contains a sensitive surveillance mechanism that prevents misfolded proteins from transiting the secretory pathway. The efficiency of protein-folding reactions not only depends on appropriate environmental and genetic factors, but also on metabolic conditions, such as diet. Conditions that disrupt protein folding threaten cell survival with decrease in viability and longevity. Accumulation of unfolded proteins in ER lumen initiates activation of an adaptive signaling cascade known as the unfolded protein response (UPR) (Malhotra and Kaufman 2007). Major transducers of the UPR include PKR-like ER kinase (PERK), inositol-requiring enzyme 1 (IRE1), activating transcription factor 6 (ATF6), The activation of these factors transmits signals from the ER to the cytoplasm or nucleus, and activate three pathways: (a) suppression of protein translation to avoid the generation of more unfolded proteins (Harding et al. 2000); (b) induction of genes encoding ER molecular chaperones to facilitate protein folding (Li et al. 2000); and (c) activation of ER-associated degradation (ERAD) to reduce unfolded protein accumulation in the ER (Ng et al. 2000). If these strategies fail, the cells are unable to maintain ER homeostasis and undergo apoptosis due to increase in ER stress (Urano et al. 2000), which activates metabolic pathways that trigger insulin resistance, activation of NF-κB and initiation of chronic inflammation. Collective evidence suggests that ER stress and the UPR are increasingly recognized as important regulators of cell function (Marciniak and Ron 2006). Conditions under which the influx of nascent, unfolded proteins exceeds the folding capacity of the ER trigger the UPR to restore homeostasis. Alterations in the protein folding status are transmitted to the nucleus by the UPR, which leads not only to ER expansion, but also elevates the protein folding capacity of the cell. Conditions under which homeostasis cannot be restored result in apoptotic cell death.

The consumption of high fat diet causes ER stress, which may lead to insulin receptor substrates (IRS) serine phosphorylation and inhibition of insulin signaling (Ozcan et al. 2004; Thaler and Schwartz 2010; Thaler et al. 2010). ER stress has also been reported to cause autophagic dysfunction in various tissues and cells (Levine and Kroemer 2008). Using autophagy-related protein 7 (Atg7) as an autophagic marker, it is shown that autophagy is highly active in the mediobasal hypothalamus of normal mice. Hypothalamic inhibition of autophagy increases energy intake and reduces energy expenditure. These metabolic changes are sufficient to increase body weight gain under normal chow feeding and exacerbate the progression of obesity along with insulin resistance under high-fat diet feeding (Meng and Cai 2011). Using brain-specific IκB kinase β knockout mice, it is shown that the effects of defective hypothalamic autophagy can be reversed by IκB kinase β inhibitors in the brain. Thus, hypothalamic autophagy is crucial for the central control of feeding, energy, and body weight balance. Decline in hypothalamic autophagy due to long term consumption of high fat diet may promote hypothalamic inflammation, leading to accelerated development of obesity (Meng and Cai 2011). Collective evidence suggests that long term consumption of high fat diet produces inflammation both in peripheral tissues and hypothalamus (Thaler et al. 2010, 2012). Unlike peripheral inflammation, which develops as a consequence of obesity and insulin resistance after weeks to months, the onset of hypothalamic inflammation occurs both in rats and mice within 1–3 days after the start of high fat and high sugar diet and prior to substantial weight gain. Hypothalamic inflammation is accompanied by reactive gliosis involving both microglial and astroglial cell populations along with increase in markers of neuron injury (TNF-α, IL1-β, and IL-6) within a week. Although these responses temporarily subside due to the onset of neuroprotective mechanisms, which may initially limit the damage, but with continuation of high fat and high sugar diet uptake, inflammation and gliosis return permanently to the mediobasal hypothalamic region (Thaler and Schwartz 2010; Thaler et al. 2012). It is also reported that high saturated fat produces ~ 50 and 20 % reduction in hypothalamic arcuate NPY and AgRP mRNA levels, respectively, compared with a low-fat or an n-3 or n-6 polyunsaturated high-fat (PUFA) diet without change in energy intake, fat mass, plasma leptin levels, and leptin receptor or POMC mRNA (Wang et al. 2002).

The presence of inflammation in the brain not only disrupts neurohormone- and neurotransmitter-mediated central regulatory functions, but also initiates and maintains propagation of obesity and related disorders in peripheral tissues following long term consumption of high fat diet (Cai 2009; Cai and Liu 2012) through the modulation of the proinflammatory axis comprising TLR4 receptor, IκB kinase-β (IKKβ) and their downstream nuclear transcription factor NF-κB (IKKβ/NF-κB signaling) in the hypothalamic neurons (Fig. 2.7). IKKβ/NF-κB-driven hypothalamic

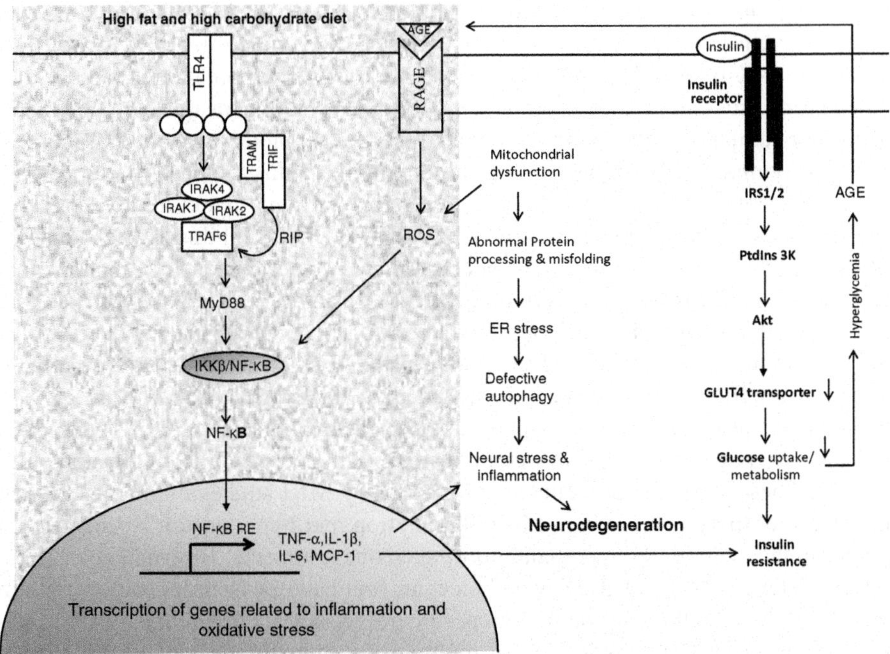

Fig. 2.7 Interactions among Toll-like receptor, receptor for advanced glycation end products, and insulin receptor. *TLR4* Toll-like receptors 4, *IRAK1,2, and 4* IL-1R-associated kinase 1, 2, and 4, *TRAF6* tumor necrosis factor receptor-associated factor adaptor protein 6, *TRAM* TRIF-related adaptor molecule, *TRIF* TIR domain-containing adaptor protein inducing IFNβ, *ROS* reactive oxygen species, *NIK* NF-κB-inducing kinase, *IKK* IκB kinase, *NF-kB* NF-kappaB, *NF-kB-RE* NF-kappaB response element, *I-κB* inhibitory subunit of NF-κB, *RIP* receptor interacting protein, *AGEs* advanced glycation end products, *RAGE* receptors for advanced glycation end products, *PtdIns 3K* phosphatidylinositol 3-kinase, *IRS1/2* insulin receptor substrate proteins1/2, *TNF-α* tumor necrosis factor-alpha, *IP-10* IFN inducible protein of 10 kDa, *MCP1* monocyte chemotactic protein-1

inflammation is also supported by increase in body weight caused by imbalance in glucose homeostasis mediated by insulin resistance (Purkayastha et al. 2011a). Activation or overexpression of IKK-β diminishes insulin signaling and induces insulin resistance whereas inhibition of IKK-β improves insulin sensitivity. IKK-β phosphorylates the inhibitor of nuclear factor kappa B (NF-κB), leading to the activation of NF-κB by the translocation of NF-κB to the nucleus, where NF-κB promotes expression of proinflammatory cytokines (TNF-α, IL-1β, and IL-6). ROS are known to activate NF-κB (Furukawa et al. 2004; Farooqui 2013), whereas antioxidants inhibit the activation of NF-κB. The finding that NF-κB deficient mice were protected from high-fat diet-induced insulin resistance suggests that NF-κB directly participates in the processes that impair insulin signaling. Recently, glial cells (astroglia and microglia) have been reported to play an important role in inducing hypothalamic inflammation in high fat diet-induced obesity (Horvath et al. 2010; Thaler et al. 2012). Hypothalamic IKKβ/NF-κB via inflammatory crosstalk between microglia and neurons has been reported to direct systemic aging by inhibiting the production of gonadotropin-releasing hormone (GnRH) and inhibition of inflammation or GnRH therapy has been reported to revert aging related degenerative symptoms at least in part (Purkayastha and Cai 2013). It is also reported that acute activation of the proinflammatory NF-κB and its upstream activator IκB kinase-β (IKK-β, encoded by Ikbkb) in the mediobasal hypothalamus rapidly elevated blood pressure in mice independently of obesity (Purkayastha et al. 2011a, b). Thus, hypothalamic inflammation-mediated increase in blood pressure is associated with the sympathetic upregulation of hemodynamics. This increase in blood pressure can be reversed by sympathetic suppression (Purkayastha et al. 2011b). This suggestion is supported by studies, which show that NF-κB inhibition in the mediobasal hypothalamus can normalize obesity-related hypertension in a manner that can be dissociated from changes in body weight. In addition, it is also indicated that pro-opiomelanocortin (POMC) neurons are crucial for the hypertensive effects of the activation of hypothalamic IKK-β and NF-κB, which underlie obesity-related hypertension. It is speculated that obesity-mediated activation of IKK-β and NF-κB in the mediobasal hypothalamus—particularly in the hypothalamic POMC neurons— is a primary pathogenic link between obesity and hypertension (Purkayastha et al. 2011b). These studies on rodents are supported by MRI studies in humans, which indicate that there is an increase in inflammation and hypertension is linked with gliosis in the mediobasal hypothalamus of aged obese humans (Bowman et al. 2012, 2013; Virtanen et al. 2013). Several mechanisms have been proposed to explain high fat-induced inflammation in both peripheral tissues and hypothalamus, including activation of TLR4, induction of endoplasmic reticulum stress, IKKβ/NF-κB signaling, and induction of SOCS3 along with other intracellular inflammatory signals associated with high levels of circulating saturated fatty acids (Thaler et al. 2012; Fessler et al. 2009; Zhang et al. 2008) that exacerbate the inflammatory response and facilitate insulin resistance. The relative contribution of these mechanisms in the onset and maintenance remains uncertain. However, much earlier onset of inflammation in hypothalamus relative to peripheral tissues suggests that different processes may be associated with the inflammation in peripheral tissues

(Thaler and Schwartz 2010). Collective evidence suggests that high-fat diet disturbs the delicate relationship between glial cells and neurons through increased endoplasmatic reticulum and oxidative stress, leading to stress–response pathways with generally cytotoxic effects (De Souza et al. 2005). The end effects of these changes are central insulin and leptin resistance and impaired hypothalamic regulation of energy balance, further favoring the development of obesity and, in turn, neurodegeneration (De Souza et al. 2005; Farooqui 2013).

These toxic effects do not stop at the level of hypothalamus, but can also affect brain areas involved in reward processing, memory formation, and cognitive decline. In recent years hippocampus has received considerable attention for its potential role in energy regulation (Davidson et al. 2007). Hippocampus is a part of a neural circuit associated with reward and energy regulation (Wang et al. 2006) and is sensitive to satiety signals involved in learning and memory (Cenquizca and Swanson 2006). In the brain, PLA_2 insoforms play critical roles in cellular growth, lipid homeostasis, and second messenger generation (Ong et al. 2010). Genetic ablation of $iPLA_2\gamma$ not only causes alterations in hippocampal phospholipid metabolism, but also induces changes in mitochondrial phospholipid homeostasis resulting in enlarged and degenerating mitochondria leading to autophagy and cognitive dysfunction (Mancuso et al. 2009). Shotgun lipidomics studies have indicated that $iPLA_2$ deficient mice ($iPLA_2\gamma^{-/-}$) mice display a markedly increase in hippocampal cardiolipin content with alterations in molecular species composition; changes in both choline and ethanolamine containing phospholipids, including a decrease in ethanolamine plasmalogen content; increase in oxidized phosphatidylethanolamine molecular species; and an increased content of ceramides (Mancuso et al. 2009). Electron microscopic studies show the presence of enlarged heteromorphic lamellar structures undergoing degeneration accompanied by the presence of ubiquitin positive spheroid inclusion bodies-derived from degenerating mitochondria. Based on these studies, it is suggested that the $iPLA_2\gamma$ not only plays an important role in neuronal mitochondrial lipid metabolism, but its genetic ablation may contribute to the pathogenesis of neurodegenerative diseases by introducing abnormalities in autophagy, and cognitive dysfunction (Mancuso et al. 2009). As stated above, the consumption of palatable high-fat diets promote excessive food intake and weight gain and interfere with hippocampal functioning associated with learning and memory. This suggestion is also supported by epidemiological studies that link diets high in saturated fat with weight gain and memory deficits (El-Gharbawy et al. 2006). High-fat diet causes maternal obesity leading to alterations in fetal hippocampal development, impairment in offspring hippocampal brain-derived neurotrophic factor (BDNF) production, and reduction in hippocampal neurogenesis during the early life of their offspring (Niculescu and Lupu 2009; Tozuka et al. 2010). Furthermore, the consumption of high fat diet in adult male rats also results in impairment of hippocampal neurogenesis (Lindqvist et al. 2006). The molecular mechanisms that link high fat diet with hippocampal neurogenesis are not known. However, one potential mechanism may involve leptin, a hormone secreted by adipose tissue, which produces its effects by interacting with leptin receptors in the hypothalamus. Leptin suppresses appetite by increasing energy expenditure, and reducing body weight gain (Halaas et al. 1995).

Leptin induces a complex regulatory response by integrating the control of bodyweight and energy expenditure. Levels of leptin are positively correlated with fasting insulin, fasting glucose, and triacylglycerols in non-diabetic men (Haffner et al. 1999). The leptin-mediated neurogenesis mainly results from increased cell proliferation, because leptin shows no significant effect on cell differentiation and survival. Leptin signaling involves the PtdIns 3K/Akt and Jak2/STAT3 pathways in adult hippocampal stem/progenitor cells, and inhibition of these signal transduction pathways leads to the attenuation of the actions of leptin on proliferation of adult hippocampal progenitor cells, supporting the view that a mechanism dependent on Akt and STAT3 activation is closely associated with the action of leptin (Garza et al. 2008). Leptin not only enhances cognition, but leptin receptor-deficient animals have impaired hippocampal LTP and poor spatial memory (Figlewicz 2004; Wayner et al. 2004).

Another mechanism of impairment of hippocampal neurogenesis by high fat diet may involve decrease in levels of BDNF in the hippocampus (Park et al. 2010). High fat diet also increases the levels of malondialdehyde (MDA) and ROS in the hippocampus (Park et al. 2010). Based on these results it is suggested that a high fat diet impairs hippocampal neurogenesis and neural progenitor cell proliferation through increase in lipid peroxidation and decrease in BDNF (Park et al. 2010). Collective evidence suggests that leptin affects brain function in several ways. Leptin acts as a trophic factor to neural progenitor cells and promotes glial as well as neuronal development (Udagawa et al. 2006). It is anti-epileptic and can directly affect calcium and potassium channels (Harvey 2007; Xu et al. 2008). In the feeding circuitry, leptin facilitates the production of melanocortin in proopiomelanocortin neurons and suppresses the production of neuropeptide Y and agouti-related protein in neuropeptide Y neurons of the arcuate nucleus of the hypothalamus (Schwartz 2001). Both proinflammatory and anti-inflammatory effects of leptin have been shown (Ahmed et al. 2007; Pinteaux et al. 2007; Lin et al. 2007). The central leptin system, therefore, might be involved in neuroinflammation after the onset of the metabolic syndrome and hyperleptinemia.

As stated above, in the brain consumption of high-fat diet produces changes in the hypothalamus and hippocampus, brain regions, which are critical for the maintenance of energy homeostasis and integration of peripheral hormonal changes associated with neuronal signaling that modulate satiety, nutritional status, body weight, and cognition (Berthoud 2002; López et al., 2007; Heyward et al. 2012). The mediobasal hypothalamus senses circulating metabolic signals such as leptin, insulin, and nutrients, and instructs to the downstream neurohormonal networks regulate various aspects of metabolic physiology related to satiety and nutritional status. In addition, hypothalamic neurons can project to the autonomic sites in the brain to modulate the sympathetic and parasympathetic nervous systems that control metabolic activities. High fat-mediated changes in hypothalamus include an increase in markers for oxidative stress, neuroinflammation, and endoplasmic reticulum (ER) stress. In addition, high fat consumption also induces autophagy defect and produces changes in the rate of apoptotic cell death (Cai 2009; Williams 2012). Inhibition of neuroinflammation and oxidative stress may prevent or reverse the metabolic dysfunction, development of insulin resistance, and obesity (Williams 2012). Long term consumption of high-fat diet also induces changes in the hippocampus, a brain region involved in learning and memory formation (Pintana et al. 2012). High fat diet-mediated hippocampal changes include mitochondrial dysfunction and changes in neural membrane plasticity and impairment of cognitive function in normal rats (Granholm et al. 2008; Stranahan et al. 2008). High fat diet-mediated changes in hippocampal morphology/plasticity are of considerable interest because this region is involved in learning and memory formation. Studies on the effect of high fat diet consumption in normal C57BL/6 mice have indicated that high fat diet not only induces loss of working memory, but also increases in APP processing (Li et al. 2007; Thirumangalakudi et al. 2008). At the molecular level, high fat reduces the expression of hippocampal SIRT1 mRNA expression and decreases SIRT1 protein (Wu et al. 2006) implicating SIRT1 in learning and memory (Gao et al. 2010; Michan et al. 2010). SIRT1 is known to upregulate IGF-1 by either derepressing IGFBP-1 (Lemieux et al. 2005) or deacetylating insulin receptor substrate-2 (IRS-2) (Zhang 2007; Li et al. 2008). The reduction of IGF-1 signaling may decrease the downstream mitogen-activated protein kinase (MAPK), ERK1/2 and PtdIns 3K, which are important for various brain functions (Zhang 2007). Mice with reduced IGF-1 levels have impaired spatial learning and this effect is partially reversed by IGF-1 replenishment (Trejo et al. 2007).

High fat diet-mediated neurochemical alterations in hypothalamic and hippocampal regions may accelerate onset of age-related disorders, including diabetes, cancer and neurological disorders (stroke, Alzheimer disease, and depression) (Fig. 2.8). Anti-diabetic drugs (Vildagliptin and sitagliptin, dipeptidyl-peptidase-4 inhibitors), long term consumption of n-3 fatty acids, and phytochemicals have been reported not only to slow down hippocampal and hypothalamic mitochondrial dysfunction,

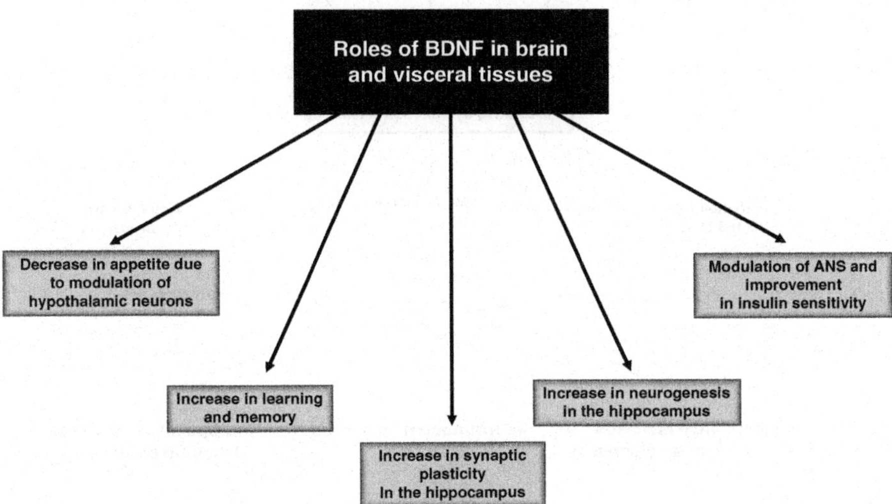

Fig. 2.8 Role of BDNF in brain and visceral tissues

but also improve peripheral insulin sensitivity and cognitive function in high-fat diet consuming rats (Pintana et al. 2012, 2013; Farooqui 2013). Based on these studies it can be suggested that high fat consumption-mediated alterations in dentate gyrus of hippocampus and hypothalamus may contribute to increase in appetite, alterations in energy homeostasis, reduction in neurogenesis and impairment in cognitive function leading to changes in learning and memory.

2.5   High Fat Diet-Mediated Changes in BDNF Signaling

BDNF is a neurotrophin with important functions in neuronal survival and differentiation (Horch 2004). However, beyond its classical neurotrophic role, BDNF also modulates and controls the neuronal activity and synaptic plasticity as a neuromodulator (Mattson 2008; Santos et al. 2010). This growth factor is also responsible for axon targeting, neuron growth, maturation of synapses during development, and synaptic plasticity (Fig. 2.9) (Kanoski et al. 2007). In the CNS, BDNF is synthesized and secreted mainly by neurons and its levels are markedly increased in response to exercise and dietary energy restriction (Mattson et al. 2004). The transcription factors CREB (cyclic AMP response element binding protein)

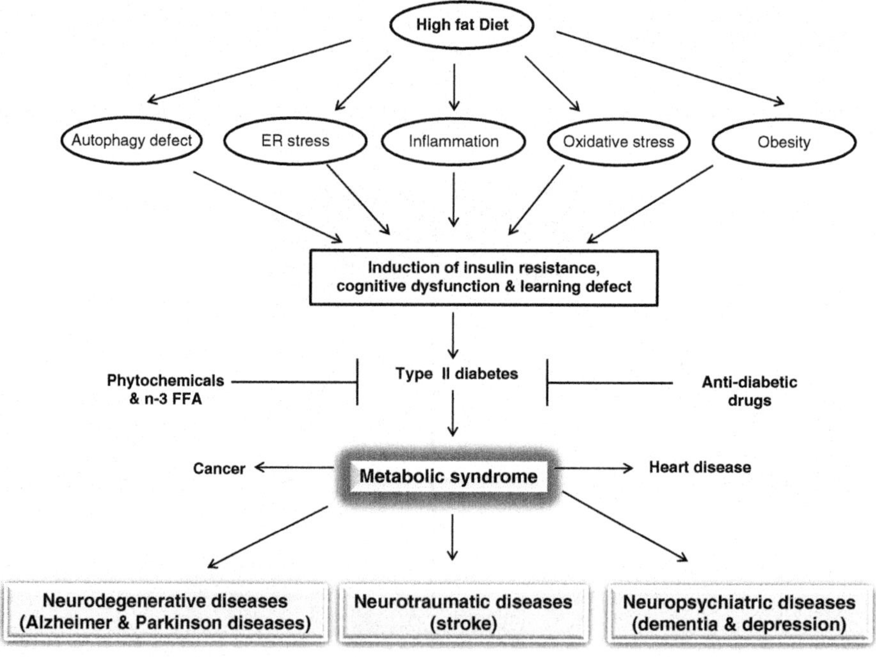

Fig. 2.9 Metabolic consequences and diseases associated with long term consumption of high fat diet

and NF-κB have been reported to induce the expression of BDNF in response to neuronal activity and metabolic stress (Lipsky et al. 2001). Studies on heterozygous BDNF knockout (BDNF+/−) have indicated that BDNF plays an important role in regulating energy metabolism (Kernie et al. 2000) through the modulation of hypothalamic neurons that are closely associated with the control appetite. In obese diabetic phenotype of BDNF+/− mice an alternate day fasting dietary restriction regimen increases BDNF expression in the brain (Duan et al. 2003). Functions of BDNF are mediated by two receptor systems: TrkB and p75^{NTR}. However, the majority of BDNF-induced functions have been attributed to signaling through TrkB. BDNF interacts with to a high-affinity receptor tyrosine kinase called trkB which is expressed by neurons and glial cells throughout the nervous system (Reichardt 2006). The binding of BDNF to TrkB elicits various intracellular signaling pathways, including mitogen-activated protein kinase/extracellular signal-regulated protein kinase (MAPK/ERK), phospholipase Cγ (PLCγ), and phosphoinositide 3-kinase (PtdIns 3K) pathways. Pro-BDNF, which preferentially binds p75^{NTR}, activates a different set of intracellular signaling cascades including nuclear factor-kappa B (NF-κB), c-jun kinase and sphingomyelin hydrolysis (Teng et al. 2005). Activation of p75^{NTR} by pro-neurotrophins has been linked to the activation of apoptotic signaling and initiation of N-methyl-D-aspartic acid (NMDA) receptor-dependent synaptic depression in the hippocampus (Lu et al. 2005). During development of the brain, the formation of synapses between neurons is modulated by their electrical activity, and neurotrophins secreted by target cells play a pivotal role in the development of neural cell networks and the structural organization. The communication between neurons is supported by interactions between neurotransmitters and neurotrophins signaling pathways. Among neurotrophins, BDNF as well as neurotrophin-3 (NT-3) have emerged as having key roles in the neurobiological mechanisms related to learning and memory. BDNF contributes to adaptive responses in neural networks through interplay between glutamate and BDNF in the adult brain. Optimal brain health throughout the lifespan is maintained by metabolic stress caused by exercise, cognitive stimulation, and dietary energy restriction (Rothman and Mattson 2013). At the molecular level, such metabolic stress to neurons not only results in the synthesis of BDNF, but also in production of proteins involved in neurogenesis, learning and memory and neuronal survival. These processes are supported by increased expression of protein chaperone GRP-78, antioxidant enzymes, the cell survival protein Bcl-2, and the DNA repair enzyme APE1. These proteins are involved in mitochondrial biogenesis, protein quality control, and resistance of cells to oxidative, metabolic and proteotoxic stress (Rothman and Mattson 2013). Intra-cerebroventricular injections of BDNF decrease energy intake and body weight (Pelleymounter et al. 1995), and it reverses the hyperphagic and obese phenotype of BDNF heterozygous mutant mice (Kernie et al. 2000). The disruption of the regulatory locus of the BDNF gene results in decrease in BDNF levels in wild types of mice (Sha et al. 2007). These animals show elevation in body weight and adiposity; hepatic steatosis; increase in levels of serum LDL cholesterol, insulin, and leptin; impairment in glucose tolerance along with age-related hyperglycemia. Similarly, human patients

Table 2.2 Neurological disorders associated with perturbed BDNF signaling

Neurological disorders	Reference
Alzheimer disease (J20 transgenic mice)	Nagahara et al. (2009)
Alzheimer disease (AD)	Ventriglia et al. (2013)
Frontotemporal dementia (FTD)	Ventriglia et al. (2013)
Vascular dementia (VAD)	Ventriglia et al. (2013)
Lewy body dementia (LBD)	Ventriglia et al. (2013)
Parkinson disease (PD)	Ventriglia et al. (2013)
Huntington disease (HD)	Lynch et al. (2007)
Epilepsy	He et al. (2004)
Mood disorders	Strauss et al. (2005)
Autism Spectrum Disorders (ASD)	Abuhatzira et al. (2007)
Sudden infant death syndrome (SIDS)	Tang et al. (2012)

with mutations in the BDNF gene (Gray et al. 2006) and the BDNF receptor (TrkB) signal transduction pathway (Gray et al. 2007) also show obesity. These observations support the view that BDNF is necessary for maintaining normal body weight and may be protective against obesity and related diseases. Collective evidence suggests that BDNF affects neurons positively or negatively through various intracellular signaling pathways triggered by activation of TrkB or p75. Long term consumption of high fat diet disrupts BDNF signaling producing changes in maturation of synapses during development, and synaptic plasticity (Kanoski et al. 2007). Perturbation of BDNF signaling is associated with many neurological disorders, such as epilepsy, mood disorders, Huntington Disease, Alzheimer Disease; neurodevelopmental diseases including Autism Spectrum Disorders (ASD), and sudden infant death syndrome (SIDS). Rett Syndrome and Tuberous sclerosis represent two of the best examples of diseases involving disrupted BDNF signaling and often presenting with ASD (Table 2.2).

2.6   High Fat Diet Mediated Changes in Neudesin Signaling

BDNF also regulates neudesin (neuron-derived neurotrophic factor; NENF) in the hypothalamus. Neudesin (neuron-derived neurotrophic factor; NENF), a 21 kDa secreted protein with 171 amino acids, is abundantly expressed in the brain, and its neurotrophic activity is mediated through MAPK and PtdIns 3K pathways (Kimura et al. 2005, 2013). Neudesin activity is inhibited by the inhibitor pertussis toxin for Gi/Go-protein but not by inhibitors for receptor tyrosine kinases (Kimura et al. 2005). Neudesin is a member of the membrane-associated progesterone receptor (MAPR) family and shares key structural motifs with the cytochrome b5-like heme/ steroid-binding domain. The "classical" mechanism by which progesterone elicits its effects is via the progesterone receptor (PR), which, like the estrogen receptor (ER), has classically been described as a nuclear transcription factor, acting through specific progesterone response elements (PRE) within the promoter region of target

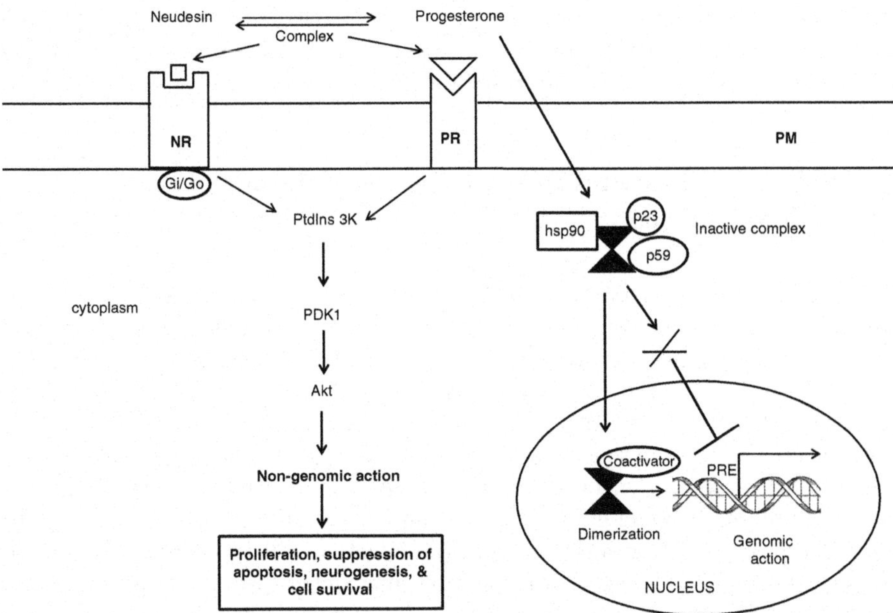

Fig. 2.10 Genomic and non-genomic effects of progesterone in the brain. *PM* plasma membrane, *NR* neudesin receptor, *PR* progesterone receptor, *PtdIns 3K* phosphatidylinositol 3-kinase, *PDK1* phosphoinositide-dependent kinase1, *Akt* protein kinase B, *hsp90* heat shock protein 90, *p23 and p59* novel proteins of inactive progesterone receptor complex

genes to regulate transcription. Non-genomic actions of progesterone are accompanied by interactions of progesterone with progesterone receptor membrane component 1 (PGRMC1). The non-genomic responses are mediated via neudesin and are not related to the nuclear progesterone receptors (PRs) (Singh et al. 2013). Based on several investigations, it is proposed that neudesin induces its effects through rapid non-genomic actions of progesterone (Fig. 2.10) (Kimura et al. 2013). Robust neudesin expression in hypothalamic nuclei regulates food intake, and its expression is altered under the diet-induced obese condition relative to the fed state (Byerly et al. 2013). Hypothalamic neudesin mRNA is regulated by BDNF, which itself is an important regulator of appetite. Delivery of purified recombinant BDNF into the lateral cerebral ventricle decreases hypothalamic neudesin expression, while pharmacological inhibition of trkB signaling increases neudesin mRNA expression. Furthermore, recombinant neudesin administration via an intracerebroventricular cannula decreases food intake and body weight and increases hypothalamic *Pomc* and *Mc4r* mRNA expression. Importantly, the appetite-suppressing effect of neudesin can be abrogated in obese mice consuming a high-fat diet supporting a diet-dependent modulation of neudesin function. Based on these results, it is revealed that neudesin is an important central modulator of food intake (Byerly et al. 2013; Kimura et al. 2013). In addition to controlling the food intake, neudesin also modulates anxiety behavior mainly through the dorsal dentate gyrus ventral hippocampus and altered dopaminergic activity (Novais et al. 2013). The modulatory role of neudesin in anxiety may be associated with its putative neurotrophic role but why and how this effect is specific for the anxiety circuits is still an open question? More studies are urgently needed on the role of neudesin in anxiety.

2.7   High Fat Diet and Induction of Mitochondrial Abnormalities

The consumption of high fat diet for 16 weeks produces non-alcoholic steatohepatitis (NASH)-like pathology, which is accompanied by elevation in triacylglycerols, increase in CYP2E1 (cytochrome P450 2E1) and iNOS (inducible nitric oxide synthase) protein, and significant enhancement in hypoxia in the pericentral region of the liver (Mantena et al. 2009). Mitochondria from the high fat diet consuming group show increased in sensitivity to NO-dependent inhibition of respiration compared with control group. In addition, accumulation of 3-nitrotyrosine (3-NT) is paralleled by the hypoxia gradient *in vivo* and 3-NT levels are increased in mitochondrial proteins. Liver mitochondria from mice fed with the high fat diet for 16 weeks exhibit depressed state 3 respiration, uncoupled respiration, alterations in cytochrome *c* oxidase activity, and mitochondrial membrane potential (Fig. 2.11)

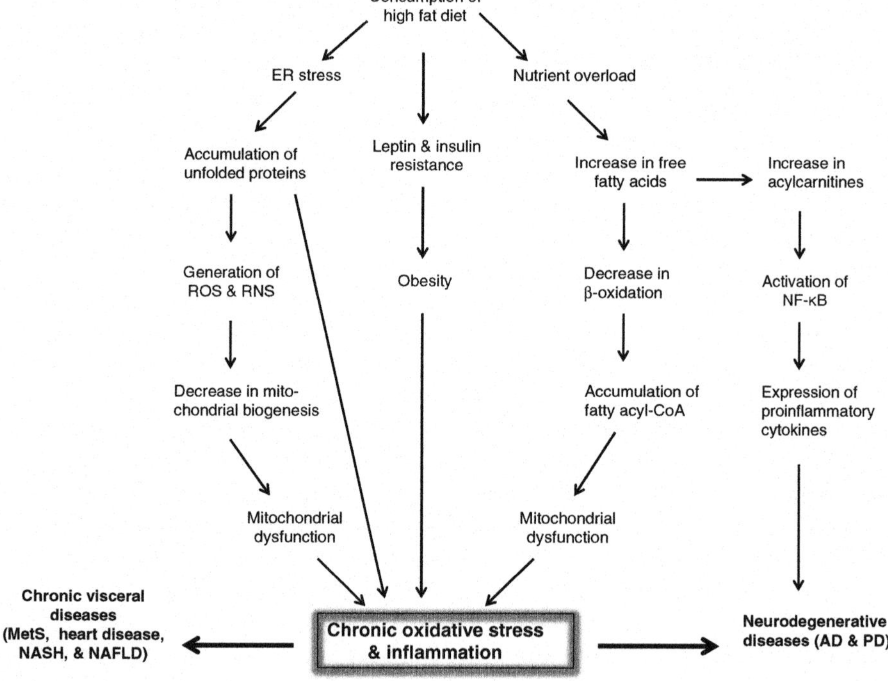

Fig. 2.11 Relationship between high fat diet and development of visceral and neurological disorders. *MetS* Metabolic syndrome, *NASH* non-alcoholic steatohepatitis, *NAFLD* nonalcoholic fatty liver disease, *AD* Alzheimer disease, *PD* parkinson disease

(Mantena et al. 2009). These findings indicate that chronic exposure to a high fat diet negatively affects the bioenergetics of liver mitochondria and this probably contributes to hypoxic stress and deleterious NO-dependent modification of mitochondrial proteins resulting in oxidative stress and insulin resistance. Increasing evidence indicates that mitochondrial dysfunction is a key player in the pathogenesis of nonalcoholic fatty liver disease (NAFLD) and NASH. These studies have been supported by other high fat diet feeding studies, which indicate that 16 weeks consumption of high fat diet decreases mitochondrial respiration and cytochrome *c* oxidase activity and increases sensitivity to NO-dependent inhibition of mitochondrial respiration when compared with measures in liver mitochondria from the control mice. Similar alterations in mitochondrial bioenergetics and NO–mitochondria signaling occur in alcohol-induced fatty liver disease (Venkatraman et al. 2003, 2004). Based on these studies it is proposed that high fat diet-induced alterations in liver NO bioavailability may contribute to mitochondrial dysfunction and oxidative stress (Clementi and Nisoli 2005). Understanding the impact of fatty liver on NO metabolism is very important because NO has been implicated as a critical physiological regulator for how mitochondria and cells respond to stress in multiple organ systems (Shiva et al. 2005; Hill et al. 2010). As stated above, the consumption of high fat diet produces alterations in signal transduction processes resulting in induction of neuroinflammation, oxidative stress, and insulin and leptin resistance.

2.8 High Fat Diet and Induction of Oxidative Stress and Neuroinflammation

In the brain, high calorie diet induced mitochondrial dysfunction and oxidative stress, which is accompanied by excessive generation and/or insufficient removal of highly reactive molecules called as reactive oxygen species (ROS) and reactive nitrogen species (RNS). The major sources of oxidative stress are mitochondrial respiratory chain, xanthine/xanthine oxidase, myeloperoxidase in cytoplasm, oxidation of ARA by cyclooxygenase (COX) and lipoxygenase (LOX) in cytoplasm, and NADPH oxidase in plasma membranes (Farooqui 2010) (Fig. 2.12). Under physiological conditions the antioxidant defense systems (superoxide dismutase, catalase, transferrin, and glutathione peroxide, glutathione, and vitamin C) within the body can easily neutralize the amount of ROS produced through three cellular antioxidative defense systems. However, disruption of tight control by high ROS production may result in the severe oxidative stress, which is closely associated with neural cell injury (Farooqui 2010). Low levels of ROS are not only needed for fundamental cellular functions, such as growth and adaptation responses, but also for optimal functioning of the immune system, which is involved in the defense against the penetrating microorganisms. As stated above, generation of high ROS in neural cells facilitates the translocation of transcription factor (NF-κB) from cytoplasm to the nucleus where it facilitates the expression of proinflammatory enzymes (PLA$_2$, COX, and nitric oxide synthase), proinflammatory cytokines and chemokines (TNF-α, IL-1β, COX-2, iNOS, VCAM-1 and ICAM-1

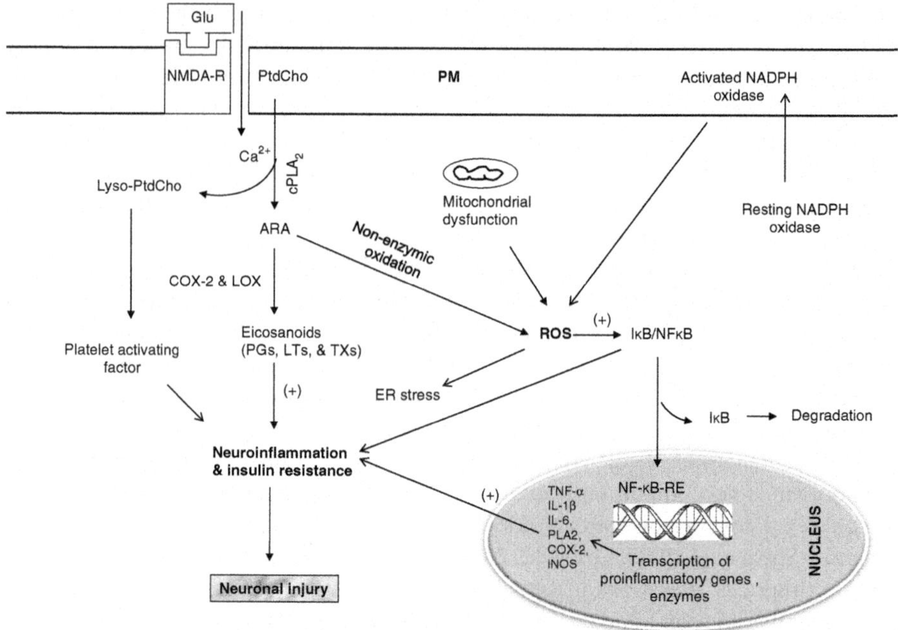

Fig. 2.12 Sources contributing to reactive oxygen species (ROS) production. *PM* plasma membrane, *NMDA-R* N-methyl-D-aspartate receptor, *Glu* glutamate, *PtdCho* phosphatidylcholine, *cPLA₂* cytosolic phospholipase A₂, *COX-2* cyclooxygenase, *LOX* lipoxygenase, *NOS* inducible nitric oxide synthase, *NADPH oxidase* nicotinamide adenine dinucleotide phosphate-oxidase, *ARA* arachidonic acid, *ROS* reactive oxygen species, *NF-κB* nuclear factor kappaB, *NF-κB-RE* nuclear factor kappaB response element, *IκB* inhibitory subunit of NF-κB, *TNF-α* tumor necrosis factor-α, *IL-1β* Interleukin-1 beta, *IL-6* Interleukin-6, *PGs* prostaglandins, *LTs* leukotrienes, *TXs* thromboxanes

and adhesion molecules), growth factors, cell cycle regulatory molecules, and adhesion molecules (Figs. 1.3 and 2.12).

The long term consumption of high fat diet produces significant decrease in cortical glutathione peroxidase activity in mice (Freeman et al. 2013). This reduction in glutathione peroxidase activity is supported by increase in total ROS and superoxide levels measured by electron paramagnetic resonance (EPR) spectroscopic studies in the cerebrocortex of high fat diet consuming mice. However, no significant differences have been observed for superoxide dismutase in the cortex or hippocampus. Based on the EPR data, it is suggested that an increase in oxidative stress in animals fed the high fat diet may be due to the reduction in glutathione peroxidase activity levels (Freeman et al. 2013). Another system, which modulates oxidative stress is transcription factor Nrf2, which is a master regulator of multiple antioxidant and detoxification pathways (Lee and Johnson 2004). The consumption of the high fat diet (high fat lard-based diet) not only causes a significant decrease in Nrf2 DNA binding activity, but also produces reduction in the Nrf2 responsive pathway proteins (heme oxygenase-1 and NAD(P)H dehydrogenase, quinone 1), and decreases the expression of Nrf2 protein expression (Morrison et al. 2010). In an elderly population, ROS production is not only closely associated with poorer cognitive function and loss of ability to perform daily activities, but also with institutionalization, as well as depressive symptoms.

2.9   High Fat Diet and Neurological Disorders

Molecular mechanisms involved in the pathogenesis of high fat diet-mediated neurological disorders remain elusive. Causes of neurodegeneration in neurological disorders include reduction in cellular antioxidant defenses (activities of superoxide dismutase, glutathione peroxidase, catalase, and glutathione reductase), increased production of ROS, increase in expression of proinflammatory cytokines, and accumulation of peroxidized lipids, proteins and DNA oxidative products supporting the view that neurodegeneration in neurological disorders is a multifactorial process involving genetic, environmental, and endogenous factors (Farooqui 2010). Endogenous factors that contribute to neurological disorders include excitotoxicity, oxidative stress, neuroinflammation, abnormal protein dynamics with defective protein degradation and aggregation related to the ubiquitin-proteasomal system resulting in generation and accumulation of misfolded proteins, autoimmunity, and mitochondrial dysfunction leading to increase in Ca^{2+} levels, and impairment in energy metabolism (Farooqui and Horrocks 2007; Farooqui 2010; Jellinger 2009).

Considerable information is available on the effect of high fat diet-mediated obesity and dementia, a major cause of disability, which is clinically-defined not only by memory deficits, and disturbances of other higher cortical functions, but also by deterioration in emotional control and social behavior (Sonnen et al. 2009). Two major types of dementia have been identified. Generalized atrophy in the cortical area of the brain results in dementia associated with Alzheimer disease and that due to vascular dementia mainly due to stroke (Farooqui 2010). Vascular dementia constitutes the second most common type of dementia, accounting for 15–20 % of all cases of dementia (Ruitenberg et al. 2001). Noteworthy, "vascular dementia" is a concept rather than a nosological entity. It encompasses a variety of conditions and dementia mechanisms including ischaemic, ischemic-hypoxic or haemorrhagic brain lesions as a result of cerebrovascular disease and cardiovascular pathological changes (Roman 2002).

Very little information is available on the direct link between high fat diet and neurological disorders. However, high fat diet-mediated increase in ROS may damage proteins essentials for important neurovascular mechanisms. For example, patients with Alzheimer disease have been shown to have high levels of oxidized soluble form of the low-density lipoprotein receptor related protein 1, which under normal conditions is the key endogenous Aβ chaperone protein in plasma (Sagare et al. 2007). 14-3-3 ζ and γ isoforms, proteins that are involved in cell growth, survival and differentiation, are also found to be highly oxidized in brain extracts from AD and cerebral amyloid angiopathy patients (Santpere et al. 2007). It is also reported

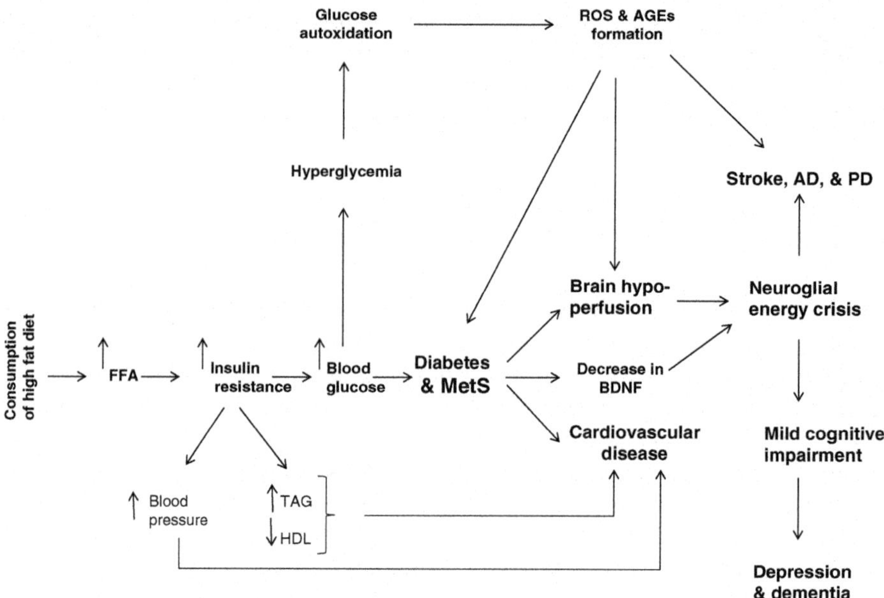

Fig. 2.13 Relationship between consumption of high fat diet and neurological disorders. *FFA* free fatty acids, *TAG* triacylglycerol, *HDL* high density lipoprotein protein, *MetS* metabolic syndrome, *AD* Alzheimer disease, *PD* parkinson disease

that in Tg2576 mice overexpressing Swedish mutant APP695 that vascular oxidative stress precedes parenchymal oxidative stress (Park et al. 2004), suggesting that ROS-mediated vascular insults may be an early event in the development of AD-like pathology in this model. Accumulating evidence suggests that high fat diet increases the risk of developing sporadic AD (Maesako et al. 2012).

As stated above, high fat diet-mediated neuroinflammation in hippocampus may cause impairment in the cognitive function (Fig. 2.13). This cognitive impairment may affect memory, judgment, speech, comprehension, execution, orientation, and learning (Qaseem et al. 2008). High fat consumption-mediated increase in saturated fatty acids has been reported to modulate presenilin-1, an important determinant of γ-secretase activity necessary for generation of Aβ in neuroblastoma cells (Liu et al. 2004). Furthermore, palmitic and stearic fatty acids induce Alzheimer disease-like hyperphosphorylation of Tau in primary rat cortical neurons (Patil and Chan 2005; Amtul et al. 2011a). These findings have been corroborated *in vivo* by using a transgenic mouse model of early-onset alzheimer disease that expresses the double-mutant form of human APP, which is the precursor protein responsible for the synthesis of Aβ peptide. Decrease in levels of Aβ peptide and less accumulation in the form of amyloid plaques has been observed in the brains of mice nourished with a diet enriched in DHA (Amtul et al. 2011a). Not only extraneously supplied but endogenously synthesized DHA can suppress the synthesis of Aβ peptide and the formation of amyloid plaques (Amtul et al. 2011a). ARA aggravates AD neuropathology by increasing the synthesis of Aβ peptide (Amtul et al. 2012), whereas oleic acid inhibits the production of Aβ peptide and amyloid plaques both *in vitro* and *in vivo* (Amtul et al. 2011b). Collective evidence suggests that neuropathology of AD is modulated by fatty acid composition of the brain. Long-term intervention studies using two specific multi-nutrients for the treatment and prevention of AD in 11–12 months old AβPPswe-PS1dE9 mice have indicated that feeding of either a control diet, the DHA + EPA + UMP (DEU) diet enriched with uridine monophosphate (UMP), the n-3 fatty acids (DHA and EPA) or the Fortasyn® Connect (FC) diet enriched with the DEU diet plus phospholipids, choline, folic acid, vitamins and antioxidants indicate that both diets are equally effective in changing brain fatty acid and cholesterol profiles (Jansen et al. 2013; Zerbi et al. 2014). However, these diets differentially affect AD-related pathologies and behavioral measures, supporting the view that the effectiveness of specific nutrients may depend on the dietary components. The FC diet is more effective than the DEU diet in counteracting neurodegenerative aspects of Alzheimer disease and enhancing processes involved in neuronal maintenance and repair (Jansen et al. 2013; Zerbi et al. 2014). Both diets increase interleukin-1β mRNA levels in AβPP-PS1 and wild-type mice. The FC diet additionally restores neurogenesis in AβPP-PS1 mice, decreases hippocampal levels of unbound choline-containing compounds in wild-type and AβPP-PS1 animals, suggesting diminished membrane turnover, and decrease in anxiety-related behavior in the open field behavior. In another study, the consumption of control diet in APPswe/PS1dE9 mice show not only decrease in cerebral blood flow (CBF), but also induced changes in brain water diffusion, in accordance with observations of hypoperfusion, axonal disconnection and neuronal loss in patients with AD. Both multinutrient diets increase cortical CBF in APPswe/PS1dE9 mice and Fortasyn reduce water diffusivity, particularly in the dentate gyrus and in cortical regions. We suggest that a specific diet intervention has the potential to slow Alzheimer disease progression, by simultaneously improving cerebrovascular health and enhancing neuroprotective mechanisms. These studies suggest that specific multi-nutrient diets can modulate water diffusivity, particularly in the dentate gyrus, and other Alzheimer disease-related etiopathogenic processes resulting in slow progression of AD (Jansen et al. 2013; Zerbi et al. 2014).

A major risk factor for dementia is advancing age. After the age of 65, the prevalence and onset of dementia double every 5 years (Alzheimer's Disease International 2010). Although, environmental factors are known to contribute to the pathogenesis and development of the dementia syndrome, but little is understood about the underlying mechanisms (World Health Organization 2006). Other risk factors such as (a) cardiovascular problems; (b) excessive alcohol consumption; (c) social isolation; (d) head injury; and (e) having one or two copies of the APOEε4 genetic variant also contribute to the pathogenesis of dementia syndrome. Among above mentioned factors, cerebral vascular abnormalities are also an important contributing factor for dementia syndrome as well as Alzheimer disease (Sato and Morishita 2013; Sudduth et al. 2013). It is estimated that as many as 40 % of Alzheimer disease patients actually have a mixed dementia of Alzheimer disease and Vascular dementia

(Kammoun et al. 2000). It is interesting to note that moderate alcohol consumption and estrogen reduce the risk of developing dementia syndrome (Alzheimer 2011).

Insulin and leptin resistance are present in most patients with type II diabetes and Alzheimer disease. This can contribute to compensatory hyperinsulinemia, which is one of the suggested mechanisms to explain the increased risk of Alzheimer disease in diabetic subjects (Qiu and Folstein 2006; Morley and Banks, 2010). There is evidence from prospective studies that there is a lower risk of Alzheimer disease development in individuals with high leptin levels (Holden et al. 2009; Lieb et al. 2009). Moreover, leptin levels are also significantly reduced in murine models (APPSwe; PSIM146V) of Alzheimer disease supporting the view that impairment in leptin system may be closely associated with the pathogenesis of Alzheimer disease (Fewlass et al. 2004).

2.10 Conclusion

The hypothalamus is the master regulator of energy balance. It governs many physiological processes including feeding, energy expenditure, body weight, and glucose metabolism.

Long term consumption of high fat diet induces alterations in hypothalamus signaling leading to neuroinflammation and oxidative stress. High fat diet-mediated alterations in hypothalamus and hippocampal signaling are associated with ER stress and mitochondrial dysfunctions, and insulin resistance, a condition that produces an increase in the secretion of insulin from the pancreas. Thus, ER is a vital organelle not only for protein synthesis and maturation, but also for quality control, and secretion. A variety of factors can disturb the proper functioning of the ER, leading to ER stress and inflammation as well as the induction of synthesis of pro-inflammatory cytokines synthesis. High fat diet is a major cause of obesity which is closely linked to a variety of health issues, including coronary heart disease, stroke, high blood pressure, fatty liver disease, diabetes, certain cancers, and neurological disorders. High fat diet may not only generate free radicals, but also contributes to the development of systemic inflammation and insulin resistance in these diseases. In addition, high saturated fats in circulation, derived mainly from diets or even from lipolysis of fat depots, lead to fatty acids and glucose competing for uptake and metabolism in tissues. Persistent increase in saturated FFA induces a lipotoxic state that is detrimental to not only neural cells, but also cells of visceral tissues through the induction of oxidative stress leading to reduction in insulin synthesis and secretion. The consumption of high fat diet activates lipid-sense nuclear factors such as peroxisome proliferator-activated receptors (PPARs) and liver X receptors, which play critical roles in cellular fatty acid and carbohydrate metabolism as well as cell proliferation. Collective evidence suggests that major biological mechanisms for high fat diet-mediated cognitive dysfunction include insulin resistance, developmental disturbances, altered membrane functioning, oxidative stress, inflammation, and altered vascularization (Farooqui 2013, 2014).

High fat-mediated changes in hippocampus have negative impact on the cognitive function not only due to vascular defects and impaired insulin metabolism, but also due to the defect in glucose transport mechanisms in brain. High fat diet-mediated reduction in mitochondrial biogenesis in hypothalamus increases oxidative stress leading to abnormalities in mitochondrial dysfunction. The resultant mitochondrial dysfunction, in turn, increases ROS production, resulting in a vicious cycle. Several mechanisms may be associated with high fat diet-mediated inflammation in hypothalamus and peripheral tissues. These mechanisms include the activation of TLR4 receptors, induction of ER stress, and activation of IKKβ. Although, the relative contribution made by these mechanisms in the induction of inflammation remains unknown, but early onset of inflammation in hypothalamus relative to that in peripheral tissues suggests that different processes may cause inflammation in peripheral tissues and hypothalamus. At the molecular level inflammation is not only accompanied by elevation in levels of ARA and its lipid mediators (PGs, LTs, and TXs) and increase in platelet activating factor, but also with increase in the expression of proinflammatory genes including genes for cytokines (TNF-α, IL-1β, and IL-6) and proinflammatory enzymes (secretory phospholipase A_2, cyclooxygenase-2, and nitric oxide synthase). In contrast, consumption of EPA and DHA-enriched diet produces anti-inflammatory effects that are partly supported by repression of genes that code for inflammatory cytokines. Overall, conclusion of information presented here is that high fat diet attenuates brain reward system, which is stimulated by multiple types of stimuli, including palatable foods and psychostimulants. Exogenously supplied and endogenously produced fatty acids alter gene transcription in liver and brain. These actions of fatty acids not only change key transcriptional networks controlling neural and hepatic carbohydrate, lipid, and protein metabolism, but also modulate inflammation and oxidative stress. Although some effects of fatty acids are beneficial (e.g., n-3 PUFA-mediated attenuation of inflammation in the brain and liver), but others are not (saturated fatty acid-mediated modulation of TLR and oxidative and ER stress) supporting the view that dietary fatty acids not only regulate several physiological functions, but changes in their levels increase or decrease the risk of visceral and brain diseases.

References

Abuhatzira L, Makedonski K, Kaufman Y, Razin A, Shemer R (2007) MeCP2 deficiency in the brain decreases BDNF levels by REST/CoREST-mediated repression and increases TRKB production. Epigenetics 2:214–222

Ahima RS (2005) Central actions of adipocyte hormones. Trends Endocrinol Metab 16:307–313

Ahmed M, Shaban Z, Yamaji D, Okamatsu-Ogura Y, Soliman M, Abd EM, Ishioka K, Makondo K, Saito M, Kimura K (2007) Induction of proinflammatory cytokines and caspase-1 by leptin in monocyte/macrophages from holstein cows. J Vet Med Sci 69:509–514

Ajuwon KM, Spurlock ME (2005) Palmitate activates the NF-kappaB transcription factor and induces IL-6 and TNFalpha expression in 3 T3-L1 adipocytes. J Nutr 135:1841–1846

Alzheimer Scotland (2011) Risk factors in dementia. Alzheimer Scotland, Edinburgh

Alzheimer's Disease International (2010) Alzheimer's disease, Senile dementia. http://www. Dementia/pages.prodigy.net/naturedoctor/alzheimers.html. Accessed 8 Jan 2010

Amtul Z, Uhrig M, Rozmahel RF, Beyreuther K (2011a) Structural insight into the differential effects of omega-3 and omega-6 fatty acids on the production of Abeta peptides and amyloid plaques. J Biol Chem 286:6100–6107

Amtul Z, Westaway D, Cechetto DF, Rozmahel RF (2011b) Oleic acid ameliorates amyloidosis in cellular and mouse models of Alzheimer's disease. Brain Pathol 21:321–329

Amtul Z, Uhrig M, Wang L, Rozmahel RF, Beyreuther K (2012) Detrimental effects of arachidonic acid and its metabolites in cellular and mouse models of Alzheimer's disease: structural insight. Neurobiol Aging 33:831

Andrews ZB, Liu ZW, Walllingford N, Erion DM, Borok E, Friedman JM, Tschöp MH, Shanabrough M, Cline G, Shulman GI, Coppola A, Gao XB, Horvath TL, Diano S (2008) UCP2 mediates ghrelin's action on NPY/AgRP neurons by lowering free radicals. Nature 454:846–851

Averill MM, Bornfeldt KE (2009) Lipids versus glucose in inflammation and the pathogenesis of macrovascular disease in diabetes. Curr Diab Rep 9:18–25

Banks WA, Coon AB, Robinson SM, Moinuddin A, Shultz JM, Nakaoke R, Morley JE (2004) Triglycerides induce leptin resistance at the blood-brain barrier. Diabetes 53:1253–1260

Bazan NG (2009) Neuroprotectin D1-mediated anti-inflammatory and survival signaling in stroke, retinal degenerations, and Alzheimer's disease. J Lipid Res 50:S400–S405

Berthoud HR (2002) Multiple neural systems controlling food intake and body weight. Neurosci Biobehav Rev 26:393–428

Bourre JM, Bonneil M, Chaudiere J, Clément M, Dumont O, Durand G, Lafont H, Nalbone G, Pascal G, Piciotti M (1992) Structural and functional importance of dietary polyunsaturated fatty acids in the nervous system. Adv Exp Med Biol 318:211–229

Bowman GL, Silbert LC, Howieson D, Dodge HH, Traber MG, Frei R, Kaye JA, Shanon J, Quinn JF (2012) Nutrient biomarker patterns, cognitive function, and MRI measures of brain aging. Neurology 78:241–249

Bowman GL, Dodge HH, Mattek N, Barbey AK, Silbert LC, Shinto L, Howieson DB, Kaye JA, Quinn JF (2013) Plasma omega-3 PUFA and white matter mediated executive decline in older adults. Front Aging Neurosci 5:92

Brüning JC, Gautam D, Burks DJ, Gillette J, Schubert M, Orban PC, Klein R, Krone W, Müller-Wieland D, Kahn CR (2000) Role of brain insulin receptor in control of body weight and reproduction. Science 289:2122–2125

Buckley JD, Howe PRC (2009) Anti-obesity effects of long-chain omega-3 polyunsaturated fatty acids. Obes Rev 10:648–659

Byerly MS, Swanson RD, Semsarzadeh NN, McCulloh PS, Kwon K, Aja S, Moran TH, Wong GW, Blackshaw S (2013) Identification of hypothalamic neuron-derived neurotrophic factor (NENF) as a novel factor modulating appetite. Am J Physiol Regul Integr Comp Physiol 305:R522–R533

Cagniard B, Balsam PD, Brunner D, Zhuang X (2006) Mice with chronically elevated dopamine exhibit enhanced motivation, but not learning, for a food reward. Neuropsychopharmacology 31:1362–1370

Cai D (2009) NFkappaB-mediated metabolic inflammation in peripheral tissues versus central nervous system. Cell Cycle 8:2542–2548

Cai D, Liu T (2012) Inflammatory cause of metabolic syndrome via brain stress and NF-kappaB. Aging (Albany NY) 4:98–115

Carvalheira JBC, Ribeiro EB, Guimarães RB, Telles MM, Torsoni M, Gontijo JA, Velloso LA, Saad MJ (2003) Selective impairment of insulin signalling in the hypothalamus of obese Zucker rats. Diabetologia 46:1629–1640

Cazanave SC, Elmi NA, Akazawa Y, Bronk SF, Mott JL, Gores GJ (2010) CHOP and AP-1 cooperatively mediate PUMA expression during lipoapoptosis. Am J Physiol Gastrointest Liver Physiol 299:G236–G243

Cenquizca LA, Swanson LW (2006) Analysis of direct hippocampal cortical field CA1 axonal projections to diencephalon in the rat. J Comp Neurol 497:101–114

Chandalia M, Abate N (2007) Metabolic complications of obesity: inflated or inflamed? J Diabetes Complications 21:128–136

Chen H, Montagnani M, Funahashi T, Shimomura I, Quon MJ (2003) Adiponectin stimulates production of nitric oxide in vascular endothelial cells. J Biol Chem 278:45021–45026

Clementi E, Nisoli E (2005) Nitric oxide and mitochondrial biogenesis: a key to long-term regulation of cellular metabolism. Comp Biochem Physiol A Mol Integr Physiol 142:102–110

Cordain L, Eaton SB, Sebastian A, Mann N, Lindeberg S, Watkins BA, O'Keefe JH, Brand-Miller J (2005) Origins and evolution of the Western diet: health implications for the 21st century. Am J Clin Nutr 81:341–354

Davidson TL, Kanoski SE, Schier LA, Clegg DJ, Benoit SC (2007) A potential role for the hippocampus in energy intake and body weight regulation. Curr Opin Pharmacol 7:613–616

de Oliveira JL, Oyama LM, Hachul AC, Biz C, Ribeiro EB, Oller do Nascimento CM, Pisani LP (2011) Hydrogenated fat intake during pregnancy and lactation caused increase in TRAF-6 and reduced AdipoR1 in white adipose tissue, but not in muscle of 21 days old offspring rats. Lipids Health Dis 13:22

De Souza CT, Araujo EP, Bordin S, Ashimine R, Zollner RL, Boschero AC, Saad MJ, Velloso LA (2005) Consumption of a fat-rich diet activates a proinflammatory response and induces insulin resistance in the hypothalamus. Endocrinology 146:4192–4199

Diano S, Farr SA, Benoit SE, McNay EC, da Silva I, Horvath B, Gaskin FS, Nonaka N, Jaeger LB, Banks WA, Morley JE, Pinto S, Sherwin RS, Xu L, Yamada KA, Sleeman MW, Tschop MH, Horvath TL (2006) Ghrelin controls hippocampal spine synapse density and memory performance. Nat Neurosci 9:381–388

Duan W, Guo Z, Jiang H, Ware M, Mattson MP (2003) Reversal of behavioral and metabolic abnormalities, and insulin resistance syndrome, by dietary restriction in mice deficient in brain-derived neurotrophic factor. Endocrinology 144:2446–2453

Dyck DJ (2009) Adipokines as regulators of muscle metabolism and insulin sensitivity. Appl Physiol Nutr Metab 34:396–402

El-Gharbawy AH, Adler-Wailes DC, Mirch MC, Theim KR, Ranzenhofer L, Tanofsky-Kraff M, Yanovski JA (2006) Serum brain-derived neurotrophic factor concentrations in lean and overweight children and adolescents. J Clin Endocrinol Metab 91:3548–3552

Elias MF, Elias PK, Sullivan LM, Wolf PA, D'Agostino RB (2003) Lower cognitive function in the presence of obesity and hypertension: the Framingham heart study. Int J Obes Relat Metab Disord 27:260–268

Eller LK, Saha DC, Shearer J, Reimer RA (2013) Dietary leucine improves whole-body insulin sensitivity independent of body fat in diet-induced obese Sprague–Dawley rats. J Nutr Biochem 12:1285–1294

Esterbauer H, Schaur RJ, Zollner H (1991) Chemistry and biochemistry of 4-hydroxynonenal, malonaldehyde and related aldehydes. Free Radic Biol Med 11:81–128

Farooqui AA (2009) Beneficial effects of fish oil on human brain. Springer, New York

Farooqui AA (2010) Neurochemical aspects of neurotraumatic and neurodegenerative diseases. Springer, New York

Farooqui AA (2011) Lipid mediators and their metabolism in the brain. Springer, New York

Farooqui AA (2013) Metabolic syndrome: an important risk factor for stroke, Alzheimer disease, and depression. Springer, New York

Farooqui AA (2014) Inflammation and oxidative stress in neurological disorders. Springer, New York

Farooqui AA, Horrocks LA (2007) Glycerophospholipids in the brain: phospholipases A_2 in neurological disorders. Springer, New York

Farooqui AA, Horrocks LA, Farooqui T (2000a) Glycerophospholipids in brain: their metabolism, incorporation into membrane, functions and involvement in neurological disorders. Chem Phys Lipids 106:1–29

Farooqui AA, Farooqui T, Horrocks LA (2000b) Molecular species of phospholipids during brain development: their occurrence, separation and roles. In: Skinner ER, Bittar EE (eds) Brain lipids and disorders in biological psychiatry. Elsevier, Amsterdam, pp 147–158

Farooqui AA, Farooqui T, Horrocks LA (2008) Metabolism and function of bioactive ether lipids in the brain. Springer, New York

Farooqui AA, Farooqui T, Panza F, Frisardi V (2012) Metabolic syndrome as a risk factor for neurological disorders. Cell Mol Life Sci 69:741–762

Farr SA, Yamada KA, Butterfield DA, Abdul HM, Xu L, Miller NE, Banks WA, Morley JE (2008) Obesity and hypertriglyceridemia produce cognitive impairment. Endocrinology 149: 2628–2636

Fedor D, Kelley DS (2009) Prevention of insulin resistance by n-3 polyunsaturated fatty acids. Curr Opin Clin Nutr Metab Care 12:138–146

Fessler MB, Rudel LL, Brown JM (2009) Toll-like receptor signaling links dietary fatty acids to the metabolic syndrome. Curr Opin Lipidol 20:379–385

Fewlass DC, Noboa K, Pi-Sunyer FX, Johnston JM, Yan SD, Tezapsidis N (2004) Obesity-related leptin regulates Alzheimer's Abeta. FASEB J 18:1870–1878

Figlewicz DP (2004) Intraventricular insulin and leptin reverse place preference conditioned with high-fat diet in rats. Behav Neurosci 118:479–487

Fisher M, Levine PH (1991) Effects of dietary omega 3 fatty acid supplementation on leukocyte free radical production. World Rev Nutr Diet 66:245–249

Franciosi S, Gama Sosa MA, English DF, Oler E, Oung T, Janssen WG, De Gasperi R, Schmeidler J, Dickstein DL, Schmitz C, Gandy S, Hof PR, Buxbaum JD, Elder GA (2009) Novel cerebro-vascular pathology in mice fed a high cholesterol diet. Mol Neurodegener 4:42

Franken IH, Muris P (2005) Individual differences in reward sensitivity are related to food craving and relative body weight in healthy women. Appetite 45:198–201

Frederich M, Balschi JA (2002) The relationship between AMP-activated protein kinase activity and AMP concentration in the isolated perfused rat heart. J Biol Chem 277:1928–1932

Freeman LR, Zhang L, Nair A, Dasuri K, Francis J, Fernandez-Kim SO, Bruce-Keller AJ, Keller JN (2013) Obesity increases cerebrocortical reactive oxygen species and impairs brain function. Free Radic Biol Med 56:226–233

Furukawa S, Fujita T, Shimabukuro M, Iwaki M, Yamada Y, Nakajima Y, Nakayama O, Makishima M, Matsuda M, Shimomura I (2004) Increased oxidative stress in obesity and its impact on metabolic syndrome. J Clin Invest 114:1752–1761

Gao J, Wang WY, Mao YW, Gräff J, Guan JS, Pan L, Mak G, Kim D, Su SC, Tsai LH (2010) A novel pathway regulates memory and plasticity via SIRT1 and miR-134. Nature 466:1105–1109

Garza JC, Guo M, Zhang W, Lu XY (2008) Leptin increases adult hippocampal neurogenesis in vivo and in vitro. J Biol Chem 283:18238–18247

Gaskin FS, Farr SA, Banks WA, Kumar VB, Morley JE (2003) Ghrelin-induced feeding is dependent on nitric oxide. Peptides 24:913–918

Giordano P, Del Vecchio GC, Cecinati V, Delvecchio M, Altomare M, De Palma F, De Mattia D, Cavallo L, Faienza MF (2011) Metabolic, inflammatory, endothelial and haemostatic markers in a group of Italian obese children and adolescents. Eur J Pediatr 13:845–850

González-Périz A, Horrillo R, Ferré N, Gronert K, Dong B, Morán-Salvador E, Titos E, Martínez-Clemente M, López-Parra M, Arroyo V, Clària J (2009) Obesity-induced insulin resistance and hepatic steatosis are alleviated by omega-3 fatty acids: a role for resolvins and protectins. FASEB J 23:1946–1957

Granholm AC, Bimonte-Nelson HA, Moore AB, Nelson ME, Freeman LR, Sambamurti K (2008) Effects of a saturated fat and high cholesterol diet on memory and hippocampal morphology in the middle-aged rat. J Alzheimers Dis 14:133–145

Gray J, Yeo GS, Cox JJ, Morton J, Adlam AL, Keogh JM, Yanovski JA, El Gharbawy A, Han JC, Tung YC, Hodges JR, Raymond FL, O'Rahilly S, Farooqi IS (2006) Hyperphagia, severe obesity, impaired cognitive function, and hyperactivity associated with functional loss of one copy of the brain-derived neurotrophic factor (BDNF) gene. Diabetes 55:3366–3371

Gray J, Yeo G, Hung C, Keogh J, Clayton P, Banerjee K, McAulay A, O'Rahilly S, Farooqi IS (2007) Functional characterization of human NTRK2 mutations identified in patients with severe early-onset obesity. Int J Obes (Lond) 31:359–364

Greenwood CE, Young SN (2001) Dietary fat intake and the brain: a developing frontier in biological psychiatry. J Psychiatry Neurosci 26:182–184

Haberzettl P, Hill BG (2013) Oxidized lipids activate autophagy in a JNK-dependent manner by stimulating the endoplasmic reticulum stress response. Redox Biol 1:56–64

Haffner SM, D'Agostino R Jr, Mykkänen L, Tracy R, Howard B, Rewers M, Selby J, Savage PJ, Saad MF (1999) Insulin sensitivity in subjects with type 2 diabetes. Relationship to cardiovascular risk factors: the Insulin Resistance Atherosclerosis Study. Diabetes Care 22:562–568

Hajer GR, van Haeften TW, Visseren FL (2008) Adipose tissue dysfunction in obesity, diabetes, and vascular diseases. Eur Heart J 13:2959–2971

Halaas JL, Gajiwala KS, Maffei M, Cohen SL, Chait BT, Rabinowitz D, Lallone RL, Burley SK, Friedman JM (1995) Weight-reducing effects of the plasma protein encoded by the obese gene. Science 269:543–546

Harding HP, Novoa I, Zhang H, Wek R, Schapira M, Ron D (2000) Regulated translation initiation controls stress-induced gene expression in mammalian cells. Mol Cell 6:1099–1108

Harnack L, Lee S, Schakel SF, Duval S, Luepker RV, Arnett DK (2003) Trends in the trans-fatty acid composition of the diet in a metropolitan area: the Minnesota Heart Survey. J Am Diet Assoc 103:1160–1166

Harvey J (2007) Leptin regulation of neuronal excitability and cognitive function. Curr Opin Pharmacol 7:643–647

Harvey KA, Arnold T, Rasool T, Antaites C, Miller SJ, Siddiqui RA (2008) Trans-fatty acids induce pro-inflammatory responses and endothelial cell dysfunction. Br J Nutr 99:723–731

He XP, Kotloski R, Nef S, Luikart BW, Parada LF, McNamara JO (2004) Conditional deletion of TrkB but not BDNF prevents epileptogenesis in the kindling model. Neuron 43:31–42

Heyward FD, Walton RG, Carle MS, Coleman MA, Garvey WT, Sweatt JD (2012) Adult mice maintained on a high-fat diet exhibit object location memory deficits and reduced hippocampal SIRT1 gene expression. Neurobiol Learn Mem 98:25–32

Hill BG, Dranka BP, Bailey SM, Lancaster JR Jr, Darley-Usmar VM (2010) What part of NO don't you understand? Some answers to the cardinal questions in nitric oxide biology. J Biol Chem 285:19699–19704

Hirosumi J, Tuncman G, Chang L, Gorgun CZ, Uysal KT, Maeda K, Karin M, Hotamisligil GS (2002) A central role for JNK in obesity and insulin resistance. Nature 420:333–336

Holden KF, Lindquist K, Tylavsky FA, Rosano C, Harris TB, Yaffe K (2009) Serum leptin level and cognition in the elderly: findings from the Health ABC Study. Neurobiol Aging 30:1483–1489

Horch HW (2004) Local effects of BDNF on dendritic growth. Rev Neurosci 15:117–129

Horvath TL, Sarman B, Garcia-Caceres C, Enriori PJ, Sotonyi P, Shanabrough M, Borok E, Argente J, Chowen JA, Perez-Tilve D (2010) Synaptic input organization of the melanocortin system predicts diet-induced hypothalamic reactive gliosis and obesity. Proc Natl Acad Sci U S A 107:14875–14880

Hotamisligil GS (2006) Inflammation and metabolic disorders. Nature 444:860–867

Hoyda TD, Fry M, Ahima RS, Ferguson AV (2007) Adiponectin selectively inhibits oxytocin neurons of the paraventricular nucleus of the hypothalamus. J Physiol 585:805–816

Hunter JE (2006) Dietary trans fatty acids: review of recent human studies and food industry responses. Lipids 41:967–992

Hwang IK, Kim IY, Kim DW, Yoo KY, Kim YN, Yi SS, Won MH, Lee IS, Yoon YS, Seong JK (2008) Strain-specific differences in cell proliferation and differentiation in the dentate gyrus of C57BL/6 N and C3H/HeN mice fed a high fat diet. Brain Res 1241:1–6

Jansen D, Zerbi V, Arnoldussen IA, Wiesmann M, Rijpma A, Fang XT, Dederen PJ, Mutsaers MP, Broersen LM, Lütjohann D, Miller M, Joosten LA, Heerschap A, Kiliaan AJ (2013) Effects of specific multi-nutrient enriched diets on cerebral metabolism, cognition and neuropathology in AβPPswe-PS1dE9 mice. PLoS One 8:e75393

Jellinger KA (2009) Recent advances in our understanding of neurodegeneration. J Neural Transm 116:1111–1162

Jump DB (2008) *N*-3 polyunsaturated fatty acid regulation of hepatic gene transcription. Curr Opin Lipidol 19:242–247

Jump DB, Tripathy S, Depner CM (2013) Fatty acid-regulated transcription factors in the liver. Annu Rev Nutr 33:249–269

Kalant D, Cianflone K (2004) Regulation of fatty acid transport. Curr Opin Lipidol 15:309–314

Kammoun S, Gold G, Bouras C, Giannakopoulos P, McGee W, Herrmann F, Michel JP (2000) Immediate causes of death of demented and non-demented elderly. Acta Neurol Scand Suppl 176:96–99

Kanoski SE, Meisel RL, Mullins AJ, Davidson TL (2007) The effects of energy-rich diets on discrimination reversal learning and on BDNF in the hippocampus and prefrontal cortex of the rat. Behav Brain Res 182:57–66

Kaplan RJ, Greenwood CE (1998) Dietary saturated fatty acids and brain function. Neurochem Res 23:615–626

Katan MB (1998) Health effects of Trans fatty acids. Eur J Clin Invest 28:257–258

Keita AV, Salim SY, Jiang T, Yang PC, Franzen L, Soderkvist P, Magnusson KE, Soderholm JD (2008) Increased uptake of non-pathogenic *E. coli* via the follicle-associated epithelium in longstanding ileal Crohn's disease. J Pathol 215:135–144

Kennedy A, Martinez K, Chuang CC, LaPoint K, McIntosh M (2009) Saturated fatty acid-mediated inflammation and insulin resistance in adipose tissue: mechanisms of action and implications. J Nutr 139:1–4

Kernie SG, Liebl DJ, Parada LF (2000) BDNF regulates eating behavior and locomotor activity in mice. EMBO J 19:1290–1300

Kettenmann H, Hanisch UK, Noda M, Verkhratsky A (2011) Physiology of microglia. Physiol Rev 91:461–553

Kim EK, Miller I, Landree LE, Borisy-Rudin FF, Brown P, Tihan T, Townsend CA, Witters LA, Moran TH, Kuhajda FP, Ronnett GV (2002) Expression of FAS within hypothalamic neurons: a model for decreased food intake after C75 treatment. Am J Physiol Endocrinol Metab 283:E867–E879

Kimura I, Yoshioka M, Konishi M, Miyake A, Itoh N (2005) Neudesin, a novel secreted protein with a unique primary structure and neurotrophic activity. J Neurosci Res 79:287–294

Kimura I, Nakayama Y, Zhao Y, Konishi M, Itoh N (2013) Neurotrophic effects of neudesin in the central nervous system. Front Neurosci 7:111

Kubota N, Yano W, Kubota T, Yamauchi T, Itoh S, Kumagai H, Kozono H, Takamoto I, Okamoto S, Shiuchi T, Suzuki R, Satoh H, Tsuchida A, Moroi M, Sugi K, Noda T, Ebinuma H, Ueta Y, Kondo T, Araki E (2007) Adiponectin stimulates AMP-activated protein kinase in the hypothalamus and increases food intake. Cell Metab 6:55–68

Kufer TA, Kremmer E, Adam AC, Philpott DJ, Sansonetti PJ (2008) The pattern-recognition molecule Nod1 is localized at the plasma membrane at sites of bacterial interaction. Cell Microbiol 10:477–486

Lam TK, Pocai A, Gutierrez-Juarez R, Obici S, Bryan J, Aguilar-Bryan L, Schwartz GJ, Rossetti L (2005) Hypothalamic sensing of circulating fatty acids is required for glucose homeostasis. Nat Med 11:320–327

Lee JM, Johnson JA (2004) An important role of Nrf2-ARE pathway in the cellular defense mechanism. J Biochem Mol Biol 37:139–143

Lee JY, Plakidas A, Lee WH, Heikkinen A, Chanmugam P, Bray G, Hwang DH (2003) Differential modulation of toll-like receptors by fatty acids preferential inhibition by n-3 polyunsaturated fatty acids. J Lipid Res 44:479–486

Lee JY, Zhao L, Youn HS, Weatherill AR, Tapping R, Feng L, Lee WH, Fitzgerald KA, Hwang DH (2004) Saturated fatty acid activates but polyunsaturated fatty acid Inhibits toll-like receptor 2 dimerized with toll-like receptor 6 or 1. J Biol Chem 279:16971–16979

Lee JS, Pinnamaneni SK, Eo SJ, Cho IH, Pyo JH, Kim CK, Sinclair AJ, Febbraio MA, Watt MJ (2006) Saturated, but not n-6 polyunsaturated, fatty acids induce insulin resistance: role of intramuscular accumulation of lipid metabolites. J Appl Physiol 12:1467–1474

Lemieux ME, Yang X, Jardine K, He X, Jacobsen KX, Staines WA, Harper ME, McBurney MW (2005) The Sirt1 deacetylase modulates the insulin-like growth factor signaling pathway in mammals. Mech Ageing Dev 126:1097–1105

Levine B, Kroemer G (2008) Autophagy in the pathogenesis of disease. Cell 132:27–42

Ley RE, Backhed F, Turnbaugh P, Lozupone CA, Knight RD, Gordon JI (2005) Obesity alters gut microbial ecology. Proc Natl Acad Sci U S A 102:11070–11075

Li M, Baumeister P, Roy B, Phan T, Foti D, Luo S, Lee AS (2000) ATF6 as a transcription activator of the endoplasmic reticulum stress element. Thapsigargin stress-induced changes and synergistic interactions with NF-Y and YY1. Mol Cell Biol 20:5096–5106

Li ZG, Zhang W, Sima AA (2007) Alzheimer-like changes in rat models of spontaneous diabetes. Diabetes 56:1817–1824

Li Y, Xu W, McBurney MW, Longo VD (2008) SirT1 inhibition reduces IGF-I/IRS-2/Ras/ERK1/2 signaling and protects neurons. Cell Metab 8:38–48

Lieb W, Beiser AS, Vasan RS, Tan ZS, Au R, Harris TB, Roubenoff R, Auerbach S, DeCarli C, Wolf PA, Seshadri S (2009) Association of plasma leptin levels with incident Alzheimer disease and MRI measures of brain aging. JAMA 302:2565–2572

Lin J, Yan GT, Xue H, Hao XH, Zhang K, Wang LH (2007) Leptin protects vital organ functions after sepsis through recovering tissue myeloperoxidase activity: an anti-inflammatory role resonating with indomethacin. Peptides 28:1553–1560

Lindqvist A, Mohapel P, Bouter B, Frielingsdorf H, Pizzo D, Brundin P, Erlanson-Albertsson C (2006) High-fat diet impairs hippocampal neurogenesis in male rats. Eur J Neurol 13:1385–1388

Lipsky RH, Xu K, Zhu D, Kelly C, Terhakopian A, Novelli A, Marini AM (2001) Nuclear factor kappaB is a critical determinant in N-methyl-D-aspartate receptor-mediated neuroprotection. J Neurochem 78:254–264

Liu Y, Yang L, Conde-Knape K, Beher D, Shearman MS, Shachter NS (2004) Fatty acids increase presenilin-1 levels and γ-secretase activity in PSwt-1 cells. J Lipid Res 45:2368–2376

López M, Tovar S, Vázquez MJ, Williams LM, Diéguez C (2007) Peripheral tissue-brain interactions in the regulation of food intake. Proc Nutr Soc 66:131–155

Lovejoy JC (2002) The influence of dietary fat on insulin resistance. Cur Diab Repo 2:435–440

Lu B, Pang PT, Woo NH (2005) The yin and yang of neurotrophin action. Nat Rev Neurosci 6:603–614

Lukiw WJ, Cui J, Marcheselli VL, Bodker M, Botkjaer A, Gotlinger K, Serhan CN, Bazan NG (2005) A role for docosahexaenoic acid-derived neuroprotectin D1 in neural cell survival and Alzheimer disease. J Clin Invest 115:2774–2783

Lynch G, Kramar EA, Rex CS, Jia Y, Chappas D, Gall CM, Simmons DA (2007) Brain derived neurotrophic factor restores synaptic plasticity in a knock-in mouse model of Huntington's disease. J Neurosci 27:4424–4434

Maesako M, Uemura K, Kubota M, Kuzuya A, Sasaki K, Hayashida N, Asada-Utsugi M, Watanabe K, Uemura M, Kihara T, Takahashi R, Shimohama S, Kinoshita A (2012) Exercise is more effective than diet control in preventing high fat diet-induced β-amyloid deposition and memory deficit in amyloid precursor protein transgenic mice. J Biol Chem 287:23024–23033

Magdeldin S, Elewa Y, Ikeda T, Ikei J, Zhang Y, Xu B, Nameta M, Fujinaka H, Yoshida Y, Yaoita E, Yamamoto T (2009) Dietary supplementation with arachidonic acid but not eicosapentaenoic or docosahexaenoic acids alter lipids metabolism in C57BL/6 J mice. Gen Physiol Biophys 12:266–275

Maitra U, Deng H, Glaros T, Baker B, Capelluto DG, Li Z, Li L (2012) Molecular mechanisms responsible for the selective and low-grade induction of proinflammatory mediators in murine macrophages by lipopolysaccharide. J Immunol 189:1014–1023

Malhotra JD, Kaufman RJ (2007) The endoplasmic reticulum and the unfolded protein response. Semin Cell Dev Biol 18:716–731

Mancuso DJ, Sims HF, Han X, Jenkins CM, Guan SP, Yang K, Moon SH, Pietka T, Abumrad NA, Schlesinger PH, Gross RW (2007) Genetic ablation of calcium-independent phospholipase A$_2$gamma leads to alterations in mitochondrial lipid metabolism and function resulting in a deficient mitochondrial bioenergetic phenotype. J Biol Chem 282:34611–34622

Mancuso DJ, Kotzbauer P, Wozniak DF, Sims HF, Jenkins CM, Guan S, Han X, Yang K, Sun G, Malik I, Conyers S, Green KG, Schmidt RE, Gross RW (2009) Genetic ablation of calcium-independent phospholipase A2{gamma} leads to alterations in hippocampal cardiolipin content and molecular species distribution, mitochondrial degeneration, autophagy, and cognitive dysfunction. J Biol Chem 284:35632–35644

Mancuso DJ, Sims HF, Yang K, Kiebish MA, Su X, Jenkins CM, Guan S, Moon SH, Pietka T, Nassir F, Schappe T, Moore K, Han X, Abumrad NA, Gross RW (2010) Genetic ablation of calcium-independent phospholipase A$_2$gamma prevents obesity and insulin resistance during high fat feeding by mitochondrial uncoupling and increased adipocyte fatty acid oxidation. J Biol Chem 285:36495–36510

Mantena SK, Vaughn DP, Andringa KK, Eccleston HB, King AL, Abrams GA, Doeller JE, Kraus DW, Darley-Usmar VM, Bailey SM (2009) High fat diet induces dysregulation of hepatic oxygen gradients and mitochondrial function *in vivo*. Biochem J 417:183–193

Marcheselli VL, Hong S, Lukiw WJ, Tian XH, Gronert K, Musto A, Hardy M, Gimenez JM, Chiang N, Serhan CN, Bazan NG (2003) Novel docosanoids inhibit brain ischemia-reperfusion-mediated leukocyte infiltration and pro-inflammatory gene expression. J Biol Chem 278:43807–43817

Marciniak SJ, Ron D (2006) Endoplasmic reticulum stress signaling in disease. Physiol Rev 86:1133–1149

Martinez de Morentin PB, Varela L, Ferno J, Nogueiras R, Dieguez C, Lopez M (2010) Hypothalamic lipotoxicity and the metabolic syndrome. Biochim Biophys Acta 1801:350–361

Mattson MP (2008) Glutamate and neurotrophic factors in neuronal plasticity and disease. Ann N Y Acad Sci 1144:97–112

Mattson MP, Maudsley S, Martin B (2004) BDNF and 5-HT: a dynamic duo in age-related neuronal plasticity and neurodegenerative disorders. Trends Neurosci 27:589–594

McGowan MK, Andrews KM, Grossman SP (1992) Chronic intrahyphothalamic infusions of insulin or insulin antibodies alter body weight and food intake in the rat. Physol Behav 51:753–766

Meng Q, Cai D (2011) Defective hypothalamic autophagy directs the central pathogenesis of obesity via the IkappaB kinase beta (IKKbeta)/NF-kappaB pathway. J Biol Chem 286:32324–32332

Mennitti LV, Oyama LM, de Oliveira JL, Hachul AC, Santamarina AB, de Santana AA, Okuda MH, Ribeiro EB, do Nascimento CM, Pisani LP (2014) Oligofructose supplementation during pregnancy and lactation impairs offspring development and alters the intestinal properties of 21-d-old pups. Lipids Health Dis 13:26

Micha R, Mozaffarian D (2009) Trans fatty acids: effects on metabolic syndrome, heart disease and diabetes. Nat Rev Endocrinol 5:335–344

Michan S, Li Y, Chou MMH, Parrella E, Ge H, Long JM, Allard JS et al (2010) SIRT1 is essential for normal cognitive function and synaptic plasticity. J Neurosci 30:9695–9707

Milanski M, Degasperi G, Coope A, Morari J, Denis R, Cintra DE, Tsukumo DM, Anhe G, Amaral ME, Takahashi HK (2009) Saturated fatty acids produce an inflammatory response predominantly through the activation of TLR4 signaling in hypothalamus: implications for the pathogenesis of obesity. J Neurosci 29:359–370

Minokoshi Y, Alquier T, Furukawa N, Kim YB, Lee A, Xue B, Mu J, Foufelle F, Ferré P, Birnbaum MJ, Stuck BJ, Kahn BB (2004) AMP-kinase regulates food intake by responding to hormonal and nutrient signals in the hypothalamus. Nature 428:569–574

Morley JE, Banks WA (2010) Lipids and cognition. J Alzheimers Dis 20:737–747

Morrison CD, Pistell PJ, Ingram DK, Johnson WD, Liu Y, Fernandez-Kim SO, White CL, Purpera MN, Uranga RM, Bruce-Keller AJ, Keller JN (2010) High fat diet increases hippocampal oxidative stress and cognitive impairment in aged mice: implications for decreased Nrf2 signaling. J Neurochem 114:1581–1589

Mozaffarian D (2006) Trans fatty acids—effects on systemic inflammation and endothelial function. Atheroscler Suppl 7:29–32

Mozaffarian D, Willett WC (2007) Trans fatty acids and cardiovascular risk: a unique cardiometabolic imprint? Curr Atheroscler Rep 9:486–493

Muhlhausler BS, Cook-Johnson R, James M, Miljkovic D, Duthoit E, Gibson R (2010) Opposing effects of omega-3 and omega-6 long chain polyunsaturated fatty acids on the expression of lipogenic genes in omental and retroperitoneal adipose depots in the rat. J Nutr Metab 2010:927836

Muurling M, Jong MC, Mensink RP, Hornstra G, Dahlmans VE, Pijl H, Voshol PJ, Havekes LM (2002) A low-fat diet has a higher potential than energy restriction to improve high-fat diet-induced insulin resistance in mice. Metabolism 51:695–701

Nagahara AH, Merrill DA, Coppola G, Tsukada S, Schroeder BE, Shaked GM, Wang L, Blesch A, Kim A, Conner JM, Rockenstein E, Chao MV, Koo EH, Geschwind D, Masliah E, Chiba AA, Tuszynski MH (2009) Neuroprotective effects of brainderived neurotrophic factor in rodent and primate models of Alzheimer's disease. Nat Med 15:331–337

Nakajima K, Kohsaka S, Tohyama Y, Kurihara T (2003) Activation of microglia with lipopolysaccharide leads to the prolonged decrease of conventional protein kinase c activity. Brain Res Mol Brain Res 110:92–99

Narazaki M, Fujimoto M, Matsumoto T, Morita Y, Saito H, Kajita T, Yoshizaki K, Naka T, Kishimoto T (1998) Three distinct domains of SSI-1/SOCS-1/JAB protein are required for its suppression of interleukin 6 signalling. Proc Natl Acad Sci U S A 95:13130–13134

Ng DT, Spear ED, Walter P (2000) The unfolded protein response regulates multiple aspects of secretory and membrane protein biogenesis and endoplasmic reticulum quality control. J Cell Biol 150:77–88

Niculescu MD, Lupu DS (2009) High fat diet-induced maternal obesity alters fetal hippocampal development. Int J Dev Neurosci 27:627–633

Novais A, Ferreira AC, Marques F, Pêgo JM, Cerqueira JJ, David-Pereira A, Campos FL, Dalla C, Kokras N, Sousa N, Palha JA, Sousa JC (2013) Neudesin is involved in anxiety behavior: structural and neurochemical correlates. Front Behav Neurosci 7:119

Nuernberg K, Breier BH, Jayasinghe SN, Bergmann H, Thompson N, Nuernberg G, Dannenberger D, Schneider F, Renne U, Langhammer M, Huber K (2011) Metabolic responses to high-fat diets rich in n-3 or n-6 long-chain polyunsaturated fatty acids in mice selected for either high body weight or leanness explain different health outcomes. Nutr Metab (Lond) 12:56

Obici S, Feng Z, Karkanias G, Baskin DG, Rossetti L (2002) Decreasing hypothalamic insulin receptors causes hyperphagia and insulin resistance in rats. Nat Neurosci 5:566–572

Obici S, Feng Z, Arduini A, Conti R, Rossetti L (2003) Inhibition of hypothalamic carnitine palmitoyltransferase-1 decreases food intake and glucose production. Nat Med 9:756–761

Olefsky JM, Glass CK (2010) Macrophages, inflammation, and insulin resistance. Annu Rev Physiol 72:219–246

Ong WY, Farooqui T, Farooqui AA (2010) Involvement of cPLA$_2$ and iPLA$_2$ in neurodegenerative and neuropsychiatric conditions. Curr Med Chem 17:2746–2763

Oresic M, Hänninen VA, Vidal-Puig A (2008) Lipidomics: a new window to biomedical frontiers. Trends Biotechnol 26:647–652

Osso FS, Moreira ASB, Teixeira MT, Pereira RO, Tavares do Carmo MG, Moura AS (2008) *Trans* fatty acids in maternal milk lead to cardiac insulin resistance in adult offspring. Nutrition 13:727–732

Ouchi N, Kihara S, Arita Y, Okamoto Y, Maeda K, Kuriyama H, Hotta K, Nishida M, Takahashi M, Muraguchi M, Ohmoto Y, Nakamura T, Yamashita S, Funahashi T, Matsuzawa Y (2000) Adiponectin, an adipocyte-derived plasma protein, inhibits endothelial NF-κB signaling through a cAMP-dependent pathway. Circulation 102:1296–1301

Ozcan U, Cao Q, Yilmaz E, Lee AH, Iwakoshi NN, Ozdelen E, Tuncman G, Görgün C, Glimcher LH, Hotamisligil GS (2004) Endoplasmic reticulum stress links obesity, insulin action, and type 2 diabetes. Science 306:457–461

Ozcan L, Ergin AS, Lu A, Chung J, Sarkar S, Nie D, Myers MG Jr, Ozcan U (2009) Endoplasmic reticulum stress plays a central role in development of leptin resistance. Cell Metab 9:35–51

Pannacciulli N, Del Parigi A, Chen K, Le DS, Reiman EM, Tataranni PA (2006) Brain abnormalities in human obesity: a voxel-based morphometric study. Neuroimage 31:1419–1425

Park L, Anrather J, Forster C, Kazama K, Carlson GA, Iadecola C (2004) Abeta-induced vascular oxidative stress and attenuation of functional hyperemia in mouse somatosensory cortex. J Cereb Blood Flow Metab 24:334–342

Park HR, Park M, Choi J, Park KY, Chung HY, Lee J (2010) A high-fat diet impairs neurogenesis: involvement of lipid peroxidation and brain-derived neurotrophic factor. Neurosci Lett 482:235–239

Patil S, Chan C (2005) Palmitic and stearic fatty acids induce Alzheimer-like hyperphosphorylation of tau in primary rat cortical neurons. Neurosci Lett 384:288–293

Pelleymounter MA, Cullen MJ, Wellman CL (1995) Characteristics of BDNF-induced weight loss. Exp Neurol 131:229–238

Phillis JW, Horrocks LA, Farooqui AA (2006) Cyclooxygenases, lipoxygenases, and epoxygenases in CNS: their role and involvement in neurological disorders. Brain Res Rev 52:201–243

Phivilay A, Julien C, Tremblay C, Berthiaume L, Julien P, Giguère Y, Calon F (2009) High dietary consumption of trans fatty acids decreases brain docosahexaenoic acid but does not alter amyloid-beta and tau pathologies in the 3xTg-AD model of Alzheimer's disease. Neuroscience 159:296–307

Pimentel GD, Lira FS, Rosa JC, Oliveira JL, Losinskas-Hachul AC, Souza GI, das Graças T, do Carmo M, Santos RV, de Mello MT, Tufik S, Seelaender M, Oyama LM, Nascimento CM O d, Watanabe RH, Ribeiro EB, Pisani LP (2012) Intake of trans fatty acids during gestation and lactation leads to hypothalamic inflammation via TLR4/NFkappaBp65 signaling in adult offspring. J Nutr Biochem 23:822–828

Pintana H, Apaijai N, Pratchayasakul W, Chattipakorn N, Chattipakorn SC (2012) Effects of metformin on learning and memory behaviors and brain mitochondrial functions in high fat diet induced insulin resistant rats. Life Sci 91:409–414

Pintana H, Apaijai N, Chattipakorn N, Chattipakorn SC (2013) DPP-4 inhibitors improve cognition and brain mitochondrial function of insulin-resistant rats. J Endocrinol 218:1–11

Pinteaux E, Inoue W, Schmidt L, Molina-Holgado F, Rothwell NJ, Luheshi GN (2007) Leptin induces interleukin-1beta release from rat microglial cells through a caspase 1 independent mechanism. J Neurochem 102:826–833

Pisani LP, Oyama LM, Bueno AA, Biz C, Albuquerque KT, Ribeiro EB, Oller do Nascimento CM (2008a) Hydrogenated fat intake during pregnancy and lactation modifies serum lipid profile and adipokine mRNA in 21-day-old rats. Nutrition 13:255–261

Pisani LP, do Nascimento CMO, Bueno AA, Biz C, Albuquerque KT, Ribeiro EB, Oyama LM (2008b) Hydrogenated fat diet intake during pregnancy and lactation modifies the PAI-1 gene expression in white adipose tissue of offspring in adult life. Lipids Health Dis 13:13

Plum L, Ma X, Hampel B, Balthasar N, Coppari R, Munzberg H, Shanabrough M, Burdakov D, Rother E, Janoschek R, Alber J, Belgardt BF, Koch L, Seibler J, Schwenk F, Fekete C, Suzuki A, Mak TW, Krone W, Horvath TL, Ashcroft FM, Bruning JC (2006) Enhanced PIP3 signaling in POMC neurons causes KATP channel activation and leads to diet-sensitive obesity. J Clin Invest 116:1886–1901

Poggi M, Bastelica D, Gual P, Iglesias MA, Gremeaux T, Knauf C, Peiretti F, Verdier M, Juhan-Vague I, Tanti JF, Burcelin R, Alessi MC (2007) C3H/HeJ mice carrying a toll-like receptor 4 mutation are protected against the development of insulin resistance in white adipose tissue in response to a high-fat diet. Diabetologia 50:1267–1276

Polozova A, Salem N Jr (2007) Role of liver and plasma lipoproteins in selective transport of n-3 fatty acids to tissues: a comparative study of 14C-DHA and 3H-oleic acid tracers. J Mol Neurosci 33:56–66

Purkayastha S, Cai D (2013) Disruption of neurogenesis by hypothalamic inflammation in obesity or aging. Rev Endocr Metab Disord 14:351–356

Purkayastha S, Zhang H, Zhang G, Ahmed Z, Wang Y, Cai D (2011a) Neural dysregulation of peripheral insulin action and blood pressure by brain endoplasmic reticulum stress. Proc Natl Acad Sci U S A 108:2939–2944

Purkayastha S, Zhang G, Cai D (2011b) Uncoupling the mechanisms of obesity and hypertension by targeting hypothalamic IKK-β and NF-κB. Nat Med 17:883–887

Qaseem A, Snow V, Cross J Jr, Forciea MA, Hopkins R Jr, Shekelle P, Adelman A, Mehr D, Schellhase K, Campos-Outcalt D, Santaguida P, Owens DK, American College of Physicians/ American Academy of Family Physicians Panel on Dementia (2008) Current pharmacologic treatment of dementia: a clinical practice guideline from the American College of Physicians and the American Academy of Family Physicians. Ann Intern Med 148:370–378

Qiu WQ, Folstein MF (2006) Insulin, insulin-degrading enzyme and amyloid-beta peptide in Alzheimer's disease: review and hypothesis. Neurobiol Aging 27:190–198

Rader DJ (2000) Inflammatory markers of coronary risk. N Engl J Med 343:1179–1182

Rapoport SI (2005) In vivo approaches and rationale for quantifying kinetics and imaging brain lipid metabolic pathways. Prostaglandins Other Lipid Mediat 77:185–196

Reichardt LF (2006) Neurotrophin-regulated signalling pathways. Philos Trans R Soc Lond B Biol Sci 361:1545–1564

Roman GC (2002) Vascular dementia revisited: diagnosis, pathogenesis, treatment, and prevention. Med Clin North Am 86:477–499

Rondinone CM (2006) Adipocyte-derived hormones, cytokines, and mediators. Endocrine 29:81–90

Rothman SM, Mattson MP (2013) Activity-dependent, stress-responsive BDNF signaling and the quest for optimal brain health and resilience throughout the lifespan. Neuroscience 239:228–240

Ruitenberg A, Ott A, van Swieten JC, Hofman A, Breteler MM (2001) Incidence of dementia: does gender make a difference? Neurobiol Aging 22:575–580

Sagare A, Deane R, Bell RD, Johnson B, Hamm K, Pendu R, Marky A, Lenting PJ, Wu Z, Zarcone T, Goate A, Mayo K, Perlmutter D, Coma M, Zhong Z, Zlokovic BV (2007) Clearance of amyloid-beta by circulating lipoprotein receptors. Nat Med 13:1029–1031

Salvati S, Attorri L, Avellino C, Di Biase A, Sanchez M (2000) Diet, lipids and brain development. Dev Neurosci 22:481–487

Santos AR, Comprido D, Duarte CB (2010) Regulation of local translation at the synapse by BDNF. Prog Neurobiol 92:505–516

Santpere G, Puig B, Ferrer I (2007) Oxidative damage of 14-3-3 zeta and gamma isoforms in Alzheimer's disease and cerebral amyloid angiopathy. Neuroscience 146:1640–1651

Sato N, Morishita R (2013) Roles of vascular and metabolic components in cognitive dysfunction of Alzheimer disease: short- and long-term modification by non-genetic risk factors. Front Aging Neurosci 5:64

Savage DB, Petersen KF, Shulman GI (2007) Disordered lipid metabolism and the pathogenesis of insulin resistance. Physiol Rev 87:507–520

Scheele C, Nielsen S, Kelly M, Broholm C, Nielsen AR, Taudorf S, Pedersen M, Fischer CP, Pedersen BK (2012) Satellite cells derived from obese humans with type 2 diabetes and differentiated into myocytes in vitro exhibit abnormal response to IL-6. PLoS One 7:e39657

Schroder M, Kaufman RJ (2005) The mammalian unfolded protein response. Annu Rev Biochem 74:739–789

Schwartz MW (2001) Brain pathways controlling food intake and body weight. Exp Biol Med 226:978–981

Seeley RJ, Woods SC (2003) Monitoring of stored and available fuel by the CNS: implications for obesity. Nat Rev Neurosci 4:901–909

Seo T, Qi K, Chang C, Liu Y, Worgall TS, Ramakrishnan R, Deckelbaum RJ (2005) Saturated fat-rich diet enhances selective uptake of LDL cholesteryl esters in the arterial wall. J Clin Invest 115:2214–2222

Serhan CN, Chiang N, Van Dyke TE (2008) Resolving inflammation: dual anti-inflammatory and pro-resolution lipid mediators. Nat Rev Immunol 8:349–361

Serhan CN, Fredman G, Yang R, Karamnov S, Belayev LS, Bazan NG, Zhu M, Winkler JW, Petasis NA (2011) Novel proresolving aspirin-triggered DHA pathway. Chem Biol 18:976–987

Sha H, Xu J, Tang J, Ding J, Gong J, Ge X, Kong D, Gao X (2007) Disruption of a novel regulatory locus results in decreased BDNF expression, obesity, and type 2 diabetes in mice. Physiol Genomics 31:252–263

Shahidi F, Zhong Y (2010) Lipid oxidation and improving the oxidative stability. Chem Soc Rev 39:4067–4079

Shelton RC, Miller AH (2010) Eating ourselves to death (and despair): the contribution of adiposity and inflammation to depression. Prog Neurobiol 91:275–299

Shiva S, Oh JY, Landar AL, Ulasova E, Venkatraman A, Bailey SM, Darley-Usmar VM (2005) Nitroxia: the pathological consequence of dysfunction in the nitric oxide-cytochrome c oxidase signaling pathway. Free Radic Biol Med 38:297–306

Shoelson S, Lee J, Goldfine A (2006) Inflammation and insulin resistance. J Clin Invest 116:1793–1801

Simopoulos AP (2008) The importance of the omega-6/omega-3 fatty acid ratio in cardiovascular disease and other chronic diseases. Exp Biol Med (Maywood) 233:674–688

Simopoulos AP (2011) Evolutionary aspects of diet: the omega-6/omega-3 ratio and the brain. Mol Neurobiol 44:203–215

Singh M, Su C, Ng S (2013) Non-genomic mechanisms of progesterone action in the brain. Front Neurosci 7:1–7

Skurk T, Hauner H (2004) Obesity and impaired fibrinolysis: role of adipose production of plasminogen activator inhibitor-1. Int J Obes Relat Metab Disord 13:1357–1364

Sonnen JA, Larson EB, Haneuse S, Woltjer R, Li G, Crane PK, Craft S, Montine TJ (2009) Neuropathology in the adult changes in thought study: a review. J Alzheimers Dis 18:703–711

Soodini GR, Hamdy O (2004) Adiponectin and leptin in relation to insulin sensitivity. Metab Syndr Relat Disord 2:114–123

Stender S, Astrup A, Dyerberg J (2008) Ruminant and industrially produced trans fatty acids: health aspects. Food Nutr Res 52:10

Stranahan AM, Norman ED, Lee K, Cutler RG, Telljohann RS, Egan JM, Mattson MP (2008) Diet-induced insulin resistance impairs hippocampal synaptic plasticity and cognition in middle-aged rats. Hippocampus 18:1085–1088

Strauss J, Barr CL, George CJ, Devlin B, Vetro A, Kiss E, Baji I, King N, Shaikh S, Lanktree M, Kovacs M, Kennedy JL (2005) Brain-derived neurotrophic factor variants are associated with childhood-onset mood disorder: confirmation in a Hungarian sample. Mol Psychiatry 10:861–867

Sudduth TL, Powell DK, Smith CD, Greenstein A, Wilcock DM (2013) Induction of hyperhomocysteinemia models vascular dementia by induction of cerebral microhemorrhages and neuroinflammation. J Cereb Blood Flow Metab 33:708–715

Suganami T, Tanimoto-Koyama K, Nishida J, Itoh M, Yuan X, Mizuarai S, Kotani H, Yamaoka S, Miyake K, Aoe S, Kamei Y, Ogawa Y (2007) Role of the toll-like receptor 4/NF-κB pathway in saturated fatty acid-induced inflammatory changes in the interaction between adipocytes and macrophages. Arterioscler Thromb Vasc Biol 27:84–91

Sundaresan S, Vijayagopal P, Mills N, Imrhan V, Prasad C (2011) A mouse model for nonalcoholic steatohepatitis. J Nutr Biochem 12:979–984

Tang S, Machaalani R, Waters KA (2012) Expression of brain-derived neurotrophic factor and TrkB receptor in the sudden infant death syndrome brainstem. Respir Physiol Neurobiol 180:25–33

Tanti J, Gual P, Gremeaux T, Gonzalez T, Barres R, Le Marchand-Brustel Y (2004) Alteration in insulin action: role of IRS-1 serine phsophorylation in the retroregulation of insulin signalling. Ann Endocrinol 65:43–48

Teng HK, Teng KK, Lee R, Wright S, Tevar S, Almeida RD, Kermani P, Torkin R, Chen ZY, Lee FS, Kraemer RT, Nykjaer A, Hempstead BL (2005) ProBDNF induces neuronal apoptosis via activation of a receptor complex of p75NTR and sortilin. J Neurosci 25:5455–5463

Thaler JP, Schwartz MW (2010) Minireview: inflammation and obesity pathogenesis: the hypothalamus heats up. Endocrinology 151:4109–4115

Thaler JP, Choi SJ, Schwartz MW, Wisse BE (2010) Hypothalamic inflammation and energy homeostasis: resolving the paradox. Front Neuroendocrinol 31:79–84

Thaler JP, Yi CX, Schur EA, Guyenet SJ, Hwang BH, Dietrich MO, Zhao X, Sarruf DA, Izgur V, Maravilla KR, Nguyen HT, Fischer JD, Matsen ME, Wisse BE, Morton GJ, Horvath TL, Baskin DG, Tschöp MH, Schwartz MW (2012) Obesity is associated with hypothalamic injury in rodents and humans. J Clin Invest 122:153–162

Thirumangalakudi L, Prakasam A, Zhang R, Bimonte-Nelson H, Sambamurti K, Kindy MS, Bhat NR (2008) High cholesterol-induced neuroinflammation and amyloid precursor protein processing correlate with loss of working memory in mice. J Neurochem 106:475–485

Thomas T, Pfeiffer AF (2011) Foods for the prevention of diabetes: how do they work? Diabetes Metab Res Rev 28:25–49

Tishinsky JM (2013) Modulation of adipokines by n-3 polyunsaturated fatty acids and ensuing changes in skeletal muscle metabolic response and inflammation. Appl Physiol Nutr Metab 38:361

Tozuka Y, Kumon M, Wada E, Onodera M, Mochizuki H, Wada K (2010) Maternal obesity impairs hippocampal BDNF production and spatial learning performance in young mouse offspring. Neurochem Int 57:235–247

Trejo JL, Piriz J, Llorens-Martin MV, Fernandez AM, Bolos M, LeRoith D, Nunez A, Torres-Aleman I (2007) Central actions of liver-derived insulin-like growth factor I underlying its pro-cognitive effects. Mol Psychiatry 12:1118–1128

Tsukumo DM, Carvalho-Filho MA, Carvalheira JB, Prada PO, Hirabara SM, Schenka AA, Araujo EP, Vassallo J, Curi R, Velloso LA, Saad MJ (2007) Loss-of-function mutation in Toll-like receptor 4 prevents diet-induced obesity and insulin resistance. Diabetes 56:1986–1998

Turnbaugh PJ, Ley RE, Mahowald MA, Magrini V, Mardis ER, Gordon JI (2006) An obesity-associated gut microbiome with increased capacity for energy harvest. Nature 444:1027–1031

Udagawa J, Hashimoto R, Suzuki H, Hatta T, Sotomaru Y, Hioki K, Kagohashi Y, Nomura T, Minami Y, Otani H (2006) The role of leptin in the development of the cerebral cortex in mouse embryos. Endocrinology 147:647–658

Urano F, Wang X, Bertolotti A, Zhang Y, Chung P, Harding HP, Ron D (2000) Coupling of stress in the ER to activation of JNK protein kinases by transmembrane protein kinase IRE1. Science 287:664–666

US Department of Health and Human Services (2005) Dietary guidelines for Americans. USDA, Washington, DC

van Thuijl H, Kola B, Korbonits M (2008) Appetite and metabolic effects of ghrelin and cannabinoids: Involvement of AMP-activated protein kinase. Vitam Horm 77:121–148

Venkatraman A, Shiva S, Davis AJ, Bailey SM, Brookes PS, Darley-Usmar V (2003) Chronic alcohol consumption increases the sensitivity of rat liver mitochondrial respiration to inhibition by nitric oxide. Hepatology 38:141–147

Venkatraman A, Landar A, Davis AJ, Chamlee L, Sanderson T, Kim H, Page G, Pompilius M, Ballinger S, Darley-Usmar V, Bailey SM (2004) Modification of the mitochondrial proteome in response to the stress of ethanol-dependent hepatotoxicity. J Biol Chem 279:22092–22101

Ventriglia M, Zanardini R, Bonomini C, Zanetti O, Volpe D, Pasqualetti P, Gennarelli M, Bocchio-Chiavetto L (2013) Serum brain-derived neurotrophic factor levels in different neurological diseases. Biomed Res Int 2013:901082

Virtanen JK, Siscovick DS, Lemaitre RN, Longstreth WT, Spiegelman D, Rimm EB, King IB, Mozaffarian D (2013) Circulating omega-3 polyunsaturated fatty acids and subclinical brain abnormalities on MRI in older adults: the Cardiovascular Health Study. J Am Heart Assoc 2:e000305

Vladykovskaya E, Sithu SD, Haberzettl P, Wickramasinghe NS, Merchant ML, Hill BG, McCracken J, Agarwal A, Dougherty S, Gordon SA, Schuschke DA, Barski OA, O'Toole T, D'Souza SE, Bhatnagar A, Srivastava S (2012) Lipid peroxidation product 4-hydroxy-*trans*-2-nonenal causes endothelial activation by inducing endoplasmic reticulum stress. J Biol Chem 287:11398–11409

Wang H, Storlien LH, Huang XF (2002) Effects of dietary fat types on body fatness, leptin, and ARC leptin receptor, NPY, and AgRP mRNA expression. Am J Physiol Endocrinol Metab 282:E1352–E1359

Wang G-J, Yang J, Volkow ND, Telang F, Ma Y, Zhu W, Wong CT, Tomasi D, Thanos PK, Fowler JS (2006) Gastric stimulation in obese subjects activates the hippocampus and other regions involved in brain reward circuitry. Proc Natl Acad Sci U S A 103:15641–15645

Watanabe RLH, Andrade IS, Telles MM, Albuquerque KT, Nascimento CM, Oyama LM, Casarini DE, Ribeiro EB (2010) Long-term consumption of fish oil-enriched diet impairs serotonin hypophagia in rats. Cell Mol Neurobiol 30:1025–1033

Wayner MJ, Armstrong DI, Phelix CF, Oomura Y (2004) Orexin-A (hypocretin-1) and leptin enhance LTP in the dentate gyrus of rats *in vivo*. Peptides 25:991–996

Weisberg SP, Hunter D, Huber R, Lemieux J, Slaymaker S, Vaddi K, Charo I, Leibel RL, Ferrante AW Jr (2006) CCR2 modulates inflammatory and metabolic effects of high-fat feeding. J Clin Invest 116:115–124

Williams LM (2012) Hypothalamic dysfunction in obesity. Proc Nutr Soc 71:521–533

Winocur G, Greenwood CE (2005) Studies of the effects of high fat diets on cognitive function in a rat model. Neurobiol Aging 26(Suppl 1):46–49

Wisse BE, Schwartz MW (2009) Does hypothalamic inflammation cause obesity? Cell Metab 10:241–242

Wolfgang MJ, Cha SH, Sidhaye A, Chohnan S, Cline G, Shulman GI, Lane MD (2007) Regulation of hypothalamic malonyl-CoA by central glucose and leptin. Proc Natl Acad Sci U S A 104(49):19285–19290

Woods SC (1996) Insulin and the brain: a mutual dependency. Prog Psychobiol Physiol Psychol 16:53–81

World Health Organization (2006) Neurological disorders: public health challenges. WHO, Geneva

Wu A, Ying Z, Gomez-Pinilla F (2006) Oxidative stress modulates Sir2α in rat hippocampus and cerebral cortex. Eur J Neurosci 23:2573–2580

Xu L, Rensing N, Yang XF, Zhang HX, Thio LL, Rothman SM, Weisenfeld AE, Wong M, Yamada KA (2008) Leptin inhibits 4-aminopyridine- and pentylenetetrazole-induced seizures and AMPAR-mediated synaptic transmission in rodents. J Clin Invest 118:272–280

Yehuda S, Rabinovitz S, Carasso RL, Mostofsky DI (2002) The role of polyunsaturated fatty acids in restoring the aging neuronal membrane. Neurobiol Aging 23:843–853

Zammit VA, Arduini A (2008) The AMPK-malonyl-CoA-CPT1 axis in the control of hypothalamic neuronal function. Cell Metab 8:175, author reply 6

Zerbi V, Jansen D, Wiesmann M, Fang X, Broersen LM, Veltien A, Heerschap A, Kiliaan AJ (2014) Multinutrient diets improve cerebral perfusion and neuroprotection in a murine model of Alzheimer's disease. Neurobiol Aging 35:600–613

Zhang J (2007) The direct involvement of SirT1 in insulin-induced insulin receptor substrate-2 tyrosine phosphorylation. J Biol Chem 282:34356–34364

Zhang X, Zhang G, Zhang H, Karin M, Bai H, Cai D (2008) Hypothalamic IKKβ/NF-κB and ER stress link overnutrition to energy imbalance and obesity. Cell 135:61–73

Chapter 3
Neurochemical Effects of Long Term Consumption of Simple Carbohydrates

3.1 Introduction

Carbohydrates, a broad group of naturally occurring substances, are important macronutrient components of our diet. Carbohydrates provide energy for the body. They are not only consumed as a naturally occurring component of foods, but are also added to the diet during processing, preparation, or at the table. Although the body can generate energy from other sources, like fats and even proteins, the consumption of carbohydrates is special because it can provide energy without the use of oxygen. Carbohydrates are the preferred energy source for the brain, and the energy derived from carbohydrates is necessary for the burning of fats. However, consumption of high amounts of simple carbohydrates is responsible for obesity, insulin resistance, diabetes, metabolic syndrome, a pathological condition which is a risk factor for stroke, Alzheimer disease and depression (Farooqui et al. 2012). Dietary carbohydrates include simple sugars such as sucrose, lactose, maltose, glucose and fructose, and more complex carbohydrates, such as starch and fiber in our diet. Sucrose (glucose + fructose) is found in sugar cane, sugar beets, honey, and corn syrup; lactose (glucose + galactose) is found in milk products; and maltose (glucose + glucose) is found in malt. The most common naturally occurring monosaccharides are glucose and fructose (found in fruits). Complex carbohydrates (polysaccharides) in diet are found in vegetables and whole grains. When complex carbohydrates are digested and hydrolyzed in the digestive tract, they are converted to glucose, which provides energy. A healthy diet contains at least some amount of naturally occurring above mentioned sugars (Murphy and Johnson 2003). As stated above, carbohydrates are the main source of energy for the body and daily nutrient recommendations are based on the Dietary Reference Intakes (DRIs) by age and gender. According to Dietary Guidelines about half (45–65 %) of daily calories should be derived from carbohydrates (starches, fiber, and sugars) (Trumbo et al. 2002). However, estimates indicate that the total caloric intake exceeds DRIs regardless of energy needs. This increases the risk of carbohydrates-supported pathological

© Springer International Publishing Switzerland 2015

A.A. Farooqui, *High Calorie Diet and the Human Brain*,

DOI 10.1007/978-3-319-15254-7_3

conditions, such as insulin resistance, diabetes, stroke, and cardiovascular disease (Oba et al. 2010). Thus significant information has been published on the relationship between intake of sugars and cardiovascular health after the last American Heart Association (AHA) scientific statement was published in 2002 (Howard and Wylie-Rosett 2002). In 2006, AHA published revised diet and lifestyle recommendations that propose minimizing the intake of beverages and foods with added sugars (Lichtenstein et al. 2006). It has been suggested that both quantity and quality of carbohydrate affect metabolic health (Gaesser 2007; Farooqui 2013). Food with high glycemic index (GI) and glycemic load (GL) has been shown to increase the risk of obesity, insulin resistance, and hypertension, but the results have been inconsistent in different human populations (Farooqui 2013).

These days high dietary carbohydrates are consumed mainly in the form of high fructose corn syrup (HFCS) and sucrose, which contribute to many unhealthy effects, including insulin resistance, lactic acidosis, lipogenesis, hypertriglyceridemia, obesity, diabetes and hypertension (Farooqui 2013). In United States, HFCS is used in processed foods such as jams, jellies, dairy products, baked desserts, cereals, canned fruits, candies, soft drinks, and ice creams (Bray et al. 2004; Bray 2007, 2010; White 2008). It should be noted that like HFCS, apple and pear juice have >66 % fructose; asparagus, raspberries, spinach, and watermelon have 56–65 % fructose; and most fruits and nuts have 42–55 % fructose. Recently European Union has put a production quota of foods containing HCFS. In 2005, this quota was set at 303,000 t. In contrast, European Union produced an average of 18.6 million tons of sugar annually between 1999 and 2001.

The increased consumption of pure fructose elevates plasma free fatty acids, leptin, adiponectin, abdominal adipose tissue resulting in impaired insulin sensitivity (Alzamendi et al. 2009; Melanson et al. 2008), leptin resistance, and exacerbated obesity (Shapiro et al. 2008). Based on above mentioned studies, it is postulated that increased consumption of fructose through food may be closely associated with stimulation of lipogenesis, high plasma levels of triacylglycerols (TAGs), obesity, insulin resistance, and metabolic syndrome that leads to cardiovascular disease (CVD) (Bray et al. 2004; Teff et al. 2004).

3.2 Differences Among Glucose, Sucrose and Fructose Metabolism in Visceral Tissues

Glucose is metabolized by all body tissues through glycolysis and TCA cycle for energy production and used for cell survival (Fig. 3.1). Two major pathways are known to mediate glucose absorption in small intestine. At low concentrations, the classic pathway facilitates an active absorption of glucose through the Na^+-glucose cotransporter 1 (SGLT1) (Tavakkolizadeh et al. 2001). When the glucose level is >30 mM in the lumen after a meal and SGLT1 is fully saturated, the absorption of glucose is mediated by the glucose transporter glucose transporter 2 (GLUT2) (Cheeseman 1993). It should be noted that SGLT1 itself is an important mediator

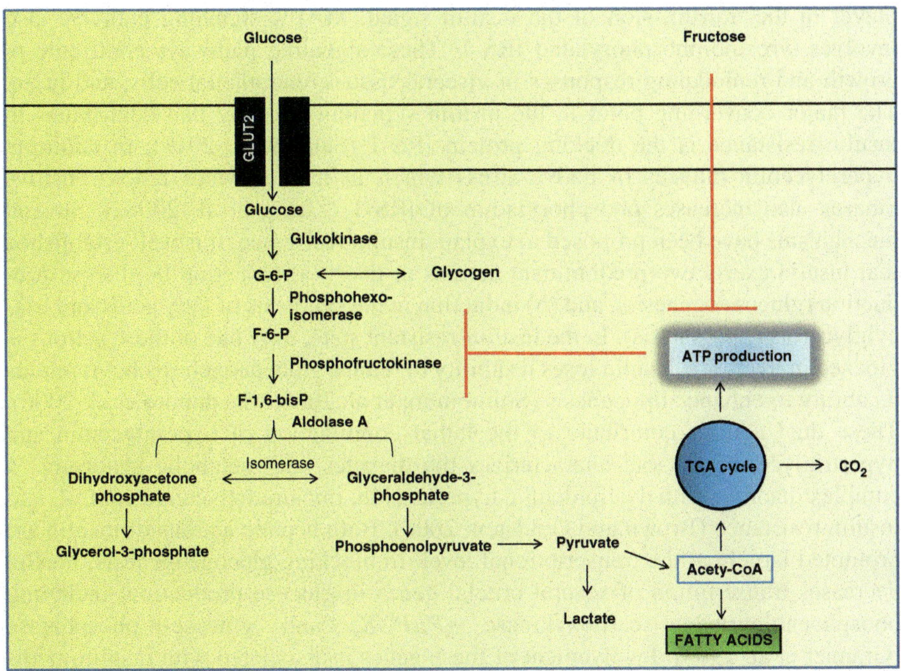

Fig. 3.1 Metabolism of glucose in the brain and visceral tissues

that is needed for the uptake of glucose by GLUT2 providing the induction of signal to generate additional transport capacity for glucose absorption through rapid translocation of GLUT2 from cytoplasmic vesicles into the apical membrane. This translocation of GLUT2 increases markedly the capacity of glucose uptake by the enterocyte (Au et al. 2002). In visceral tissues such as muscle or liver, insulin lowers blood glucose levels through the involvement of the insulin receptor. Insulin signaling occurs through two different pathways: phosphatidylinositol 3 kinase (PtdIns 3K)/protein kinase B (Akt) signaling pathway eliciting mainly metabolic responses and the mitogen-activated protein kinase (MAPK) signaling pathway eliciting growth factor-like responses (Saha et al. 2011). In PtdIns 3K/protein kinase B (Akt) pathway, the interaction between insulin and its receptor results in tyrosine phosphorylation of insulin receptor substrate 1 (IRS-1) (Sykiotis and Papavassiliou 2001). This provides a docking site for proteins with Src Homology 2 domains, such as PtdIns 3K (Saltiel and Kahn 2001). Activation of the catalytic subunit of PtdIns 3K catalyzes conversion of phosphatidylinositol 4,5-bisphosphate (PtdIns-4,5-P_2) to phosphatidylinositol (3,4,5)-trisphosphate (PtdIns(3,4,5) P_3), at the plasma membrane. This allows proteins that contain pleckstrin-homology domains, such as phosphoinositide-dependent kinase-1 (PDK-1) and Akt, to be activated (Khan and Pessin 2002). It is widely recognized that Akt is a crucial player in the transmission of the insulin signal. MAPK signaling pathway also involves tyrosine-phosphorylated IRS-1. These signaling pathways contribute to growth and remodeling responses in visceral tissues, endothelial cells, and brain. The major converging point in the insulin signaling pathway that contributes to insulin resistance is the docking protein IRS-1 (Saha et al. 2011). In addition, hyperglycemia induces oxidative stress which in turn activates redox-sensitive kinases also increases phosphorylation of IRS-1 (Zhang et al. 2008a). Several mechanisms have been proposed to explain insulin resistance. It is well established that insulin exerts two predominant actions in liver: (a) reduction in glucose production (gluconeogenesis), and (b) induction in the synthesis of fatty acids and triacylglycerols (lipogenesis). In the insulin-resistant state, only one of these actions is blocked in liver. The insulin loses its ability to reduce gluconeogenesis but it retains its ability to enhance lipogenesis (Shimomura et al. 2000; Matsumoto et al. 2006). These dual actions contribute to the lethal combination of hyperglycemia and hypertriglyceridemia that characterizes the diabetes and metabolic syndrome, a complex disorder with dyslipidemia, hypertension, impaired glycemic control, and insulin resistance (Brown and Goldstein 2008). Both hepatic actions of insulin are promoted largely at the transcriptional level. In blocking gluconeogenesis, insulin decreases transcription of several crucial genes in glucose production, including phosphoenolpyruvate carboxykinase (*PEPCK*) and glucose-6-phosphatase (Granner et al. 1983). Involvement of these genes is associated with insulin-mediated phosphorylation of the transcription factor FoxO1, a process that results in the exclusion of FoxO1 from the nucleus (Nakae et al. 2002). In activating hepatic lipogenesis, insulin increases transcription of genes encoding acetyl-CoA carboxylase, fatty acid synthase, glycerol-3-phosphate acyltransferase, and others. These actions are supported by an insulin-mediated increase in the active nuclear fragment of sterol regulatory element-binding protein-1c (SREBP-1c) (Horton et al. 2002). This increase occurs largely because insulin not only increases the transcription of *SREBP-1c* (Ferre et al. 2001), but also enhances the proteolytic processing of the membrane-bound SREBP-1c precursor, allowing its entry into the nucleus (Hegarty et al. 2005). The mechanism associated with insulin-mediated enhancement of transcription of *SREBP-1c* is not fully understood, but the stimulation requires the participation of liver X receptors (LXR), a transcription factor, which is known to elevate feed-forward stimulation (Chen et al. 2004). Intracellular accumulation of long-chain fatty acyl-CoA molecules during lipogenesis activates serine/threonine kinase signal transduction cascades (potentially involving PKCθ and/or IKKβ) that dampen insulin signal transduction via serine phosphorylation of the insulin receptor and IRS proteins (Shulman 2000). According to another mechanism FFAs induce insulin resistance by initially disrupting the phosphorylation process in the insulin-signaling pathway and consequently reducing glucose oxidation and glycogen synthesis (Roden et al. 1996). In addition, reduction in glucose oxidation and glycogen synthesis increase FFA oxidation, which causes an increase in and accumulation of glucose-6-phosphate, inhibiting the action of hexokinase II in the glycogen synthesis pathway (Roden et al. 1996). Such inhibitory effects increase the glucose level in the cells, prompting glucose uptake to halt, leading to an enhancement of the glucose levels in the bloodstream. This will eventually lead to insulin resistance and diabetes, as a long-term impact (Roden et al. 1996). It is also suggested that JNK and TNF-α may be important mediators of ROS-induced insulin resistance (Houstis et al. 2006).

In contrast to glucose, fructose is mainly metabolized by the liver. Fructose alone is poorly absorbed, but enhanced by glucose in the gut, thus accounting for the rapid and complete absorption of both fructose and glucose when ingested as sucrose. In addition, fructose is absorbed further down the intestine whereas circulating glucose releases insulin from the pancreas (Vilsboll et al. 2003). Fructose stimulates insulin synthesis but does not release it (Le and Tappy 2006). The release of insulin by glucose modifies food intake by inhibiting eating (Schwartz et al. 2000) and increasing leptin release (Saad et al. 1998), which also inhibits food intake. The uptake of glucose is regulated by a variety of glucose transporters that differ in their tissue distribution, regulation and kinetics (Thorens and Mueckler 2010). The consumption of glucose associated with the release of insulin, which further promotes glucose uptake by increasing the translocation of glucose transporter4 (GLUT4) in striated muscle (skeletal and cardiac) and adipose tissues to further facilitate its uptake. Once in the cell, glucose is phosphorylated by glucokinase into glucose-6-phosphate. This metabolite is transformed into fructose 1,6-bisphosphate through the action of phosophofructokinase, a multisubunit allosteric enzyme, which is inhibited by ATP and citrate (Fig. 3.1). Fructose 1,6-bisphosphate is metabolized to triose-phosphate and pyruvate by aldolase A. Pyruvate is then decarboxylated to acetyl coenzyme A, and enters the tricarboxylic acid cycle (TCA cycle), a reaction sequence for energy production. The net equations for glucose breakdown (glycolysis and TCA cycle) are as follows.

1. $C_6H_{12}O_6 + 2\ ADP + 2\ [P]i + 2\ NAD^+ \longrightarrow 2$ pyruvate $+ 2\ ATP + 2\ NADH$
2. Acetyl CoA $+ 3\ NAD^+ + Q + GDP + [P]i + 2\ H_2O \longrightarrow CoA\text{-}SH + 3\ NADH + 3\ H^+ + QH2 + GTP + 2\ CO_2$

Liver and gut are major sites for the metabolism of fructose. Absorption of fructose from the intestine into the portal blood is facilitated by glucose transporter5 (GLUT5) at the brush border and basolateral membranes of the jejunum. Fructose through the portal vein enters liver, where it is phosphorylated by fructokinase (2-ketohexokinase) producing fructose-1-phosphate (Hayward and Bonthron 1998; Diggle et al. 2009; Asipu et al. 2003). Fructokinase is not regulated by energy status. In contrast to glucokinase pathway, the fructokinase pathway bypasses phosphofructokinase, an enzyme that tightly regulates glycolysis, and transforms fructose into fructose-1-phosphate. Thus, while glucose metabolism is negatively regulated by phosphofructokinase, fructose can continuously enter the glycolytic pathway via fructose-1-phosphate formation resulting in depletion of ATP. Two forms of fructokinase (fructokinase-A and C) are known to occur in liver and intestine (Diggle et al. 2009; Ishimoto et al. 2012). The enzymic properties of purified recombinant fructokinase-A and fructokinase-C differ markedly. In particular, fructokinase-A has a tenfold higher *Km* for fructose (8 mM), suggesting that it phosphorylates fructose poorly at physiological concentrations (Diggle et al. 2009).

Unlike fructokinase-C, the phosphorylation of fructose by fructokinase-A is slow with only minimal ATP consumption. The induction of obesity, fatty liver, and insulin resistance that occurs in mice consuming fructose can be retarded in mice lacking both fructokinase-C and fructokinase-A. However, these pathological conditions can be exacerbated in fructokinase-A knockout despite similar overall energy intake (Ishimoto et al. 2012). The molecular mechanism involved in this process is not known. It is suggested that a lack of fructokinase-A results in more delivery of fructose to the liver where it is metabolized by fructokinase-C supporting the view that hepatic fructokinase-C plays an important role in the induction of insulin resistance. Fructokinase-C has no negative feedback system, and adenosine triphosphate is utilized as a source of phosphorylation leading to intracellular phosphate depletion and the rapid generation of uric acid. These processes not only lead to transient interruption of protein synthesis (Maenpaa et al. 1968), but also result in the stimulation of AMP deaminase, an enzyme that catalyzes transformation of AMP to IMP ultimately facilitating the generation of uric acid inside the cell (See below) (Kim et al. 2009; Cirillo et al. 2009). Generation of uric acid and its metabolism are complex processes involving various factors that regulate hepatic production, and renal and gut excretion of this compound. The exogenous pool of uric acid varies significantly with diet, and animal proteins contribute significantly to this purine pool. The endogenous production of uric acid is mainly from the liver, intestines and other tissues like muscles, kidneys and the vascular endothelium.

Fructose-1-phosphate is subsequently converted to dihydroxyacetone-phosphate and D-glyceraldehyde by the action of the aldolase B (Jianghai et al. 2012). D-glyceraldehyde is phosphorylated and metabolized to pyruvate by the glycolysis pathway. The metabolism of fructose by glycolysis pathway is catalyzed by phosphoglycerate and pyruvate kinases, which utilize two ATP molecules (Murray et al. 2003). The conversion of pyruvate to acetyl-CoA occurs in mitochondria, which exports citrate to the cytosol for fatty acid synthesis (Mayes 1993). Fructose not only controls the activity of glucokinase, but is also a potent and acute regulator of liver glucose uptake and glycogen synthesis. Inclusion of catalytic quantities of fructose in a carbohydrate meal improves glucose tolerance. This improvement is primarily mediated by the activation of hepatic glucokinase resulting in improved liver glucose uptake. The consumption of fructose also leads to larger increase in the circulating lactate than does the consumption of a comparable amount of glucose. Low-dose fructose has also been found to restore the ability of hyperglycemia to regulate hepatic glucose production.

The consumption of high fructose diet decreases insulin receptor activation and the phosphorylation of insulin receptor substrate-1in skeletal muscle of rodents (Eiffert et al. 1991; Le et al. 2006). Fructose also produces oxidative stress and mitochondrial dysfunction, resulting in a stimulation of peroxisome proliferator-activated receptor gamma coactivator 1-α and β (PGC1-α and PGC-1β) that drive both insulin resistance and lipogenesis. Fructose mediated lipogenesis alters the activity of key lipogenic enzymes and transcription factors in the liver, such as pyruvate dehydrogenase kinase and sterol-regulatory-element-binding protein-1c (SREBP-1c), the principal inducer of hepatic lipogenesis (Matsuzaka et al. 2004).

Fructose consumption dramatically induces SREBP1c expression, compared with feeding with regular chow. This effect can be retarded by treating fructose fed rats with PGC1β antisense oligonucleotides (Nagai et al. 2009). The decrease in SREBP1c expression produces a decrease in induction of lipogenic enzymes such as fatty acid synthase, which in turn may result in reduction in accumulation of di- and triacylglycerol within the livers of fructose fed rats. This decrease in tissue lipid content is accompanied by the improvement in insulin action. Another protein, which is associated with the regulation of fructose-mediated lipogenesis is the X-box binding protein (XBP)1. This protein regulates the expression of many proteins that are involved in endoplasmic reticulum membrane expansion, including the lipogenic enzymes (Glimcher 2010). XBP1 protein expression in mice can be increased after fructose consumption and is associated with the induction of critical genes involved in fatty acid synthesis (Lee et al. 2008). In contrast, deletion of XBP1 lowers expression of SREBP1 and key lipogenic enzymes, decreases rates of hepatic de novo lipogenesis and cholesterol synthesis and, *in vivo*, decreases plasma triglyceride concentration and secretion.

Consumption of fructose prevents adaptive increases in serum levels of 1,25-dihydroxyvitamin D_3 ($1,25(OH)_2D_3$) (vinh quốc Lương and Nguyen 2013; Dekker et al. 2010). This fructose-mediated decrease in serum $1,25(OH)_2D_3$ concentrations is not associated with a reduction in serum levels of its precursor $25(OH)D_3$, supporting the view that fructose has little effect on the initial steps of vitamin D synthesis. Instead, consumption of fructose produces a specific effect at the renal step of either $1,25(OH)_2D_3$ synthesis or degradation (Douard et al. 2012). Fructose-mediated decrease in $1,25(OH)_2D_3$ levels is always associated with a fructose-induced decrease in CYP27B1 expression and less consistent with increase in CYP24A1 expression. Based on these observations, It is suggested that chronic fructose intake reduces over time normal circulating levels of $1,25(OH)_2D_3$ independently of any increase in Ca^{2+} requirement and these processes are closely associated with metabolic diseases, such as hypertension, chronic kidney disease, metabolic syndrome, insulin resistance, and obesity.

Fructose metabolism in the liver differs from glucose indicating that glucose and fructose are metabolized by different metabolic processes (Bray et al. 2004). Thus, increased metabolism of fructose in the liver not only increases hepatic acetyl-CoA leading to increased production of very low-density lipoprotein and triglycerides (TAGs) (Park et al. 1992), but also elevates uric acid production by increasing ATP degradation to AMP, a precursor of uric acid (Choi et al. 2005) (Figs 3.2 and 3.3). In liver, phosphorylation of fructose and depletion of phosphate limits the regeneration of ATP from ADP, which in turn serves as the substrate for the catabolic pathway to uric acid formation (Fox et al. 1987) (Fig. 3.3). Thus, minutes after an infusion of fructose, plasma uric acid concentrations are significantly increased. To compensate purine nucleotide depletion, rates of de novo purine synthesis is also increased to potentiate uric acid production (Raivio et al. 1975). Increase in uric acid due to fructose consumption is accompanied by increase in levels of leptin and leptin resistance leading to high food intake and development of obesity (Shapiro et al. 2008). The mechanism underlying the fructose-induced leptin resistance is not

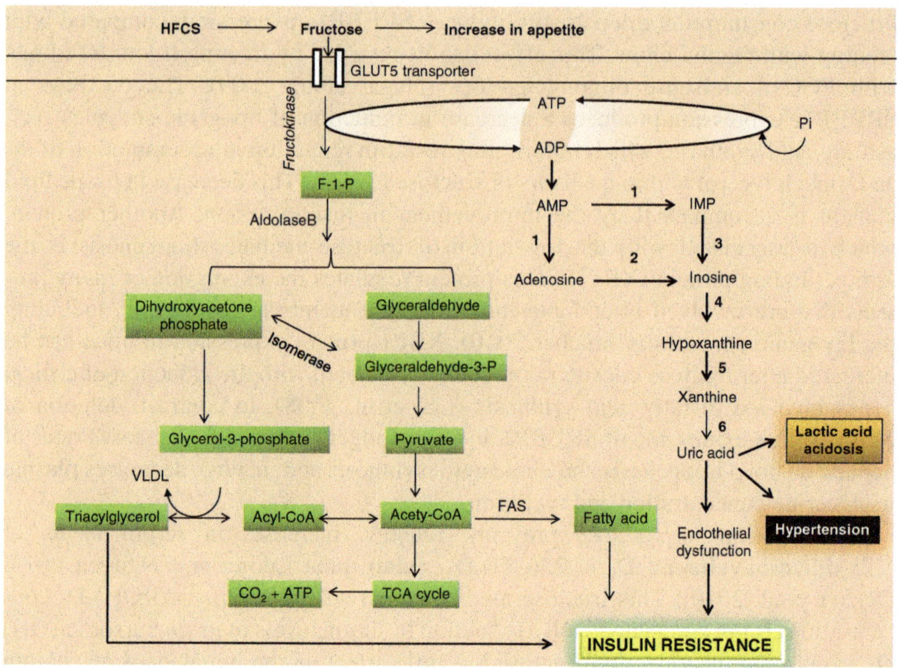

Fig. 3.2 Metabolism of fructose and generation of uric acid in visceral tissues. AMP deaminase (1); adenosine deaminase (2); 5′-nucleotidase (3); purine nucleoside phosphorylase (4); xanthine oxidase (5); *F-1-P* fructose-1-phosphate, *FAS* fatty acid synthase

fully understood. Induction of central leptin resistance involves reduction in activation of JAK-mediated STAT3 phosphorylation, defective leptin receptor signaling, and/or a subsequent failure to stimulate downstream signaling events (Scarpace and Zhang 2007). In addition, impaired leptin transport across the blood–brain barrier may also contribute to leptin resistance.

Increase in uric acid levels is also promoted not only by enhancement in renal uric acid reabsorption via stimulation of uric acid-anion exchanger (Enomoto et al. 2002) and/or the Na⁺-dependent anion co-transporter in brush border membranes of the renal proximal tubule (Choi et al. 2005), but also by reduction in the renal excretion of uric acid (Ter Maaten et al. 1997). In addition, an increase in fructose consumption often leads to positive energy balance, which may contribute to increase in body weight (Farooqui 2013). Increase in body weight is associated with a higher concentration of non-esterified fatty acids (McGarry 1994), which may reduce insulin sensitivity by increasing the intramyocellular lipid content in muscle cells where insulin receptors are located (Wu et al. 2004). A close association between serum uric acid levels and individual components of the metabolic syndrome has been reported (Farooqui 2013). Some of these 3 carbon molecules are either converted into glucose through gluconeogenesis, or transformed into triacylglycerol (TAG) (Fig. 3.2). In liver, TAG interferes with insulin signaling (Kim et al. 2007).

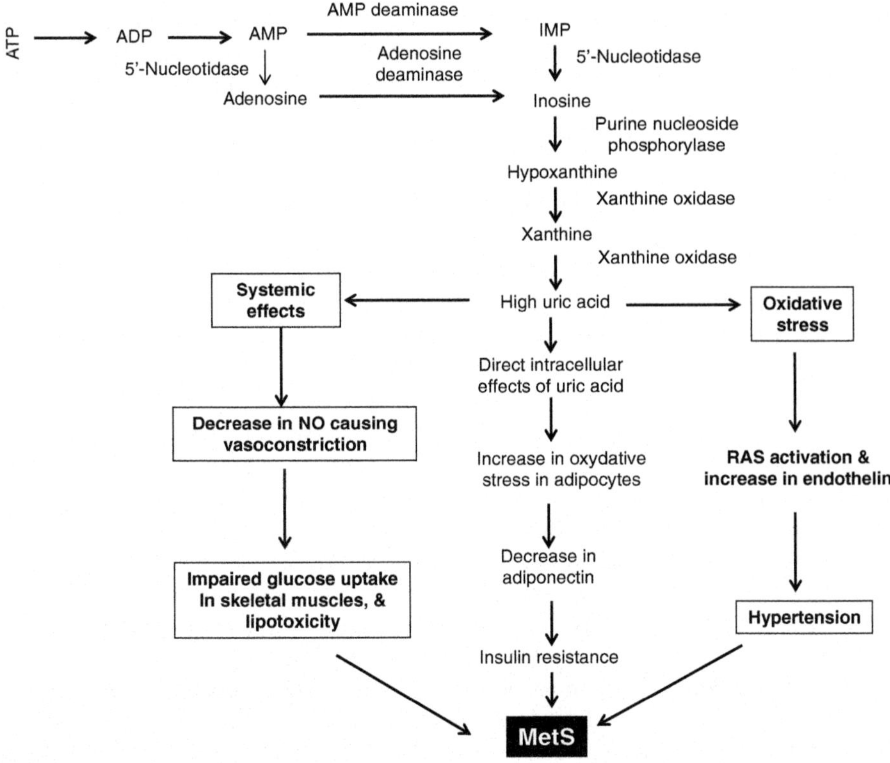

Fig. 3.3 Generation of uric acid from ATP and its metabolic consequences

Based on the above information, it is proposed that a high fructose diet is analogous to a high fat diet in many metabolic ways (Havel 2005). It is also reported that consumption of high calorie diet enriched in fructose results in the generation of glycerol-3-phosphate, which causes fixation of fat in the central adipose tissues. Fructose tricks the body into gaining weight by turning off body's appetite-control system. Fructose does not suppress ghrelin (the "hunger hormone") and doesn't stimulate leptin (the "satiety hormone"), which together results in feeling hungry all the time, even though one has eaten (Havel 2005). Thus, consumption of diet enriched in fructose contributes to overeating and obesity in rodents due to increase in leptin secretion (Teff et al. 2002; Havel 2005). Consumption of fructose also regulates mitochondrial enzymes. Thus, fructose consumption also results in an increase flux of acetyl CoA through the TCA cycle with a concomitant increase in cellular energy status (increase in ATP/ADP ratio and NADH/NAD$^+$ ratio). A high NADH/NAD$^+$ ratio in the mitochondria results in substrate-mediated inhibition of isocitrate dehydrogenase in the TCA cycle, modulating not only an increase in the export of citrate to the cytosol, but also activates acetyl-CoA carboxylase and increases in the synthesis of malonyl-CoA. This metabolite is the precursor for fatty acid synthesis (Locke et al. 2008; Cox et al. 2012). Elevated cytosolic concentrations of malonyl-CoA is known to inhibit the carnitine shuttle via carnitine palmitoyl transferase (CPT), resulting in reduced entry of fatty acids into the mitochondria, decreased fatty acid oxidation, and increased fatty acid levels, which lead to insulin resistance (Locke et al. 2008; Cox et al. 2012). Increase in consumption of fructose enriched high calorie diet stimulates the expression of protein tyrosine phosphatase-1B (Li et al. 2010). Fructose fed rats exhibit impaired c-Jun NH_2-terminal kinase (JNK) and mitogen-activated protein kinase signaling and increase the expression of FOXO1 due to SOCS3 expression (Vila et al. 2008). In turn, these changes lead to decrease in peroxisome proliferator-activated receptor alpha (PPARα), reduction in fatty acid oxidation, and accumulation of TAG in the liver. In addition, fructose feeding also increases hepatic ceramide levels supporting the view that incomplete fatty acid oxidation due to PPARα impairments provides substrate for ceramide synthesis (Vila et al. 2008). This may not only also result in activation of protein phosphatase-2A, but may also contribute to the deficiency in leptin signaling leading to metabolic disease (Vila et al. 2008). In metabolic diseases, adipocyte hypertrophy is accompanied by the accumulation of multinucleated macrophages, which attempt to protect the contents of dysfunctional adipocytes from further glycation, fructation, and oxidation (Seneff et al. 2011). The exposure of macrophages to advanced glycated end products (AGEs) results in the formation of dysfunctional macrophages, which try to enter into the artery wall and initiate the formation of plaque formation leading thrombosis (Seneff et al. 2011). Collective evidence suggests that a high-fructose diet produces insulin resistance not only by significantly reducing the protein expression of insulin receptor, insulin receptor substrate-1 and reducing circulating insulin and leptin levels, but also by altering expression of Akt. Fructose consumption is associated with a positive feedback system in which fructose up-regulates its transporter (GLUT5) as well as fructokinase. Experimentally, fructose administration has been shown to upregulate GLUT5 and fructokinase in the rat intestine, liver, and kidney (Korieh and Crouzoulon 1991). Although fructose does not promote increase in insulin levels, chronic exposure seems to indirectly cause hyperinsulinemia and obesity through other mechanisms. One proposed mechanism involves GLUT5, a fructose transporter that is found to have significantly higher expression levels in young Zucker obese rats compared to lean controls. As the rats age and become diabetic, GLUT5 abundance and activity is compromised, causing an even more marked insulin resistance over lean rats, implying a possible role of GLUT5 receptors in the pathology of metabolic syndrome associated with fructose feeding and insulin resistance (Litherland et al. 2004). In rats fed 66 % fructose for 2 weeks, insulin receptor mRNA, and subsequent insulin receptor numbers in skeletal muscle and liver are significantly decreased compared to rats fed a standard chow diet. In addition, blood pressure and plasma TAG are increased in the fructose-fed rats, even though there is no change in plasma insulin, glucose, or body weight (Catena et al. 2003). Another mechanism of induction of insulin resistance may involve fructose feeding-mediated reduction in insulin autophosphorylation and significant decreases in insulin induced IRS (1/2) phosphorylation in both the liver and muscle of the fructose fed rats. The consumption of fructose increases mammalian target of rapamycin (mTOR) and mitogen-activated protein kinase (p38-MAPK) activities in hepatocytes (Kyriakis and Avruch 2012). Two mechanisms may be responsible for fructose-mediated metabolic burden: increase in activity of protein phosphatase A_2 (Vilà et al. 2011) and the presence of bacterial toxins in blood, as a result of fructose-related alteration of the intestinal barrier permeability (Kanuri et al. 2011). Furthermore, increase in p38-MAPK activity may be linked with increase in mTOR activity. The mTOR signaling pathway not only transduces information from growth factors, amino acids and energy overload of the cell (Yap et al. 2008), but also causes mTOR-mediated increase in the splicing of X-Box Binding Protein-1 (XBP-1) (Pfaffenbach et al. 2010), which interacts with Forkhead box O1 (FoxO1) transcription factor to direct it to proteasome-mediated degradation, but also plays an important role in ER folding capacity and increasing insulin sensitivity. These biochemical changes are accompanied by production of polyol and AGEs, which may contribute to the insulin resistance along with numerous diabetes complications (Giacco and Brownlee 2010) (see below). Fructose feeding also produces ventricular dilatation, ventricular hypertrophy, decrease in ventricular contractile function, infiltration of inflammatory cells in heart and hepatic steatosis (Bray et al. 2004; Patel et al. 2009; Chang et al. 2007). Collectively, these studies suggest that long term consumption of high carbohydrate containing diet increases levels of inflammatory biomarkers such as C-reactive protein (Liu et al. 2002) and haptoglobin (Engstrom et al. 2003), which may be associated with the risk of diabetes and heart disease.

In mice, a chronic moderate consumption of fructose results in increased expression of lipocalin-2, a 25-kDa secretory glycoprotein, which plays important roles in regulation of immune responses (Yang et al. 2002), and in binding of small lipophilic substances (arachidonic acid, iron, and lipids) (Flower 1996). The increase in lipocalin-2 expression is not only associated with inflammation, but also with the mitochondrial dysfunction (Alwahsh et al. 2014). It is stated that lipocalin-2 may be a potential link among obesity, inflammation, and obesity-associated metabolic dysfunction such as insulin resistance (Wang et al. 2007).

Dietary sucrose has been shown to reduce the mechanical strength and bone mineral contents of rat bones (Tjaderhane and Larmas 1998), and excessive consumption of soft drinks laden with sugars may also decrease bone mineral density in humans (Tucker et al. 2006; McGartland et al. 2003). However, the mechanisms underlying the detrimental effect of sugars on bone remain completely unexplored. It is suggested that excessive fructose consumption may affect Ca^{2+} transport, 1,25-$(OH)_2D_3$ levels, and bone quality. Since fructose induces a deficiency in 1,25-$(OH)_2D_3$ without perturbing 25-$(OH)D_3$ levels, it may be important to distinguish in humans the effect of fructose on levels of both metabolites. Future work is needed on the effects of fructose consumption on bones.

Accumulating evidence suggests that chronic consumption of fructose not only stimulates ROS production, lowers antioxidant power and initiates proinflammatory processes, but also causes dysregulation of adipokines (Carvalho et al. 2010). ROS induce direct damage to cellular components by oxidizing lipids, proteins, or DNA. 4-Hydroxy-2,3-nonenal (4-HNE) and 4-hydroxy-2,3-alkenals of different chain length originate as a consequence of peroxidation of lipids. These molecules form aldehyde-protein adducts by reacting with reactive groups of cysteine, lysine, and histidine which produce damage to cells and tissues (Chaiswing et al. 2004). Increase in ROS also activates nuclear factor kappa B (NF-κB) and c-Jun N-terminal kinases (JNK), a member of mitogen activated protein kinase (MAPK) family which plays a major role in inflammatory signaling (Shen and Liu 2006).

3.3   Induction of Type II Diabetes Complications by High Levels of Glucose

Type II diabetes is a multifactorial disease. Several hypotheses have proposed to explain the origin of the pathology; that is, that it is an abnormality of the anterior hypothalamus and the endocrine pancreas induced by progressive ischemia or that there is abnormal islet innervation. Based on many studies it is also proposed that type II diabetes mellitus develops due to a complex interaction between genetic predisposition and lifestyle. The actual manifestations of the disease are preceded by a phase of impairment in glucose regulation promoted by the increase in the vascular risk factor leading to heart disease (Schulze and Hu 2005). However, the most important risk factors for the development of type II diabetes mellitus are consumption of unhealthy diet, obesity and physical inactivity. Diabetic patients often develop secondary complications (heart disease, blindness, end-stage renal failure, stroke, Alzheimer disease and depression), which are mainly caused not only by increased sorbitol production, oxidative-nitrosative stress, and endogenous antioxidant depletion, but also by enhanced lipid peroxidation, and alterations in hormonal responses (Fig. 3.4) (Brownlee 2005). These processes are promoted by increase in

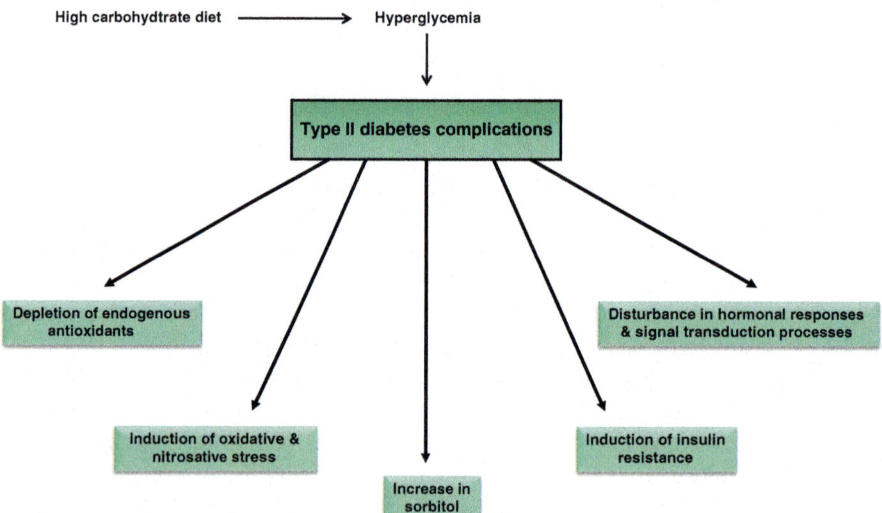

Fig. 3.4 Consumption of high carbohydrate diet and type II diabetes complication

intracellular glucose level and its downstream metabolic metabolism. Excessive levels of glucose are known to disrupt the electron transport chain in the mitochondria, leading to overproduction of superoxide anions (Nishikawa et al. 2000). High glucose can also stimulate oxidative stress via the auto-oxidation of glucose and through non-enzymatic glycation (see below). The production of superoxide in mitochondria overwhelms the capacity of MnSOD to dismutase superoxide to H_2O_2. These reactive oxygen species trigger DNA single-strand breakage to induce a rapid activation of poly (ADP-ribose) polymerase (PARP), which in turn reduces the activity of glyceraldehyde-3-phosphate dehydrogenase (GAPDH) to increase all the glycolytic intermediates that are upstream of GAPDH (Fig. 3.5). Inhibition of GAPDH also leads to the modification ADP-ribose polymerase (PARP) (Du et al. 2003; Sawa et al. 1997). Specific inhibitors of PARP block the inhibition of GAPDH. Increased amounts of glyceraldehyde-3 phosphate also activate two pathways including the generation of advanced glycation end products (see below) and diacylglycerol-mediated activation of protein kinase (PKC). Another upstream metabolite of fructose

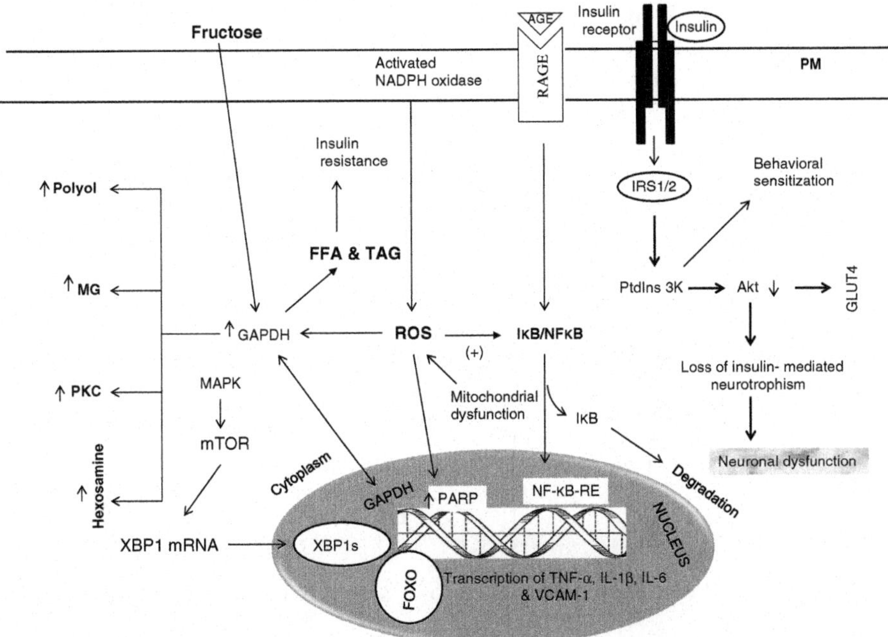

Fig. 3.5 Interactions between AGE and RAGE receptors. *PM* plasma membrane, *AGEs* advanced glycation products, *RAGE* receptor for advanced glycation end products, *BBB* blood brain barrier, *PtdCho* phosphatidylcholine, *cPLA2* cytosolic phospholipase A_2, *COX-2* cyclooxygenase, *ARA* arachidonic acid, *ROS* reactive oxygen species, *NF-κB* nuclear factor kappaB, *NF-κB-RE* nuclear factor kappaB response element, *IκB* inhibitory subunit of NFκB, *TNF-α* tumor necrosis factor-α, *IL-1β* interleukin-1 beta, IL-6 interleukin-6, *PGs* prostaglandins, *LTs* leukotrienes, *TXs* thromboxanes; *MG* methylglyoxal, *MAPK* mitogen-activated protein kinase, *PtdIns 3K* phosphatidylinositol 3 kinase, *PKC* protein kinase C, *NADP oxidase* nicotinamide adenine dinucleotide phosphate oxidase, *PARP* poly(ADP-ribose)polymerase, *GAPDH* glyceraldehydes 3-phosphate dehydrogenase

metabolism (fructose-6 phosphate) enters the hexosamine pathways (Fig. 3.5). Fructose-6 phosphate-mediated increase in the hexosamine pathway leads to a greater production of UDP (uridine diphosphate) N-acetylglucosamine, which often results in pathologic changes in gene expression such as increased expression of transforming growth factor β1 and plasminogen activator inhibitor-1.

In type II diabetes and MetS patients, alterations in glucose metabolism due to insulin resistance result in increased generation of methylglyoxal (MG), a major source of intracellular advanced glycation end-products (AGEs) (Beisswenger et al. 2003; Thornalley 1993). Methylglyoxal, an important glycating agent is a highly reactive α-oxoaldehyde. It is formed primarily from the intermediates of glycolysis (dihydroxyacetone phosphate, and glyceraldehyde-3-phosphate) in cells (Fig. 3.6). Nonenzymically, MG is spontaneously formed from dihydroxyacetone phosphate in vascular endothelial cells and smooth muscle cells (Wu 2006). It is interesting to note that triose phosphate pool is regulated by cellular levels of fructose, not by glucose and plasma levels of fructose are more involved in MG overproduction than glucose (Wang et al. 2008) (Fig. 3.6). MG reacts with arginine groups to form such AGEs as 5-hydro-5-methylimidazolone and the fluorescent AGE, argpyrimidine (Westwood and Thornalley 1995). MG levels are regulated by two enzymes, glyoxalase I and glyoxalase II. These enzymes "detoxify" MG and

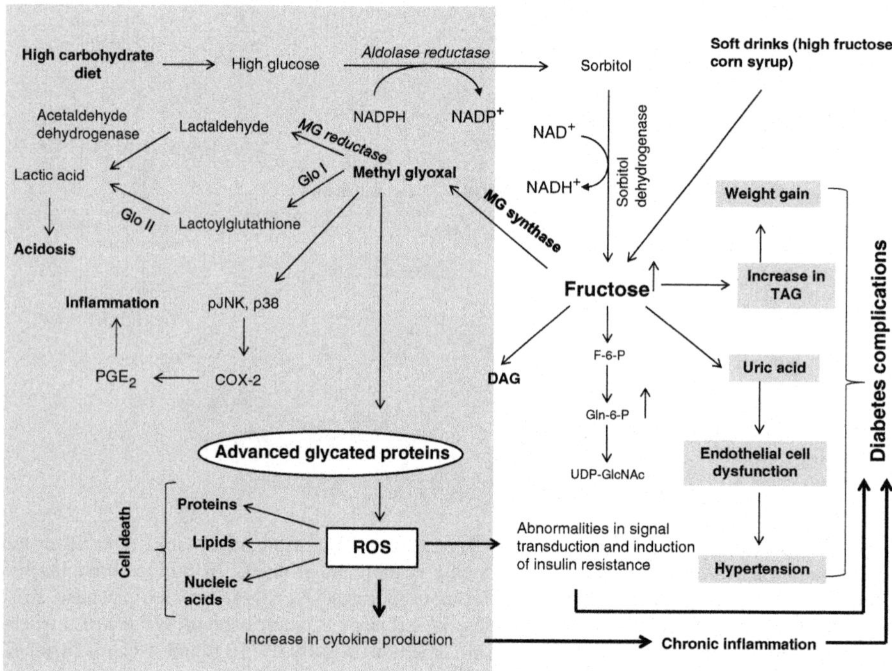

Fig. 3.6 Consumption of high carbohydrate and generation of methylglyoxal from glucose. *ROS* reactive oxygen species, *PKC* protein kinase C, *JNK* c-Jun N-terminal kinase, *TAG* triacylglycerol; *COX-2* cyclooxygenase-2

convert it to D-lactate (Phillips and Thornalley 1993). When the level of activity of glyoxalases is impaired, MG levels rise, thereby favoring further AGE production (Thornalley 2003). MG reduces activity of antioxidant enzymes like glutathione reductase and glutathione peroxidase, resulting in further increase in oxidative stress, which contribute to pathophysiological changes in diabetes, hypertension, and aging (Farooqui 2013).

Under normal conditions, glucose is metabolized via glycolysis and TCA cycle for energy production. However, during insulin resistance, abnormalities in insulin function may result in down regulation of GAPDH, slowing glucose metabolism and increasing glucose metabolism via the polyol pathway (Thornalley 1993), a glucose metabolic shunt that is defined by two enzymic reactions catalyzed by aldose reductase (AR) and sorbitol dehydrogenase (SDH), respectively (Giacco and Brownlee 2010). Biochemically, AR catalyzes the rate-limiting reduction of glucose to sorbitol, with the aid of cofactor NADPH. SDH converts sorbitol to fructose using NAD^+. Under normal conditions AR/polyol pathway represents a minor source of glucose utilization, accounting for less than 3 % of glucose consumption. However, during hyperglycemia, the activity of AR is substantially increased, representing up to 30 % of total glucose consumption (Chung et al. 2003). AR/the polyol pathway has been reported to play important roles in the development and progression of diabetic complications in a number of tissues including kidney, retina, lens, and peripheral neuron tissues (Giacco and Brownlee 2010). In the liver, the expression of AR is relatively low under normal physiological conditions, but the hepatic expression of sorbitol dehydrogenase is quite high (Alexiou et al. 2009). In liver, the hepatic AR is dynamically regulated under a variety of conditions. For instance, in rats fed with fructose, hepatic AR is significantly upregulated, which is associated with impaired activation of STAT3 and suppressed activity of PPARα in the liver (Roglans et al. 2007).

Under pathological conditions, such as type II diabetes and MetS, inhibition of AR caused significant dephosphorylation of hepatic PPARα results in the activation of this transcriptional factor as well significant reduction in serum TG levels (Qiu et al. 2008). In addition, in type II diabetes and MetS, high levels of glucose and fructose initiate AGE formation by attaching themselves with the α-amino group of either the amino terminus of proteins or lysine residues via nucleophilic attack forming a Schiff base, which undergoes an Amadori rearrangement forming ketoamines (Singh et al. 2001). These ketoamines (Amadori products) through an oxidative or nonoxidative pathway form irreversible AGE (Singh et al. 2001). The binding of AGEs with AGE receptors (RAGEs), a group of cell surface receptor, which belongs to the immunoglobulin superfamily (Neeper et al. 1992; Ahmed 2005) results in the induction of oxidative stress, largely through the NADPH oxidase system (Wautier et al. 2001). This oxidative stress facilitates more AGE formation leading to an environment, which favors more AGE formation. Thus, it is not surprising that in mice devoid of RAGE, levels of oxidative stress and AGEs were lower compared to wild-type RAGE-expressing animals (Reiniger et al. 2010). The activation of the AGE pathway can damage cells by three mechanisms: first, these compounds modify intracellular proteins, especially those involved in gene transcription regulation; second, these compound can diffuse to the extracellular space and modify extracellular proteins such as laminin and fibronectin to disturb signaling between the matrix and the cells; and finally, these compounds modify blood proteins such as albumin, causing them to bind to AGE receptors on macrophages/mesangial cells and increasing the production of growth factors and proinflammatory cytokines (Brownlee 2005). Collective evidence suggests that hyperglycemia in type II diabetes and MetS causes excessive glycation of proteins found in serum (e.g., albumin, hemoglobin, and LDL) and in the vessel wall (e.g., collagen, fibronectin, vitronectin, and laminin) (Singh et al. 2001). Glycation results of impairment in cellular function and glycated proteins, which become highly susceptible to oxidative damage. Glycated proteins are also resistant to degradation by lysosomal enzymes. For example, glycated haemoglobin is used as a marker for type II diabetes and glycated LDL are poorly recognized by lipoprotein receptors and scavenger receptors (Zimmermann et al. 2001). Glycation may also promote protein aggregation, for instance, glycation of lens crystallins may lead to cataract formation in type II diabetes and in old age (Crabbe et al. 2003). Over time, these proteins and their debris accumulate in the blood serum and along arterial walls (Giacco and Brownlee 2010). Glycated proteins play a critical role in aging, type II diabetes, atherosclerosis, cardiovascular disease, autoimmune diseases, cancer, and neurodegenerative diseases. The binding of glycated proteins with RAGEs is closely associated with the pathogenesis of endothelial cell dysfunction (Funk et al. 2012). Several lines of evidence suggest that chronic endothelial dysfunction plays a pivotal role not only in visceral diseases, but also in neurological disorders. In neurological disorders, endothelial cell dysfunction may be responsible for breakdown of the BBB, impaired cerebrovascular blood flow, cerebral amyloid angiopathy, atherosclerosis, or other important secondary-related inflammatory phenomena (Bell and Zlokovic 2009). Thus, intracellular production of AGE and glycated proteins can damage visceral and neural cells by three general mechanisms: (a) AGE-mediated modification of intracellular proteins causes alterations in their functions, (b) AGE may produce alterations in extracellular matrix components and with matrix receptors (integrins) on the surface of cells, and (c) Plasma proteins modified by AGE precursors may bind to RAGE on cells such as, macrophages, vascular endothelial cells and vascular smooth muscle cells mediating the production of ROS, which in turn activates the pleiotropic transcription factor, nuclear factor kappa B (NF-κB) causing multiple pathological changes in gene expression (VCAM-1, E-selectin, cytokines) (Fig. 3.5) (Goldin et al. 2006). AGEs and their precursors can diffuse out of the cell and modify extracellular matrix molecules nearby (McLellan et al. 1994), which changes signaling between the matrix and the cell, inducing cellular dysfunction (Charonis et al. 1990). Collective evidence suggests that fructose-derived AGEs production contributes to insulin resistance by a variety of mechanisms, including generation of tumor necrosis factor-α direct modification of the insulin molecule, thereby leading to its impaired action by generating oxidative stress, and impairing mitochondrial function. In contrast, acute temporary ingestion of fructose produces beneficial effects under some circumstances. Thus, short-term treatment of astroglial C6 cells with fructose has been reported to protect astroglial C6 cells against peroxide-induced stress (Spasojević et al. 2009a). Fructose has been shown to prevent apoptosis induced by reoxygenation in rat hepatocytes by decreasing the level of ROS (Frenzel et al. 2002; Valeri et al. 1996; MacAllister et al. 2011). It has been demonstrated that fructose and its phosphorylated derivatives (fructose-1,6-bisphosphate) have significantly higher antioxidant capacities against ROS than other carbohydrates (Spasojević et al. 2009a). Based on these observations, it is suggested that acute infusion or ingestion of fructose can produce cytoprotective effects in disorders related to oxidative stress (Spasojević et al. 2009b). The molecular mechanisms associated with beneficial effects of fructose are not fully understood. However, defensive effect of fructose may involve: (a) its iron binding ability resulting in the prevention of the Fenton reaction (Spasojević et al. 2009a; MacAllister et al. 2011); (b) stabilization of the glutathione pool in the cell (Frenzel et al. 2002); (c) upregulation of the pentose phosphate pathway producing NADPH (Spasojević et al. 2009b); and (d) production of fructose-1,6-bisphosphate, a compound well known for its cytoprotective and antioxidant effects (Muntané 2006; Marangos et al. 1998; Lopes et al. 2006).

3.4 Metabolism of Fructose, Generation of Uric Acid, and Development of Hypertension

Unlike glucose, which is utilized by all organs, fructose is solely metabolized in the liver. As stated above, metabolism of fructose by liver results in ATP depletion in animal models of diabetes and humans patients with type II diabetes (Fig. 3.2). The ATP depletion activates enzymes of purine metabolism (AMP deaminase-1), which degrades adenine nucleotides to uric acid via xanthine oxidoreductase with the development of hyperuricemia (Hallfrisch 1990; Nakagawa et al. 2005). Uric acid synthesis and metabolism are complex processes involving various factors that regulate its hepatic production, and renal and gut excretion of this compound. Uric acid is the end product of an exogenous pool of purines and endogenous purine metabolism. The exogenous pool varies significantly with diet, and animal proteins contribute significantly to this purine pool. The endogenous production of uric acid is mainly from the liver, and intestine. However, other tissues like muscles, kidneys and the vascular endothelium also contribute to uric acid production. Uric acid formation from purine catabolism occurs through a series of enzymic steps involving the enzyme xanthine oxidase. An intermediate product of this metabolism is inosine. This intermediate is converted by the purine nucleoside phosphorylase to hypoxanthine. Xanthine oxidase converts hypoxanthine to xanthine and subsequently to uric acid (Khitan and Kim 2013; Jalal et al. 2013) (Fig. 3.7). Under normal conditions, two thirds of the produced uric acid is eliminated in the urine and one third is removed by the biliary tree. In the kidney, uric acid is readily filtered by the glomerulus and subsequently reabsorbed by the proximal tubular cells of the kidney; the normal fractional excretion of uric acid is approximately 10 %. Low levels of uric acid have antioxidant properties, but high

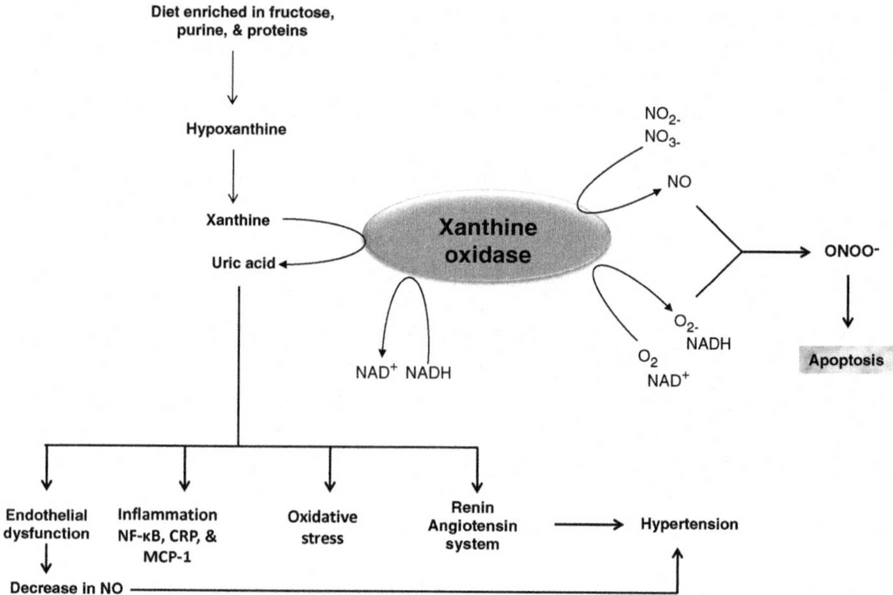

Fig. 3.7 Consumption of high fructose diet and metabolic changes caused by the high levels of uric acid. *NO* nitric oxide, *NF-κB* nuclear factor kappaB, *CRP* C-reactive protein, *MCP-1* monocyte chemotactic protein-1

serum uric acid levels are associated with other cardiovascular risk factors and predict cardiovascular events in adults (Gao et al. 2007; Gagliardi et al. 2009). Biologically, high levels of uric acid play an important role in worsening of insulin resistance in animal models of diabetes by inhibiting the bioavailability of nitric oxide, a multifunctional molecule that possesses both concentration-dependent and cell-specific cytoprotective and cytoxic properties (Wink et al. 2001). It is produced by nitric oxide synthase (NOS). This enzyme transforms L-arginine to L-citrulline. NOS isoenzymes comprise inducible NOS (iNOS or NOS2), endothelial NOS (eNOS or NOS3), and neuronal NOS (nNOS or NOS1) (Aliev et al. 2009). Nitric oxide (NO) plays a crucial role in mitochondrial respiration (Moncada and Erusalimsky 2002), since even low (nanomolar) concentrations of NO have been found to reversibly inhibit the mitochondrial respiratory chain enzyme cytochrome oxidase (complex IV) and compete with molecular oxygen. Inhibition of cytochrome oxidase by nitric oxide results in the reduction of the electron-transport chain, and favors the formation of the superoxide radical anions (O_2^-). The reaction of NO with superoxide radical anion results in the formation of peroxynitrite ($ONOO^-$), which is more cytotoxic than NO itself (Pacher et al. 2007). Enhanced production of ROS and impaired production and bioavailability of NO play a central role in endothelial dysfunction, arterial stiffness, and impaired diastolic function. The molecular mechanism involved in the influence of NO on insulin sensitivity has been intensely studied. In experimental studies, reduction in NO synthesis prevents glucose transportation in skeletal muscle cells (Balon and Nadler 1997). It is also reported that **arginine** improves glucose metabolism in skeletal muscle via the NO/c-GMP cascade (de Castro et al. 2013). Additionally, it has also been demonstrated that l-arginine may increase the activity of glucokinase in isolated liver cells of rats (Monti et al. 2000). Collectively, these studies indicate that NO is essential for insulin-stimulated glucose uptake (Khosla et al. 2005). NO may also promote oxidative damage by reacting with superoxide anion to form peroxynitrite (Pacher et al. 2007). Alternatively, NO has been shown to be neuro protective by overexpressing heme oxygenase, promoting CREB and Akt survival and protecting excitotoxicity through S-nitrosylation (Iadecola 1997).

High levels of serum uric acid levels are also associated not only with its deleterious effects on endothelial function, platelet adhesion and aggregation, but also with increased risk of hypertension, cardiovascular disease, kidney disease, gout, metabolic syndrome, stroke, and vascular dementia, (Fig. 3.8) (Alper et al. 2005). Chronic increase in uric acid levels is linked to reduction in adiponectin and elevation in E-selectin, in parallel with positive effects such as reduction in nitrotyrosine and increase in total antioxidant capacity (Bo et al. 2008). Uric acid also reduces NO bioavailability in endothelial cells, adipocytes, and vascular smooth muscle cells (Khosla et al. 2005; Corry et al. 2008; Zharikov et al. 2008) through several mechanisms including induction of oxidative oxidative stress (Corry et al. 2008),

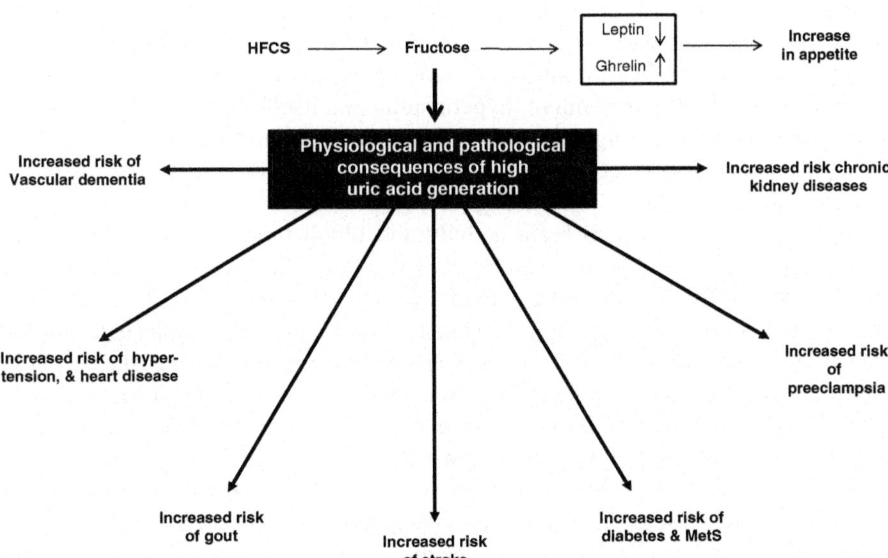

Fig. 3.8 Diseases associated with high levels of uric acid

and the direct scavenging of NO by uric acid (Gersch et al. 2008). Uric acid also stimulates vascular smooth muscle cells by entering cells via a specific organic anion transport pathway with the stimulation of mitogen-activated kinases (p38 and ERK) and nuclear transcription factors (nuclear factor-κB and activator protein-1). These processes result in platelet-derived growth factor-dependent proliferation, cyclooxygenase-2-dependent thromboxane production, monocyte chemotactic protein 1 (MCP-1) and C-reactive protein synthesis, and stimulation of angiotensin II (Corry et al. 2008; Kang et al. 2002; Watanabe et al. 2002). Uric acid not only activates the cytoplasmic phospholipase A_2 leading to the inhibition of proximal tubular cellular proliferation (Han et al. 2007), but also stimulates fructokinase expression through the activation of the carbohydrate response element binding protein (ChREBP), a transcription factor that in turn results in the transcriptional activation of fructokinase by the binding to a specific sequence within its promoter (Lanaspa et al. 2012). Uric acid blocks endothelial cell proliferation and migration (Kang et al. 2005). Finally, uric acid has potent effects on proximal tubular cells (stimulating MCP-1 production) as well as adipocytes (inducing oxidative stress, stimulating oxidized lipids, and lowering NO levels) (Sautin et al. 2007). Uric acid not only increases blood pressure through the inhibition of NOS, but also speeds metabolism of high carbohydrate diet. Several other mechanisms may also contribute to increase in blood pressure. These mechanisms include increase in sympathetic nervous system activity (Farah et al. 2006; Verma et al. 1999), insulin resistance, increased oxidative stress, elevation in circulating catecholamines (Tran et al. 2009), decrease in NO bioavailability, enhancement in renin-angiotensin system activity and angiotensin II levels (a potent vasoconstrictor) (Tran et al. 2009), increase in sodium reabsorption (De Fronzo 1981), impaired endothelium-dependent relaxation (Katakam et al. 1998), and increase in secretion of endothelin-1 (ET-1) (Juan et al. 1998). These factors may also contribute to an increase in vascular tone and impairment in endothelial function. Furthermore, hyperinsulinemia itself may cause hypertension as it is well known that long-term insulin administration causes an increase in blood pressure in rats (Meehan et al. 1994), which is reversed upon discontinuation of the insulin (Hsieh and Huang 1993). High blood pressure impairs functional hyperemia, the process by which brain activity and blood flow are coordinated. This impairment is mediated by dysregulation of vasoactive mediators such as NO and endothelin-1, which not only induce oxidative stress, but also cause structural alteration of the blood vessels through inadequate cerebral autoregulation (Iadecola and Davisson 2008). All of these processes are linked with insulin resistance (Craft 2009). In addition, increase in uric acid is also associated with elevated circulating levels of systemic inflammatory mediators such as monocyte chemoattractant protein-1, NF-κB, interleukin-1β, interleukin-6, and tumor necrosis factor-α, and vascular smooth muscle proliferation (Johnson et al. 2003; Kanellis et al. 2004). Collective evidence suggests that elevation in serum uric acid contributes to impairment in NO production/endothelial dysfunction, increase in vascular stiffness, inappropriate activation of the renin-angiotensin-aldosterone system, enhancement in oxidative stress, and maladaptive immune and inflammatory responses (Corry et al. 2008; Gersch et al. 2008).

3.5 Metabolism of Glucose, Sucrose and Fructose Metabolism in the Brain

Glucose homeostasis is maintained through interplay between central and peripheral control mechanisms which are utilized by cells for storing excess glucose following food consumption and mobilizing glucose from stores (glycogen) during periods of fasting. Central control of glucose homeostasis is achieved by a system of glucose sensing elements, which are integrative neural networks (Watts and Donovan 2010). Glucose is transported across the cell membrane by two types of glucose transporters: (a) sodium dependent glucose transporters (SGLTs) which transport glucose against its concentration gradient and (b) sodium independent glucose transporters (GLUTs), which transport glucose by facilitative diffusion in its concentration gradient (Jurcovicova 2014). Brain contains several SGLTs including GLUT1, GLUT2, and GLUT3. These transporters are found throughout the brain. GLUT5 is a predominantly fructose transporter. In brain GLUT5 is only found in microglia, where its function and regulation is not fully understood (Jurcovicova 2014). Brain synthesizes its own insulin and contains insulin receptors. In brain, insulin helps neurons in glucose-uptake and the regulation of neurotransmitters, such as acetylcholine, which are crucial for memory and learning. This is why induction of insulin insensitivity and insulin resistance in the brain impairs cognitive function (Farooqui 2013). Detailed investigations have indicated that insulin plays a key modulatory role in afferent (and efferent) hypothalamic pathways governing energy intake (Isganaitis and Lustig 2005). The antagonism of insulin signaling in the brain is known to impair the ability of circulating insulin to inhibit glucose production (Obici et al. 2002a). On the other hand, there is evidence that increased sympathetic activity results in insulin resistance due to α_1-adrenoceptor activation causing a reduction in blood flow and therefore a reduction in glucose delivery to skeletal muscle (Rattigan et al. 1999). The net outcome of this process is a continuous stimulation of sympathetic nervous system due to compensatory hyperinsulinemia in response to the insulin resistance. This proposal is supported by studies on the blockade of the sympathetic nervous system, by chemical sympathectomy (Verma et al. 1999) or decreasing sympathetic outflow (Pénicaud et al. 1998). These processes prevent not only the development of hyperinsulinemia, but also elevation in the blood pressure of fructose-fed rats. This would imply that the development of hyperinsulinemia and high blood pressure in this model depends on an intact sympathetic nervous system (Verma et al. 1999).

It is well known that he hypothalamus is involved in the modulation of feeding behavior and ability to detect changes in blood glucose. Receptors for both insulin (Kenner and Heidenreich 1991) and leptin (Elmquist et al. 1998; Leshan et al. 2006) are widely expressed throughout the brain. The medial hypothalamus, a key center for the regulation of energy homeostasis and coordination of metabolic events, is a major target for both insulin and leptin action (Mirshamsi et al. 2004; Niswender et al. 2003). Studies utilizing antisense oligonucleotides against the insulin receptor and conditional, localized knockout of the insulin receptor have indicated the contribution of the brain insulin receptor to energy homeostasis and glucose homeostasis (Koch et al. 2008; Obici and Rossetti 2003). Similarly, leptin receptors are also extensively expressed in the brain, with the 'signaling' form leptin receptors having the major role in leptin action. Collectively, these studies indicate that both insulin and leptin play important role in energy homeostasis in the brain. The metabolic interactions between glial cells and neurons may require for glucose sensing in the brain a short time after feeding. Thus, the glycolytic metabolism of glucose by glial cells and transfer of lactate to neighboring glucose sensing neurons may lead to enhanced production of ATP synthesis along with closure of ATP-sensitive K^+ (K_{ATP}) channels leading to membrane depolarization (Ainscow et al. 2002), whereas in glucose-inhibited neurons, the inhibitory effect of an elevated glucose concentration is mediated by an ATP-independent K^+ channel. These glucose-sensing neurons are involved in the control of neuroendocrine function, nutrient metabolism and energy metabolism (Levin et al. 2004).

Like visceral tissues, in brain glucose is metabolized for energy production through glycolysis and TCA cycle. Brain contains insulin receptors, which are linked with the stimulation of PtdIns 3K. This enzyme in turn activates Akt, a regulator of the mammalian target of rampamycin (mTOR). mTOR activation results in protein synthesis leading to mitosis through S6k1 and 4E-BP-1, and then cell growth. The activation of PtdIns 3K by insulin in the mediobasal hypothalamus inhibits food intake through the involvement of neuronal PtdIns 3K signaling (Carvalheira et al. 2003; Niswender et al. 2003). Since weight gain is associated with insulin resistance at the level of PtdIns 3K signaling in the peripheral tissues (Carvalheira et al. 2003); therefore, it is hypothesized that a similar defect in insulin-stimulated PtdIns 3K activation also occurs in the hypothalamus and thereby favors further weight gain. It is recently reported that diet-induced obesity in rodents induces inflammatory genes and cytokine expression in the hypothalamus and then these effects are associated with diminished hypothalamic insulin signal transduction (De Souza et al. 2005).

As stated above, brain does not utilize fructose for energy production. However, hyperphysiological doses of fructose administered directly to the brain has been reported to induce increase in feeding. In contrast, administration of glucose causes a decrease in food consumption (Miller et al. 2002). These effects may be due to differential induction of malonyl-CoA by these two sugars. Glucose administration in the brain is known to activate the glycolytic cycle, which leads to increase in the neuronal ATP levels, decreasing AMP production. Since AMP is an activator of AMP kinase (AMPK); therefore, reduction in AMP results in a decrease in AMPK phosphorylation/activity. AMPK activity, in turn, catalyzes the phosphorylation/activation of acetyl-CoA carboxylase (ACC). Glucose blocks this activation and this leads to increased malonyl-CoA and decreased food intake. In contrast, central administration of fructose decreases ATP levels leading to decrease in hypothalamic malonyl-CoA levels (Cha et al. 2008; Wolfgang et al. 2007). This process may be associated with a decrease in satiety.

It is well known that the consumption of high carbohydrate diet results in hyperglycemia. Under these conditions glucose can be converted to fructose inside the cells via the AR/polyol pathway (Chung et al. 2003). An increased flux of glucose through the AR/polyol pathway in hyperglycemic conditions causes tissue damage through different mechanisms, including an osmotic imbalance due to sorbitol accumulation, an imbalance of the pyridine nucleotide redox status that decreases the antioxidant cell ability, and an induction in the advanced glycated end products. These processes may be involved in the pathogenesis of diabetic complications such as cataracts and nephropathy (Yabe-Nishimura 1998). In addition, there is also a possible link between activation/deactivation of AR/the polyol pathway and altered regulation the in lipid metabolism. It has been reported that in diabetic patients with dyslipidemia, there is significant increase in plasma (serum) and urinary sorbitol and fructose, indicating that the increased flux in the polyol pathway is concomitant with diabetic dyslipidemia (Yoshii et al. 2001). Based on evidence mentioned above, it is proposed that high carbohydrate diet is a high fat diet because overconsumption of carbohydrates results in their conversion into fat not only in visceral tissues, but also to some extent in the brain (Havel 2005). High carbohydrate-mediated decrease in leptin or insulin signaling in the hypothalamus is closely associated with obesity and type 2 diabetes (Burks et al. 2000; Obici et al. 2002b; Balthasar et al. 2004). In obesity and type II diabetes insulin and leptin levels in the cerebrospinal fluid (CSF) are elevated indicating a chronic state of central insulin and leptin resistance. Central administration of insulin or leptin consistently compromises the ability to control food intake in animals during the development of dietary obesity, supporting the view that hypothalamic (or central) leptin and insulin resistance not only contribute to the pathophysiology of obesity and type II diabetes, but are also closely associated with the pathogenesis of neurological disorders, such as stroke, Alzheimer disease, and depression (Fig. 3.9) (Farooqui et al. 2012). Recent research has also indicated that over-nutrition specifically due to high carbohydrate consumption directly blunts central insulin and leptin sensitivity before the onset of obesity (Woods et al. 2004). The molecular mechanism involved in this process in not fully understood. However, it is proposed that innate immunity regulator, IKKβ/NF-κB, which is enriched in the hypothalamic neurons plays an important role. This pathway is activated in the hypothalamic neurons by the chronic overnutrition with carbohydrates and fats. Persistent signals to the brain during overnutrition with carbohydrate and fat may stimulate IKKβ/NF-κB pathway in the hypothalamic neurons and this process contribute to the neuronal regulation of energy balance (Fig. 3.10). It is suggested that IKKβ/NF-κB pathway is a master-switch and central regulator of innate immunity and related functions (Zhang et al. 2008b). In the quiescent state, NF-κB remains inactive in the cytoplasm through binding to the inhibitory protein IκB. Activation of IKKβ by phosphorylation at S177 and S181 induces phosphorylation of its substrate IκBα at S32 and S36, ubiqitination, and subsequent proteosomal degradation. The liberation of IκBα from NF-κB/IκBα complex results in the translocation of NF-κB into the nucleus where it mediates the transcription of proinflammatory cytokines, such as TNFα, interleukin-1, and interleukin-6. These cytokines have been implicated in the pathogenesis of insulin resistance and may represent a causal link among neuroinflammation, insulin resistance, type II diabetes, and neurological disorders (Farooqui et al. 2012; Farooqui 2013).

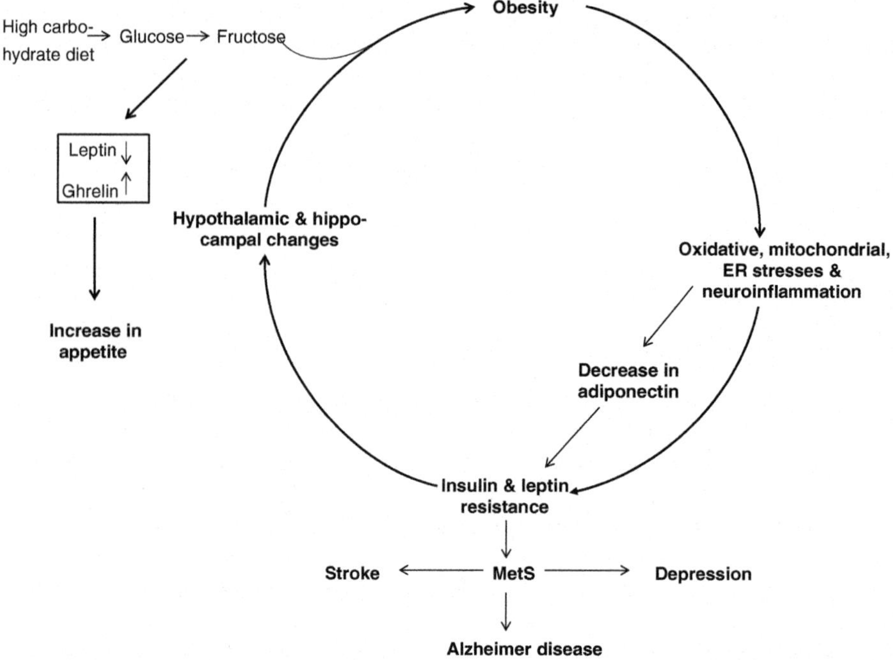

Fig. 3.9 Consumption of high fructose corn syrup (HFCS) and induction of insulin and leptin resistance

IKKβ/NF-κB-driven hypothalamic inflammation, ER stress, and autophagic defect is closely associated with imbalance of glucose homeostasis, mediated by impaired insulin secretion, insulin resistance and glucose intolerance (Meng and Cai 2011; Posey et al. 2009). Recently, non-neuronal cell types like astroglia and microglia have been shown to act as additional platforms of inducing hypothalamic inflammation in diet-induced obesity and insulin resistance (Thaler et al. 2012). Collective evidence suggests that hypothalamic inflammation in the brain involves the IKKβ/NF-κB, which may lead to metabolic as well as neurological diseases. Under these conditions, functions of inflamed neurons are generally compromised causing dysregulation of neuronal signaling. However, implication and scope of these developments are still in early stage. To date, only a few molecules have been identified with the activation of neural IKKβ/NF-κB pathway. For example, suppressor of cytokine signaling-3 (SOCS3) is an important molecule, which inhibits both leptin and insulin signaling (Howard and Flier 2006). Studies have shown that overnutrition-induced IKKβ/NF-κB activation can cause upregulation of hypothalamic SOCS3 gene expression to induce hypothalamic leptin and insulin resistance (Zhang et al. 2008b). Genetic mouse models have shown that SOCS3 knockout in hypothalamic neurons can improve central leptin signaling and reduce obesity (Reed et al. 2010; Mori et al. 2004). Protein-tyrosine phosphatase 1B (PTP1B) is another protein which inhibits insulin and leptin signaling and is known to contribute to

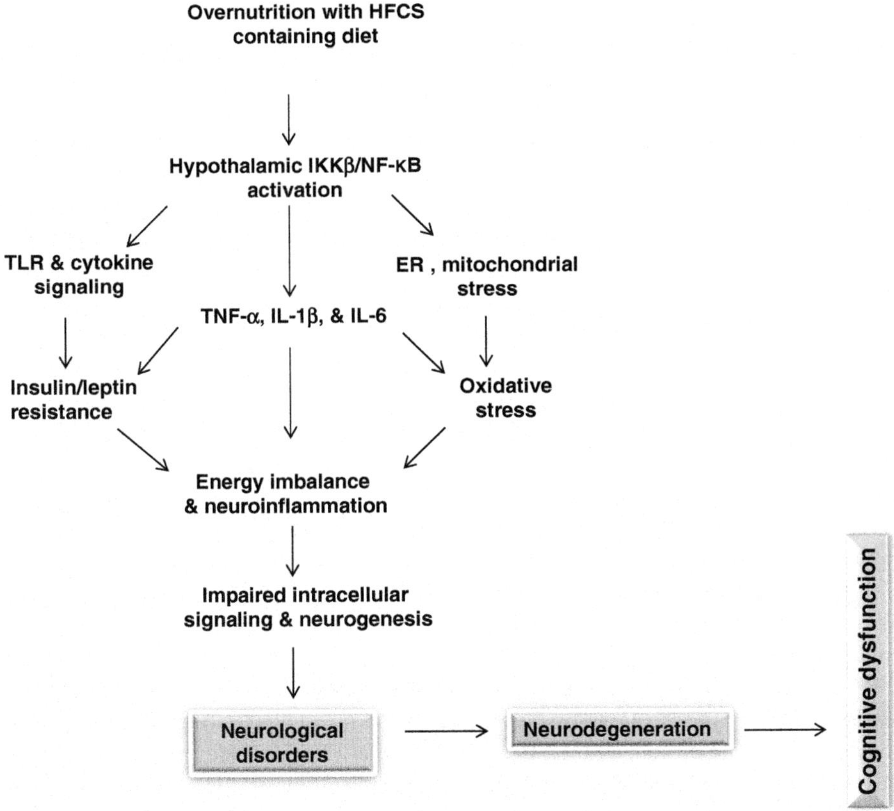

Fig. 3.10 Neurochemical events associated with cognitive dysfunction caused by overconsumption of diet enriched in high fructose corn syrup. *ER* endoplasmic reticulum, *NF-κB* nuclear factor kappaB, *TNF-α* tumor necrosis factor-α, *IL-1β* interleukin-1 beta, *IL-6* interleukin-6, *TLR* toll like receptor

PTP1B-mediated IKKβ/NF-κB inflammatory pathway. In liver, PTP1B is tethered to the endoplasmic reticulum (ER) via its hydrophobic C-terminal targeting sequence. PTP1B inhibits insulin signaling by dephosphorylating the insulin receptor (IR) and possibly insulin receptor substrate 1 (Salmeen et al. 2000). Complete absence of PTP1B in mice (PTP1B$^{-/-}$) results in increased in systemic insulin sensitivity, improvement in glucose tolerance, and enhanced muscle and liver IR phosphorylation, supporting the view that PTP1B is a physiologically important IR phosphatase (Klaman et al. 2000). Furthermore, mice with neuronal-, muscle-, or liver-specific PTP1B-deficiency also display improved insulin sensitivity (Delibegovic et al. 2009). PTP1B likely contributes to the pathogenesis of insulin resistance since it is over-expressed in many rodent and human models of obesity and insulin resistance (Goldstein 2001; Zabolotny et al. 2008). Little is known about the factors that modulate PTP1B expression in carbohydrate-mediated obesity. Potential factors may include insulin, leptin, glucose, free fatty acids, glucocorticoids, and pro-inflammatory cytokines (Pickup and Crook 1998). Insulin, leptin, and glucose induce modest increases in PTP1B expression in cultured cells (Lam et al. 2004). TNF-α, a proinflammatory cytokine and gene product of IKKβ/NF-κB, increases hypothalamic PTP1B expression (Zabolotny et al. 2008) and neural PTP1B inhibition counteracts overnutrition-induced leptin resistance, obesity and disorders of glucose metabolism (Banno et al. 2010; Picardi et al. 2008). Collectively, these studies suggest that Protein-tyrosine phosphatase 1B (PTP1B) inhibits insulin and leptin signaling. Recently brain PTP1B has been linked to AD in genetic mouse models (Liao et al. 2009), and thus may represent a connection between neurodegeneration and central mechanism of metabolic diseases.

Long term consumption of high carbohydrate diet results in high levels of glucose (hyperglycemia), which may be transformed into fructose through polyol pathway. High fructose is far more damaging to neural cells than glucose because of rapid AGE generation and glycation of various proteins. For example, studies involving feeding rats controlled diets containing either fructose or glucose indicate that the fructose-fed rats are worse off on many indicators of glycation damage (Levi and Werman 1998). Collectively, these studies suggest that a high carbohydrate diet produces insulin resistance in brain and visceral tissues not only by significantly reducing the protein expression of insulin receptor, insulin receptor substrate-1, and dysregulated secretion of adipokines/cytokines, but also by reducing circulating insulin and leptin levels and altering expression of Akt (Farooqui 2013). Consumption of high carbohydrate diet results in generation of AGEs and glycation of various proteins results in impairment in cellular function. Glycation may also promote protein aggregation, which plays a critical role in aging, diabetes, atherosclerosis, cardiovascular disease, MetS, autoimmune diseases, cancer, and neurodegenerative diseases (Fig. 3.11). Many proteins, such as apolipoproteins are regulated by glycation. Among apolipoproteins, Apolipoprotein E (ApoE; mol mass ~34 kDa) is a major cholesterol carrier that supports lipid transport and injury repair in the brain (Mahley and Rall 2000). *APOE* polymorphic alleles are the main genetic determinants of Alzheimer disease (AD) risk. Individuals carrying the ε4 allele are at increased risk of AD compared with those carrying the more common ε3 allele, whereas the ε2 allele decreases risk of AD (Mahley and Rall 2000). ApoB, the main apolipoprotein in LDL, is rich in lysine, an amino acid that is especially susceptible to glycation (Younis et al. 2008). Furthermore, glycation or fructation, of lysine is increased in type II diabetes, an important risk factor for AD (Karachalias et al. 2003). Small LDL particles are more susceptible to glycation than large ones. ApoE4 is the major risk factor for late-onset AD, whereas apoE3, the common isoform, is neutral with respect to this disease. This is tempting to speculate that glycation of apolipoproteins may contribute to the pathogenesis of AD and retardation of the formation of glycated ApoE may restore lipid homeostasis in AD brain. Although ApoE-based therapies are still in the early stages of development, they offer great promises for the potential therapy of AD. It is also shown that mice consuming high carbohydrate diet gain more body weight and develop glucose intolerance, hyperinsulinemia, and hypercholesterolemia (Cao et al. 2007). These metabolic alterations are associated with the exacerbation of memory impairment and a 2–3-fold increase

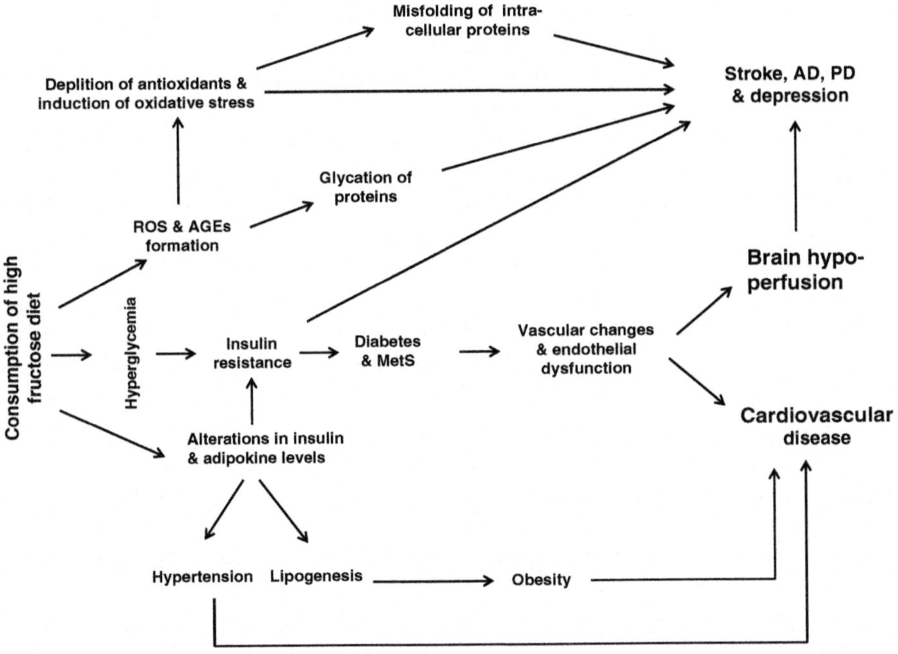

Fig. 3.11 Fructose consumption and pathogenesis of diabetes, stroke, Alzheimer disease, Parkinson disease, and depression. *ROS* reactive oxygen species, *AGEs* advanced glycation end products, *MetS* metabolic syndrome, *AD* Alzheimer disease, *PD* parkinson disease

in insoluble amyloid-beta protein (Aβ) levels and deposition in the brain. The consumption of high carbohydrate diet results in a 2.5-fold increase in brain apoE levels. Based on these observations, it is suggested that the up-regulation of apoE accelerates the aggregation of Aβ, resulting in the exacerbation of cerebral amyloidosis in high carbohydrate-fed mice (Cao et al. 2007).

Amyloid-β protein precursor (APP) is a large transmembrane protein, which has been implicated in neuroprotection and as a regulator of neuronal cell growth, cell-cell, or cell-matrix interactions and synaptic plasticity (Storey and Cappai 1999). However, its degradation through the amyloidogenic pathway can contribute to neurotoxicity. Indeed, soluble monomeric Aβ fragments are normally produced in the human body, but they can aggregate into various sized oligomers and insoluble fibrils, which subsequently form neuritic plaques. Aβ monomers are generated in neuronal and non-neuronal cells. However, neuronal cells seem to generate greater amounts of Aβ than other cell types (Fukumoto et al. 1999), indicating that the Aβ peptide might play an important role in the normal physiology of the CNS. Thus, Aβ contributes to synaptic dysfunction and loss associated with cognitive impairment (changes in perception, attention, memory, decision making, and language comprehension) in AD. Aβ is also produced in the peripheral cells by a number of different cells from where is transported across the BBB via receptor-mediated transcytosis. Thus, in the normal brain, levels of the Aβ peptide are regulated by its production from APP as well as its influx into the brain across the BBB. The influx of Aβ not only occurs via the receptor for advanced glycation end products (RAGE), but also through clearance from the brain via the low-density lipoprotein receptor-related protein-1 (LRP1) and enzymic degradation within brain (Selkoe 2001; Deane et al. 2003, 2009). Thus, impairment of these regulatory mechanisms may result in the accumulation and deposition of excessive amounts of *Aβ* peptide in the brain of individual with AD. RAGE is the key receptor that transports Aβ from the blood into the brain (Deane et al. 2003), which occurs at a rate that is five to six-fold lower than the rate determined for the transport of large neutral amino acids (Segal et al. 1990). LRP1 is the major cell surface Aβ clearance receptor that transports Aβ out of the brain across the BBB (Bell et al. 2007; Shibata et al. 2000) and promotes Aβ clearance on vascular smooth muscle cells (Urmoneit et al. 1997). Aβ is not only cleared from the brain interstitial fluid (ISF) (Bell et al. 2007) as a soluble peptide, but it can also be transported by its chaperone proteins in the ISF, such as apolipoprotein E (apoE), apolipoprotein J, and α2-macroglobulin (Zlokovic 1996). In brain, Aβ is a specific ligand for RAGEs. These receptors are expressed in the brain in neurons, microglia, and astrocytes (Lue et al. 2001; Sasaki et al. 2001; Yan et al. 1996). They interact with the N-terminal domain of RAGE (Fig. 3.12) (Chaney et al. 2005).

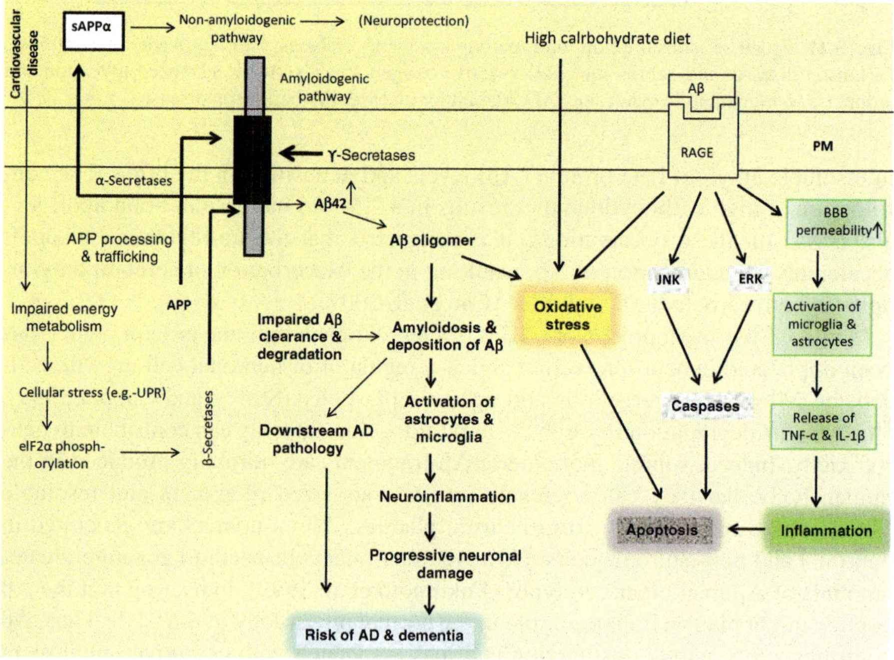

Fig. 3.12 Generation of beta-amyloid and its interactions with RAGE receptors. *ER* endoplasmic reticulum, *RAGE* receptor for AGE, *APP* amyloid precursor protein, *Aβ* beta amyloid, *JNK* c-Jun NH(2)-terminal kinase, *ERK* extracellular signal-regulated kinase, *TNF-α* tumor necrosis factor-α, *IL-1β* Interleukin-1 beta, *BBB* blood brain barrier

Aβ levels and RAGE expression are elevated in AD pathology-enriched brain regions, including hippocampus and inferior frontal cortex, when compared to cerebellum where AD pathology is limited. The interactions of RAGE with Aβ mediate multiple physiological and pathological functions, including inflammation, oxidative stress, and neurodegeneration (Ramasamy et al. 2005; Meneghini et al. 2010). These processes involve NF-κB, MAPKs, PtdIns 3K/Akt, Rho GTPases, Jak/STAT, and Src family kinases (Fang et al. 2010; Kislinger et al. 1999).

3.6 Conclusion

The consumption of high carbohydrate diet alters the electrophysiological properties of neurons, reduces the density of synaptic inputs, induces gliosis, and impairs insulin signaling in hypothalamic neurons controlling energy balance. In the liver, metabolism of fructose bypasses the two highly regulated steps in glycolysis (glucokinase and phosphofructokinase). Both steps are inhibited by increasing concentrations of their byproducts. Fructose is metabolized by fructokinase, which is not regulated by the negative feedback mechanism. Utilization of ATP in fructokinase reaction results in intracellular phosphate depletion and in the rapid generation of uric acid due to activation of AMP deaminase. The consumption of fructose in high calorie diet has increased dramatically in recent years and correlates closely with the rise in insulin resistance, obesity, type II diabetes as well as neurological disorders. Fructose is a highly lipogenic sugar. It stimulates triglyceride synthesis, and increases fat deposition in the liver, likely mediated in part by increasing fatty acyl coenzyme A and diacylglycerol. The consumption of high fructose diet decreases insulin receptor activation and also decline phosphorylation of insulin receptor substrate-1in skeletal muscle of rodents. Fructose also produces oxidative stress and mitochondrial dysfunction, resulting in a stimulation of peroxisome proliferator-activated receptor gamma coactivator 1-α and β (PGC1-α and PGC-1β) that drive both insulin resistance and lipogenesis. Fructose mediated lipogenesis alters the activity of key lipogenic enzymes and transcription factors in the liver, such as pyruvate dehydrogenase kinase and sterol-regulatory-element-binding protein-1c (SREBP-1c), the principal inducer of hepatic lipogenesis. Fructose-mediated uric acid generation not only causes hypertension, but is also linked with endothelial dysfunction, and insulin and leptin resistance. It is becoming increasingly evident that the metabolic effects of sugar may matter as much as its energy content. The functional resistance to leptin and insulin in the hypothalamus is a consequence of HFCS-mediated activation of inflammatory signaling, specifically in this region of the brain leading to the molecular impairment of leptin and insulin signal transduction by several distinct mechanisms including induction of SOCS3 expression, activation of JNK and I kappa kinase (IKK), and induction of PTP1B. These mechanisms impair leptin and insulin signaling in the hypothalamus and are closely associated with disruption of the main satietogenic and adipostatic routes that are involved in the stabilization of body mass.

References

Ahmed N (2005) Advanced glycation endproducts—role in pathology of diabetic complications. Diabetes Res Clin Pract 67:3–21

Ainscow EK, Mirshamsi S, Tang T, Ashford ML, Rutter GA (2002) Dynamic imaging of free cytosolic ATP concentration during fuel sensing by rat hypothalamic neurones: evidence for ATP-independent control of ATP-sensitive K(+) channels. J Physiol 544:429–445

Alexiou P, Pegklidou K, Chatzopoulou M, Nicolaou I, Demopoulos VJ (2009) Aldose reductase enzyme and its implication to major health problems of the 21st century. Curr Med Chem 16:734–752

Aliev G, Palacios HH, Lipsitt AE, Fischbach K, Lamb BT, Obrenovich ME, Morales L, Gasimov E, Bragin V (2009) Nitric oxide as an initiator of brain lesions during the development of Alzheimer disease. Neurotox Res 16:293–305

Alper AB Jr, Chen W, Yau L, Srinivasan SR, Berenson GS, Hamm LL (2005) Childhood uric acid predicts adult blood pressure: the Bogalusa Heart Study. Hypertension 45:34–38

Alwahsh SM, Xu M, Seyhan HA, Ahmad S, Mihm S, Ramadori G, Schultze FC (2014) Diet high in fructose leads to an overexpression of lipocalin-2 in rat fatty liver. World J Gastroenterol 20:1807–1821

Alzamendi A, Giovambattista A, Raschia A, Madrid V, Gaillard RC, Rebolledo O, Gagliardino JJ, Spinedi E (2009) Fructose-rich diet-induced abdominal adipose tissue endocrine dysfunction in normal male rats. Endocrine 35:227–232

Asipu A, Hayward BE, O'Reilly J, Bonthron DT (2003) Properties of normal and mutant recombinant human ketohexokinases and implications for the pathogenesis of essential fructosuria. Diabetes 52:2426–2432

Au A, Gupta A, Schembri P, Cheeseman CI (2002) Rapid insertion of GLUT2 into the rat jejunal brush-border membrane promoted by glucagon-like peptide 2. Biochem J 367:247–254

Balon TW, Nadler JL (1997) Evidence that nitric oxide increases glucose transport in skeletal muscle. J Appl Physiol 82:359–363

Balthasar N, Coppari R, McMinn J, Liu SM, Lee CE, Tang V, Kenny CD, McGovern RA, Chua SC Jr, Elmquist JK, Lowell BB (2004) Leptin receptor signaling in POMC neurons is required for normal body weight homeostasis. Neuron 42:983–991

Banno R, Zimmer D, De Jonghe BC, Atienza M, Rak K, Yang W, Bence KK (2010) PTP1B and SHP2 in POMC neurons reciprocally regulate energy balance in mice. J Clin Invest 120:720–734

Beisswenger PJ, Howell SK, Smith K, Szwergold BS (2003) Glyceraldehyde-3-phosphate dehydrogenase activity as an independent modifier of methylglyoxal levels in diabetes. Biochim Biophys Acta 1637:98–106

Bell RD, Zlokovic BV (2009) Neurovascular mechanisms and blood-brain barrier disorder in Alzheimer's disease. Acta Neuropathol 118:103–113

Bell RD, Sagare AP, Friedman AE et al (2007) Transport pathways for clearance of human Alzheimer's amyloid beta-peptide and apolipoproteins E and J in the mouse central nervous system. J Cereb Blood Flow Metab 27:909–918

Bo S, Gambino R, Durazzo M, Ghione F, Musso G, Gentile L (2008) Associations between serum uric acid and adipokines, markers of inflammation, and endothelial dysfunction. J Endocrinol Invest 31:499–504

Bray GA (2007) How bad is fructose? Am J Clin Nutr 12:895–896

Bray GA (2010) Soft drink consumption and obesity: it is all about fructose. Curr Opin Lipidol 21:51

Bray GA, Nielsen SJ, Popkin BM (2004) Consumption of high-fructose corn syrup in beverages may play a role in the epidemic of obesity. Am J Clin Nutr 79:537–543

Brown MS, Goldstein JL (2008) Selective *vs* total insulin resistance: a pathogenic paradox. Cell Metab 7:95–96

Brownlee M (2005) The pathobiology of diabetic complications: a unifying mechanism. Diabetes 54:1615–1625

Burks DJ, de Font MJ, Schubert M, Withers DJ, Myers MG, Towery HH, Altamuro SL, Flint CL, White MF (2000) IRS-2 pathways integrate female reproduction and energy homeostasis. Nature 407:377–382

Cao D, Lu H, Lewis TL, Li L (2007) Intake of sucrose-sweetened water induces insulin resistance and exacerbates memory deficits and amyloidosis in a transgenic mouse model of Alzheimer disease. J Biol Chem 282:36275–36282

Carvalheira JB, Ribeiro EB, Araujo EP, Guimaraes RB, Telles MM, Torsoni M, Gontijo JA, Velloso LA, Saad MJ (2003) Selective impairment of insulin signalling in the hypothalamus of obese Zucker rats. Diabetologia 46:1629–1640

Carvalho CR, Bueno AA, Mattos AM et al (2010) Fructose alters adiponectin, haptoglobin and angiotensinogen gene expression in 3T3-L1 adipocytes. Nutr Res 30:644–649

Catena C, Giacchetti G, Novello M, Colussi G, Cavarape A, Sechi LA (2003) Cellular mechanisms of insulin resistance in rats with fructose-induced hypertension. Am J Hypertens 16:973–978

Cha SH, Wolfgang M, Tokutake Y, Chohnan S, Lane MD (2008) Differential effects of central fructose and glucose on hypothalamic malonyl-CoA and food intake. Proc Natl Acad Sci U S A 105:16871–16875

Chaiswing L, Cole MP, St. Clair DK, Ittarat W, Szweda LI, Oberley TD (2004) Oxidative damage precedes nitrative damage in adriamycin-induced cardiac mitochondrial injury. Toxicol Pathol 32:536–547

Chaney MO, Stine WB, Kokjohn TA, Kuo YM, Esh C, Rathman A, Luehr DC, Schmidt AM, Stern D, Yan SD, Roher AE (2005) RAGE and amyloid β interactions: atomic force microscopy and molecular modeling. Biochim Biophys Acta 1741:199–205

Chang KC, Liang JT, Tseng CD, Wu C-D, Hsu K-L, Wu M-S, Lin Y-T, Tseng T (2007) Aminoguanidine prevents fructose-induced deterioration in left ventricular-arterial coupling in Wistar rats. Br J Pharmacol 151:341–346

Charonis AS, Reger LA, Dege JE, Kouzi-Koliakos K, Furcht LT, Wohlhueter RM, Tsilibary EC (1990) Laminin alterations after in vitro nonenzymatic glycosylation. Diabetes 39:807–814

Cheeseman CI (1993) GLUT2 is the transporter for fructose across the rat intestinal basolateral membrane. Gastroenterology 105:1050–1056

Chen G, Liang G, Ou J, Goldstein JL, Brown MS (2004) Central role for liver X receptor in insulin-mediated activation of *Srebp-1c* transcription and stimulation of fatty acid synthesis in liver. Proc Natl Acad Sci U S A 101:11245–11250

Choi HK, Mount DB, Reginato AM (2005) Pathogenesis of gout. Ann Intern Med 143:499–516

Chung SS, Ho EC, Lam KS, Chung SK (2003) Contribution of polyol pathway to diabetes-induced oxidative stress. J Am Soc Nephrol 14:S233–S236

Cirillo P, Gersch MS, Mu W, Scherer PM, Kim KM, Gesualdo L, Henderson GN, Johnson RJ, Sautin YY (2009) Ketohexokinase-dependent metabolism of fructose induces proinflammatory mediators in proximal tubular cells. J Am Soc Nephrol 20:545–553

Corry DB, Eslami P, Yamamoto K, Nyby MD, Makino H, Tuck ML (2008) Uric acid stimulates vascular smooth muscle cell proliferation and oxidative stress via the vascular renin-angiotensin system. J Hypertens 26:269–275

Cox CL, Stanhope KL, Schwarz JM, Graham JL, Hatcher B, Griffen SC, Bremer AA, Berglund L, McGahan JP, Havel PJ, Keim NL (2012) Consumption of fructose-sweetened beverages for 10 weeks reduces net fat oxidation and energy expenditure in overweight/obese men and women. Eur J Clin Nutr 66:201–208

Crabbe MJC, Cooper LR, Corne DW (2003) Use of essential and molecular dynamics to study γB-crystallin unfolding after non-enzymic post-translational modifications. Comput Biol Chem 27:507–510

Craft S (2009) The role of metabolic disorders in Alzheimer disease and vascular dementia: two roads converged. Arch Neurol 66:300–305

de Castro BT, Jiang LQ, Zierath JR, Nunes MT (2013) l-Arginine enhances glucose and lipid metabolism in rat L6 myotubes via the NO/c-GMP pathway. Metabolism 62:79–89

De Fronzo RA (1981) The effect of insulin on renal sodium metabolism. A review with clinical implications. Diabetologia 21:165–171

De Souza CT, Araujo EP, Bordin S, Ashimine R, Zollner RL, Boschero AC, Saad MJ, Velloso LA (2005) Consumption of a fat-rich diet activates a proinflammatory response and induces insulin resistance in the hypothalamus. Endocrinology 146:4192–4199

Deane R, Du Yan S, Submamaryan RK, LaRue B, Jovanovic S, Hogg E, Welch D, Manness L, Lin C, Yu J, Zhu H, Ghiso J, Frangione B, Stern A, Schmidt AM, Armstrong DL, Arnold B, Liliensiek B, Nawroth P, Hofman F, Kindy M, Stern D, Zlokovic B (2003) RAGE mediates amyloid-beta peptide transport across the blood-brain barrier and accumulation in brain. Nat Med 9:907–913

Deane R, Bell RD, Sagare A, Zlokovic BV (2009) Clearance of amyloid-β peptide across the blood-brain barrier: implication for therapies in Alzheimer's disease. CNS Neurol Disord Drug Targets 8:16–30

Dekker MJ, Su Q, Baker C, Rutledge AC, Adeli K (2010) Fructose: a highly lipogenic nutrient implicated in insulin resistance, hepatic steatosis, and the metabolic syndrome. Am J Physiol Endocrinol Metab 299:E685–E694

Delibegovic M, Zimmer D, Kauffman C, Rak K, Hong EG, Cho YR, Kim JK, Kahn BB, Neel BG, Bence KK (2009) Liver-specific deletion of protein-tyrosine phosphatase 1B (PTP1B) improves metabolic syndrome and attenuates diet-induced endoplasmic reticulum stress. Diabetes 58:590–599

Diggle CP, Shires M, Leitch D, Brooke D, Carr IM, Markham AF, Hayward BE, Asipu A, Bonthron DT (2009) Ketohexokinase: expression and localization of the principal fructose-metabolizing enzyme. J Histochem Cytochem 57:763–774

Douard V, Suzuki T, Sabbagh Y, Lee J, Shapses S, Lin S, Ferraris RP (2012) Dietary fructose inhibits lactation-induced adaptations in rat 1,25-(OH)(2)D(3) synthesis and calcium transport. FASEB J 26:707–721

Du X, Matsumura T, Edelstein D, Rossetti L, Zsengeller Z, Szabo C, Brownlee M (2003) Inhibition of GAPDH activity by poly(ADP-ribose) polymerase activates three major pathways of hyperglycemic damage in endothelial cells. J Clin Invest 112:1049–1057

Eiffert KC, McDonald RB, Stern JS (1991) High sucrose diet and exercise: effects on insulin-receptor function of 12- and 24-mo-old Sprague-Dawley rats. J Nutr 121:1081–1089

Elmquist JK, Bjorbaek C, Ahima RS, Flier JS, Saper CB (1998) Distribution of leptin receptor mRNA isoforms in the rat brain. J Comp Neurol 395:535–547

Engstrom G, Stavenow L, Hedblad B, Lind P, Tydén P, Janzon L, Lindgärde F (2003) Inflammation-sensitive plasma proteins and incidence of myocardial infarction in men with low cardiovascular risk. Arterioscler Thromb Vasc Biol 23:2247–2251

Enomoto A, Kimura H, Chairoungdua A, Shigeta Y, Jutabha P, Cha SH, Hosoyamada M, Takeda M, Sekine T, Igarashi T, Matsuo H, Kikuchi Y, Oda T, Ichida K, Hosoya T, Shimokata K, Niwa T, Kanai Y, Endou H (2002) Molecular identification of a renal urate anion exchanger that regulates blood urate levels. Nature 417:447–452

Fang F, Lue LF, Yan S, Xu H, Luddy JS, Chen D, Walker DG, Stern DM, Yan S, Schmidt AM, Chen JX, Yan SS (2010) RAGE-dependent signaling in microglia contributes to neuroinflammation, Abeta accumulation, and impaired learning/memory in a mouse model of Alzheimer's disease. FASEB J 24:1043–1055

Farah V, Elased KM, Chen Y, Key MP, Cunha TS, Ingoyen M, Morris M (2006) Nocturnal hypertension in mice consuming a high fructose diet. Auton Neurosci 130:41–50

Farooqui AA (2013) Metabolic syndrome: an important risk factor for stroke, Alzheimer disease, and depression. Springer, New York

Farooqui AA, Farooqui T, Panza F, Frisardi V (2012) Metabolic syndrome as a risk factor for neurological disorders. Cell Mol Life Sci 69:741–762

Ferre P, Foretz M, Azzout-Marniche D, Bécard D, Foufelle F (2001) Sterol-regulatory-element-binding protein 1c mediates insulin action on hepatic gene expression. Biochem Soc Trans 29:547–552

Flower DR (1996) The lipocalin protein family: structure and function. Biochem J 318:1–14

Fox IH, Palella TD, Kelley WN (1987) Hyperuricemia: a marker for cell energy crisis. N Engl J Med 317:111–112

Frenzel J, Richter J, Eschrich K (2002) Fructose inhibits apoptosis induced by reoxygenation in rat hepatocytes by decreasing reactive oxygen species via stabilization of the glutathione pool. Biochim Biophys Acta 1542:82–94

Fukumoto H, Tomita T, Matsunaga H, Ishibashi Y, Saido TC, Iwatsubo T (1999) Primary cultures of neuronal and non-neuronal rat brain cells secrete similar proportions of amyloid β peptides ending at Aβ40 and Aβ42. Neuro Report 10:2965–2969

Funk SD, Yurdogul A Jr, Orr W (2012) Hyperglycemia and endothelial dysfunction in atherosclerosis: lessons from type 1 diabetes. Int J Vasc Med 2012:569654

Gaesser GA (2007) Carbohydrate quantity and quality in relation to body mass index. J Am Diet Assoc 107:1768–1780

Gagliardi ACM, Miname MH, Santos RD (2009) Uric acid: a marker of increased cardiovascular risk. Atherosclerosis 202:11–17

Gao X, Qi L, Qiao N, Choi HK, Curhan G, Tucker KL, Ascherio A (2007) Intake of added sugar and sugar-sweetened drink and serum uric acid concentration in US men and women. Hypertension 50:306–312

Gersch C, Palii SP, Kim KM, Angerhofer A, Johnson RJ, Henderson GN (2008) Inactivation of nitric oxide by uric acid. Nucleosides Nucleotides Nucleic Acids 27:967–978

Giacco F, Brownlee M (2010) Oxidative stress and diabetic complications. Circ Res 107:1058–1070

Glimcher LH (2010) XBP1: the last two decades. Ann Rheum Dis 69(Suppl 1):i67–i71

Goldin A, Beckman JA, Schmidt AM, Creager MA (2006) Advanced glycation end products: sparking the development of diabetic vascular injury. Circulation 114:597–605

Goldstein BJ (2001) Protein-tyrosine phosphatase 1B (PTP1B): a novel therapeutic target for Type 2 diabetes mellitus, obesity and related states of insulin resistance. Curr Drug Targets 1:265–275

Granner D, Andreone T, Sasaki K, Beale E (1983) Inhibition of transcription of the phosphoenolpyruvate carboxykinase gene by insulin. Nature 305:549–551

Hallfrisch J (1990) Metabolic effects of dietary fructose. FASEB J 4:2652–2660

Han HJ, Lim MJ, Lee YJ, Lee JH, Yang IS, Taub M (2007) Uric acid inhibits renal proximal tubule cell proliferation via at least two signaling pathways involving PKC, MAPK, cPLA$_2$, and NF-kappaB. Am J Physiol Renal Physiol 292:F373–F381

Havel PJ (2005) Dietary fructose: implications for dysregulation of energy homeostasis and lipid/carbohydrate metabolism. Nutr Rev 63:133–157

Hayward BE, Bonthron DT (1998) Structure and alternative splicing of the ketohexokinase gene. Eur J Biochem 257:85–91

Hegarty BD, Bobard A, Hainault I, Ferré P, Bossard P, Foufelle F (2005) Distinct roles of insulin and liver X receptor in the induction and cleavage of sterol regulatory element-binding protein-1c. Proc Natl Acad Sci U S A 102:791–796

Horton JD, Goldstein JL, Brown MS (2002) SREBPs: activators of the complete program of cholesterol and fatty acid synthesis in the liver. J Clin Invest 109:1125–1131

Houstis N, Rosen ED, Lander ES (2006) Reactive oxygen species have a causal role in multiple forms of insulin resistance. Nature 440:944–948

Howard JK, Flier JS (2006) Attenuation of leptin and insulin signaling by SOCS proteins. Trends Endocrinol Metab 17:365–371

Howard BV, Wylie-Rosett J (2002) Sugar and cardiovascular disease: a statement for healthcare professionals from the Committee on Nutrition of the Council on Nutrition, Physical Activity, and Metabolism of the American Heart Association. Circulation 106:523–527

Hsieh PS, Huang WC (1993) Chemical sympathectomy attenuates hyperinsulinemia-induced hypertension in conscious rats. Nutr Metab Cardiovasc Dis 3:173–178

Iadecola C (1997) Bright and dark sides of nitric oxide in ischemic brain injury. Trends Neurosci 20:132–139

Iadecola C, Davisson RL (2008) Hypertension and cerebrovascular dysfunction. Cell Metab 7:476–484

Isganaitis E, Lustig RH (2005) Fast food, central nervous system insulin resistance, and obesity. Arterioscler Thromb Vasc Biol 25:2451–2462

Ishimoto T, Lanaspa MA, Le MT, Garcia GE, Diggle CP, Maclean PS, Jackman MR, Asipu A, Roncal-Jimenez CA, Kosugi T, Rivard CJ, Maruyama S, Rodriguez-Iturbe B, Sánchez-Lozada LG, Bonthron DT, Sautin YY, Johnson RJ (2012) Opposing effects of fructokinase C and A isoforms on fructose-induced metabolic syndrome in mice. Proc Natl Acad Sci U S A 109:4320–4325.

Jalal DI, Chonchol M, Chen W, Targher G (2013) Uric acid as a target of therapy in CKD. Am J Kidney Dis 61:134–146

Jianghai L, Wang R, Desai K, Wu L (2012) Up-regulation of aldolase B and overproduction of methylglyoxal in vascular tissues from rats with metabolic syndrome. Cardiovasc Res 92:494–503

Johnson R, Kang D, Feig D, Kivlighn S, Kanellis J, Watanabe S, Tuttle KR, Rodriguez-Iturbe B, Herrera-Acosta J, Mazzali M (2003) Is there a pathogenetic role for uric acid in hypertension and cardiovascular and renal disease? Hypertension 41:1183–1189

Juan CC, Fang VS, Hsu YP, Huang DB, Hsia DB, Yu PC, Kwok CF, Ho LT (1998) Overexpression of vascular endothelin-1 and endothelin-A receptors in a fructose-induced hypertensive rat model. J Hypertens 16:1775–1782

Jurcovicova J (2014) Glucose transport in brain—effect of inflammation. Endocr Regul 48:35–48

Kanellis J, Feig D, Johnson R (2004) Does asymptomatic hyperuricaemia contribute to the development of renal and cardiovascular disease? An old controversy renewed. Nephrology (Carlton) 9:394–399

Kang DH, Nakagawa T, Feng L, Watanabe S, Han L, Mazzali M, Truong L, Harris R, Johnson RJ (2002) A role for uric acid in the progression of renal disease. J Am Soc Nephrol 13:2888–2897

Kang DH, Park SK, Lee IK, Johnson RJ (2005) Uric acid-induced C-reactive protein expression: implication on cell proliferation and nitric oxide production of human vascular cells. J Am Soc Nephrol 16:3553–3562

Kanuri G, Spruss A, Wagnerberger S, Bischoff SC, Bergheim I (2011) Role of tumor necrosis factor α (TNFα) in the onset of fructose-induced nonalcoholic fatty liver disease in mice. J Nutr Biochem 22:527–534

Karachalias N, Babaei-Jadidi R, Ahmed N, Thornalley PJ (2003) Accumulation of fructosyl-lysine and advanced glycation end products in the kidney, retina and peripheral nerve of streptozotocin-induced diabetic rats. Biochem Soc Trans 31:1423–1425

Katakam PV, Ujhelyi MR, Hoenig ME, Miller AW (1998) Endothelial dysfunction precedes hypertension in diet-induced insulin resistance. Am J Physiol 275:R788–R792

Kenner KA, Heidenreich KA (1991) Insulin and insulin-like growth factors stimulate in vivo receptor autophosphorylation and tyrosine phosphorylation of a 70K substrate in cultured fetal chick neurons. Endocrinology 129:301–311

Khan AH, Pessin JE (2002) Insulin regulation of glucose uptake: a complex interplay of intracellular signalling pathways. Diabetologia 45:1475–1483

Khitan Z, Kim DH (2013) Fructose: a key factor in the development of metabolic syndrome and hypertension. J Nutr Metab 2013:682673

Khosla UM, Zharikov S, Finch JL, Nakagawa T, Roncal C, Mu W, Krotova K, Block ER, Prabhakar S, Johnson RJ (2005) Hyperuricemia induces endothelial dysfunction. Kidney Int 67:1739–1742

Kim DS, Jeong SK, Kim HR, Kim DS, Chae SW, Chae HJ (2007) Effects of triglyceride on ER stress and insulin resistance. Biochem Biophys Res Commun 363:140–145

Kim KM, Henderson GN, Ouyang X, Frye RF, Sautin YY, Feig DI, Johnson RJ (2009) A sensitive and specific liquid chromatography-tandem mass spectrometry method for the determination of intracellular and extracellular uric acid. J Chromatogr B Analyt Technol Biomed Life Sci 877:2032–2038

Kislinger T, Fu C, Huber B, Qu W, Taguchi A, Du Yan S, Hofmann M, Yan SF, Pischetsrieder M, Schmidt AM (1999) N(epsilon)-(carboxymethyl)lysine adducts of proteins are ligands for receptor for advanced glycation end products that activate cell signaling pathways and modulate gene expression. J Biol Chem 274:31740–31749

Klaman LD, Boss O, Peroni OD, Kim JK, Martino JL, Zabolotny JM, Moghal N, Lubkin M, Kim YB, Sharpe AH, Stricker-Krongrad A, Shulman GI, Neel BG, Kahn BB (2000) Increased energy expenditure, decreased adiposity, and tissue-specific insulin sensitivity in protein-tyrosine phosphatase 1B-deficient mice. Mol Cell Biol 20:5479–5489

Koch L, Wunderlich FT, Seibler J, Konner AC, Hampel B, Irlenbusch S, Brabant G, Kahn CR, Schwenk F, Bruning JC (2008) Central insulin action regulates peripheral glucose and fat metabolism in mice. J Clin Invest 118:2132–2147

Korieh A, Crouzoulon G (1991) Dietary regulation of fructose metabolism in the intestine and in the liver of the rat. Duration of the effects of a high fructose diet after the return to the standard diet. Arch Int Physiol Biochim Biophys 99:455–460

Kyriakis JM, Avruch J (2012) Mammalian MAPK signal transduction pathways activated by stress and inflammation: a 10-year update. Physiol Rev 92:689–737

Lam NT, Lewis JT, Cheung AT, Luk CT, Tse J, Wang J, Bryer-Ash M, Kolls JK, Kieffer TJ (2004) Leptin increases hepatic insulin sensitivity and protein tyrosine phosphatase 1B expression. Mol Endocrinol 18:1333–1345

Lanaspa MA, Sanchez-Lozada LG, Cicerchi C, Li N, Roncal-Jimenez CA, Ishimoto T, Le M, Garcia GE, Thomas JB, Rivard CJ, Andres-Hernando A, Hunter B, Schreiner G, Rodriguez-Iturbe B, Sautin YY, Johnson RJ (2012) Uric acid stimulates fructokinase and accelerates fructose metabolism in the development of fatty liver. PLoS One 7:e47948

Le KA, Tappy L (2006) Metabolic effects of fructose. Curr Opin Clin Nutr Metab Care 9:469–475

Le KA, Faeh D, Stettler R, Ith M, Kreis R, Vermathen P, Boesch C, Ravussin E, Tappy L (2006) A 4-wk high-fructose diet alters lipid metabolism without affecting insulin sensitivity or ectopic lipids in healthy humans. Am J Clin Nutr 84:1374–1379

Lee AH, Scapa EF, Cohen DE, Glimcher LH (2008) Regulation of hepatic lipogenesis by the transcription factor XBP1. Science 320:1492–1496

Leshan RL, Bjornholm M, Munzberg H, Myers MG Jr (2006) Leptin receptor signaling and action in the central nervous system. Obesity 14(Suppl):208S–212S

Levi B, Werman MJ (1998) Long-term fructose consumption accelerates glycation and several age-related variables in male rats. J Nutr 128:1442–1449

Levin BE, Routh VH, Kang L, Sanders NM, Dunn-Meynell AA (2004) Neuronal glucosensing: what do we know after 50 years? Diabetes 53:2521–2528

Li JM, Li YC, Kong LD, Hu QH (2010) Curcumin inhibits hepatic protein-tyrosine phosphatase 1B and prevents hypertriglyceridemia and hepatic steatosis in fructose-fed rats. Hepatology 51:1555–1566

Liao G, Cheung S, Galeano J, Ji AX, Qin Q, Bi X (2009) Allopregnanolone treatment delays cholesterol accumulation and reduces autophagic/lysosomal dysfunction and inflammation in Npc1−/− mouse brain. Brain Res 1270:140–151

Lichtenstein AH, Appel LJ, Brands M, Carnethon M, Daniels S, Franch HA, Franklin B, Kris-Etherton P, Harris WS, Howard B, Karanja N, Lefevre M, Rudel L, Sacks F, Van Horn L, Winston M, Wylie-Rosett J (2006) Diet and lifestyle recommendations revision 2006: a scientific statement from the American Heart Association Nutrition Committee. Circulation 114:82–96

Litherland GJ, Hajduch E, Gould GW, Hundal HS (2004) Fructose transport and metabolism in adipose tissue of Zucker rats: diminished GLUT5 activity during obesity and insulin resistance. Mol Cell Biochem 261:23–33

Liu S, Manson JE, Buring JE, Stampfer MJ, Willett WC, Ridker PM (2002) Relation between a diet with a high glycemic load and plasma concentrations of high-sensitivity C-reactive protein in middle-aged women. Am J Clin Nutr 75:492–498

Locke GA, Cheng D, Witmer MR, Tamura JK, Haque T, Carney RF, Rendina AR, Marcinkeviciene J (2008) Differential activation of recombinant human acetyl-CoA carboxylases 1 and 2 by citrate. Arch Biochem Biophys 475:72–79

Lopes RP, Lunardelli A, Preissler T, Leite CE, Alves-Filho JC, Nunes FB, de Oliveira JR, Bauer ME (2006) The effects of fructose-1,6-bisphosphate and dexamethasone on acute inflammation and T-cell proliferation. Inflamm Res 55:354–358

Lue LF, Walker DG, Brachova L, Beach TG, Roger J, Schmidt AM, Stern DM, Yan SD (2001) Involvement of microglial receptor for advanced glycation endproducts (RAGE) in Alzheimer's disease: identification of a cellular activation mechanism. Exp Neurol 171:29–45

MacAllister SL, Choi J, Dedina L, O'Brien PJ (2011) Metabolic mechanisms of methanol/formaldehyde in isolated rat hepatocytes: carbonyl-metabolizing enzymes versus oxidative stress. Chem Biol Interact 191:308–314

Maenpaa PH, Raivio KO, Kekomaki MP (1968) Liver adenine nucleotides: fructose-induced depletion and its effect on protein synthesis. Science 161:1253–1254

Mahley RW, Rall SC Jr (2000) Apolipoprotein E: far more than a lipid transport protein. Annu Rev Genomics Hum Genet 1:507–537

Marangos PJ, Fox AW, Riedel BJ, Royston D, Dziewanowska ZE (1998) Potential therapeutic applications of fructose-1,6-diphosphate. Expert Opin Investig Drugs 7:615–623

Matsumoto M, Han S, Kitamura T, Accili D (2006) Dual role of transcription factor FoxO1 in controlling hepatic insulin sensitivity and lipid metabolism. J Clin Invest 116:2464–2472

Matsuzaka T, Shimano H, Yahagi N, Amemiya-Kudo M, Okazaki H, Tamura Y, Iizuka Y, Ohashi K, Tomita S, Sekiya M, Hasty A, Nakagawa Y, Sone H, Toyoshima H, Ishibashi S, Osuga J, Yamada N (2004) Insulin-independent induction of sterol regulatory element-binding protein-1c expression in the livers of streptozotocin-treated mice. Diabetes 53:560–569

Mayes PA (1993) Intermediary metabolism of fructose. Am J Clin Nutr 58:754S–765S

McGarry JD (1994) Disordered metabolism in diabetes: have we underemphasized the fat component? J Cell Biochem 55(suppl):29–38

McGartland C, Robson PJ, Murray L, Cran G, Savage MJ, Watkins D, Rooney M, Boreham C (2003) Carbonated soft drink consumption and bone mineral density in adolescence: the Northern Ireland Young Hearts project. J Bone Miner Res 18:1563–1569

McLellan AC, Thornalley PJ, Benn J, Sonksen PH (1994) Glyoxalase system in clinical diabetes mellitus and correlation with diabetic complications. Clin Sci (Lond) 87:21–29

Meehan WP, Buchanan TA, Hsueh W (1994) Chronic insulin administration elevates blood pressure in rats. Hypertension 23:1012–1017

Melanson KJ, Angelopoulos TJ, Nguyen V, Zukley L, Lowndes J, Rippe JM (2008) High-fructose corn syrup, energy intake, and appetite regulation. Am J Clin Nutr 88:1738S–1744S

Meneghini V, Francese MT, Carraro L, Grilli M (2010) A novel role for the receptor for advanced glycation end-products in neural progenitor cells derived from adult subventricular zone. Mol Cell Neurosci 45:139–150

Meng Q, Cai D (2011) Defective hypothalamic autophagy directs the central pathogenesis of obesity via the IkappaB kinase beta (IKKbeta)/NF-kappaB pathway. J Biol Chem 286:32324–32332

Miller CC, Martin RJ, Whitney ML, Edwards GL (2002) Intracerebroventricular injection of fructose stimulates feeding in rats. Nutr Neurosci 5:359–362

Mirshamsi S, Laidlaw HA, Ning K, Anderson E, Burgess LA, Gray A, Sutherland C, Ashford ML (2004) Leptin and insulin stimulation of signaling pathways in arcuate nucleus neurons: PI3K dependent actin reorganization and KATP channel activation. BMC Neurosci 5:54

Moncada S, Erusalimsky JD (2002) Does nitric oxide modulate mitochondrial energy generation and apoptosis? Nat Rev Mol Cell Biol 3:214–220

Monti LD, Valsecchi G, Costa S (2000) Effects of endothelin 1 and nitric oxide on glucokinase activity in isolated rat hepatocytes. Metabolism 49:73–80

Mori H, Hanada R, Hanada T, Aki D, Mashima R, Nishinakamura H, Torisu T, Chien KR, Yasukawa H, Yoshimura A (2004) Socs3 deficiency in the brain elevates leptin sensitivity and confers resistance to diet-induced obesity. Nat Med 10:739–743

Muntané J (2006) Multiple beneficial actions of fructose 1,6-biphosphate in sepsis-associated liver injury. Crit Care Med 34:927–929

Murphy S, Johnson R (2003) The scientific basis of recent US guidance on sugars intake. Am J Clin Nutr 78:827S–833S

Murray R, Bender D, Botham KM, Kennelly PJ, Rodwell V, Weil PA (2003) Harper's illustrated biochemistry. McGraw-Hill, New York, pp 55–65

Nagai Y, Yonemitsu S, Erion DM, Iwasaki T, Stark R, Weismann D, Dong J, Zhang D, Jurczak MJ, Löffler MG, Cresswell J, Yu XX, Murray SF, Bhanot S, Monia BP, Bogan JS, Samuel V, Shulman GI (2009) The role of peroxisome proliferator-activated receptor gamma coactivator-1 beta in the pathogenesis of fructose-induced insulin resistance. Cell Metab 9:252–264

Nakae J, Biggs WH 3rd, Kitamura T, Cavenee WK, Wright CV, Arden KC, Accili D (2002) Regulation of insulin action and pancreatic β-cell function by mutated alleles of the gene encoding forkhead transcription factor Foxo1. Nat Genet 32:245–253

Nakagawa T, Tuttle KR, Short RA, Johnson RJ (2005) Hypothesis: fructose-induced hyperuricemia as a causal mechanism for the epidemic of the metabolic syndrome. Nat Clin Pract Nephrol 1:80–86

Neeper M, Schmidt AM, Brett J, Yan SD, Wang F, Pan YC, Elliston K, Stern D, Shaw A (1992) Cloning and expression of a cell surface receptor for advanced glycosylation end products of proteins. J Biol Chem 267:14998–15004

Nishikawa T, Edelstein D, Du XL, Yamagishi S, Matsumura T, Kaneda Y, Yorek MA, Beebe D, Oates PJ, Hammes HP, Giardino I, Brownlee M (2000) Normalizing mitochondrial superoxide production blocks three pathways of hyperglycaemic damage. Nature 404:787–790

Niswender KD, Morrison CD, Clegg DJ, Olson R, Baskin DG, Myers MG Jr, Seeley RJ, Schwartz MW (2003) Insulin activation of phosphatidylinositol 3-kinase in the hypothalamic arcuate nucleus: a key mediator of insulin-induced anorexia. Diabetes 52:227–231

Oba S, Nagata C, Nakamura K, Fujii K, Kawachi T, Takatsuka N, Shimizu H (2010) Dietary glycemic index, glycemic load, and intake of carbohydrate and rice in relation to risk of mortality from stroke and its subtypes in Japanese men and women. Metabolism 59:1574–1582

Obici S, Rossetti L (2003) Minireview: nutrient sensing and the regulation of insulin action and energy balance. Endocrinology 144:5172–5178

Obici S, Zhang BB, Karkanias G, Rossetti L (2002a) Hypothalamic insulin signaling is required for inhibition of glucose production. Nat Med 8:1376–1382

Obici S, Feng Z, Karkanias G, Baskin DG, Rossetti L (2002b) Decreasing hypothalamic insulin receptors causes hyperphagia and insulin resistance in rats. Nat Neurosci 5:566–572

Pacher P, Beckman JS, Liaudet L (2007) Nitric oxide and peroxynitrite in health and disease. Physiol Rev 87:315–424

Park OJ, Cesar D, Faix D, Wu K, Shackleton CH, Hellerstein MK (1992) Mechanisms of fructose-induced hypertriglyceridaemia in the rat. Activation of hepatic pyruvate dehydrogenase through inhibition of pyruvate dehydrogenase kinase. Biochem J 282:753–757

Patel J, Iyer A, Brown L (2009) Evaluation of the chronic complications of diabetes in a high fructose diet in rats. Indian J Biochem Biophys 46:66–72

Pénicaud L, Berthault MF, Morin J, Dubar M, Ktorza A, Ferre P (1998) Rilmenidine normalizes fructose-induced insulin resistance and hypertension in rats. J Hypertens Suppl 16:S45–S49

Pfaffenbach KT, Nivala AM, Reese L, Ellis F, Wang D, Wei Y, Pagliassotti MJ (2010) Rapamycin inhibits postprandial-mediated X-box-binding protein-1 splicing in rat liver. J Nutr 140:879–884

Phillips SA, Thornalley PJ (1993) Formation of methylglyoxal and D-lactate in human red blood cells in vitro. Biochem Soc Trans 21:163S

Picardi PK, Calegari VC, Prada PO, Moraes JC, Araújo E, Marcondes MC, Ueno M, Carvalheira JB, Velloso LA, Saad MJ (2008) Reduction of hypothalamic protein tyrosine phosphatase improves insulin and leptin resistance in diet-induced obese rats. Endocrinology 149:3870–3880

Pickup JC, Crook MA (1998) Is type II diabetes mellitus a disease of the innate immune system? Diabetologia 41:1241–1248

Posey KA, Clegg DJ, Printz RL, Byun J, Morton GJ, Vivekanandan-Giri A, Pennathur S, Baskin DG, Heinecke JW, Woods SC (2009) Hypothalamic proinflammatory lipid accumulation, inflammation, and insulin resistance in rats fed a high-fat diet. Am J Physiol –. Endocrinol Metab 296:E1003–E1012

Qiu L, Wu X, Chau JFL (2008) Aldose reductase regulates hepatic peroxisome proliferator-activated receptor α phosphorylation and activity to impact lipid homeostasis. J Biol Chem 283:17175–17183

Raivio KO, Becker A, Meyer LJ, Greene ML, Nuki G, Seegmiller JE (1975) Stimulation of human purine synthesis de novo by fructose infusion. Metabolism 24:861–869

Ramasamy R, Vannucci SJ, Yan SS, Herold K, Yan SF, Schmidt AM (2005) Advanced glycation end products and RAGE: a common thread in aging, diabetes, neurodegeneration, and inflammation. Glycobiology 15:16R–28R

Rattigan S, Clark MG, Barrett EJ (1999) Acute vasoconstriction-induced insulin resistance in rat muscle in vivo. Diabetes 48:564–569

Reed AS, Unger EK, Olofsson LE, Piper ML, Myers MG Jr, Xu AW (2010) Functional role of suppressor of cytokine signaling 3 upregulation in hypothalamic leptin resistance and long-term energy homeostasis. Diabetes 59:894–906

Reiniger N, Lau K, McCalla D, Eby B, Cheng B, Lu Y, Qu W, Quadri N, Ananthakrishnan R, Furmansky M, Rosario R, Song F, Rai V, Weinberg A, Friedman R, Ramasamy R, D'Agati V, Schmidt AM (2010) Deletion of the receptor for advanced glycation endproducts reduces glomerulosclerosis and preserves renal function in the diabetic OVE26 mouse. Diabetes 59:2043–2054

Roden M, Price TB, Perseghin G, Petersen KF, Rothman DL, Cline GW, Shulman GI (1996) Mechanism of free fatty acid-induced insulin resistance in humans. J Clin Invest 97:2859–2865

Roglans N, Vila L, Farre M, Alegret M, Sanchez RM, Vazquez-Carrera M, Laguna JC (2007) Impairment of hepatic Stat-3 activation and reduction of PPARα activity in fructose-fed rats. Hepatology 45:778–788

Saad MF, Riad-Gabriel MG, Khan A, Sharma A, Michael R, Jinagouda SD, Boyadjian R, Steil GM (1998) Diurnal and ultradian rhythmicity of plasma leptin: effects of gender and adiposity. J Clin Endocrinol Metab 83:453–459

Saha AK, Xu XJ, Balon TW, Brandon A, Kraegen EW, Ruderman NB (2011) Insulin resistance due to nutrient excess: is it a consequence of AMPK downregulation? Cell Cycle 10:3447–3451

Salmeen A, Andersen JN, Myers MP, Tonks NK, Barford D (2000) Molecular basis for the dephosphorylation of the activation segment of the insulin receptor by protein tyrosine phosphatase 1B. Mol Cell 6:1401–1412

Saltiel AR, Kahn CR (2001) Insulin signalling and the regulation of glucose and lipid metabolism. Nature 414:799–806

Sasaki N, Toki S, Chowei H, Saito T, Nakano N, Hayashi Y, Takeuchi M, Makita Z (2001) Immunohistochemical distribution of the receptor for advanced glycation end products in neurons and astrocytes in Alzheimer's disease. Brain Res 888:256–262

Sautin YY, Nakagawa T, Zharikov S, Johnson RJ (2007) Adverse effects of the classical antioxidant uric acid in adipocytes: NADPH oxidase-mediated oxidative/nitrosative stress. Am J Physiol Cell Physiol 293:C584–C596

Sawa A, Khan AA, Hester LD, Snyder SH (1997) Glyceraldehyde-3-phosphate dehydrogenase: nuclear translocation participates in neuronal and nonneuronal cell death. Proc Natl Acad Sci U S A 94:11669–11674

Scarpace PJ, Zhang Y (2007) Elevated leptin: consequence or cause of obesity? Front Biosci 12:3531–3544

Schulze MB, Hu FB (2005) Primary prevention of diabetes: what can be done and how much can be prevented? Annu Rev Public Health 26:445–467

Schwartz MW, Woods SC, Porte D Jr, Seeley RJ, Baskin DG (2000) Central nervous system control of food intake. Nature 404:661–671

Segal MB, Preston JE, Collis CS, Zlokovic BV (1990) Kinetics and Na independence of amino acid uptake by blood side of perfused sheep choroid plexus. Am J Physiol 258:F1288–F1294

Selkoe DJ (2001) Clearing the brain's amyloid cobwebs. Neuron 32:177–180

Seneff S, Wainwright G, Mascitelli L (2011) Is the metabolic syndrome caused by a high fructose, and relatively low fat, low cholesterol diet? Arch Med Sci 7:8–20

Shapiro A, Mu W, Roncal C, Cheng KY, Johnson RJ, Scarpace PJ (2008) Fructose-induced leptin resistance exacerbates weight gain in response to subsequent high-fat feeding. Am J Physiol Regul Integr Comp Physiol 295:R1370–R1375

Shen H-M, Liu Z-G (2006) JNK signaling pathway is a key modulator in cell death mediated by reactive oxygen and nitrogen species. Free Radic Biol Med 40:928–939

Shibata M, Yamada S, Kumar SR, Calero M, Bading J, Frangione B, Holtzman DM, Miller CA, Strickland DK, Ghiso J, Zlokovic BV (2000) Clearance of Alzheimer's amyloid-beta, (1–40) peptide from brain by LDL receptor-related protein-1 at the blood-brain barrier. J Clin Invest 106:1489–1499

Shimomura I, Matsuda M, Hammer RE, Bashmakov Y, Brown MS, Goldstein JL (2000) Decreased IRS-2 and increased SREBP-1c lead to mixed insulin resistance and sensitivity in livers of lipodystrophic and *ob/ob* mice. Mol Cell 6:77–86

Shulman GI (2000) Cellular mechanisms of insulin resistance. J Clin Invest 106:171–176

Singh R, Barden A, Mori T, Beilin L (2001) Advanced glycation end-products: a review. Diabetologia 44:129–146

Spasojević I, Mojović M, Blagojević D, Spasić SD, Jones DR, Nikolić-Kokić A, Spasić MB (2009a) Relevance of the capacity of phosphorylated fructose to scavenge the hydroxyl radical. Carbohydr Res 344:80–84

Spasojević I, Bajić A, Jovanović K, Spasić M, Andjus P (2009b) Protective role of fructose in the metabolism of astroglial C6 cells exposed to hydrogen peroxide. Carbohydr Res 344:1676–1681

Storey E, Cappai R (1999) The amyloid precursor protein of Alzheimer's disease and the Aβ peptide. Neuropathol Appl Neurobiol 25:81–97

Sykiotis GP, Papavassiliou AG (2001) Serine phosphorylation of insulin receptor substrate-1: a novel target for the reversal of insulin resistance. Mol Endocrinol 15:1864–1869

Tavakkolizadeh A, Berger UV, Shen KR, Levitsky LL, Zinner MJ, Hediger MA, Ashley SW, Whang EE, Rhoads DB (2001) Diurnal rhythmicity in intestinal SGLT-1 function, V(max), and mRNA expression topography. Am J Physiol Gastrointest Liver Physiol 280:G209–G215

Teff K, Elliot S, Tschoep MR (2002) Consuming high fructose meals reduces 24 hour plasma insulin and leptin concentrations, does not sup- press circulating ghrelin, and increases postprandial and fasting triglycerides in women. Diabetes 52:A408

Teff KL, Elliott SS, Tschop M, Kieffer TJ, Rader D, Heiman M, Townsend RR, Keim NL, D'Alessio D, Havel PJ (2004) Dietary fructose reduces circulating insulin and leptin, attenuates postprandial suppression of ghrelin, and increases triglycerides in women. J Clin Endocrinol Metab 89:2963–2972

Ter Maaten JC, Voorburg A, Heine RJ, Ter Wee PM, Donker AJ, Gans RO (1997) Renal handling of urate and sodium during acute physiological hyperinsulinaemia in healthy subjects. Clin Sci (Lond) 92:51–58

Thaler JP, Yi CX, Schur EA, Guyenet SJ, Hwang BH, Dietrich MO, Zhao X, Sarruf DA, Izgur V, Maravilla KR (2012) Obesity is associated with hypothalamic injury in rodents and humans. J Clin Invest 122:153–162

Thorens B, Mueckler M (2010) Glucose transporters in the 21ˢᵗ Century. Am J Physiol Endocrinol Metab 298:E141–E145

Thornalley PJ (1993) Modification of the glyoxalase system in disease processes and prospects for therapeutic strategies. Biochem Soc Trans 1:531–534

Thornalley PJ (2003) Glyoxalase I: structure, function and a critical role in the enzymatic defence against glycation. Biochem Soc Trans 31:1343–1348

Tjaderhane L, Larmas M (1998) A high sucrose diet decreases the mechanical strength of bones in growing rats. J Nutr 128:1807–1810

Tran LT, MacLeod KM, McNeill JH (2009) Chronic etanercept treatment prevents the development of hypertension in fructose-fed rats. Mol Cell Biochem 330:219–228

Trumbo P, Schlicker S, Yates AA, Poos M (2002) Dietary reference intakes for energy, carbohydrate, fiber, fat, fatty acids, cholesterol, protein and amino acids. J Am Diet Assoc 102:1621–1630

Tucker KL, Morita K, Qiao N, Hannan MT, Cupples LA, Kiel DP (2006) Colas, but not other carbonated beverages, are associated with low bone mineral density in older women: The Framingham Osteoporosis Study. Am J Clin Nutr 84:936–942

Urmoneit B, Prikulis I, Wihl G, D'Urso D, Frank R, Heeren J, Beisiegel U, Prior R (1997) Cerebrovascular smooth muscle cells internalize Alzheimer amyloid beta protein via a lipoprotein pathway: implications for cerebral amyloid angiopathy. Lab Invest 77:157–166

Valeri F, Boess F, Wolf A, Göldlin C, Boelsterli UA (1996) Fructose and tagatose protect against oxidative cell injury by iron chelation. Free Radic Biol Med 22:257–268

Verma S, Bhanot S, McNeill JH (1999) Sympathectomy prevents fructose-induced hyperinsulinemia and hypertension. Eur J Pharmacol 373:R1–R4

Vila L, Roglans N, Alegret M, Sanchez RM, Vazquez-Carrera M, Laguna JC (2008) Suppressor of cytokine signaling-3 (SOCS-3) and a deficit of serine/threonine (Ser/Thr) phosphoproteins involved in leptin transduction mediate the effect of fructose on rat liver lipid metabolism. Hepatology 48:1506–1516

Vilà L, Roglans N, Perna V, Sánchez RM, Vázquez-Carrera M, Alegret M, Laguna JC (2011) Liver AMP/ATP ratio and fructokinase expression are related to gender differences in AMPK activity and glucose intolerance in rats ingesting liquid fructose. J Nutr Biochem 22:741–751

Vilsboll T, Krarup T, Madsbad S, Holst JJ (2003) Both GLP-1 and GIP are insulinotropic at basal and postprandial glucose levels and contribute nearly equally to the incretin effect of a meal in healthy subjects. Regul Pept 114:115–121

vinh quốc Lương K, Nguyen LT (2013) The beneficial role of vitamin D in obesity: possible genetic and cell signaling mechanisms. Nutr J 12:89

Wang Y, Lam KS, Kraegen EW, Sweeney G, Zhang J, Tso AW, Chow WS, Wat NM, Xu JY, Hoo RL, Xu A (2007) Lipocalin-2 is an inflammatory marker closely associated with obesity, insulin resistance, and hyperglycemia in humans. Clin Chem 53:34–41

Wang X, Jia X, Chang T, Desai K, Wu L (2008) Attenuation of hypertension development by scavenging methylglyoxal in fructose-treated rats. J Hypertens 26:765–772

Watanabe S, Kang DH, Feng L, Nakagawa T, Kanellis J, Lan H, Mazzali M, Johnson RJ (2002) Uric acid, hominoid evolution, and the pathogenesis of salt-sensitivity. Hypertension 40:355–360

Watts AG, Donovan CM (2010) Sweet talk in the brain: glucosensing, neural networks, and hypoglycemic counterregulation. Front Neuroendocrinol 31:32–43

Wautier MP, Chappey O, Corda S, Stern DM, Schmidt AM, Wautier JL (2001) Activation of NADPH oxidase by AGEs links oxidant stress to altered gene expression via RAGE. Am J Physiol Endocrinol Metab 280:E685–E694

Westwood ME, Thornalley PJ (1995) Molecular characteristics of methylglyoxal-modified bovine and human serum albumins. Comparisons with glucose-derived advanced glycation endproduct-modified serum albumins. J Protein Chem 14:359–372

White JS (2008) Straight talk about high-fructose corn syrup: what it is and what it ain't. Am J Clin Nutr 88:1716S–1721S

Wink DA, Miranda KM, Espey MG, Pluta RM, Hewett SJ, Colton C, Vitek M, Feelisch M, Grisham MB (2001) Mechanisms of the antioxidant effects of nitric oxide. Antioxid Redox Signal 3:203–213

Wolfgang MJ, Cha SH, Sidhaye A, Chohnan S, Cline G, Shulman GI, Lane MD (2007) Regulation of hypothalamic malonyl-CoA by central glucose and leptin. Proc Natl Acad Sci U S A 104:19285–19290

Woods SC, D'Alessio DA, Tso P, Rushing PA, Clegg DJ, Benoit SC, Gotoh K, Liu M, Seeley RJ (2004) Consumption of a high-fat diet alters the homeostatic regulation of energy balance. Physiol Behav 83:573–578

Wu L (2006) Is methylglyoxal a causative factor for hypertension development? Can J Physiol Pharmacol 84:129–139

Wu T, Giovannucci E, Pischon T, Hankinson SE, Ma J, Rifai N, Rimm EB (2004) Fructose, glycemic load, and quantity and quality of carbohydrate in relation to plasma C-peptide concentrations in US women. Am J Clin Nutr 80:1043–1049

Yabe-Nishimura C (1998) Aldose reductase in glucose toxicity: a potential target for the prevention of diabetic complications. Pharmacol Rev 50:21–33

Yan SD, Chen X, Fu J, Chen M, Zhu H, Roher A, Slattery T, Zhao L, Nagashima M, Moser J, Migheli A, Nawroth P, Stern D, Schmidt AM (1996) RAGE and amyloid-β peptide neurotoxicity in Alzheimer's disease. Nature 382:685–691

Yang J, Goetz D, Li JY, Wang W, Mori K, Setlik D, Du T, Erdjument-Bromage H, Tempst P, Strong R, Barasch J (2002) An iron delivery pathway mediated by a lipocalin. Mol Cell 10:1045–1056

Yap TA, Garrett MD, Walton MI, Raynaud F, de Bono JS, Workman P (2008) Targeting the PI3K-AKT-mTOR pathway: progress, pitfalls, and promises. Curr Opin Pharmacol 8:393–412

Yoshii H, Uchino H, Ohmura C, Watanabe K, Tanaka Y, Kawamori R (2001) Clinical usefulness of measuring urinary polyol excretion by gas-chromatography/mass-spectrometry in type 2 diabetes to assess polyol pathway activity. Diabetes Res Clin Pract 51:115–123

Younis N, Sharma R, Soran H, Charlton-Menys V, Elseweidy M, Durrington PN (2008) Glycation as an atherogenic modification of LDL. Curr Opin Lipidol 19:378–384

Zabolotny JM, Kim YB, Welsh LA, Kershaw EE, Neel BG, Kahn BB (2008) Protein-tyrosine phosphatase 1B expression is induced by inflammation in vivo. J Biol Chem 283:14230–14241

Zhang J, Gao Z, Yin J, Quon MJ, Ye J (2008a) S6K directly phosphorylates IRS-1 on Ser-270 to promote insulin resistance in response to TNF-(alpha) signaling through IKK2. J Biol Chem 283:35375–35382

Zhang X, Zhang G, Zhang H, Karin M, Bai H, Cai D (2008b) Hypothalamic IKKbeta/NF-kappaB and ER stress link overnutrition to energy imbalance and obesity. Cell 135:61–73

Zharikov SI, Krotova K, Hu H, Baylis C, Johnson RJ, Block ER, Patel JM (2008) Uric acid decreases NO production and increases arginase activity in cultured pulmonary artery endothelial cells. Am J Physiol Cell Physiol 295:C1183–C1190

Zimmermann R, Panzenböck U, Wintersperger A, Levak-Frank S, Graier W, Glatter O, Fritz G, Kostner GM, Zechner R (2001) Lipoprotein lipase mediates the uptake of glycated LDL in fibroblasts, endothelial cells, and macrophages. Diabetes 50:1643–1653

Zlokovic BV (1996) Cerebrovascular transport of Alzheimer's amyloid beta and apolipoproteins J and E: possible anti-amyloidogenic role of the blood-brain barrier. Life Sci 59:1483–1497

Chapter 4
Effects of Long Term Consumption of Animal Proteins in the High Calorie Diet

4.1 Introduction

Protein is an essential macronutrient needed by the human body for growth and maintenance of metabolic functions. Proteins are major functional and structural components of all the cells of the body. Proteins occur in seafood, meat, poultry, eggs, milk and milk products, beans, peas, soy products, nuts, and seeds. The chemical composition and physical structure of dietary and body proteins varies considerably. However, all proteins are composed of amino acids, which contain a requisite amino group. Amino acids function as precursors of many coenzymes, hormones, and nucleic acids in addition to the role of the diverse protein structures in the body. Twenty amino acids have been identified in proteins of animal origin. They all are needed for human growth and metabolism. Twelve of these amino acids (eleven in children) are termed nonessential, meaning that they can be synthesized by our body and do not need to be consumed in the diet. The remaining amino acids (phenylalanine, valine, threonine, tryptophane, methionine, leucine, isoleucine, lysine, and histidine) cannot be synthesized in the body and are described as essential meaning that they need to be consumed in our diets. The absence of any of these amino acids will compromise the ability of tissue to grow, be repaired or be maintained. The current recommended dietary allowance (RDA) for protein is 0.8 g protein/kg body weight/day for adults, 1.5 g protein/kg body weight/day for children, and 1.0 g protein/kg body weight/day for adolescents (Fulgoni 2008). Protein is the most important macronutrient, which is vital for human metabolism as a source of essential amino acids. Proteins of animal origin contain all essential amino acids and are enriched in branched chain amino acids (leucine, isoleucine, and tyrosine) whereas vegetable proteins generally lack one or more of the essential amino acids (methionine, lysine, and tryptophan). In adults, approximately ~ 32–46 g of high-quality dietary protein/day is required in maintaining protein balance (Food and Nutrition Board 2005). However, Americans adults consume ~ 65–$100+$ g/day of dietary protein (Fulgoni 2008). Animal sources of dietary protein, despite providing a complete

© Springer International Publishing Switzerland 2015
A.A. Farooqui, *High Calorie Diet and the Human Brain*,
DOI 10.1007/978-3-319-15254-7_4

protein and numerous vitamins and minerals, have some health professionals concerned about the amount of saturated fat common in these foods compared to vegetable sources. Dietary proteins regulate body weight by modulating four targets for body weight regulation: satiety, thermogenesis, energy efficiency, and body composition (Westerterp-Plantenga et al. 2009a, b; Keller 2011). The consumption of proteins results in higher ratings of satiety than equicaloric amounts of carbohydrates or fat. The effect of proteins on satiety is mainly due to oxidation of amino acids fed in excess (Keller 2011). Mechanisms associated with protein-induced satiety are nutrient-specific, and consist mainly of synchronization with elevated amino acid concentrations. Different proteins produce different nutrient related responses through the release of glucagon-like peptide-1 (GLP-1) (Veldhorst et al. 2008).

4.2 Metabolic Changes Following High Protein Consumption

The consumption of high protein diet dramatically increases the amino acid delivery to the body by increasing the pathways involved in the elimination of ammonia and maintenance of the nitrogen balance (Jackson 1999). In rats and humans, consumption of high protein diet increases oxidation of proteins, reduces carbohydrate oxidation, and elevates lipid oxidation, (Petzke et al. 2007; Tentolouris et al. 2008; Leidy et al. 2007). Consumption of high protein diet also increases the degradation of amino acids (Morens et al. 2003), along with the activation of urea cycle enzymes (Colombo et al. 1990). High protein intake also leads to a negative fat balance (Westerterp-Plantenga et al. 2009a, b), a reduction in the expression of lipogenic enzymes (Blouet et al. 2006) and reduction in glucose disposal by adipose tissues (Pichon et al. 2006), which may be due to decrease in insulin response following a decrease in carbohydrate to protein ratio. It is well known that amino acids can act as fuel for gluconeogenesis when delivered in abundance even in the fed state (Azzout-Marniche et al. 2007). For these reasons, high protein diets have been reported to have positive effects on glucose homoeostasis in rats (Blouet et al. 2006) and humans (Bowen et al. 2006).

Among body tissues, kidney and the liver play a central role in protein metabolism. Synthesis of proteins mainly occurs in the liver, whereas protein breakdown and excretion are handled through an intricate interaction between liver and kidney. Thus, onset of liver and kidney diseases invariably result in clinically relevant disturbances of protein metabolism. Conversely, metabolic processes regulated by these two organs are directly affected by dietary protein intake. According to The Institute of Medicine guidelines for an acceptable macronutrient distribution range (AMDR) for proteins is 5–35 % of daily calories (depending on age), with a special emphasis that there is insufficient data on the long-term safety of the upper limit of this range (Food and Nutrition Board IoM 2002). A major concern in relation to potential deleterious effects of high protein diet is the increased risk of renal dysfunction (Martin et al. 2005). High protein intake triggers renal hyper-filtration and therefore may cause renal damage (Brenner et al. 1996). Animal and human studies on consumption

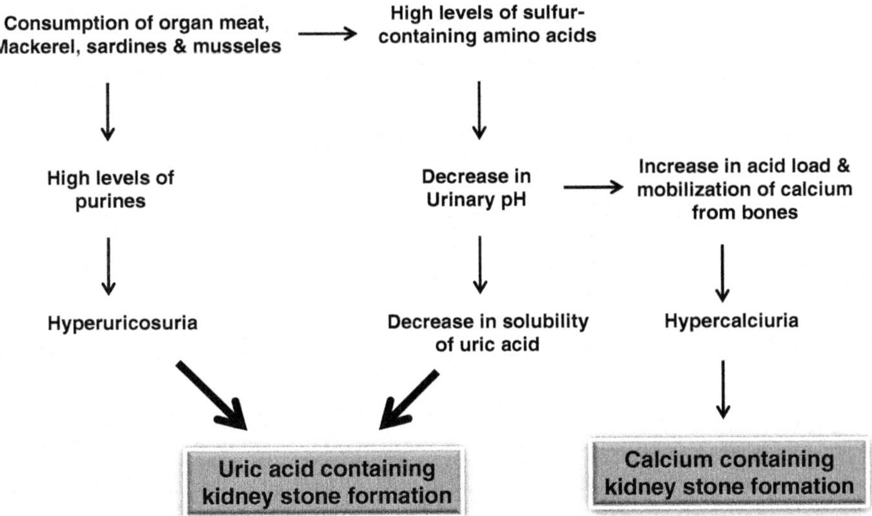

Fig. 4.1 Potential molecular mechanisms associated with kidney stone formation

of high protein diet have indicated an acceleration of chronic kidney disease (CKD) and increase in albuminuria and diuresis, natriuresis, and kaliuresis (Friedman 2004; Schwingshackl and Hoffmann 2014). The effect of high protein diet on induction of renal diseases is a controversial topic. Some studies have indicated that the consumption of high protein diet not only lowers urinary pH and increases calcium excretion (Calvez et al. 2012), but also increases chances of kidney stones formation and induction of other renal diseases (Fig. 4.1).

Protein constitutes roughly 50 % of the volume of bone and about one-third its mass (Heaney 2007). The bone protein matrix undergoes continuous turnover and remodeling. The bone mass at any particular time reflects the balance between bone formation and resorption. At the cellular level, osteoblast number and activity is decreased while osteoclast number and activity is increased with aging (Cao et al. 2005). Because the cross-linking of collagen molecules in bone involves post-translational modifications of amino acids (including hydroxylation of lysine and proline), many of the collagen fragments released during proteolysis of bone are not utilized for building new bone matrix. Accordingly, a daily supply of dietary protein is required for bone maintenance. High protein diet consumption promotes weight loss, bone growth, and retards bone loss and low-protein diet is associated with demineralization of bones and higher risk of hip fractures (Calvez et al. 2012). Consumption of a balanced variety of protein foods of plant origin can contribute to improved nutrient intake and health benefits. For example, moderate evidence indicates that eating proteins of plant origin (chickpeas, lentils, peanuts, walnuts, almonds, and pistachios) reduces risk factors for cardiovascular disease when consumed as part of a diet that is nutritionally adequate and within calorie needs (Krajcovicova-Kudlackova et al. 2005).

4.3 Consumption of High Protein Diet and Induction of Type II Diabetes

Increase in dairy and organ meat consumption is a major risk factor for the development of type II diabetes (Aune et al. 2009; Pounis et al. 2010; Melnik 2012). The biochemical mechanism of high protein consumption-mediated type II diabetes is not fully understood. However, it is proposed that the presence of leucine and other branched chain amino acids in meat and dairy products may contribute to the pathogenesis of type II diabetes (Melnik 2012) through the over-stimulation of mammalian target of rapamycin complex 1 (mTORC1). mTOR is an atypical serine/threonine protein kinase that belongs to the phosphoinositide 3-kinase (PtdIns 3K)-related kinase family and interacts with several proteins to form two distinct complexes named mTOR complex 1 (mTORC1) and mTOR complex 2 (mTORC2). The characteristic components of mTORC1 are Raptor (regulatory-associated protein of mTOR) and PRAS40 (proline-rich Akt substrate 40 kDa). Components of mTORC2 include Rictor (rapamycin-insensitive companion of TOR), mSin1 (mammalian stress-activated MAP kinase-interacting protein 1) and Protor-1 and 2 (proteins observed with Rictor 1 and 2) (Laplante and Sabatini 2012). The mTORC1 pathway integrates inputs from at least five major intracellular and extracellular cues—growth factors, stress, energy status, oxygen, and amino acids—to control many major processes, including protein and lipid synthesis and autophagy. Collective evidence suggests that mTORC1is an important master regulator of cell growth. It is predominantly activated by branched amino acids (leucine, and isoleucine). The consumption of high protein diet increases levels of insulin-like growth factor-1 (IGF-1), which binds to its receptor and stimulates the phosphoinositide-3 kinase (PtdIns 3K)/Akt pathway (Fig. 4.2). PtdIns 3K activation then leads to phosphorylation of Akt. Akt can translocate into the nucleus to activate mTOR (mammalian target of rapamycin) promoting increased cell growth and metabolism. In contrast, nutrient deprivation triggers other upstream pathways, such as the LKB1/AMPK pathway, which ultimately inhibits mTOR activity and limits cell growth and metabolism. The downstream target of mTORC1is the kinase S6K1. Leucine, and isoleucine have been reported not only to stimulate insulin synthesis and insulin secretion of pancreatic β-cells (McDaniel et al. 2002), but also activate the translational regulators 4E-BP1 and the kinase S6K1 in an mTORC1-dependent manner (Xu et al. 2001). Detailed investigations have revealed that leucine stimulates protein synthesis in skeletal muscle through both insulin-dependent and independent mechanisms. The insulin-dependent mechanism involves signaling through mTOR to 4E-BP1 and S6K1, whereas the insulin-independent effect is mediated by an unknown mechanism that may involve phosphorylation of eIF4G and/or its association with eIF4E. Leucine produces its stimulatory effect on assembly of the eIF4F complex, a key component in the mRNA binding step in translation initiation, as assessed by the phosphorylation status of the eIF4E binding protein (4E-BP1) and by the association of eIF4E with 4E-BP1 and eIF4G in skeletal muscle. Leucine also stimulates the phosphorylation status of eIF4G as well as ribosomal protein S6

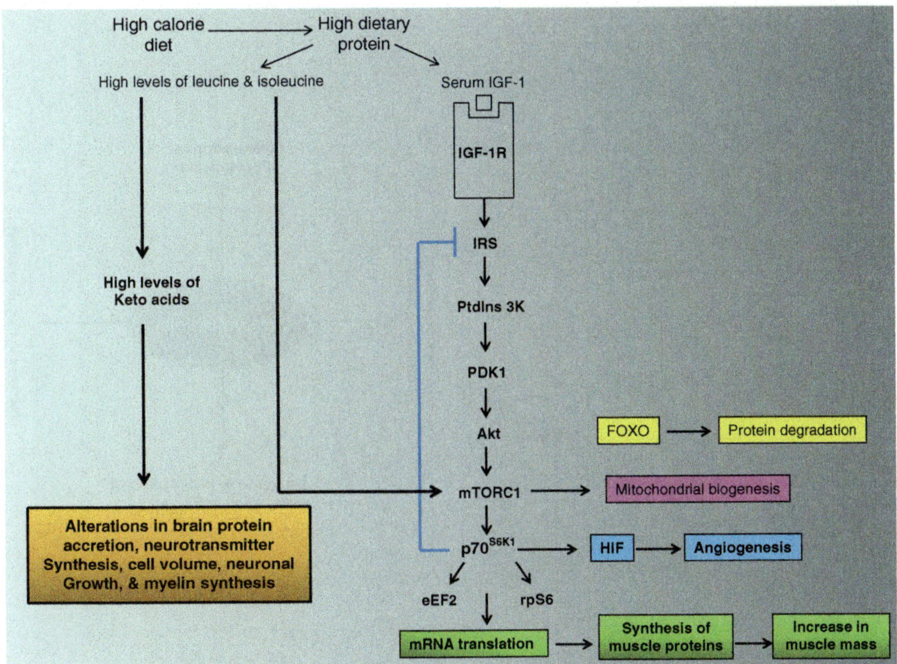

Fig. 4.2 Effect of high protein consumption on mTOR. *IGF-1* insulin-like growth factor 1, *IGF-1R* insulin-like growth factor 1 receptor, *IRS* insulin receptor substrate, *PtdIns 3K* phosphatidylinositol 3-kinase, *PDK1* phosphoinositide-dependent protein kinase 1, *Akt* protein kinase B, *mTORC1* mammalian target of rapamycin, *p70S6* p70S6 kinase, *FOXO* Forkhead box O; mTOR signaling to its downstream effectors: *S6K1* S6 kinase 1, *eEF2* eukaryotic elongation factor 2

kinase (S6K1) and its downstream substrate S6. Because phosphorylation of 4E-BP1, eIF4G, and S6K1 is mediated in part by the mTOR, these results suggest that leucine stimulates a signaling pathway involving serine/threonine protein kinase in skeletal muscle. Kinase S6K1 initiates insulin resistance through the phosphorylation of insulin receptor substrate-1, thereby increases the metabolic burden of β-cells. In addition, leucine-induced mTORC1-S6K1-signaling plays an important role in adipogenesis, further increasing the risk of obesity-mediated insulin resistance (Melnik 2012) (Fig. 4.3a). High consumption of leucine-rich proteins increases the mTORC1-dependent insulin secretion, increases in β-cell growth, and β-cell proliferation, promoting an early onset of replicative β-cell senescence with subsequent induction of β-cell apoptosis (Melnik 2012). These processes are hallmarks of type II diabetes and insulin resistance. Onset of type II diabetes is an important risk factor for stroke, Alzheimer disease, and depression (Farooqui 2013). In addition, high leucine and its keto acid (α-ketoisocaproic acid) have been reported to produce toxic effects in the brain because increased plasma levels of these compounds are associated with the appearance of neurological symptoms (Mello et al. 1999; Coitinho et al. 2001). High levels of leucine and its keto acids disturb protein accretion, neurotransmitter synthesis, cell volume, neuronal growth, and myelin

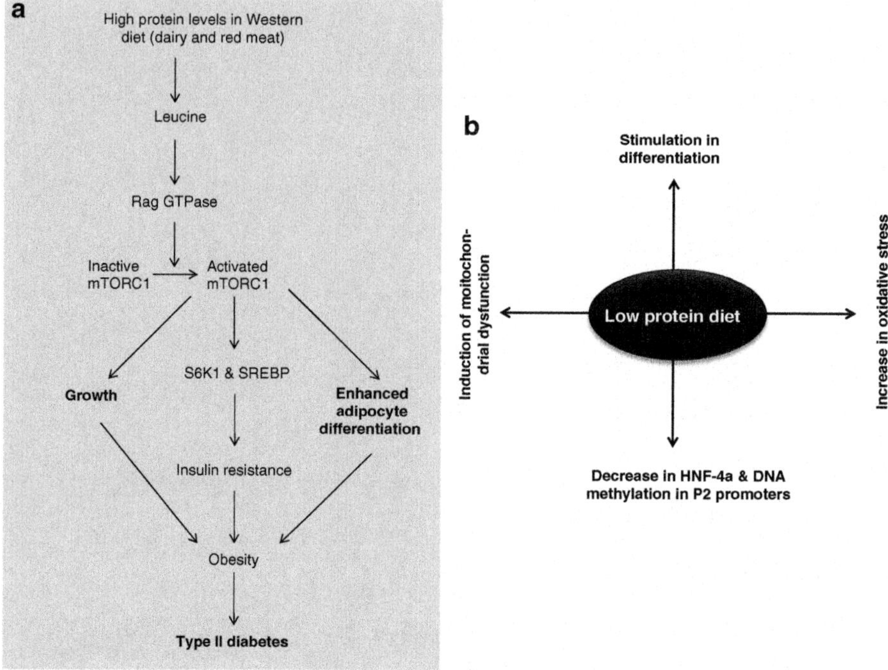

Fig. 4.3 Effects of high and low calorie diets on pathogenesis of diabetes. *mTORC1* mammalian target of rapamycin, *p70S6* p70S6 kinase; mTOR signaling to its downstream effectors: *S6K1* S6 kinase 1

synthesis in the brain. The neurotoxicity of leucine stems in part from its ability to interfere with the transport of other large neutral amino acids across the blood brain barrier, reduction in the supply of tryptophane, methionine, tyrosine, phenylalanine, histidine, valine, and threonine (Mello et al. 1999; Coitinho et al. 2001). Cerebral amino acid deficiency has adverse consequences for brain growth and the synthesis of neurotransmitters, such as dopamine, serotonin, norepinephrine, and histamine (Mello et al. 1999; Coitinho et al. 2001). Furthermore, coadministration of lipopolysaccharide (LPS) and high concentrations of branched-chain amino acids (BCAA) has been reported to alter BBB and to cause changes in matrix metalloproteinases (MMP-2 and MMP-9). Thus, the coadministration of branched chain amino acids and LPS not only induces breakdown of the BBB, but also increases the levels of MMP-2 and MMP-9 in the hippocampus of these rats. These studies support the view that MMP-mediated neuroinflammation is closely associated with BBB breakdown (Scaini et al. 2014).

Among branched chain amino acids, leucine is the most effective amino acid in inducing protein synthesis by stimulating the intracellular energy-sensing mammalian target of mTORC1 pathway (Proud 2007; Laplante and Sabatini 2012; Andre and Cota 2012). Thus, it critically affects energy balance regulation (Reiter et al. 2004; Blouet et al. 2009; Blouet and Schwartz 2012; Ropelle et al. 2008) through the regulation of the mTORC1-signaling pathway in neurons, in vivo (Lynch et al. 2000; Macotela et al. 2011).

Neurodegenerative diseases are accompanied by the malfunctioning of proteostasis control leading in the accumulation of misfolded and aggregated proteins. Aging in particular is characterized by decreasing proteostasis capacity and increasing protein damage which are in combination a major challenge for the cell due to the inability to maintain metastable proteins in folded states (Ben-Zvi et al. 2009). To counteract a cascade of protein destabilization caused by metastable proteins which escaped the PQC the proteostasis network including molecular chaperones, the UPS and autophagic pathways must manage the increasing burden of protein misfolding to maintain proteome stability (Gidalevitz et al. 2006). The induction of protein aggregation involves a crystallization-like seeding mechanism by which a specific protein is structurally corrupted by its misfolded conformer. Recent studies have indicated that once formed, proteopathic seeds can spread from one location to another via cellular uptake, transport, and release. Impeding this process may represent a unified therapeutic strategy for slowing the progression of a wide range of neurodegenerative disorders. It is well known that aging is the major risk factor neurodegenerative diseases (Kern and Behl 2009) the age-dependent loss of proteostasis may be an important contributor to the pathology of neurodegenerative diseases (Farooqui 2010, 2013).

The consumption of low protein diet also contributes to type II diabetes. The mechanisms responsible for the insulin resistance and type II diabetes remain unclear. However, it is reported that low protein diet may cause mitochondrial dysfunction and increase oxidative stress along with fibrosis (Tarry-Adkins et al. 2010), decrease in HNF4a expression with increased DNA methylation in P2 promoters (Sandovici et al. 2011). These processes may cause—cellular dysfunction and consequently increase the incidence of T2DM in postnatal life (Fig. 4.3b).

Interestingly, a recent study indicates that reduction in protein intake and, more specifically, methionine replacement in diet by 80 % increases life expectancy in rats due to significant reduction in ROS and oxidative damage (Pamplona and Barja 2006). This observation correlates well with the oxidative damage in brains of laboratory animals associated with oxidation of methionine in its proteins. It is also reported that calorie restriction reduces metabolic rate and oxidative damage, improves markers of type II diabetes such as insulin sensitivity resulting in increased longevity. The locomotor activity and brain dopamine levels of methionine sulfoxide reductase knockout ($Msr^{-/-}$) mice are considerably altered, unless the $Msr^{-/-}$ are on caloric restriction (Oien et al. 2010). From a therapeutic standpoint, the focus on caloric restriction in neural aging has resulted in the identification of several potential targets, such as the sirtuins, BDNF, FoxO and PPAR (Contestabile 2009).

4.4 Consumption of High Protein Diet and Renal and Bone Diseases

Body tissues respond to high protein diet challenge in several ways. Kidneys respond by increasing acid excretion, and bones respond to acid environment through demineralization leading to leaching of calcium from bones and promoting osteoporosis (Kerstetter et al. 1999; Barzel and Massey 1998; Calvez et al. 2012). Under these conditions, the protein-induced hypercalciuria may lead to the formation of calcium kidney stones (Fig. 4.1) (Goldfarb 1988). Furthermore, animal proteins are enriched in purines, which are precursors of uric acid. The solubility of uric acid largely depends upon the urinary pH. As the pH falls below 5.5–6.0, the solubility of uric acid decreases, and uric acid precipitates, even in the absence of hyperuricosuria (Rodman et al. 1996). Thus, consumption of high protein diet for 6 weeks results in a marked acid load on kidneys and increases the risk for stone formation by several folds (Fig. 4.1) (Reddy et al. 2002).

Long term consumption of high protein diet is also associated with chronic kidney diseases (CKD), which are complex diseases due to cardiovascular complications, oxidative stress, and high morbidity. As stated above, CKD is characterized either by kidney damage or a decline in renal function due to decrease in glomerular filtration rate (GFR) for three or more months (Levey et al. 2003). High BP is the second leading cause of CKD and accounts for approximately 30 % of all cases in the U.S. (Palmer 2004). High BP-mediated mechanisms in the progression of renal damage involve the magnitude of increase in systemic BP and the degree to which the elevation in systemic BP is transmitted to the renal microvasculature (Palmer 2001). It is well known that in the healthy kidney, renal autoregulation facilitates the maintenance of a constant renal blood flow and intraglomerular capillary pressure despite fluctuations in systemic BP between 80 and 170 mmHg (Palmer 2001). This involves the modulation of a myogenic reflex inherent to the kidney, wherein the preglomerular vasculature constricts or dilates in response to increase or decrease in systemic blood pressure. When systemic blood pressure increases, the afferent arteriole constricts, thereby limiting transmission of increased pressure to glomerular capillaries (Palmer 2001). Kidney damage is accompanied by the bluntness in myogenic reflex, impairment in renal autoregulation, and partial loss of glomerular circulation (Palmer 2001). These processes result in intraglomerular pressure-mediated changes in systemic arterial pressure along with increase in renin-angiotensin-aldosterone system and oxidative stress (Palmer 2001, 2004). Based on this information, it is suggested that glomerular capillary-mediated high BP is closely associated with the development of glomerular sclerosis and progressive kidney failure (Weir and Dworkin 1998). Nuclear factor erythroid-2 related factor 2 (Nrf2), a member of Cap 'n' collar/basic-leucine zipper family, is an important transcription factor that regulates the expression of many genes encoding for antioxidant proteins (superoxide dismutase, catalase, glutathione peroxidase, glutathione reductase, γ-glutamine cysteine ligase, heme oxygenase-1), thiol molecules and their generating enzymes, detoxifying enzymes, and stress response proteins. Nrf2 is present in cytosol in form of complex with Keap1, which functions as adaptor for the culin 3-based E3 ligase. Under normal unstressed condition, Nrf2 is ubiquitinated and rapidly degraded (half life ~20 min) by ubiquitin-proteasome system (McMahon et al. 2006). Under conditions of oxidative stress by either reactive electrophiles, or ARE inducers, the interaction between Nrf2 and Keap1 is disrupted and Nrf2 translocates to the nucleus. In the nucleus, it binds to small Maf proteins that increase the transcription rate of ARE-driven genes. The exact mechanism involved in disruption of the Keap1-Nrf2 complex remains elusive but it is suggested that ARE inducers may directly modify cysteine thiols groups in Keap1 (Dinkova-Kostova et al. 2002) that leads to the release of Nrf2, thereby increasing Nrf2 activity (Yamamoto et al. 2008). Furthermore, activation of several kinases may participate in this process through the phosphorylation of Nrf2 at serine and threonine residues that could mediate/modify translocation of Nrf2 to nucleus (Kobayashi and Yamamoto 2006). There is considerable experimental evidence supporting the view that Nrf2 signaling plays a protective role in renal injuries caused by CKD (Fig. 4.4) (Choi et al. 2014). Impairments in Nrf2 activity and consequent target gene repression have been reported to occur in animal models of CKD. Based on several studies, it is proposed that a pharmacological intervention activating Nrf2 signaling may be beneficial in protecting against kidney dysfunction in CKD (Choi et al. 2014).

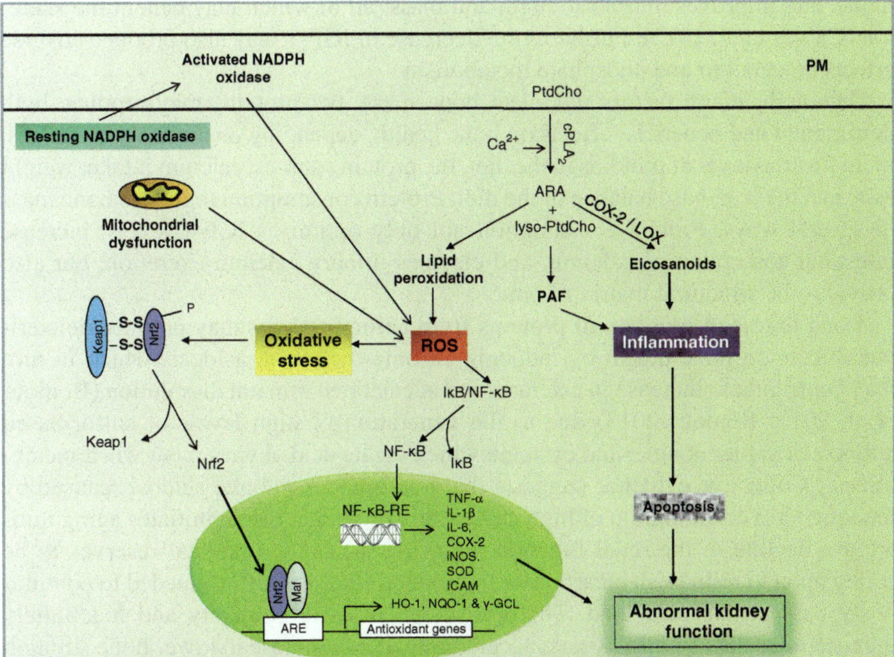

Fig. 4.4 Nrf2 inhibits reactive oxygen species and inflammatory pathways that lead to kidney dysfunction. *Glu* Glutamate, *NMDA-R* NMDA receptor, *PLA₂* phospholipase A₂, *PtdCho* phosphatidylcholine, *lyso-PtdCho* lyso-phosphatidylcholine, *ARA* arachidonic acid, *COX* cyclooxygenase, *PAF* platelet activating factor, *ROS* reactive oxygen species, *NF-kB* NF-kappaB, *NF-kB-RE* NF-kappaB response element, *I-κB* inhibitory subunit of NF-κB, *TNF-α* tumor necrosis factor-alpha, *IL-1β* interleukin-1beta, *IL-6* interleukine-6, *MCP1* monocyte chemotactic protein-1, *COX-2* cyclooxygenase-2, *iNOS* inducible nitric oxide synthase, *NO* nitric oxide, *SOD* superoxide dismutase, *Nrf2* nuclear factor-erythroid-2-related factor 2, *Keap1* kelch-like erythroid Cap 'n' Collar homologue-associated protein 1, *ARE* antioxidant response-element, *Maf* musculo-aponeurotic fibrosarcoma protein, *γ-GCL* gamma-glutamylcysteine ligase, *NQO1* NAD(P)H:quinone oxidoreductase-1, *HO-1* heme oxygenase-1, *SOD* superoxide dismutase, *GPS* glutathione peroxidase

Furthermore, since IGFs respond to renal hemodynamics both directly and indirectly by interacting with the renin-angiotensin system (Bach and Hale 2014), high protein consumption-mediated changes in the IGF system have been implicated in a number of kidney diseases. IGF activity is increased in early diabetic nephropathy and polycystic kidneys, whereas IGF resistance is associated with chronic kidney failure (Bach and Hale 2014).

Proteins are key nutrients that are necessary for bone health along with calcium and an adequate vitamin D supply. However, molecular mechanisms through which proteins impact bone health are still controversial. It is proposed that high protein intake results in bone loss due to protein catabolism, and the leaching of calcium from bones (Barzel and Massey 1998; Kerstetter et al. 1999). This opinion is supported by the observation that high protein diet not only increases calcium content in the urine (Kerstetter and Allen 1990), but also increases IGF-1, a key mediator of bone health, increases intestinal calcium absorption, suppresses parathyroid hormone, and improves muscle strength and mass, all of which may benefit the skeleton (Calvez et al. 2012). Furthermore, decrease in IGF-1 may also produce adverse effects on calcium and phosphate metabolism.

Although, many factors modulate bone mass, but proteins may produce both detrimental and beneficial effects on bone health, depending on a variety of factors, including the level of proteins in the diet, the protein sources, calcium intake, weight loss, and the acid/base balance of the diet. Protein consumption regulates bone mass in several ways. Protein consumption not only optimizes IGF-1 levels, increase intestinal absorption of calcium, and elevates urinary calcium excretion, but also provides the structural matrix of bone.

Long term consumption of proteins from animal sources may produce deleterious effects on bone health by inducing chronic metabolic acidosis which in turn may contribute to increase in calciuria and accelerated mineral dissolution (Bonjour et al. 2013; Bonjour 2013) due to the generation of high levels of sulfur-based amino acids, methionine, and cysteine which create acid environment when metabolized . Collective evidence suggests that low grade metabolic acidosis caused by the long term consumption of high calorie diet of animal origin initiates aging initiating a decline in the renal function requiring the body's skeletal reserves to be called upon to relinquish bicarbonate to produce alkaline buffers needed to continuously balance the acid load. Therefore bone mass is gradually and indefinitely reduced after the age of 30 years, accelerating at menopause to lower bone strength and mineral density. Long term consumption of high protein diet of animal origin also increases levels of iron coming from haem containing proteins of animal origin. Recently iron has been reported to promote atherosclerosis due to its remarkable capacity to generate free radicals. In contrast, proteins of plant origin produce alkalinogenic effects due to high mineral content in the form of salts of organic acids. Alkalinogenic environment produces a number of benefits such as prevention of cancers, infections, allergies, obesity, and osteoporosis due to high loads of phosphate and nonvolatile acid (Vallianou et al. 2013). Furthermore, proteins of plant origin have low in iron supporting the view that it may not generate as many free radicals as animal protein diet and may not induce atherosclerosis.

4.5 Conclusion

Short term consumption of high protein diet promotes weight loss and prevents weight (re)gain. This is because high protein diet not only induces satiety, increases secretion of gastrointestinal hormones, and increases diet-mediated thermogenesis, but also induces adaptations of the metabolic pathways involved in protein and energy metabolism. Proteins produce unique effects in the body depending on characteristics related to their sources, amino acid composition, rate of absorption, and protein/food texture. These characteristics may be important factors that may determine the metabolic effects. However, long term consumption of high-protein diets may produce detrimental effects on human health. Five branched chain and aromatic amino acids (isoleucine, leucine, valine, tyrosine, and phenylalanine) produce insulin resistance and markedly increase the risk of type II diabetes, cardiovascular disease, and chronic kidney disease.

References

Andre C, Cota D (2012) Coupling nutrient sensing to metabolic homoeostasis: the role of the mammalian target of rapamycin complex 1 pathway. Proc Nutr Soc 71:502–510

Aune D, Ursin G, Veierod MB (2009) Meat consumption and the risk of type 2 diabetes: a systematic review and meta-analysis of cohort studies. Diabetologia 52:2277–2287

Azzout-Marniche D, Gaudichon C, Blouet C, Bos C, Mathe V, Huneau JF, Tomé D (2007) Liver glyconeogenesis: a pathway to cope with postprandial amino acid excess in high-protein fed rats? Am J Physiol Regul Integr Comp Physiol 292:R1400–R1407

Bach LA, Hale LJ (2015) Insulin-like growth factors and kidney disease. Am J Kidney Dis 65(2):327–336, pii: S0272-6386(14)01000-2

Barzel US, Massey LK (1998) Excess dietary protein can adversely affect bone. J Nutr 128:1051–1053

Ben-Zvi A, Miller EA, Morimoto RI (2009) Collapse of proteostasis represents an early molecular event in *Caenorhabditis elegans* aging. Proc Natl Acad Sci U S A 106:14914–14919

Blouet C, Schwartz GJ (2012) Brainstem nutrient sensing in the nucleus of the solitary tract inhibits feeding. Cell Metab 16:579–587

Blouet C, Mariotti F, Azzout-Marniche D, Bos C, Mathe V, Tomé D, Huneau JF (2006) The reduced energy intake of rats fed a high-protein low-carbohydrate diet explains the lower fat deposition, but macronutrient substitution accounts for the improved glycemic control. J Nutr 136:1849–1854

Blouet C, Jo YH, Li X, Schwartz GJ (2009) Mediobasal hypothalamic leucine sensing regulates food intake through activation of a hypothalamus-brainstem circuit. J Neurosci 29:8302–8311

Bonjour JP (2013) Nutritional disturbance in acid-base balance and osteoporosis: a hypothesis that disregards the essential homeostatic role of the kidney. Br J Nutr 110:1168–1677

Bonjour JP, Kraenzlin M, Levasseur R, Warren M, Whiting S (2013) Dairy in adulthood: from foods to nutrient interactions on bone and skeletal muscle health. J Am Coll Nutr 32:251–263

Bowen J, Noakes M, Trenerry C, Clifton PM (2006) Energy intake, ghrelin, and cholecystokinin after different carbohydrate and protein preloads in overweight men. J Clin Endocrinol Metab 91:1477–1483

Brenner BM, Lawler EV, Mackenzie HS (1996) The hyperfiltration theory: a paradigm shift in nephrology. Kidney Int 49:1774–1777

Calvez J, Poupin N, Chesneau C, Lassale C, Tomé D (2012) Protein intake, calcium balance and health consequences. Eur J Clin Nutr 66:281–295

Cao JJ, Wronski TJ, Iwaniec U, Phleger L, Kurimoto P, Boudignon B, Halloran BP (2005) Aging increases stromal/osteoblastic cell-induced osteoclastogenesis and alters the osteoclast precursor pool in the mouse. J Bone Miner Res 20:1659–1668

Choi BH, Kang KS, Kwak MK (2014) Effect of redox modulating NRF2 activators on chronic kidney disease. Molecules 19:12727–12759

Coitinho AS, de Mello CF, Lima TT, de Bastiani J, Fighera MR, Wajner M (2001) Pharmacological evidence that alpha-ketoisovaleric acid induces convulsions through GABAergic and glutamatergic mechanisms in rats. Brain Res 894:68–73

Colombo JP, Pfister U, Cervantes H (1990) The regulation of N-acetylglutamate synthetase in rat liver by protein intake. Biochem Biophys Res Commun 172:1239–1245

Contestabile A (2009) Benefits of caloric restriction on brain aging and related pathological States: understanding mechanisms to devise novel therapies. Curr Med Chem 16:350–361

Dinkova-Kostova AT, Holtzclaw WD, Cole RN, Itoh K, Wakabayashi N, Katoh Y, Yamamoto M, Talalay P (2002) Direct evidence that sulfhydryl groups of Keap1 are the sensors regulating induction of phase 2 enzymes that protect against carcinogens and oxidants. Proc Natl Acad Sci U S A 99:11908–11913

Farooqui AA (2010) Neurochemical aspects of neurotraumatic and neurodegenerative diseases. Springer, New York

Farooqui AA (2013) Metabolic syndrome: an important risk factor for neurodegenerative diseases. Springer, New York

Food and Nutrition Board (2005) Proteins and amino acids. In: Dietary reference intakes for energy, carbohydrate, fiber, fat, fatty acids, cholesterol, protein, and amino acids. The National Academies Press, Washington, DC, pp 589–768

Food and Nutrition Board IoM (2002) In: DNAP (ed) Dietary reference intakes for energy, carbohydrates, fiber, fatty acids, cholesterol, protein, and amino acids (macronutrients). National Academy Press, Washington, DC, pp 207–264

Friedman AN (2004) High-protein diets: potential effects on the kidney in renal health and disease. Am J Kidney Dis 44:950–962

Fulgoni VL (2008) Current protein intake in America: analysis of the National Health and Nutrition Examination Survey, 2003-2004. Am J Clin Nutr 87:1554S–1557S

Gidalevitz T, Ben-Zvi A, Ho KH, Brignull HR, Morimoto RI (2006) Progressive disruption of cellular protein folding in models of polyglutamine diseases. Science 311:1471–1474

Goldfarb S (1988) Dietary factors in the pathogenesis and prophylaxis of calcium nephrolithiasis. Kidney Int 34:544–555

Heaney RP (2007) Effects of protein on the calcium economy. In: Burckhardt P, Heaney RP, Dawson-Hughes B (eds) Nutritional aspects of osteoporosis 2006. Proceedings of the 6th International Symposium on Nutritional Aspects of Osteoporosis, 4–6 May 2006, Lausanne, Switzerland. Elsevier, Amsterdam, pp 191–197

Jackson AA (1999) Limits of adaptation to high dietary protein intakes. Eur J Clin Nutr 53(Suppl 1): S44–S52

Keller U (2011) Dietary proteins in obesity and in diabetes. Int J Vitam Nutri Res 81:125–133

Kern A, Behl C (2009) The unsolved relationship of brain aging and late-onset alzheimer disease. Biochim Biophys Acta 1790:1124–1132

Kerstetter JE, Allen LH (1990) Dietary protein increases urinary calcium. J Nutr 120:134–136

Kerstetter JE, Mitnick ME, Gundberg CM, Caseria DM, Ellison AF, Carpenter TO, Insogna KL (1999) Changes in bone turnover in young women consuming different levels of dietary protein. J Clin Endocrinol Metab 84:1052–1055

Kobayashi M, Yamamoto M (2006) Nrf2-Keap1 regulation of cellular defense mechanisms against electrophiles and reactive oxygen species. Adv Enzyme Regul 46:113–140

Krajcovicova-Kudlackova M, Babinska K, Valachovicova M (2005) Health benefits and risks of plant proteins. Bratisl Lek Listy 106:231–234

Laplante M, Sabatini DM (2012) mTOR signaling in growth control and disease. Cell 149: 274–293

Leidy HJ, Carnell NS, Mattes RD, Campbell WW (2007) Higher protein intake preserves lean mass and satiety with weight loss in pre-obese and obese women. Obesity (Silver Spring) 15:421–429

Levey AS, Coresh J, Balk E, Kausz AT, Levin A, Steffes MW, Hogg RJ, Perrone RD, Lau J, Eknoyan G (2003) National Kidney Foundation practice guidelines for chronic kidney disease: evaluation, classification, and stratification. Ann Intern Med 139:137–147

Lynch CJ, Fox HL, Vary TC, Jefferson LS, Kimball SR (2000) Regulation of amino acid-sensitive TOR signaling by leucine analogues in adipocytes. J Cell Biochem 77:234–251

Macotela Y, Emanuelli B, Bang AM, Espinoza DO, Boucher J, Beebe K, Gall W, Kahn CR (2011) Dietary leucine—an environmental modifier of insulin resistance acting on multiple levels of metabolism. PLoS One 6:e21187

Martin WF, Armstrong LE, Rodriguez NR (2005) Dietary protein intake and renal function. Nutr Metab 2:25

McDaniel ML, Marshall CA, Pappan KL, Kwon G (2002) Metabolic and autocrine regulation of the mammalian target of rapamycin by pancreatic β-cells. Diabetes 10:2877–2885

McMahon M, Thomas N, Itoh K, Yamamoto M, Hayes JD (2006) Dimerization of substrate adaptors can facilitate cullin-mediated ubiquitylation of proteins by a "tethering" mechanism: a two-site interaction model for the Nrf2-Keap1 complex. J Biol Chem 281:24756–24768

Mello CF, Feksa L, Brusque AM, Wannmacher CM, Wajner M (1999) Chronic early leucine administration induces behavioral deficits in rats. Life Sci 65:747–755

Melnik BC (2012) Leucine signaling in the pathogenesis of type 2 diabetes and obesity. World J Diabetes 3:38–53

Morens C, Bos C, Pueyo ME, Benamouzig R, Gausseres N, Luengo C, Tomé D, Gaudichon C (2003) Increasing habitual protein intake accentuates differences in postprandial dietary nitrogen utilization between protein sources in humans. J Nutr 133:2733–2740

Oien DB, Osterhaus GL, Lundquist BL, Fowler SC, Moskovitz J (2010) Caloric restriction alleviates abnormal locomotor activity and dopamine levels in the brain of the methionine sulfoxide reductase A knockout mouse. Neurosci Lett 468:38–41

Palmer BF (2001) Impaired renal autoregulation: implications for the genesis of hypertension and hypertension-induced renal injury. Am J Med Sci 321:388–400

Palmer BF (2004) Disturbances in renal autoregulation and the susceptibility to hypertension-induced chronic kidney disease. Am J Med Sci 328:330–343

Pamplona R, Barja G (2006) Mitochondrial oxidative stress, aging and caloric restriction: the protein and methionine connection. Biochim Biophys Acta 1757:496–508

Petzke KJ, Riese C, Klaus S (2007) Short-term, increasing dietary protein and fat moderately affect energy expenditure, substrate oxidation and uncoupling protein gene expression in rats. J Nutr Biochem 18:400–407

Pichon L, Huneau JF, Fromentin G, Tome D (2006) A high-protein, high-fat, carbohydrate-free diet reduces energy intake, hepatic lipogenesis, and adiposity in rats. J Nutr 136:1256–1260

Pounis GD, Tyrovolas S, Antonopoulou M, Zeimbekis A, Anastasiou F, Bountztiouka V, Metallinos G, Gotsis E, Lioliou E, Polychronopoulos E, Lionis C, Panagiotakos DB (2010) Long-term animal-protein consumption is associated with an increased prevalence of diabetes among the elderly: The Mediterranean islands (MEDIS) study. Diabetes Metab 36:484–490

Proud CG (2007) Amino acids and mTOR signalling in anabolic function. Biochem Soc Trans 35:1187–1190

Reddy ST, Wang C-Y, Sakhaee K, Brinkley L, Pak CYC (2002) Effect of low-carbohydrate high-protein diets on acid-base balance, stone-forming propensity, and calcium metabolism. Am J Kidney Dis 40:265–274

Reiter AK, Anthony TG, Anthony JC, Jefferson LS, Kimball SR (2004) The mTOR signaling pathway mediates control of ribosomal protein mRNA translation in rat liver. Int J Biochem Cell Biol 36:2169–2179

Rodman JS, Sosa RE, Lopez MA (1996) Diagnosis and treatment of uric acid calculi. In: Coe FL, Favus MJ, Pak CY, Parks JH, Preminger GM (eds) Kidney stones: medical and surgical management. Lippincott-Raven, New York, pp 973–989

Ropelle ER, Pauli JR, Fernandes MF, Rocco SA, Marin RM, Morari J, Souza KK, Dias MM, Gomes-Marcondes MC, Gontijo JA, Franchini KG, Velloso LA, Saad MJ, Carvalheira JB (2008) A central role for neuronal AMP-activated protein kinase (AMPK) and mammalian target of rapamycin (mTOR) in high-protein diet-induced weight loss. Diabetes 57:594–605

Sandovici I, Smith NH, Nitert MD, Ackers-Johnson M, Uribe-Lewis S, Ito Y, Jones RH, Marquez VE, Cairns W, Tadayyon M, O'Neill LP, Murrell A, Ling C, Constância M, Ozanne SE (2011) Maternal diet and aging alter the epigenetic control of a promoter-enhancer interaction at the Hnf4a gene in rat pancreatic islets. Proc Natl Acad Sci U S A 108:5449–5454

Scaini G, Morais MO, Galant LS, Vuolo F, Dall'Igna DM, Pasquali MA, Ramos VM, Gelain DP, Moreira JC, Schuck PF, Ferreira GC, Soriano FG, Dal-Pizzol F, Streck EL (2014) Coadministration of branched-chain amino acids and lipopolysaccharide causes matrix metalloproteinase activation and blood-brain barrier breakdown. Mol Neurobiol 50:358–367

Schwingshackl L, Hoffmann G (2014) Comparison of high vs. normal/low protein diets on renal function in subjects without chronic kidney disease: a systematic review and meta-analysis. PLoS One 9:e97656

Tarry-Adkins JL, Chen JH, Jones RH, Smith NH, Ozanne SE (2010) Poor maternal nutrition leads to alterations in oxidative stress, antioxidant defense capacity, and markers of fibrosis in rat islets: potential underlying mechanisms for development of the diabetic phenotype in later life. FASEB J 24:2762–2771

Tentolouris N, Pavlatos S, Kokkinos A, Perrea D, Pagoni S, Katsilambros N (2008) Diet-induced thermogenesis and substrate oxidation are not different between lean and obese women after two different isocaloric meals, one rich in protein and one rich in fat. Metabolism 57:313–320

Vallianou NG, Bountziouka VP, Georgousopoulou E, Evangelopoulos AA, Bonou MS, Vogiatzakis ED, Barbetseas JD, Avgerinos PC, Panagiotakos DB (2013) Influence of protein intake from haem and non-haem animals and plant origin on inflammatory biomarkers among apparently-healthy adults in Greece. J Health Popul Nutr 31:446–454

Veldhorst M, Smeets A, Soenen S, Hochstenbach-Waelen A, Hursel R, Diepvens K, Lejeune M, Luscombe-Marsh N, Westerterp-Plantenga M (2008) Protein-induced satiety: effects and mechanisms of different proteins. Physiol Behav 94:300–307

Weir MR, Dworkin LD (1998) Antihypertensive drugs, dietary salt, and renal protection: how low should you go and with which therapy? Am J Kidney Dis 32:1–22

Westerterp-Plantenga MS, Nieuwenhuizen A, Tomé D, Soenen S, Westerterp KR (2009a) Dietary protein, weight loss, and weight maintenance. Annu Rev Nutr 29:21–41

Westerterp-Plantenga MS, Lejeune MP, Smeets AJ, Luscombe-Marsh ND (2009b) Sex differences in energy homeostatis following a diet relatively high in protein exchanged with carbohydrate, assessed in a respiration chamber in humans. Physiol Behav 97:414–419

Xu G, Kwon G, Cruz WS, Marshall CA, McDaniel ML (2001) Metabolic regulation by leucine of translation initiation through the mTOR-signaling pathway by pancreatic beta-cells. Diabetes 10:353–360

Yamamoto T, Suzuki T, Kobayashi A, Wakabayashi J, Maher J, Motohashi H, Yamamoto M (2008) Physiological significance of reactive cysteine residues of Keap1 in determining Nrf2 activity. Mol Cell Biol 28:2758–2770

Chapter 5
Effect of Soft Drink Consumption on Human Health

5.1 Introduction

The consumption of soft drinks (sodas and fruit juices) containing natural sugars and artificial sweeteners has increased throughout the world. Sodas and fruit juices are viewed by many as a major contributor to obesity and related to health problems and have consequently been targeted as a means to help curtail the rising prevalence of obesity, particularly among children. Soft drinks have been banned from schools in Britain and France, and in the United States school systems. Soft drinks contain high fructose corn syrup (HFCS) or artificial sweeteners. At present five artificial sweeteners are approved by the US Food and Drug Administration (FDA): aspartame, sucralose, saccharin, acesulfame-K, and neotame. In addition, stevia, an herb extract of intense sweetness, and erythritol, a sugar alcohol are used to a limited extent. HFCS is manufactured by hydrolyzing corn starch into glucose, which then is partly isomerized into fructose by enzymic processes. Fructose is preferred by food and soft drink manufacturers due to the fact that fructose exerts a significantly increased perception of sweetness than glucose. Some consumers and patients with type II diabetes prefer to limit their food energy intake by replacing high energy sugar or corn syrup with natural or artificial sweeteners having little or no food energy (sugar substitutes). This allows them to eat the same foods they normally consume, while allowing them to lose weight and avoid other problems associated with excessive calorie intake (Bellisle and Drewnowski 2007). Artificial sweeteners are food additive that duplicate the effect of sugar in taste, but usually have less food energy. Artificial sweeteners are many times sweeter than sugar (FDA 2011). Animal studies have shown that artificial sweeteners may increase appetite, produce weight gain, induce brain tumors, bladder cancer and many other health hazards. Some kind of health related side effects including carcinogenicity are also noted in humans. A large number of studies have been carried out on these substances with conclusions ranging from "safe under all conditions" to "unsafe at any dose". Researchers are divided in their views on the issue of artificial sweetener safety.

© Springer International Publishing Switzerland 2015
A.A. Farooqui, *High Calorie Diet and the Human Brain*,
DOI 10.1007/978-3-319-15254-7_5

Because of low cost, being sweeter than sugar, increased shelf life of fruit juices and sodas, and aggressive marketing, the use of sweeteners and HFCS containing fruit juices and sodas has increased worldwide (Bray et al. 2004). It is reported that from 1970 to 1990, the consumption of HFCS increased more than 1,000 % and currently accounts for 40 % of all added caloric sweeteners (Bray et al. 2004; Bray 2007, 2010; Ritzkalla 2010). This translates, according to one estimate, to an average of about 40 teaspoons of added sugar per person per day (Elliott et al. 2002), and many individuals consume much more than that. Since HFCS contains virtually no vitamins, minerals, or other micronutrients, its consumption decreases overall micronutrient intake by an average of almost 20 %. In addition, fructose in HFCS is energy-dense (i.e., they provide a large number of calories in a small volume) component and contain no fiber. Because it takes a relatively large number of calories from energy-dense foods to produce a feeling of fullness, excessive intake of fructose in HFCS may lead to overeating and obesity (Anderson 2007).

The tasting of sweetness is a complex physiologic process. It involves the mouth (tongue, soft palate), which contains taste buds, the sensory organs of taste (Yarmolinsky et al. 2009). In humans, gustatory information is perceived by taste receptors on the tongue, which ascend through the thalamus and eventually terminates in the anterior insula/frontal operculum and the orbitofrontal cortex (Kobayakawa et al. 1999; Small 2006). Subpopulations of sensory cells in the taste bud respond to sweet molecules by activating local sensory neurons that project to the brain areas that process and interpret sensory information (e.g., brainstem, thalamus, cerebral cortex, and amygdala) (Yarmolinsky et al. 2009). Tasting sweetness involves the activation of pleasure-generating brain circuitry. This circuitry is the same or overlaps with which mediates the addictive nature of drugs such as alcohol and opiates (Drewnowski et al. 1995). Brain dopamine (DA) reward system plays an important role in regulating the ingestion of sugars. It is well known that DA antagonists attenuate the hedonic value of sweet-tasting nutrients and pretreatment of animals with either D_1 or D_2 type of DA receptor antagonists produce weaker sugary taste for high-concentration sucrose solutions (Geary and Smith 1985; Wise 2006). Conversely, the tasting of palatable foods increases DA levels in the nucleus accumbens of the ventral striatum (Hernandez and Hoebel 1988), a brain region implicated in food reinforcement (Kelley et al. 2005).

Mechanisms of taste perception of sugar and artificial sweeteners are similar in humans and non-human mammals. HFCS and artificial sweeteners are sensed by receptors in taste buds. Sweet taste receptors T2R and T1R are coupled through G-proteins, α-gustducin and transducin, to activate phospholipase C β2 and increase intracellular calcium concentration (Fig. 5.1) (Jang et al. 2007). Sweet-taste receptors, including the T1R family and α-gustducin, respond not only to caloric sugars such as HFCS or sucrose but also to artificial sweeteners (sucralose, aspartame, and acesulfame-K), and sweet proteins such as thaumatin and monellin (Mace et al. 2007). The T1R2+T1R3 heterodimer senses sweet taste (glucose, fructose, and sucrose) at concentrations in the 100 mM range and the artificial sweeteners (acesulfame potassium, sucralose, and saccharin) at much lesser concentrations (1–10 mM). The T1R1+T1R3 heterodimer senses the "bitter" amino acids, such as L-aspartate

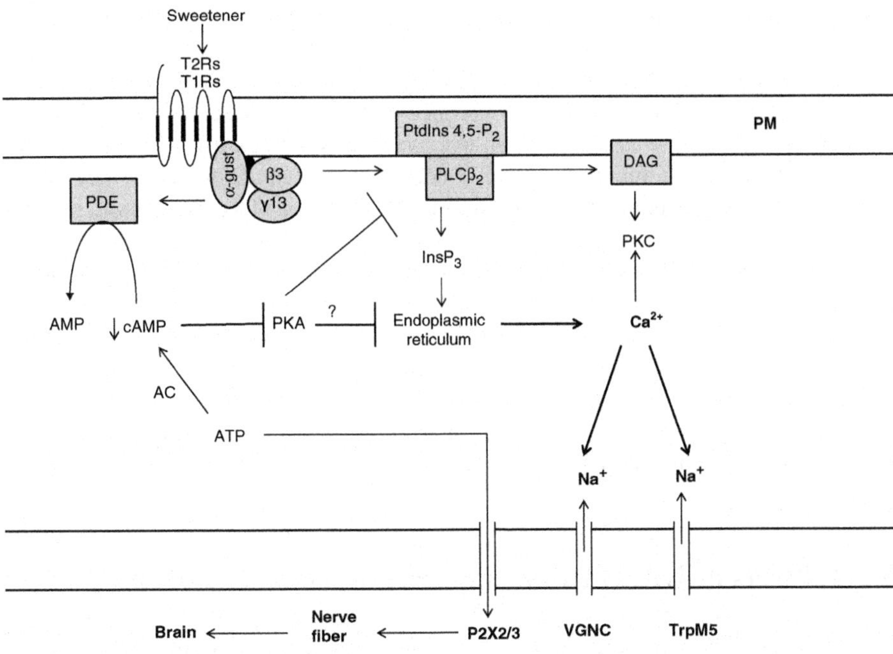

Fig. 5.1 Hypothetical diagram showing interactions between sweetener and taste receptor and downstream signaling pathway. *Gβγ* G-protein, *PLCβ2* phospholipase Cβ2, *InsP3* inositol 1,4,5-trisphosphate, *PtdInd 4,5-P₂* phosphatidylinositol 4,5-bisphosphate, *DAG* diacylglycerol, *PKA* protein kinase A, *PKC* protein kinase C, *TrpM5* Ca²⁺ dependent transient receptor potential channel5, *VGNC* depolarization, activation of voltage-gated Na⁺ channels, *P2X2/3* release of ATP through pannexin-1 hemichannels, *cAMP* cyclic AMP, *PDE* phosphodiesterase, which subsequently prevents phosphorylation and desensitization of Ca²⁺ signaling effectors

and L-glutamate. It is also reported that simple sugars and artificial sweeteners act synergistically through a T1R2 + T1R3-α-gustducin-PLC βII pathway to stimulate activation of PKC BII, which appears to be essential for apical translocation of GLUT2 (Mace et al. 2007; Jang et al. 2007). The binding of sweetener with receptor results in increase in phospholipase CβII, which increases production of the second messengers, inositol trisphosphate and diacylglycerol. This in turn leads to activation of the taste-transduction channel, TRPM5 (transient receptor potential cation channel subfamily M member 5, also known as long transient receptor potential channel 5), resulting in increased intracellular calcium and neurotransmitter release (Chandrashekar et al. 2006). These events lead to membrane depolarization, action potentials, and release of ATP as a transmitter to activate gustatory afferents. The Gα subunit, α-gustducin, activates a phosphodiesterase to decrease intracellular cAMP levels, although the precise targets of cAMP have not been identified (Kinnamon 2012). It is becoming increasingly evident that artificial sweeteners do not activate the food reward pathways in the same fashion as natural sweeteners. Lack of caloric contribution generally eliminates the post-ingestive component. Studies on functional magnetic imaging in normal weight men indicate that glucose ingestion may lead to a prolonged signal depression in the hypothalamus. This response is not observed with sucralose ingestion (Smeets et al. 2005). Natural and artificial sweeteners also activate the gustatory branch differently. As stated above, the sweet taste receptors are heterodimeric G protein coupled transmembrane receptors, which contain several ligand-binding sites. Sugar and sweeteners bind to different sites of these receptors inducing different intensity of sweetness (Cui et al. 2006). Thus, the binding of sucrose to taste receptors produces significantly greater global activation, particularly in the hunger state, than other tastes whether they were sweet (the artificial sweetener saccharin), bitter (caffeine) or another taste quality. Artificial sweeteners do not have nutritive value and, therefore, the brain may respond to sweeteners less than glucose or sucrose because the taste of sweeteners does not reward with calories (Haase et al. 2009). It is generally assumed that inclusion of artificial sweeteners with few calories does not alter glucose metabolism. However, recent animal studies have indicated that the consumption of HFCS or artificial sweeteners markedly affects carbohydrate and lipid metabolism in visceral organs and brain (Yarmolinsky et al. 2009).

5.2 Effects of HFCS Intake on Visceral Organs and Brain

As stated in Chap. 3, fructose in HFCS exerts adverse metabolic effects in the body compare to glucose. Fructose is preferentially metabolized to lipid in the liver, leading to increased hepatic de novo lipogenesis, increase in triacylglycerol, low HDL cholesterol, small, dense LDL, atherogenic dyslipidemia, and insulin resistance in rodents and other animals (Bray 2007). Growing evidence also indicates that a high-fructose diet and sodas containing HFCS produces insulin resistance not only by significantly reducing the protein expression of insulin receptor, insulin receptor substrate-1, but also by altering expression of Akt and GLUT4. Fructose feeding also produces ventricular dilatation, ventricular hypertrophy, decrease in ventricular contractile function, infiltration of inflammatory cells in heart and hepatic steatosis (Patel et al. 2009; Panchal and Brown 2011). In the liver, fructose feeding causes both microvesicular and macrovesicular steatosis with periportal fibrosis and lobular inflammation (Kawasaki et al. 2009). Collectively, these studies suggest that at the molecular level, the consumption of HFCS reduces circulating insulin and leptin levels (Teff et al. 2004) contributing to increase in body weight. Thus, fructose intake may not result in the degree of satiety that normally ensues with an equally caloric meal of glucose or sucrose. Secondly, fructose-mediated increase in triacylglycerols effectively reduces the amount of leptin crossing blood–brain barrier. These processes may interfere with the communication between leptin and hypothalamus. Lastly, fructose does not suppress ghrein (hunger hormone) leading to overeating. Consequently, the consumption of fructose-enriched drinks in rodents has been used as a model for investigating the development and pathogenesis of the type II diabetes and metabolic syndrome (MetS), a complex disorder with dyslipidemia, hypertension, impaired glycaemic control, and insulin resistance (Rayssiguier et al. 2006; Després et al. 2008; Stanhope et al. 2009; Panchal and Brown 2011).

Onset of type II diabetes and MetS not only increases the risk of developing cardiovascular (heart disease), but also increases of the chances of developing cerebrovascular diseases (stroke, Alzheimer disease (AD), and depression). Thus, the incidences of cerebrovascular diseases increase many folds in type II diabetes and MetS patients with cardiovascular diseases. The incidences of stroke are two- to fourfold higher in patients with MetS and cardiovascular diseases compared to normal subjects of the same age. Similarly, patients with MetS have a two- to threefold increased risk for developing dementia and AD. Metabolic syndrome doubles the risk of depression. The molecular mechanism underlying the mirror relationship between MetS and neurological disorders is not fully understood. However, biochemical alterations observed in MetS like induction of chronic inflammation and oxidative stress, impairment of endothelial cell function, induction of insulin and leptin resistance, hyperglycemia-related increase in advanced glycation end-products, and micro-vascular injury may represent a pathological bridge between metabolic syndrome and neurological disorders (Farooqui et al. 2012; Farooqui 2013).

As stated in Chap. 3, the fundamental problem with consumption of HFCS containing drinks is the conversion of fructose into long chain fatty acids (FFA) and generations of reactive carbonyls (RCS) and oxygen species (ROS) (Semchyshyn 2013). Increase in storage of FFA not only occurs in liver and white adipose tissue (WAT) but also skeletal muscle. FFA storage in white adipose tissue, or WAT, and myocellular fat not only contributes to inflammation, but also paradoxically releases more FFA into circulation and impairs normal uptake of glucose following consumption of meals and insulin secretion. In the liver, the accumulation of FFA results in hepatic insulin resistance in which more FFA are synthesized by de novo lipogenesis along with excess production of hepatic glucose and release into the circulation (Leclercq et al. 2007; Bocarsly et al. 2010). Long-term consumption of excessive HFCS (fructose) leads to glycoxidation, generation of ROS and RCS, and accumulation of damaged cellular constituents involved in type II diabetes, metabolic syndrome, and age-related disorders (Semchyshyn 2013).

Like liver, fructose also produces damaging effects in the brain (Mielke et al. 2006; Ross et al. 2009). It is shown that feeding 60 % fructose for 6 weeks to male Syrian hamster produces hippocampal insulin resistance. This observation is particularly significant given that the hippocampus is integral to many forms of learning and memory (Ergorul and Eichenbaum 2004) and that converging lines of evidence indicate that neural insulin signaling facilitates hippocampal-dependent memory (Park 2001) and activity-dependent synaptic plasticity, which is associated with the persistent alterations in excitatory synaptic strength. Several lines of evidence indicate that leptin also has the ability to modulate hippocampal synaptic plasticity. Indeed, leptin insensitive obese rodents (*fa/fa* rats and *db/db* mice) display deficits in hippocampal long term potentiation (LTP) and long-term depression (LTD) (Li et al. 2002; Harvey 2007). High fructose intake does not influence navigational ability. In addition, the fructose fed animals were able to learn and retain the location of the platform for short periods of time. Only impairments on the retention tests have recorded associated with high fructose intake (given 48 h after training), which suggests that the consumption of such diet may impair long-term memory storage or retrieval, or both (Stranahan et al. 2008). Most importantly, direct infusions of insulin into the hippocampus enhance performance in a variety of memory tasks, and the memory-enhancing effects of hippocampal insulin administration are not observed in diabetic rats (Moosavi et al. 2006; Ross et al. 2009). It is also reported that diet enriched with n-3 FFA prevents the detrimental effects of fructose on cognition not only under "normal" conditions, but also during metabolic stress (Agrawal and Gomez-Pinilla 2012). In animal experiments, the n-3 FFA supplementation also results in normalizing the phosphorylation of CREB and Synapsin I and Synaptophysin even in the presence of fructose, suggesting that n-3 FFA can restore the cognitive function under challenging conditions by normalizing the action of insulin resistance on synaptic plasticity (Simopoulos 2013). Based on many biochemical studies, it is postulated that increased consumption of HFCS containing soft drinks may be closely associated with the stimulation of lipogenesis, high plasma level of triacylglycerols (TAGs), obesity, oxidative stress, insulin resistance and metabolic syndrome leading to cardiovascular disease (CVD) and neurological disorders (Bray et al. 2004; Farooqui 2013), and n-3 FFA retard these effects.

As stated in Chap. 3, most fructose in HFCS is metabolized by fructokinase (ketohexokinase), which phosphorylates fructose to fructose 1-phosphate (Fig. 5.2). Unlike glucose, whose phosphorylation is tightly regulated by phosphofructokinase so that ATP levels are never depleted; the phosphorylation of fructose results in a

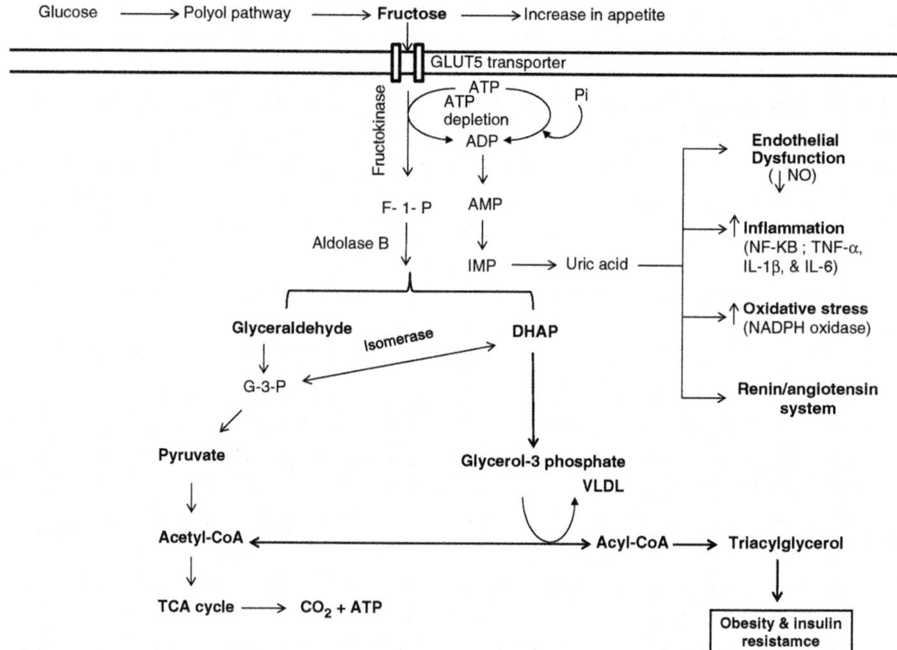

Fig. 5.2 Metabolism of fructose and generation of uric acid in the liver. *F-1-P* fructose-1-phosphate, *G-3-P* glyceraldehyde-3- phosphate, *DHAP* dihydroxyacetone phosphate, *NO* nitric oxide, *TNF-α* tumor necrosis factor-α, *IL-1β* interleukin-1beta, *VLDL* very low density protein

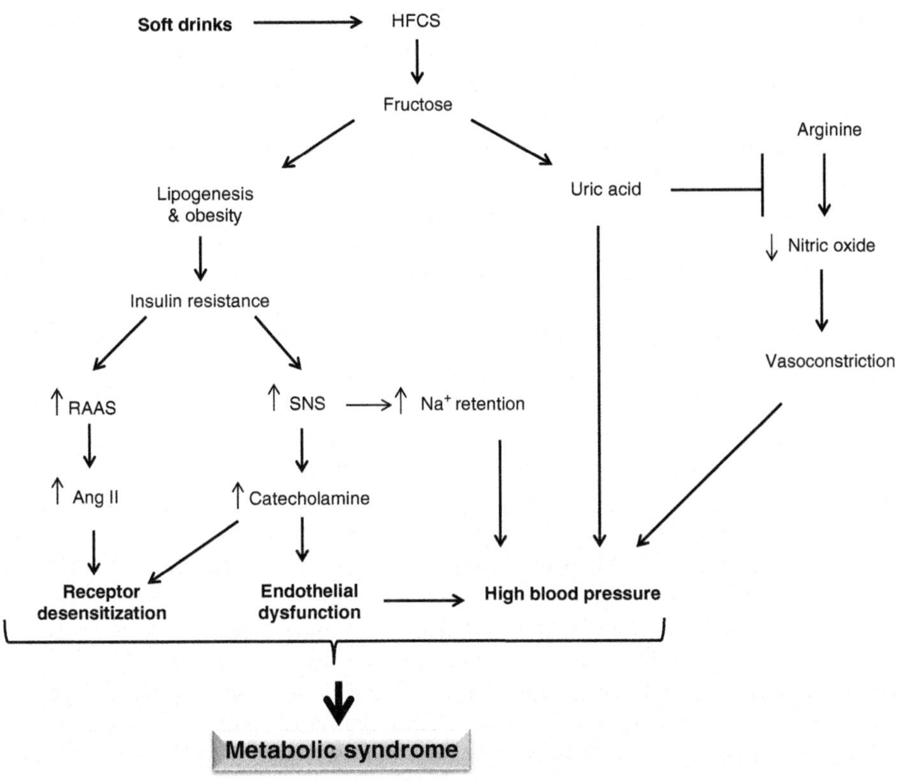

Fig. 5.3 Generation of uric acid and induction of lipogenesis from fructose consumption. *RAS* renin-angiotensin system, *Ang II* angiotensin II, *SNS* sympathetic nervous system

decrease in intracellular phosphate and ATP depletion, resulting in transient inhibition of protein synthesis. Activation of AMP deaminase generates inosine monophosphate (IMP) and eventually high levels of uric acid, a metabolite, which increases blood pressure through the inhibition of nitric oxide synthase, but also modulates several other processes (Fig. 5.3). These processes include increase in sympathetic nervous system activity (Farah et al. 2006; Verma et al. 1999), insulin resistance, increase in oxidative stress, elevation in circulating catecholamines (Tran et al. 2009), decrease in nitric oxide (NO) bioavailability, enhancement in renin-angiotensin aldosterone system activity (RAAS), and angiotensin II (Ang II), levels (a potent vasoconstrictor) (Tran et al. 2009; Wright et al. 2013) (Fig. 5.3). The brain RAAS system contains several functional components to produce the active ligands angiotensins II (AngII), angiotensin III, angiotensins (IV), angiotensin (1–7), and angiotensin (3–7). These ligands interact with several receptor proteins including AT_1, AT_2, AT_4, and Mas, which are distributed within the central and peripheral nervous systems (Wright and Harding 2010). Alterations in brain RAAS have been reported to contribute to dementia and blockade of RAAS system may delay/retard the onset of dementia, a pathological condition that is characterized by a decrease in intelligence, memory, and perception and may be caused by various diseases. Logical and critical thinking, judgement, retentive memory, and short-term memory are impaired, while remote memory (long-term memory) can remain for a long time. In addition, personality may deteriorate (Iwanami et al. 2009; Wright and Harding 2010). The precise molecular mechanism of RAAS-mediated processes and their role are not fully understood. However, it is becoming increasingly evident that octapeptide AngII disrupts learning and memory; while the hexapeptide angiotensin IV (AngIV) facilitates memory acquisition and consolidation (Wright et al. 2013). Collective evidence suggests that HFCS containing soft drinks increase fasting insulin and induce insulin resistance. Insulin resistance is known to contribute to chronic inflammation and oxidative stress, leading to irreversible protein aggregation and neuronal degeneration (Salkovic-Petrisic and Hoyer 2007; Tinahones et al. 2009). It is well known that ROS are removed by antioxidant defense system that works as a complex team. The three oxidative damage markers: ROS, TBARS and protein carbonyl are significantly increased in the brains of fructose-drinking insulin resistance rats compared with control rats (Yin et al. 2013). In antioxidant defense system, SOD converts superoxide to hydrogen peroxide, GPx and CAT convert hydrogen peroxide to water, and they are recognized as primary antioxidant enzymes, which decrease oxidative stress (Kumar and Kumar 2008). Activities of SOD, GPx, and CAT are significantly decreased in the brains of fructose-drinking insulin resistance rats supporting the view that antioxidant systems in brains of fructose-drinking insulin resistance rats convert hydrogen peroxide to water less efficiently causing the accumulation of hydrogen peroxide. The excessive presence of hydrogen peroxide induces cytotoxicity in the brain. Moreover, fructose-drinking insulin resistance rats also show significant lower levels of GSH and total antioxidant capacity in brains (Yin et al. 2013).

5.3 Effects of Aspartame Intake on Visceral Organs and Brain

Aspartame (NutraSweet or Equal) is a methyl ester of a dipeptide used as a synthetic nonnutritive sweetener in over 90 countries worldwide in over 6,000 products. Aspartame (L-aspartyl-L-phenylalanyl-methyl ester) is about 200 times sweeter than sucrose. Aspartame consists of two amino acids, phenylalanine and aspartate, linked to a methanol backbone (Fig. 5.4) (Tandel 2011). Due to the small amount ingested at a time, its caloric contribution is negligible. Many short-term animal studies have indicated that the consumption of aspartame is safe. However, a recent large and long term study in rats has indicated that aspartame increases the risk of developing lymphomas, leukemias, and transitional cell carcinomas of the pelvis, ureter, and bladder in a dose-dependent manner within ranges that are considered to be safe for human consumption (doses as low as 20 mg/kg body weight) (Soffritti et al. 2006). Short term studies on aspartame consumption in humans produce on unhealthy effects. So, large long term studies in humans are needed on the safety of aspartame consumption.

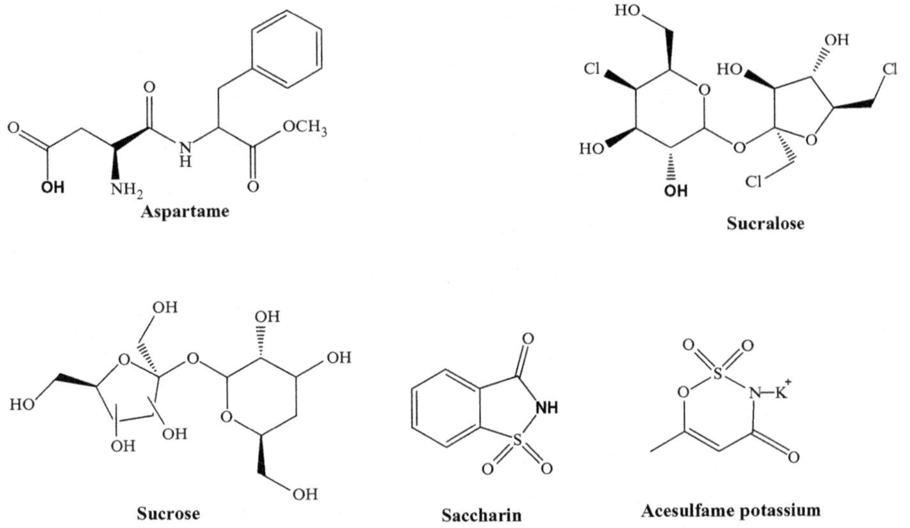

Fig. 5.4 Chemical structures of sucrose and artificial sweeteners

Studies on the effect of aspartame on neural cells indicate that astrocytes directly or indirectly contribute to the harmful effects of aspartame metabolites on neurons. The artificial sweetener is degraded into phenylalanine (50 %), aspartic acid (40 %) and methanol (10 %) during metabolism in the body. Phenylalanine and aspartic acid directly impact brain and central nervous system functions. Evidence shows that these amino acids play a role in mood disorders, memory problems and other neurological illnesses. Aspartic acid may cause excitotoxicity and neuronal damage (Finkelstein et al. 1983). In mammalian tissues, methanol is considered a poison because the enzyme alcohol dehydrogenase (ADH) converts methanol into toxic formaldehyde. Using a genome-wide analysis of the mouse brain, it is shown that an increase in blood methanol concentration led to the accumulation of mRNAs from genes primarily involved in detoxification processes and regulation of the alcohol/aldehyde dehydrogenases gene cluster (Komarova et al. 2014). However, manufacturers of aspartame claim that methanol and its byproducts are quickly excreted. Recent studies have indicated that measurable amounts of formaldehyde in the liver, kidney and brain of test subjects after ingestion of aspartame (Aune 2012). Phenylalanine is an amino acid, which is metabolized to tyrosine by the hepatic enzyme phenylalanine hydroxylase. Both phenylalanine and tyrosine are transported by neutral amino acid transporter through the blood–brain barrier. These amino acids are intimately involved in the production of several key neurotransmitters such as dopamine, norepinephrine and serotonin. The excess of phenylalanine blocks the transport of important amino acids to the brain contributing to the reduced levels of dopamine and serotonin (Rycerz and Jaworska-Adamu 2013), leading to chemical imbalances that cause depression, mood and emotional disorders, anxiety, insomnia, headaches, and memory loss (Fig. 5.5). Hyperphenylalaninemic rodents have lower brain weights (Burri et al. 1990) together with impaired myelinogenesis (Reynolds et al. 1992).

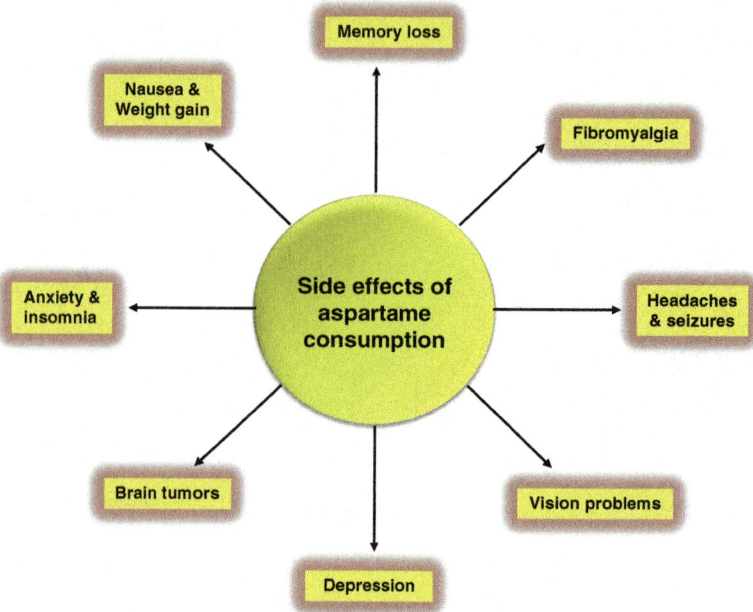

Fig. 5.5 Side effects of long term consumption of aspartame in animals

Phenylalanine is especially dangerous for people with the hereditary disease, phenylketonuria (PKU), an autosomal recessive metabolic genetic disorder characterized by a mutation in the gene for the hepatic enzyme phenylalanine hydroxylase (PAH), rendering it nonfunctional (Adler-Abramovich et al. 2012). This enzyme metabolizes phenylalanine to tyrosine. When PAH activity is reduced, phenylalanine accumulates and is converted into phenylpyruvate (also known as phenylketone), which can be detected in the urine (Gonzalez and Willis Monte 2010). Following early diagnosis, PKU patients are maintained at a strict diet for a normal life span.

As stated above, aspartic acid at high concentrations is a neurotoxin that causes hyperexcitability of neurons. It is also a precursor of other excitatory amino acid such as glutamates. The excess of glutamate and lack of astrocytic uptake induces excitotoxicity and leads to the degeneration of astrocytes and neurons. Methanol, which forms 10 % of the degraded product, is converted in the body to formate, which can either be excreted or can give rise to formaldehyde, diketopiperazine (a carcinogen) and a number of other highly toxic derivatives. It is reported that aspartame may cause leukemias/lymphomas and mammary (breast) cancer (Zwillich 2007). Diketopiperazine contributes to the formation of tumors in the brain such as gliomas, medulloblastomas and meningiomas. Glial cells are the main source of tumors, which can be caused by the sweetener in the brain. It has been proposed that consumption of aspartame may cause neurological and behavioral disturbances in sensitive individuals. Headaches, insomnia, seizures, behavioral and cognitive alterations,

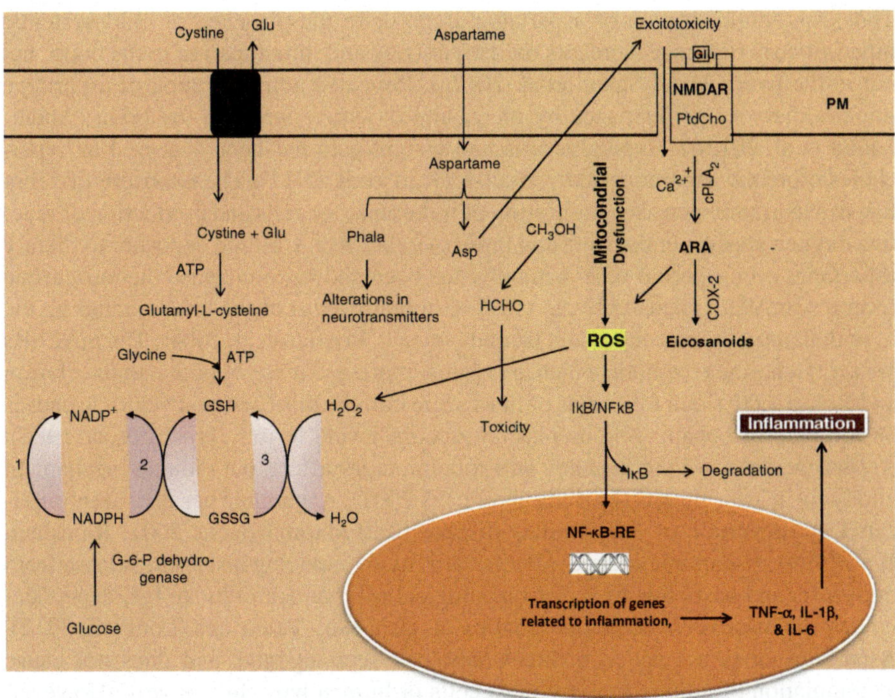

Fig. 5.6 Metabolism of aspartame and potential mechanism associated with side effects of aspartame. *NMDA-R* N-Methyl-D-aspartate receptor, *PtdCho* phosphatidylcholine, *cPLA₂* cytosolic phospholipase A₂, *ARA* arachidonic acid, *COX-2* cyclooxygenase-2, *Phala* phenylalanine, *Asp* aspartic acid, *CH₃OH* methanol, *HCHO* formaldehyde, *ROS* reactive oxygen species, *NF-κB* nuclear factor kappa-B, *NF-κB-RE* nuclear factor kappaB response element, *IκB* inhibitory subunit of NFκB, *TNF-α* tumor necrosis factor-α, *IL-1β* interleukin-1 beta, *IL-6* interleukin-6, *H₂O₂* hydrogen peroxide, *GSH* reduced glutathione, *GSSG* oxidized glutathione

as well as allergic reactions are also some of the neurological effects that have been encountered by the consumption of aspartame in sensitive subpopulations. Most of these changes can be accredited to changes in regional brain concentrations of neurotransmitters (Fig. 5.5) (Schiffman et al. 1987; Humphries et al. 2008).

The administration of aspartame increases brain lipid peroxidation, decreases reduced glutathione, and increases TNF-α. Aspartame also produces dose-dependent inhibition of brain serotonin, noradrenaline, and dopamine (Abdel-Salam et al. 2012a) (Fig. 5.6). However, aspartame has no effect on lipid peroxidation, nitrite, glutathione (GSH), aspartate aminotransferase (AST), Alanine aminotransferase (ALT), and alkaline phosphatase (ALP) in the liver. In contrast, the administration of LPS increases levels of nitrite in the brain as well as liver. LPS also decreases GSH in brain and liver, increases brain TNF-α, and glucose, and causes marked increase in brain monoamines. Administration of aspartame to LPS-treated mice increases lipid peroxidation, and levels of nitrite but mitigates the increase in monoamines. Aspartame does not alter liver TBARS, nitrite, GSH, ALT, AST, or ALP.

Thus, the administration of aspartame alone or in the presence of mild systemic inflammatory response increases oxidative stress and inflammation in the brain, but not in the liver (Abdel-Salam et al. 2012a). Repeated administration of aspartame impairs memory performance by increasing oxidative stress in the brain (Abdel-Salam et al. 2012b). Hyperglycemia and weight gain has been observed in hyper-cholesterolemic aspartame-fed zebrafish (Kim et al. 2011). The exposure of zebra fish to aspartame increases infiltration of inflammatory cells and production of reactive oxygen species in the liver and brain of zebra fish. Chronic exposure to dietary aspartame over a period of 3–4 months has been shown to increase the muscarinic receptor (mAChR) density by up to 80 % in many areas of the brain, including the hypothalamus, hippocampus and frontal cortex (Christian et al. 2004). The mAChRs are acetylcholine receptors, which are highly expressed in the hypothalamus (Quiron and Boksa 1986), and injections of muscarine into the third cerebral ventricle causes an increase in hepatic venous plasma glucose levels in rats (Iguchi et al. 1986). Aspartame does not produce any anti-inflammatory effect, but reduces mechanical allodynia in arthritic rats (LaBuda and Fuchs 2001). Aspartame has also been implicated in alleviation of autoimmune disease (LaBuda and Fuchs 2001; Ramsland et al. 1999). Aspartame induces allergic activity by inducing histamine release from mast cells and basophils by a pharmacological mechanism or by an IgE-dependent process of mast cell activation (Szucs et al. 1986; Veien and Lomholt 2012). Aspartame does not act as a direct mast cell secretagogue, and does not cause degranulation of cultured mouse mast cells or human basophils *in vitro*. However, aspartame does decrease antigen-induced histamine release from cultured mouse mast cells after long-term exposure only. It is proposed that the presence of aspartame makes mast cells less responsive to anaphylactic stimulation.

5.4   Effects of Saccharin Intake on Visceral Organs and Brain

Saccharin (ortho-sulfobenzoic Acid Imide) is an artificial sweetener, which is 300 times as sweet as sucrose by weight (Fig. 5.4). It is non-cariogenic and noncaloric but has a slightly bitter taste. It is available as a tablet, powder, or liquid form and is widely used in food products including diet sodas, pharmaceuticals, and cosmetics. It is not metabolized by the digestive tract. Although saccharin noncaloric in nature, some studies report that it may trigger the release of insulin in humans and rats, presumably as a result of its taste. Many studies have been published about the effect of saccharin in laboratory rats (Jacobson et al. 1998). Approximately 20 study groups analyzed the effect of saccharin in one generation of rats, which have been exposed to high doses of saccharin for at least 1.5 years. Saccharin at a dose of 7.5 % causes bladder cancer in 30 % of all rats (Taylor et al. 1980; Squire 1985). Because of these results, saccharin was prohibited in Canada. In the USA, since 1981, saccharin-containing products had to be labeled with a warning that saccharin can cause cancer in laboratory animals.

5.5 Effects of Sodium Cyclamate on Visceral Organs and Brain

Sodium cyclamate is an artificial low calorie sweetener, which is 30–50 times sweeter than sugar (depending on concentration) (Fig. 5.7). Health hazards associated with sodium Cyclamate includes irritation of the skin, eyes, mucous membranes and respiratory tract; diarrhea, photosensitization and birth defects like Down Syndrome and behavioral changes in the offsprings of pregnant women exposed to it. Cyclamate is transformed into cyclohexylamine, which has been reported to be rather toxic (Renwick 1986). Studies in rats and dogs have indicated that cyclohexylamine causes testicular atrophy with impairment of spermatogenesis (James et al. 1981). Long-term toxicity study with cyclamate in non-human primates indicates that animals consuming cyclamate show malignancies (Takayama et al. 2000).

5.6 Effects of Neotame on Visceral Organs and Brain

Neotame, a relatively recent approved noncaloric sweetener, is a chemical derivative of aspartame (Fig. 5.7). Neotame is manufactured by combining aspartame with 3,3-dimethylbutyraldehyde, which was added to block enzymes that break

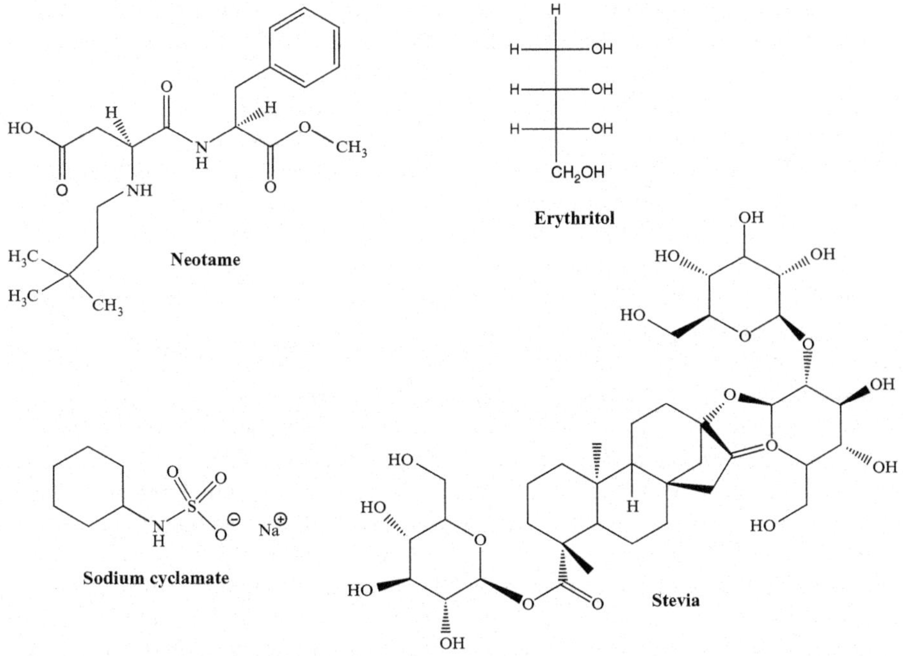

Fig. 5.7 Chemical structures of sweeteners

the peptide bond between aspartic acid and phenylalanine, thereby reducing the availability of phenylalanine. 3,3-Dimethylbutyraldehyde is a highly flammable and an irritant, which causes irritation in eyes, skin, and lungs. It received FDA approval in 2002 for use as a general purpose sweetener in selected food products (except not in meat and poultry) and flavor enhancer. Neotame is an intense non-nutritive sweetener that is not fermentable by the oral microbiota and possesses a crisp, clean taste with no detectable aftertaste. It is reported to be greater than 7,000–13,000 times more potent than sucrose on a weight basis depending on the food product and how it is prepared (Walters et al. 2000; Prakash et al. 2001). Unlike aspartame, it is safe for consumption by people with phenylketonuria. It is also heat stable in baking applications and can be safely used by diabetics and pregnant women. Neotame is stable in carbonated soft drinks, powdered soft drinks, yellow cake, yogurt, and hot-packed still drinks (Witt 1999).

5.7   Effects of Acesulfame Potassium on Visceral Organs and Brain

Acesulfame potassium is a calorie-free, non-cariogenic, and nonnutritive artificial sweetener, which is approximately 200 times sweeter than sucrose (Fig. 5.4). Like saccharin, acesulfame potassium, has a slightly bitter aftertaste and is often blended with other sweeteners to mask this property. Acesulfame potassium is stable at high cooking/baking temperatures, even under moderately acidic or basic conditions, which permits it to be used in baking or in products requiring a long shelf life. As stated above, in mammals, the sweetness is detected by the sweet taste receptors, which are coupled through G-proteins, α-gustducin and transducin, to activate phospholipase C $\beta2$ and increase intracellular calcium concentration (Jang et al. 2007; Zhao et al. 2003). The taste preference for acesulfame potassium in rodents is mainly dependent upon T1R3 subunit expression. Recent studies have indicated that the T1R3 subunit is not only expressed in peripheral tissues such as the tongue, gut and pancreas (Nakagawa et al. 2009), but also in the brain including the hypothalamus, cortex and hippocampus (Martin et al. 2010; Ren et al. 2009). It is suggested that the presence of CNS T1R3 may provide a possible mechanism by which acesulfame potassium may affect the brain, in addition to its somatic actions. Studies on the effect of acesulfame potassium indicate that consumption of this sweetener (40 weeks) by normal C57BL/6J mice results in a moderate and limited influence on metabolic homeostasis, including altering fasting insulin and leptin levels, pancreatic islet size and lipid levels, without affecting insulin sensitivity and bodyweight (Cong et al. 2013). C57BL/6J mice consuming acesulfame potassium show impaired cognitive memory functions (as evaluated by Morris Water Maze and Novel Objective Preference tests), while no differences in motor function and anxiety levels are observed. It is proposed that the generation of an acesulfame potassium-induced neurological symptoms may be associated with metabolic dysregulation (glycolysis inhibition and functional ATP depletion) and neurosynaptic abnormalities (dysregulation of TrkB-mediated BDNF and Akt/Erk-mediated cell growth/survival pathway) in hippocampal neurons. Collectively, these studies suggest that chronic use of acesulfame potassium may impair cognitive functions, potentially via altering neuro-metabolic functions in male C57BL/6J mice (Cong et al. 2013).

5.8 Effects of Sucralose on Visceral Organs and Brain

Sucralose (Splenda) is a nonnutritive, zero-calorie artificial sweetener. It is a trichlorinated disaccharide with the chemical name 1,6-dichloro-1,6-dideoxy-β-D-fructofuranosyl-4-chloro-4-deoxy-α-D-galactopyranoside. Sucralose is about 600 times as sweet as sucrose. It is produced from sucrose by the selective substitution of three of the sucrose hydroxyl groups with chlorine atoms, involving inverting the configuration at the 4-position from the gluco- to the galacto-analogue (Fig. 5.4). It is consumed in the form of power and tablets (Blendy) by diabetic and obese patients. Due to its unique sugar-like taste sucralose is used in virtually every type of food including beverages, frozen desserts, chewing gum, baked goods, and other foods. Unlike other artificial sweeteners, it is stable when heated (119 °C) and can, therefore, be used in baked and fried foods (Schiffman and Gatlin 1993; Schiffman et al. 2008). Recent studies indicate that cooking with sucralose at high temperatures (550 °C) generates chloropropanols (3-monochloropropanediol and 1,2- and 1,3-dichloropropanols), a potentially toxic class of compounds (Rahn and Yaylayan 2010). These sucralose metabolites are known to cause cancer. Although, the molecular mechanism of thermal decomposition of chloropropanol production is not known, but it is suggested that generation of hydrogen chloride from sucralose contributes to the synthesis of chloropropanols. Another important concern is that sucralose is a class of chemicals called organic chlorides, some types of which are carcinogenic. In addition, some studies have indicated that sucralose cause dizziness, head and muscle aches, stomach cramps, diarrhea, chronic inflammation and bladder issues (Fig. 5.8) (Bigal and Krymchantowski 2006; Grotz 2008). A 3-month study of 128 people with type II diabetes, in which sucralose is consumed at a dose approximately three times the maximum estimated daily intake, show no adverse effects on any measure of blood glucose control (Grotz et al. 2003). However, prolonged consumption of sucralose and other high-intensity sweeteners may produce harmful deficits in energy metabolism and balance (Swithers et al. 2013; Simon et al. 2013). These deficits in regulating energy balance may be due to the disruption of a learned signaling relationship between sweet tastes and energetic outcomes (Davidson et al. 2011), supporting the view that consuming noncaloric sweeteners may promote excessive intake and body weight gain by weakening a predictive relationship between sweet taste and the caloric consequences of eating.

As stated above, sucrose activates dopamine neurons in a region of the brain called the striatum, and the resulting release of dopamine is associated with pleasure (Velloso et al. 2008). In contrast, sucralose does not produce this effect. However, like sucrose, sucralose binds with T1R2 and T1R3 sweet taste receptors on the

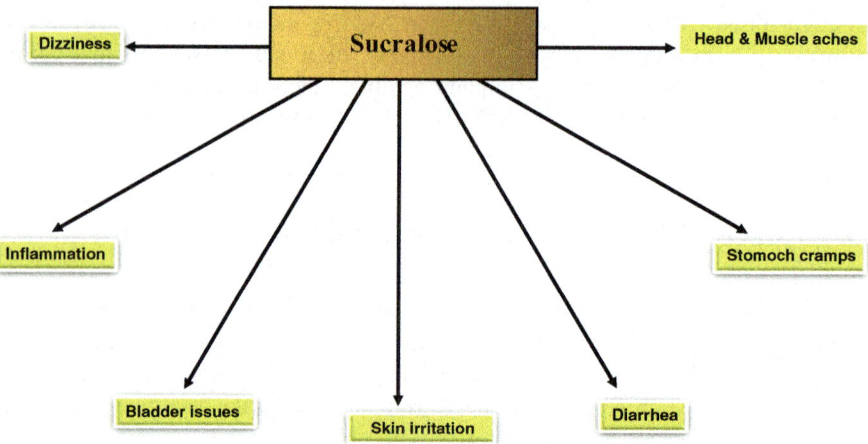

Fig. 5.8 Side effects of sucralose

mouth and interacts with chemosensors in the alimentary tract that play a role in sweet taste sensation and hormone secretion (cholecystokinin, peptide tyrosine tyrosine, neurotensin, glucagon-like peptide-1, glucagon-like peptide-2, and glucose-dependent insulinotropic peptide). Stimulation of T1R2 and T1R3 sweet taste receptors by diet soda (sucralose) as well as glucose results in increased secretion of glucagon-like peptide-1 (GLP-1) in humans and rodents. This increase in GLP-1 secretion is inhibited using the sweet taste receptor antagonist lactisole (Steinert et al. 2011). The observed increase in GLP-1 secretion from diet soda likely requires the presence of a caloric, metabolizable sugar as well because non-nutritive sweeteners alone do not appear to alter gut hormone secretion in vivo (Ma et al. 2009; Fujita et al. 2009). Several groups have demonstrated the presence of additional glucose-sensing molecules in both lingual and enteroendocrine cells that express taste receptors, including components of ATP-sensitive K^+ channels (SUR1 and Kir 6.1), GLUTs, sodium-glucose linked transporter 1, and glucokinase (Yee et al. 2011; Reimann et al. 2008). These glucose sensors may serve as a primary signal for GLP-1 secretion, with stimulation of sweet taste receptors (by caloric or nonnutritive sweeteners) serving as a secondary signal. It is also reported that in rats, sucralose ingestion increases the expression of the efflux transporter P-glycoprotein (P-gp) and two cytochrome P-450 (CYP) isozymes in the intestine. P-gp and CYP are key components of the presystemic detoxification system involved in first-pass drug metabolism (Abou-Donia et al. 2008; Schiffman and Rother 2013). Sucralose induces changes in the microbial composition in the gastrointestinal tract (GIT), with relatively greater reduction in beneficial bacteria. Early studies have claimed that sucralose is not metabolized by the GIT, but subsequent analysis indicates that some of sucralose is metabolized in the GIT, as indicated by multiple peaks found in thin-layer radiochromatographic profiles of methanolic fecal extracts after oral sucralose administration. It is also reported that both human and rodent sucralose alters glucose, insulin, and GLP-1 levels.

5.9 Effects of Polyols on Visceral Organs and Brain

Polyols are hydrogenated carbohydrates used as sweeteners. Biochemically, polyol are non-cariogenic, low-glycaemic, low-energy and low-insulinemic, low-digestible, osmotic carbohydrates. Common polyol, which have sugary taste include sorbitol, tagatose, maltitol, and xylitol. In comparison to other sugar alcohols used currently as sweeteners, erythritol has a much lower energy value ($\sim$0.2 kcal g^{-1}) than sucrose (4 kcal g^{-1}) and other sugar alcohols ($\sim$2.4 kcal g^{-1}).

Erythritol is a four-carbon polyhydric alcohol present in some fruits such as pears, melons, and grapes and fermented foods (Fig. 5.7). It is 60–70 % as sweet as table sugar. Its consumption has no affect on blood sugar (Goossens and Röper 1994). It is noncarcinogenic, noncaloric and does not cause tooth decay. It is only partially absorbed by the body and then excreted in urine and feces (Munro et al. 1998). The metabolic and safety studies in animals and humans have indicated that erythritol well tolerated and elicits no toxicological effects even at high doses. It is almost completely absorbed, not metabolized systemically and is excreted unchanged in faeces and urine. A comparison of the human and animal data indicated a high degree of similarity in the metabolism of erythritol and this finding supports the use of the animal species used to evaluate the safety of erythritol for human consumption. In the human body, erythritol does not change the insulin level; thus, it is also safe for diabetics. It can be concluded that erythritol did not produce evidence of toxicity (Munro et al. 1998; Chung and Lee 2013). Unlike other sugar alcohol erythritol is not metabolized by the digestive pathway (Hino et al. 2000; Arrigoni et al. 2005). This has resulted in requirement of labeling by the U.S. Food and Drug Administration (FDA/CFSAN 2001). Nowadays, consumption of too much fat and carbohydrates has resulted in a higher body weight and consequently many diseases such as type II diabetes, cardiovascular disease, and neurological disorders (Farooqui 2013). The easiest way to control these pathological conditions is to limit consumption of HFCS and sugars (monosaccharides and disaccharides) in the human diet. A good candidate as a substitute for sucrose is erythritol is not digested by humans. For this reason, demand for erythritol has been growing over many years.

5.10 Effects of Stevia on Visceral Organs and Brain

The South American herb *Stevia rebaudiana* (family Asteraceae) contains several sweet tasting diterpene glycosides in its leaves. Shade dried leaves and stem are used to purify glycosides, which are used as sugar substitute (Starrat et al. 2002; Kalpana and Khan Md 2008). *Stevia rebaudiana*-derived glycosides are high-intensity sweeteners ranging from 50 to 300 times sweeter than sugar, with low water solubility and high melting points (Crammer and Ikan 1987; Brahmachari et al. 2011). These molecules are also highly stable at broad pH and temperature ranges in solution. Prolonged heating of the stevioside solution produces a decrease of the stevioside in the solution by 16.7 % after 12 h at 100 °C at neutral pH and

46–54 % after 4 h in acidic solutions of pH 2.4–2.6 (Chang and Cook 1983). The instability of the stevioside under very low pH and baking has led to doubts regarding its safety (Virendra and Kalpagam 2008).

The leaves of *Stevia rebaudiana* contain at least eight sweet steviol glycosides (stevioside, rebaudioside A and rebaudiosides A, B, C, and D) (Fig. 5.7) of which the main constituents are stevioside and rebaudioside A. As a sugar substitute steviosides have several advantages for human subjects. Steviosides are not only stable and non-calorific, but also maintains good dental health. Pharmacological activities of steviosides include antitumour and anticancer, antiinflammatory, antioxidative, antihyperglycemic, antihypertensive, antidiarrheal, immunomodulatory, diuretic, and enzyme inhibitory actions (Brahmachari et al. 2011). Steviosides are not absorbed by the human gut; only bacteria of the colon degrade steviosides to steviols. Part of this steviol is absorbed by the colon and transported to the liver by portal blood. Liver transforms steviol into steviol glucuronide, which is released into the blood and filtered out by the kidneys into the urine. High levels of steviol glucuronide in the urine suggest that there is no accumulation of steviol derivatives in the human body. The steviol glucuronide still present is expected to be excreted in the urine during the next 24 h. Besides steviol glucuronide, no free steviol or any other possible steviol metabolite could be detected in blood or urine. Steviosides have also been used to control weight in obese subjects. Stevioside, rebaudioside A and related compounds have been reported to regulate plasma glucose by modulating insulin secretion and sensitivity (Lailerd et al. 2004). Moreover, steviosides have also been reported to regulate blood glucose levels by enhancing not only insulin secretion, but also insulin utilization in insulin-deficient rats; the latter is due to decreased PEPCK gene expression in rat liver by stevioside's action of slowing down gluconeogenesis (Chen et al. 2005). These antihyperglycemic, insulinotropic, and glucagonostatic effects, especially for rebaudioside A, are largely plasma glucose level dependent, requiring high glucose levels (Abudula et al. 2008). Despite the large number of studies on *Stevia* and steviol glycosides, very little is known about the cellular and molecular mechanisms underlying its effects. In C57BL6J insulin-resistant mice model, steviosides not only downregulate the NF-κB pathway, but also enhance whole-body insulin sensitivity, glucose infusion rate, and the level of the glucose-lowering effect of insulin (Wang et al. 2012). The downregulation of NF-κB retards the expression of proinflammatory cytokines including tumor necrosis factor-α (TNFα), interleukin 6 (IL6), interleukin 1β (IL1β), and interleukin 10 (IL10) supporting the view that steviosides potentiate the reduction of insulin resistance by reducing the inflammation. It is also reported that steviol glycosides act by modulating GLUT translocation through the PtdIns 3K/Akt pathway (Rizzo et al. 2013). Steviol glycosides and insulin probably share a similar mechanism in regulating glucose entry into cells supporting the view that the use of *Stevia* extracts goes beyond their sweetening power by supporting its anticancer, antiinflammatory, antioxidantive, antihyperglycemic, antihypertensive, antidiarrheal, immunomodulatory, diuretic properties (Geuns 2003; Chatsudthipong and Muanprasat 2009). Steviol glycosides are also safe for phenylketonuria patients as no aromatic amino acids are involved (Geuns 2003).

5.11   Consumption of Sugar-Sweetened and Low-Calorie Sodas and Neurological Disorders

According to US Department of Agriculture per capita soft-drink consumption has increased by almost 500 % in the past 50 years (Putnam and Allshouse 1999). Large studies of US men and women have indicated that greater consumption of sugar-sweetened and artificial sweeteners containing sodas may contribute to insulin resistance and obesity, which is caused by an imbalance between energy intake and energy expenditure. Obesity is an important risk-factor for the development of type II diabetes and metabolic syndrome (Fig. 5.9). Indeed, the incidences of stroke increase many folds in metabolic syndrome patients with cardiovascular diseases. Thus, patients with heart disease have ninefold high risk of cerebral infarction compared to the general population. Induction of neuroinflammation, increased production of free radicals, alterations in neurotrophic factors, and reduction of insulin transport into the brain has been reported in the patients with metabolic syndrome (Farooqui 2013). Similarly, long-term presence of hyperglycemia and insulin resistance may result in cerebrovascular disease, which may accelerate cognitive decline and promote dementia and AD. Furthermore, insulin and insulin-signaling mechanisms are important for neuronal survival. Insulin place an important role in regulating memory therefore long-term disturbance in insulin signaling may have a negative impact on memory formation and may promote the initiation of AD (Bernstein et al. 2012; Farooqui et al. 2012; Farooqui 2013). It is also reported that type II diabetes and depression are linked with each

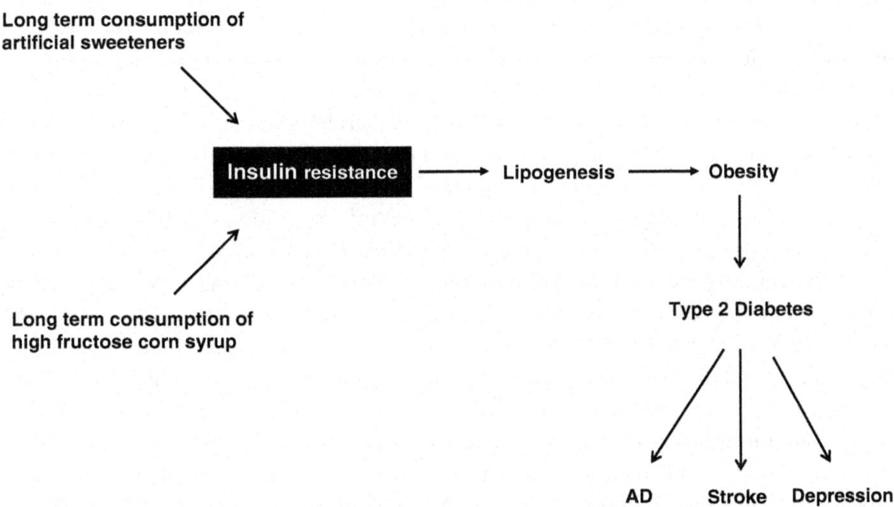

Fig. 5.9 Effects of artificial sweeteners and high fructose corn syrup on the pathogenesis of neurological disorders

other through stress, which impairs the ability of brain to regulate corticosteroid release. This may lead to hypercortisolemia. Excessive stimulation of corticosteroid receptors in hippocampus may cause hippocampal atrophy, which may lead to depression and dementia (Farooqui et al. 2012; Farooqui 2013). Collective evidence suggests that cellular and biochemical alterations observed in metabolic syndrome like elevation in lipid mediators, impairment of endothelial cell function, abnormality in essential fatty acid metabolism along with abnormal insulin/leptin signaling may represent a pathological bridge between metabolic syndrome and neurological disorders such as stroke, Alzheimer's disease (AD), and depression (Farooqui et al. 2012; Farooqui 2013).

5.12   Conclusion

HCFS (Fructose) is sweetener, which is metabolized by the body differently than glucose or sucrose. Fructose stimulates insulin synthesis but does not release it. Insulin modifies food intake by inhibiting eating and by increasing leptin release, which also can inhibit food intake. Meals enriched in HFCS can reduce circulating insulin and leptin levels, contributing to increased body weight. Thus, fructose intake may not result in the degree of satiety that normally ensues with an equally caloric meal of glucose or sucrose. Among artificial sweeteners, aspartame is composed of phenylalanine, aspartic acid, and methanol. Phenylalanine regulates neurotransmitters, whereas aspartic acid plays an important role in inducing excitotoxicity in the brain. Glutamate, asparagines and glutamine are formed from their precursor, aspartic acid. Methanol is oxidized into formaldehyde and diketopiperazine, a carcinogen, which mediates a number of other highly toxic effects. Other sugar substitutes have also been reported to cause many side effects. Thus, saccharin has been reported to cause bladder cancer in experimental rats, and consumption of aspartame is associated with cancer as well as neurological and psychiatric side effects in rodents. However, saccharin is not classified as carcinogenic to humans because of critical interspecies differences in urine composition, despite sufficient evidence of carcinogenicity in animals. Steviol, a natural extract from the Stevia plant, has been suspected to be a mutagen, but the safety of steviol glycoside as well as steviol oxidatives has been proved in many studies. Most artificial sweeteners are metabolized in the gastrointestinal tract, providing a mechanistic explanation for observed metabolic effects. Sweet-taste receptors, including the taste receptor T1R family and α-gustducin, respond not only to caloric sugars, such as sucrose, glucose, and fructose corn syrup but also to artificial sweeteners, including sucralose (Splenda™) and acesulfame-K. The consumption of an artificial sweetener in conjunction with a sugar containing food or drink may lead to more rapid sugar absorption, as well as increased GLP-1 and insulin secretion, potentially affecting weight, appetite, and glycemia. Collective evidence suggests that the safety of artificial sweeteners has been a concern, especially for their neurological effects and cancer-related risks.

References

Abdel-Salam OM, Salem NA, Hussein JS (2012a) Effect of aspartame on oxidative stress and monoamine neurotransmitter levels in lipopolysaccharide-treated mice. Neurotox Res 21:245–255

Abdel-Salam OM, Salem NA, El-Shamarka ME, Hussein JS, Ahmed NA, El-Nagar ME (2012b) Studies on the effects of aspartame on memory and oxidative stress in brain of mice. Eur Rev Med Pharmacol Sci 16:2092–2101

Abou-Donia MB, El-Masry EM, Abdel-Rahman AA, McLendon RE, Schiffman SS (2008) Splenda alters gut microflora and increases intestinal p-glycoprotein and cytochrome p-450 in male rats. J Toxicol Environ Health A 71:1415–1429

Abudula R, Matchkov VV, Jeppesen PB, Nilsson H, Aalkjær C, Hermansen K (2008) Rebaudioside A directly stimulates insulin secretion from pancreatic beta cells: a glucose-dependent action via inhibition of ATP-sensitive K$^+$-channels. Diabetes Obes Metab 10:1074–1085

Adler-Abramovich L, Vaks L, Carny O, Trudler D, Magno A, Caflisch A, Frenkel D, Gazit E (2012) Phenylalanine assembly into toxic fibrils suggests amyloid etiology in phenylketonuria. Nat Chem Biol 8:701–706

Agrawal R, Gomez-Pinilla F (2012) Metabolic syndrome in the brain: deficiency in omega-3 fatty acid exacerbates dysfunctions in insulin receptor signalling and cognition. J Physiol 12:2485–2499

Anderson GH (2007) Much ado about high-fructose corn syrup in beverages: the meat of the matter. Am J Clin Nutr 86:1577–1578

Arrigoni E, Brouns F, Amadò R (2005) Human gut microbiota does not ferment erythritol. Br J Nutr 94:643–646

Aune D (2012) Soft drinks, aspartame, and the risk of cancer and cardiovascular disease. Am J Clin Nutr 96:1249–1251

Bellisle F, Drewnowski A (2007) Intense sweeteners, energy intake and the control of body weight. Eur J Clin Nutr 61:691–700

Bernstein AM, de Koning L, Flint AJ, Rexrode KM, Willett WC (2012) Soda consumption and the risk of stroke in men and women. Am J Clin Nutr 95:1190–1199

Bigal ME, Krymchantowski AV (2006) Migraine triggered by sucralose–a case report. Headache 46:515–517

Bocarsly ME, Powell ES, Avena NM, Hoebel BG (2010) High-fructose corn syrup causes characteristics of obesity in rats: increased body weight, body fat and triglyceride levels. Pharmacol Biochem Behav 12:101–106

Brahmachari G, Mandal LC, Roy R, Mondal S, Brahmachari AK (2011) Stevioside and related compounds—molecules of pharmaceutical promise: a critical overview. Arch Pharm 344:5–19

Bray GA (2007) How bad is fructose? Am J Clin Nutr 12:895–896

Bray GA (2010) Soft drink consumption and obesity: it is all about fructose. Curr Opin Lipidol 21:51

Bray GA, Nielsen SJ, Popkin BM (2004) Consumption of high-fructose corn syrup in beverages may play a role in the epidemic of obesity. Am J Clin Nutr 79:537–543

Burri R, Matthieu JM, Vandevelde M, Lazeyras F, Posse S, Herschkowitz N (1990) Brain damage and recovery in hyperphenylalaninemic rats. Dev Neurosci 12:116–125

Chandrashekar J, Hoon MA, Ryba NJ, Zuker CS (2006) The receptors and cells for mammalian taste. Nature 444:288–294

Chang SS, Cook JM (1983) Stability studies of stevioside and rebaudioside A in carbonated beverages. J Agric Food Chem 31:409–412

Chatsudthipong V, Muanprasat C (2009) Stevioside and related compounds: therapeutic benefits beyond sweetness. Pharmacol Ther 121:41–54

Chen T-H, Chen S-C, Chan P, Chu Y-L, Yang H-Y, Cheng J-T (2005) Mechanism of the hypoglycemic effect of stevioside, a glycoside of Stevia rebaudiana. Planta Med 71:108–113

Christian B, McConnaughey K, Bethea E, Brantley S, Coffey A, Hammond L, Harrell S, Metcalf K, Muelenbein D, Spruill W, Brinson L, McConnaughey M (2004) Chronic aspartame affects T-maze performance, brain cholinergic receptors and Na+, K+-ATPase in rats. Pharmacol Biochem Behav 78:121–127

Chung YS, Lee M (2013) Genotoxicity assessment of erythritol by using short-term assay. Toxicol Res 29:249–255

Cong WN, Wang R, Cai H, Daimon CM, Scheibye-Knudsen M, Bohr VA, Turkin R, Wood WH 3rd, Becker KG, Moaddel R, Maudsley S, Martin B (2013) Long-term artificial sweetener acesulfame potassium treatment alters neurometabolic functions in C57BL/6J mice. PLoS One 8:e70257

Crammer B, Ikan R (1987) Progress in the chemistry and progress of the rebaudiosides. In: Grenby TH (ed) Developments in Sweeteners-3. Elsevier Applied Science, London, pp 45–64

Cui M, Jiang P, Maillet E, Max M, Margolskee RF, Osman DR (2006) The heterodimeric sweet taste receptor has multiple potential ligand binding sites. Curr Pharm Des 12:4591–4600

Davidson TL, Martin AA, Clark K, Swithers SE (2011) Intake of high-intensity sweeteners alters the ability of sweet taste to signal caloric consequences: implications for the learned control of energy and body weight regulation. Q J Exp Psychol (Hove) 64:1430–1441

Després JP, Lemieux I, Bergeron J, Pibarot P, Mathieu P, Larose E, Rodés-Cabau J, Bertrand OF, Poirier P (2008) Abdominal obesity and the metabolic syndrome: contribution to global cardio-metabolic risk. Arterioscler Thromb Vasc Biol 28:1039–1049

Drewnowski A, Krahn DD, Demitrack MA, Nairn K, Gosnell BA (1995) Naloxone, an opiate blocker, reduces the consumption of sweet high-fat foods in obese and lean female binge eaters. Am J Clin Nutr 61:1206–1212

Elliott SS, Keim NL, Stern JS, Teff K, Havel PJ (2002) Fructose, weight gain, and the insulin resistance syndrome. Am J Clin Nutr 76:911–922

Ergorul C, Eichenbaum H (2004) The hippocampus and memory for "what", "where", and "when". Learn Mem 11:397–405

Farah V, Elased KM, Chen Y, Key MP, Cunha TS, Ingoyen M, Morris M (2006) Nocturnal hypertension in mice consuming a high fructose diet. Auton Neurosci 130:41–50

Farooqui AA (2013) Metabolic syndrome an important risk factor for stroke, Alzheimer disease and depression. Springer, New York

Farooqui AA, Farooqui T, Panza F, Frisardi V (2012) Metabolic syndrome as a risk factor for neurological disorders. Cell Mol Life Sci 69:741–762

FDA (2011) No Calories. Sweet! http://www.fda.gov/fdac/features/2006/406_sweeteners.html. Accessed 1 Feb 2011

FDA/CFSAN (2001) Agency response letter: GRAS notice. FDA, Washington, DC

Finkelstein MW, Daabees TT, Steginik LD, Applebaum AE (1983) Correlation of aspartate dose, plasma dicarboxylic amino acid concentration, and neuronal necrosis in infant mice. Toxicology 29:109–119

Fujita Y, Wideman RD, Speck M, Asadi A, King DS, Webber TD, Haneda M, Kieffer TJ (2009) Incretin release from gut is acutely enhanced by sugar but not by sweeteners in vivo. Am J Physiol Endocrinol Metab 296:E473–E479

Geary N, Smith GP (1985) Pimozide decreases the positive reinforcing effect of sham fed sucrose in the rat. Pharmacol Biochem Behav 22:787–790

Geuns JMC (2003) Molecules of interest: stevioside. Phytochemistry 64:913–921

Gonzalez J, Willis Monte S (2010) Ivar Asbjorn Folling discovered phenylketonuria (PKU). Lab Med 41:118–119

Goossens J, Röper H (1994) Erythritol: a new bulk sweetener. Int Food Ingredients 1(2):27–33

Grotz VL (2008) Sucralose and migraine. Headache 48:164–165

Grotz VL, Henry RR, McGill JB, Prince MJ, Shamoon H, Trout JR, Pi-Sunyer FX (2003) Lack of effect of sucralose on glucose homeostasis in subjects with type 2 diabetes. J Am Diet Assoc 103:1607–1612

Haase L, Cerf-Ducastel B, Murphy C (2009) Cortical activation in response to pure taste stimuli during the physiological states of hunger and satiety. Neuroimage 44:1008–1021

Harvey J (2007) Leptin regulation of neuronal excitability and cognitive function. Curr Opin Pharmacol 7:643–647

Hernandez L, Hoebel BG (1988) Food reward and cocaine increase extracellular dopamine in the nucleus accumbens as measured by microdialysis. Life Sci 42:1705–1712

Hino H, Kasai S, Hattori N, Kenjo K (2000) A case of allergic urticaria caused by erythritol. J Dermatol 27:163–165

Humphries P, Pretorius E, Naudé H (2008) Direct and indirect cellular effects of aspartame on the brain. Eur J Clin Nutr 62:451–462

Iguchi A, Gotoh M, Matsunaga H, Yatomi A, Honmura A, Yanase M, Sakamoto N (1986) Mechanism of central hyperglycemic effect of cholinergic agonists in fasted rats. Am J Physiol 254:E431–E437

Iwanami U, Mogi M, Iwai M, Horiuchi M (2009) Inhibition of the renin-angiotensin system and target organ protection. Hypertens Res 32:229–237

Jacobson MF, Farber E, Clapp R (1998) Re: long-term feeding of sodium saccharin to nonhuman primates: implications for urinary tract cancer. J Natl Cancer Inst 90:934–936

James RW, Heywood R, Crook D (1981) Testicular responses of rats and dogs to cyclohexylamine overdosage. Food Cosmet Toxicol 19:291–296

Jang HJ, Kokrashvili Z, Theodorakis MJ, Carlson OD, Kim BJ, Zhou J, Kim HH, Xu X, Chan SL, Juhaszova M, Bernier M, Mosinger B, Margolskee RF, Egan JM (2007) Gut-expressed gustducin and taste receptors regulate secretion of glucagon-like peptide-1. Proc Natl Acad Sci U S A 104:15069–15074

Kalpana R, Khan Md K (2008) Post-harvest management of stevia leaves: a review. J Food Sci Technol 45:391–397

Kawasaki T, Igarashi K, Koeda T, Sugimoto K, Nakagawa K, Hayashi S, Yamaji R, Inui H, Fukusato T, Yamanouchi T (2009) Rats fed fructose-enriched diets have characteristics of non-alcoholic hepatic steatosis. J Nutr 139:2067–2071

Kelley AE, Schiltz CA, Landry CF (2005) Neural systems recruited by drug- and food-related cues: studies of gene activation in corticolimbic regions. Physiol Behav 86:11–14

Kim J-Y, Seo J, Cho K-H (2011) Aspartame-fed zebrafish exhibit acute deaths with swimming defects and saccharin-fed zebrafish have elevation of cholesteryl ester transfer protein activity in hypercholesterolemia. Food Chem Toxicol 49:2899–2905

Kinnamon SC (2012) Taste receptor signaling—from tongues to lungs. Acta Physiol (Oxf) 204:158–168

Kobayakawa T, Ogawa H, Kaneda H, Ayabe-Kanamura S, Endo H, Saito S (1999) Spatio-temporal analysis of cortical activity evoked by gustatory stimulation in humans. Chem Senses 24:201–209

Komarova TV, Petrunia IV, Shindyapina AV, Silachev DN, Sheshukova EV, Kiryanov GI, Dorokhov YL (2014) Endogenous methanol regulates Mammalian gene activity. PLoS One 9:e90239

Kumar P, Kumar A (2008) Prolonged pretreatment with carvedilol prevents 3-nitropropionic acid-induced behavioral alterations and oxidative stress in rats. Pharmacol Rep 60:706–715

LaBuda CJ, Fuchs PN (2001) A comparison of chronic aspartame exposure to aspirin on inflammation, hyperalgesia and open field activity following carrageenan-induced monoarthritis. Life Sci 69:443–454

Lailerd N, Saengsirisuwan V, Sloniger JA, Toskulkao C, Henriksen EJ (2004) Effects of stevioside on glucose transport activity in insulin-sensitive and insulin-resistant rat skeletal muscle. Metabolism 53:101–107

Leclercq IA, Da Silva MA, Schroyen B, Van Hul N, Geerts A (2007) Insulin resistance in hepatocytes and sinusoidal liver cells: mechanisms and consequences. J Hepatol 47:142–156

Li XL, Aou S, Oomura Y, Hori N, Fukunaga K, Hori T (2002) Impairment of long-term potentiation and spatial memory in leptin receptor-deficient rodents. Neuroscience 113:607–615

Ma J, Bellon M, Wishart JM, Young R, Blackshaw LA, Jones KL, Horowitz M, Rayner CK (2009) Effect of the artificial sweetener, sucralose, on gastric emptying and incretin hormone release in healthy subjects. Am J Physiol Gastrointest Liver Physiol 296:G735–G739

Mace OJ, Affleck J, Patel N, Kellett GL (2007) Sweet taste receptors in rat small intestine stimulate glucose absorption through apical GLUT2. J Physiol 582:379–392

Martin B, Shin YK, White CM, Ji S, Kim W, Carlson OD, Napora JK, Chadwick W, Chapter M, Waschek JA, Mattson MP, Maudsley S, Egan JM (2010) Vasoactive intestinal peptide-null mice demonstrate enhanced sweet taste preference, dysglycemia, and reduced taste bud leptin receptor expression. Diabetes 59:1143–1152

Mielke JG, Nicolitch K, Avellaneda V, Earlam K, Ahuja T, Mealing G, Messier C (2006) Longitudinal study of the effects of a high-fat diet on glucose regulation, hippocampal function, and cerebral insulin sensitivity in C57BL/6 mice. Behav Brain Res 175:374–382

Moosavi M, Naghdi N, Maghsoudi N, Zahedi Asl S (2006) The effect of intrahippocampal insulin microinjection on spatial learning and memory. Horm Behav 50:748–752

Munro IC, Berndt WO, Borzelleca JF, Flamm G, Lynch BS, Kennepohl E, Bär EA, Modderman J (1998) Erythritol: an interpretive summary of biochemical, metabolic, toxicological and clinical data. Food Chem Toxicol 36:1139–1174

Nakagawa Y, Nagasawa M, Yamada S, Hara A, Mogami H, Nikolaev VO, Lohse MJ, Shigemura N, Ninomiya Y, Kojima I (2009) Sweet taste receptor expressed in pancreatic beta-cells activates the calcium and cyclic AMP signaling systems and stimulates insulin secretion. PLoS One 4:e5106

Panchal SK, Brown L (2011) Rodent models for metabolic syndrome research. J Biomed Biotechnol 2011:351982

Park CR (2001) Cognitive effects of insulin in the central nervous system. Neurosci Biobehav Rev 25:311–323

Patel J, Iyer A, Brown L (2009) Evaluation of the chronic complications of diabetes in a high fructose diet in rats. Indian J Biochem Biophys 46:66–72

Prakash I, Bishay IE, Desai N, Eric Walters D (2001) Modifying the temporal profile of the high-potency sweetener neotame. J Agric Food Chem 49:786–789

Putnam JJ, Allshouse JE (1999) Food consumption, prices, and expenditures, 1970–97. Food and Consumers Economics Division, Economic Research Service, US Department of Agriculture, Washington, DC

Quiron R, Boksa P (1986) Autoradiographic distribution of muscarinic [3H]acetylcholine receptors in rat brain: comparison with antagonists. Eur J Pharmacol 123:170–172

Rahn A, Yaylayan VA (2010) Thermal degradation of sucralose and its potential in generating chloropropanols in the presence of glycerol. Food Chem 118:56–61

Ramsland PA, Movafagh BF, Reichlin M, Edmundson AB (1999) Interference of rheumatoid factor activity by aspartame, a dipeptide methyl ester. J Mol Recognit 12:249–257

Rayssiguier Y, Gueux E, Nowacki W, Rock E, Mazur A (2006) High fructose consumption combined with low dietary magnesium intake may increase the incidence of the metabolic syndrome by inducing inflammation. Magnes Res 19:237–243

Reimann F, Habib AM, Tolhurst G, Parker HE, Rogers GJ, Gribble FM (2008) Glucose sensing in L cells: a primary cell study. Cell Metab 8:532–539

Ren X, Zhou L, Terwilliger R, Newton S, De Araujo IE (2009) Sweet taste signaling functions as a hypothalamic glucose sensor. Front Integr Neurosci 3:1–15

Renwick AG (1986) The metabolism of intense sweeteners. Xenobiotica 16:1057–1071

Reynolds R, Burri R, Mahal S, Herschkowitz N (1992) Disturbed myelinogenesis and recovery in hyperphenylalaninemia in rats: an immunohistochemical study. Exp Neurol 115:347–367

Ritzkalla SW (2010) Health implications of fructose consumption: a review of recent data. Nutr Metab 12:82

Rizzo B, Zambonin L, Angeloni C, Leoncini E, Dalla Sega FV, Prata C, Fiorentini D, Hrelia S (2013) Steviol glycosides modulate glucose transport in different cell types. Oxid Med Cell Longev 2013:348169

Ross AP, Bartness TJ, Mielke JG, Parent MB (2009) A high fructose diet impairs spatial memory in male rats. Neurobiol Learn Mem 92:410–416

Rycerz K, Jaworska-Adamu JE (2013) Effects of aspartame metabolites on astrocytes and neurons. Folia Neuropathol 51:10–17

Salkovic-Petrisic M, Hoyer S (2007) Central insulin resistance as a trigger for sporadic Alzheimer-like pathology: an experimental approach. J Neural Transm 72:217–233

Schiffman SS, Gatlin CA (1993) Sweeteners: state of knowledge review. Neurosci Biobehav Rev 17:313–345

Schiffman SS, Rother KI (2013) Sucralose, a synthetic organochlorine sweetener: overview of biological issues. J Toxicol Environ Health B Crit Rev 16:399–451

Schiffman SS, Buckley CEI, Sampson HA, Massey EW, Baraniuk JN et al (1987) Aspartame and susceptibility to headache. N Engl J Med 317:1181–1185

Schiffman SS, Sattely-Miller EA, Bishay IE (2008) Sensory properties of neotame: comparison with other sweeteners. In: Weerasinghe DK, DuBois GE (eds) Sweetness and sweeteners: biology, chemistry and psychophysics, ACS symposium series. Oxford University Press, New York, pp 511–529

Semchyshyn HM (2013) Fructation *in vivo*: detrimental and protective effects of fructose. BioMed Res Int 2013:343914

Simon BR, Parlee SD, Learman BS, Mori H, Scheller EL, Cawthorn WP, Ning X, Gallagher K, Tyrberg B, Assadi-Porter FM, Evans CR, MacDougald OA (2013) Artificial sweeteners stimulate adipogenesis and suppress lipolysis independently of sweet taste receptors. J Biol Chem 288:32475–32489

Simopoulos AP (2013) Dietary omega-3 fatty acid deficiency and high fructose intake in the development of metabolic syndrome, brain metabolic abnormalities, and non-alcoholic fatty liver disease. Nutrients 5:2901–2923

Small DM (2006) Central gustatory processing in humans. Adv Otorhinolaryngol 63:191–220

Smeets PAM, de Graaf C, Stafleu A, van Osch MJP, van der Grond J (2005) Functional magnetic resonance imaging of human hypothalamic responses to sweet taste and calories. Am J Clin Nutr 82:1011–1016

Soffritti M, Belpoggi F, Degli ED, Lambertini L, Tibaldi E, Rigano A (2006) First experimental demonstration of the multipotential carcinogenic effects of aspartame administered in the feed to Sprague-Dawley rats. Environ Health Perspect 114:379–385

Squire RA (1985) Histopathological evaluation of rat urinary bladders from the IRDC two-generation bioassay of sodium saccharin. Food Chem Toxicol 23:491–497

Stanhope KL, Schwarz JM, Keim NL, Griffen SC, Bremer AA, Graham JL, Hatcher B, Cox CL, Dyachenko A, Zhang W, McGahan JP, Seibert A, Krauss RM, Chiu S, Schaefer EJ, Ai M, Otokozawa S, Nakajima K, Nakano T, Beysen C, Hellerstein MK, Berglund L, Havel PJ (2009) Consuming fructose-sweetened, not glucose-sweetened, beverages increases visceral adiposity and lipids and decreases insulin sensitivity in overweight/obese humans. J Clin Invest 119:1322–1334

Starrat AN, Kirby CW, Pocs R, Brandle JE (2002) Rebaudioside F, a diterpene glycoside from *Stevia rebaudiana*. Phytochemistry 59:367–370

Steinert RE, Gerspach AC, Gutmann H, Asarian L, Drewe J, Beglinger C (2011) The functional involvement of gut-expressed sweet taste receptors in glucose-stimulated secretion of glucagon-like peptide-1 (GLP-1) and peptide YY (PYY). Clin Nutr 30:524–532

Stranahan AM, Norman ED, Lee K, Cutler RG, Telljohann RS, Egan JM, Mattson MP (2008) Diet-induced insulin resistance impairs hippocampal synaptic plasticity and cognition in middle-aged rats. Hippocampus 12:1085–1088

Swithers SE, Sample CH, Davidson TL (2013) Adverse effects of high-intensity sweeteners on energy intake and weight control in male and obesity-prone female rats. Behav Neurosci 127:262–274

Szucs EF, Barrett KE, Metcalfe DD (1986) The effects of aspartame on mast cells and basophils. Food Chem Toxicol 24:171–174

Takayama S, Renwick AG, Johansson SL (2000) Long-term toxicity and carcinogenicity study of cyclamate in nonhuman primates. Toxicol Sci 53:33–39

Tandel KR (2011) Sugar substitutes: health controversy over perceived benefits. J Pharmacol Pharmacother 2:236–243

Taylor JM, Weinberger MA, Friedman L (1980) Chronic toxicity and carcinogenicity to the urinary bladder of sodium saccharin in the in utero-exposed rat. Toxicol Appl Pharmacol 54:57–75

Teff KL, Elliott SS, Tschop M, Kieffer TJ, Rader D, Heiman M, Townsend RR, Keim NL, D'Alessio D, Havel PJ (2004) Dietary fructose reduces circulating insulin and leptin, attenuates postprandial suppression of ghrelin, and increases triglycerides in women. J Clin Endocrinol Metab 89:2963–2972

Tinahones FJ, Murri-Pierri M, Garrido-Sánchez L, García-Almeida JM, García-Serrano S, García-Arnés J, García-Fuentes E (2009) Oxidative stress in severely obese persons is greater in those with insulin resistance. Obesity 17:240–246

Tran LT, MacLeod KM, McNeill JH (2009) Chronic etanercept treatment prevents the development of hypertension in fructose-fed rats. Mol Cell Biochem 330:219–228

Veien NK, Lomholt HB (2012) Systemic allergic dermatitis presumably caused by formaldehyde derived from aspartame. Contact Dermatitis 67:315–316

Velloso LA, Araujo EP, de Souza CT (2008) Diet-induced inflammation of the hypothalamus in obesity. Neuroimmunomodulation 15:189–193

Verma S, Bhanot S, McNeill JH (1999) Sympathectomy prevents fructose-induced hyperinsulinemia and hypertension. Eur J Pharmacol 373:R1–R4

Virendra VP, Kalpagam P (2008) Assessment of stevia (*Stevia rebaudiana*)—natural sweetener: a review. J Food Sci Tech Mys 45:467–473

Walters DE, Prakash I, Desai N (2000) Active conformations of neotame and other high-potency sweeteners. J Med Chem 43:1242–1245

Wang Z, Xue L, Guo C (2012) Stevioside ameliorates high-fat diet-induced insulin resistance and adipose tissue inflammation by downregulating the NF-κB pathway. Biochem Biophys Res Commun 417:1280–1285

Wise RA (2006) Role of brain dopamine in food reward and reinforcement. Philos Trans R Soc Lond B Biol Sci 361:1149–1158

Witt J (1999) Discovery and development of neotame. World Rev Nutr Diet 85:52–57

Wright JW, Harding JW (2010) The brain RAS and Alzheimer's disease. Exp Neurol 223:326–333

Wright JW, Kawas LH, Harding JW (2013) A Role for the Brain RAS in Alzheimer's and Parkinson's Diseases. Front Endocrinol (Lausanne) 4:158

Yarmolinsky DA, Zuker CS, Ryba NJ (2009) Common sense about taste: from mammals to insects. Cell 139:234–244

Yee KK, Sukumaran SK, Kotha R, Gilbertson TA, Margolskee RF (2011) Glucose transporters and ATP-gated K+ (KATP) metabolic sensors are present in type 1 taste receptor 3 (T1r3)-expressing taste cells. Proc Natl Acad Sci U S A 108:5431–5436

Yin QQ, Pei JJ, Xu S, Luo DZ, Dong SQ, Sun MH, You L, Sun ZJ, Liu XP (2013) Pioglitazone improves cognitive function via increasing insulin sensitivity and strengthening antioxidant defense system in fructose-drinking insulin resistance rats. PLoS One 8:e59313

Zhao GQ, Zhang Y, Hoon MA, Chandrashekar J, Erlenbach I, Ryba NJ, Zuker CS (2003) The receptors for mammalian sweet and umami taste. Cell 115:255–266

Zwillich T (2007) Aspartame safety study stirs emotions. Italian study shows sweetener promotes cancer in rats; FDA says it's safe. http://www.webmd.com/diet/news/20070626/aspartame-safety-studystirs-emotions. Accessed 8 Mar 2011

Chapter 6
Contribution of Salt in Inducing Biochemical Changes in the Brain

6.1 Introduction

Sodium is an essential nutrient necessary for maintaining plasma volume, acid–base balance, neurotransmission, and normal cell function (Holbrook et al. 1984). The minimum daily required intake is estimated at 1–2 g/day (He and MacGregor 2009; Campbell et al. 2012). The World Health Organization currently recommends a daily consumption of less than 5 g of salt (WHO 2010). The Departments of Agriculture and Health and Human Services recommend daily intake of less than 6 g of salt (2,300 mg of sodium), with a lower target of 3.7 g/day of salt for most adults (CDC 2009). However, average consumption of sodium is well above that needed for physiological function. Americans consume 8–10 g salt/day. Sodium is not only a main chemical component of common table salt, but is also found in foods such as soy and fish sauces, and processed foods (breads, crackers, meats, and snack foods), milk, cheeses, and shellfish. Based on large observational follow-up studies, it is suggested that a reduction in dietary sodium intake may lead not only to a reduction in blood pressure (BP), but also a reduction in the long-term risk of cardiovascular events by 25 % over a period of 10–15 years (Cook et al. 2007). Moreover, recent systematic reviews and meta-analyses have shown that there is a significant relationship between a high sodium intake and an increased risk of morbidity and mortality related to stroke and cardiovascular diseases (de Wardener and MacGregor 2002; Strazzullo et al. 2009). Response of BP to a low or high sodium diet is not uniform. Some individuals show a dramatic response to dietary sodium intake, with an increase in BP by ingesting a diet high in sodium or decreasing BP by a diet low in sodium (salt sensitive). Other individuals show no response (salt insensitive or salt resistance) or a negative response to diets low and high in sodium (Nguyen et al. 2013). However, excess sodium intake also has BP-independent effects, promoting left ventricular hypertrophy as well as fibrosis in the heart, kidneys, and arteries (Frohlich 2007).

High calorie diet enriched in sodium is not only linked with increase in BP, but also with defective insulin sensitivity and impaired glucose homeostasis

© Springer International Publishing Switzerland 2015
A.A. Farooqui, *High Calorie Diet and the Human Brain*,
DOI 10.1007/978-3-319-15254-7_6

(Lastra et al. 2010; Lanaspa et al. 2012). Increase in salt intake is known to increase calcium excretion and loss of hip bone density (He and MacGregor 2010). Excessive sodium intake induces hypertrophy of vascular smooth muscles independent of blood pressure, increases NADPH oxidase activity and oxidative stress, and reduces the availability and production of nitric oxide (Kitiyakara et al. 2003; Oberleithner et al. 2007).

In salt sensitive individuals, sensitivity of BP to sodium is not only defined by the elevation in BP, but also by endothelial dysfunction, alterations in cardiovascular structure and function, albuminuria, kidney disease progression, and cardiovascular morbidity and mortality in the general population. In contrast, salt-resistant animals or humans excrete a sodium load without changes in BP; i.e., without resorting to pressure natriuresis. Genetic factors, such as age and race, play an important role in sodium sensitivity to BP. Genetic causes of salt sensitivity include mutations in genes that promote salt retention through a defect in renal sodium handling. The gene mutations in the aldosterone synthase/11β-hydroxysteroid dehydrogenase, mineralo-corticoid receptor, β and γ subunits of the epithelial sodium channel (ENaC), the ATP-binding cassette, subfamily B, member 1 (*ABCB1*), cytochrome P450 3A5 (*CYP3A5*), and *CYP4A11* cause monogenic hypertension, which is rare, accounting for less than 1 % of the prevalence of essential hypertension. These genetic muta-tions cannot explain the high incidence of salt sensitivity, especially in normotensive humans (Iwai et al. 2007; Eap et al. 2007; Rocha et al. 1998).

Humans with salt-sensitive high BP have been reported to show not only increase in sodium transport in the renal proximal tubule and medullary thick ascending limb, but also cause sodium retention. Sodium transport is regulated by natriuretic and antinatriuretic hormones and humoral agents, such as dopamine and angiotensin, which exert their effects through G protein–coupled receptors. Activation of certain post-junctional dopamine receptor subtypes (D_1R, D_3R, D_4R, and D_5R) and the angiotensin receptor by GPCR kinase 4γ (GRK4γ) on renal proximal tubule and thick ascending limb of Henle inhibits sodium transport, whereas activation of the post-junctional D_2R and angiotensin type 1 receptor (AT_1R) increases sodium trans-port (Fig. 6.1) (Felder and Jose 2006). In salt sensitive individuals, diet-mediated insulin resistance, which is caused by high calorie diet also contributes to sodium sensitivity of BP in rodents and humans (Ogihara et al. 2002; Fuenmayor et al. 1998). In addition to excess of sodium intake, a variety of other risk factors including excess alcohol intake, weight gain, and physical inactivity also contribute to essential hyper-tension (Kaplan and Domino 2013). BP is controlled through a complex interplay among multiple organ systems in the body. The mean perfusion pressure of the sys-temic circulation is modulated by the product of the cardiac output and the peripheral vascular resistance. Thus, BP depends on a number of factors including the cardiac contractile function, the circulating blood volume, the vascular smooth muscle, and endothelial function in resistance arteries, the vascular structure, viscosity of the blood, the secretory function of endocrine glands, and the sympathetic nervous tone (Fig. 6.2). An important hallmark of high BP-mediated structural abnormalities is represented by changes in the mechanical properties of arteries, with particular regard for increased stiffness (Intengan and Schiffrin 2000). Vascular fibrosis is

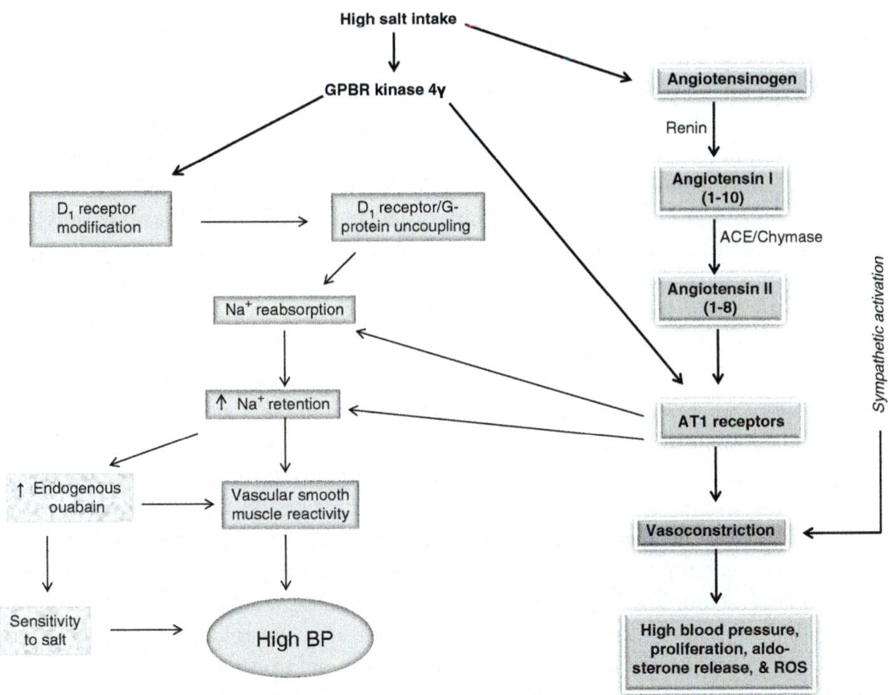

Fig. 6.1 Modulation of sodium by dopamine and angiotensin

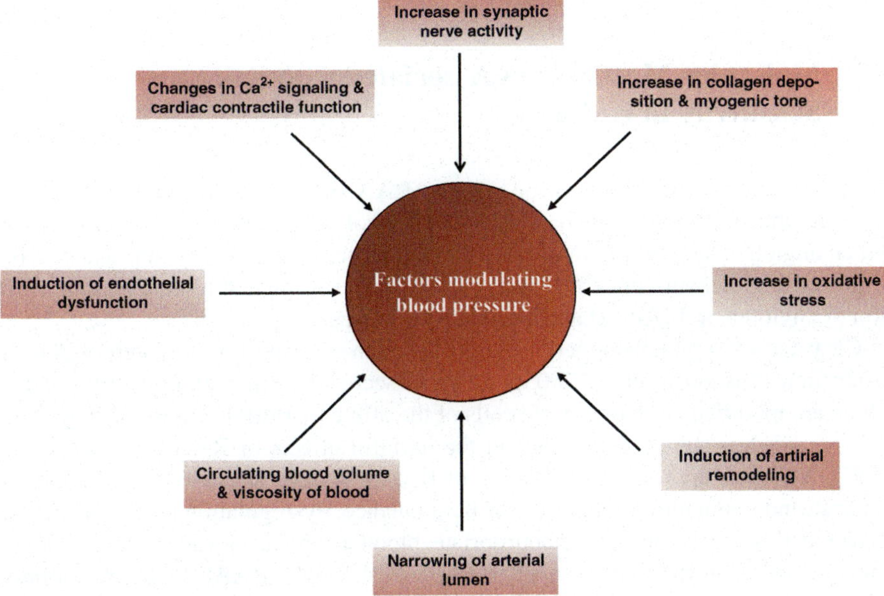

Fig. 6.2 Factors associated with modulation of blood pressure

critically important in the determining the vascular structural modifications, and it involving changes in extracellular matrix (ECM) components, such as collagen type I and III, elastin, and fibronectin (Intengan and Schiffrin 2000). All these changes are promoted by common denominators such as changes in oxidative stress and alterations in immune cell function (Harrison 2013). The presence of local organ specific renin–angiotensin aldosterone systems (RAAS), which modulates BP. It has been reported in the heart, large arteries and arterioles, kidneys, and other organs and their activation lead to structural and functional changes that are independent of those elicited by the classical RAAS (Frohlich 2007; Bader and Ganten 2008; Frohlich and Re 2009; Re 2009; Kurdi et al. 2005). However, despite extensive research on the effect of salt on BP, the specific mechanisms associated by high dietary salt-mediated elevation in BP remain elusive. Many investigators have focused on the mechanisms by which the kidneys retain salt and mediate increase in BP. However, recent studies have indicated that the brain can play an important role in sodium sensitivity. Brain has specific and discrete short-term rapid-acting and long-term slow-acting pathways that modulate BP. The slow pathway involves the activation of central aldosterone (Aldo), mineralocorticoid receptors (MR), epithelial sodium channels (EnaCs), endogenous ouabain (EO), and AT1 receptors. Chronic elevation of circulating endogenous EO amplifies existing sympathetic tone, augments the contractile function of the peripheral vasculature, suppresses nitric oxide-related signaling in the renal medulla and resets kidney function to help maintain the elevated BP (Joyner et al. 2008; Fink and Arthur 2009; Grassi 2009). Inhibition of any component in this central pathway retards salt-sensitive hypertension in experimental animals (Blaustein et al. 2012; Hamlyn and Blaustein 2013).

6.2   Molecular Mechanisms Associated with Salt Sensitivity in Kidneys

The renin–angiotensin–aldosterone system (RAAS) is a master regulator of blood pressure and fluid homeostasis. This system involves a multi-enzymic cascade in which angiotensinogen, the major substrate, is processed in a two-step reaction by renin and angiotensin-converting enzyme (ACE), resulting in the sequential generation of angiotensin I (Ang I) and angiotensin II (Ang II).

Cleavage of angiotensinogen to Ang I by renin is a rate-limiting step of Ang II production (Takahashi et al. 2005). Renin is secreted from juxtaglomerular cells, which are modified smooth muscle cells of the afferent arterioles in the kidney. Ang II is regarded as the central player in the harmful effects of RAAS (Kobori et al. 2007). Although appropriate activation of RAAS is vital for preventing circulatory collapse and maintaining intravascular fluid balance, dysregulation and/or persistent RAAS activation leading to inappropriate blood pressure elevation, target organ damage, and even reduced survival (Benigni et al. 2009). RAAS blockers not only inhibit the production or action of Ang II but also suppress the synthesis of aldosterone (Ruggenenti et al. 2010). Indeed, mineralocorticoid receptor (MR) blockers

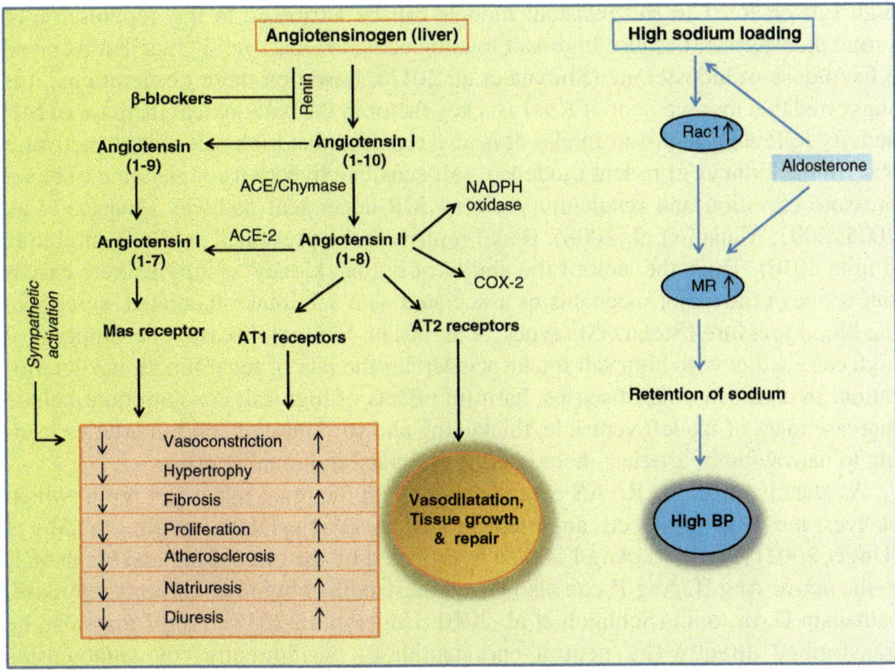

Fig. 6.3 Effect of salt loading on blood pressure in visceral tissues (kidneys). *Rac1* GTPase, *ACE* angiotensin converting enzyme, *AT1 receptor* angiotensin II type-1 receptor, *AT2 receptor* angiotensin II type-2 receptor, *ROS* reactive oxygen species, *MR* mineralocorticoid receptors, *BP* blood pressure

have been shown to protect against various kidney diseases. In rodent models of salt-sensitive hypertension, high-salt loading activates Rho-family GTPases (Rac1, a small GTP-binding protein) in the kidneys causing blood pressure elevation and renal injury via an MR-dependent pathway (Fig. 6.3) (Shibata et al. 2011). Rac-1 is not only associated with cell migration, but is also involved in NADPH oxidase activation and thus inhibition of Rac-1 reduces oxidative damage. For renal cells, it was shown that the Rac-1 inhibitor NSC23766 blocks oxalate-induced NADPH oxidase-mediated oxidative cell injury in tubular cells (Thamilselvan et al. 2012). High-salt diet also causes renal Rac1 upregulation in salt-sensitive Dahl (Dahl-S) rats. In contrast, salt-insensitive Dahl (Dahl-R) rats show downregulation of Rac 1 (Shibata et al. 2011). These observations support the view that the aldosterone/MR signaling is closely related to dietary salt intake in both health and disease. Detailed investigations have indicated that salt loading increases not only serum- and glucocorticoid-inducible kinase (Sgk1), but also nuclear MR accumulation in Dahl-S rats and that increase in MR signaling can be retarded by Rac1 inhibition. In contrast, high-salt loading down-regulates Rac1 and MR signaling in salt-resistant models, including Dahl-R and SD rats, indicating that high-salt loading potentiates Rac1-MR signaling only in salt-sensitive models, not in salt-resistant subjects. The negative action of high salt on Rac1 in salt-resistant models can be attributed to the suppression of serum aldosterone, because high-salt loading activates Rac1 in SD rats that received a fixed dose of aldosterone (Shibata et al. 2011). Based on these observations, it is suggested that involvement of Rac1 is a key factor in the paradoxical increase of MR activity following high-salt intake. It is also reported that high-salt loading activates Rac1in the kidneys of rodent models of salt-sensitive hypertension, leading to blood pressure elevation and renal injury via an MR-dependent pathway (Nagase et al. 2006, 2007; Matsui et al. 2008). Renal injury can be prevented by Rac1 inhibitors (Fujita 2010). Thus, the lack of the ability of human kidney to fully excrete excess salt is one of the major mechanisms associated with salt intake-mediated increase in the blood pressure (Stolarz-Skrzypek et al. 2011). In renal disease, consumption of high calorie diet with high salt intake accelerates the rate of renal functional deterioration. In cardiovascular diseases, harmful effects of high salt consumption include increase mass of the left ventricle, thickening and stiffening of conduit arteries leading to narrowing of arteries, including the coronary and renal arteries.

As stated above, the RAAS cascade starts with the renal release of renin, which cleaves the liver-produced angiotensinogen, a glycosylated protein, to Ang I (Unger 2002) (Fig. 6.3). Ang I is then hydrolyzed by the circulating and local ACE to the active Ang II. Ang II can also be synthesized by chymase, carboxypeptidase, cathepsin G, or tonin (Schlaich et al. 2010; Unger et al. 2011). Ang I may also be transformed directly (by neutral endopeptidase) or indirectly (by angiotensin-converting enzyme and angiotensin-converting enzyme 2 into a variety of Ang II forms, such as angiotensin 1–9, angiotensin 1–7, angiotensin 1–5. The Ang II is the most prominent regulator of blood pressure, exerting its effects through the control of vasoconstriction and regulation of water and salt balance (Audoly et al. 2000). In addition, Ang II has important effects on diverse cellular systems, including neuromodulation, cellular growth, and proliferation (Allen et al. 2000). Ang II exerts its functions through activation of type 1 (AT1) and type 2 (AT2) angiotensin receptors. AT1 receptor has two subtypes, (AT1a and AT1b) in mice. AT1 is a member of the superfamily of G protein-coupled receptors (GPCR) and activates G proteins through regions of the third intracellular loop and intracellular carboxyl-terminal tail of the receptor (Guo et al. 2001). AT_1R mediates vasoconstriction, antinatriuresis, contributes to an increase in blood pressure, thirst, and release of vasopressin and aldosterone, fibrosis, cellular growth, and migration (Fyhrquist and Saijonmaa 2008). AT_2R mediates vasodilatation, inhibits cellular growth, promotes natriuresis, and potentially lowers blood pressure (Carey and Padia 2008; Padia et al. 2008). Activation of AT_1 receptor involves several proteins. The best characterized chain of events is the rapid activation of heterotrimeric G proteins ($G\alpha_{q/11}$, $G\alpha_{12/13}$, and $G\beta\gamma$) that subsequently activate phospholipase Cβ and phospholipase Cγ to induce the generation of inositol-1,4,5-trisphosphate ($InsP_3$) and diacylglycerol (DAG), increase cytosolic Ca^{2+}, activate cytosolic kinases and transcriptional factors (Inagami et al. 1999). In addition, other mediators that are associated with activation of AT_1 receptors are cytosolic and transmembrane tyrosine kinases, phosphatases, serine-threonine kinases, and other signaling enzymes (Blume et al. 1999). At the molecular level, the stimulation of the AT_1 receptor involves the activation of the phosphoinositide pathway, increase in intracellular calcium, inhibition of adenylate cyclase (depending on the cell type and tissue), and activation of transcription of c-*fos*, c-*myc*, AP1, tissue inhibitors of metalloproteinase (TIMP), and α_2-macroglobulin promoters (Blume et al. 1999). These processes are closely associated with vasoconstriction, leading to an increase in blood pressure. AT_1 receptor activation also contributes independently to chronic disease pathology by promoting vascular growth and proliferation, and endothelial dysfunction. These negative consequences of angiotensin II are partly counteracted by angiotensin II type-2 (AT_2) receptor stimulation, which has favorable effects on tissue growth and repair processes (Unger et al. 2011). Recent studies have shown that Ang II not only produces proinflammatory effects (Ruiz-Ortega et al. 2001), but also induces endothelial cell function (Campbell 2014). In addition, Ang II is one factor that increases oxidative stress by activating NADH/NADPH oxidases leading to production of superoxide anion and hydrogen peroxide and reduction in nitric oxide bioavailability (Campbell 2014). These processes result in activation and translocation of NF-κB to the nucleus, where it increases the expression of proinflammatory cytokines (TNF-α, IL-1β, IL-6), adhesion molecules (VCAM-1, ICAM-1, and P-selectin), promoting the adhesion of monocytes and neutrophil to vascular endothelial cells (Campbell 2014). Ang II also induces vascular remodeling by increasing expression of platelet-derived growth factor, basic fibroblast growth factor, insulin-like growth factor, and transforming growth factor along with release of matrix glycoprotein and metalloproteinases (Figs. 6.4 and 6.5) (Campbell 2014). Collective evidence suggests that Ang II plays a crucial role in inducing mechanical changes of arteries, with particular regard for increase in stiffness, and in determining changes in extracellular matrix components within the vascular wall (Neves et al. 2003; Brassard et al. 2005). Indeed, an increased collagen and fibronectin depositions, together with a decrease in elastin content, have been described in the media of small arteries from Ang II-infused animals (Neves et al. 2003; Brassard et al. 2005). Furthermore, the activation of RAAS is critical for the development of cardiovascular and renal fibrosis in hypertension and type II diabetes. Although all major components of RAAS contribute to profibrotic activity, Ang II, the principal constituent peptide of RAAS, plays an important role in the development of renal fibrosis (Mezzano et al. 2001). Chronic Ang II infusion induces extensive renal fibrosis in rats (Zhao et al. 2008). The angiotensin converting enzyme inhibitor enalapril retards glomerular gene expression of extracellular matrix proteins in diabetic rats (Nakamura et al. 1995). In mesangial cell culture, Ang II stimulates the synthesis of matrix proteins such as collagen type I and fibronectin; this can be blocked by the AT1 antagonist losartan (Ray et al. 1994). Based on above mentioned studies, it is proposed that targeting the RAAS may be an effective strategy to decrease fibrogenesis in progressive renal disease (Iwano and Neilson 2004).

As a component of RAAS, aldosterone (Aldo) regulates fluid and electrolyte balance through several pathways. In the first pathway, interactions between Aldo to the MR result in translocation of mineralocorticoid receptor (MR) to the nucleus, where it regulates target gene expression (Fig. 6.6) (Booth et al. 2002; Yang and Fuller 2012; Yang et al. 2011). In the second pathway, Aldo elicits rapid "nongenomic"

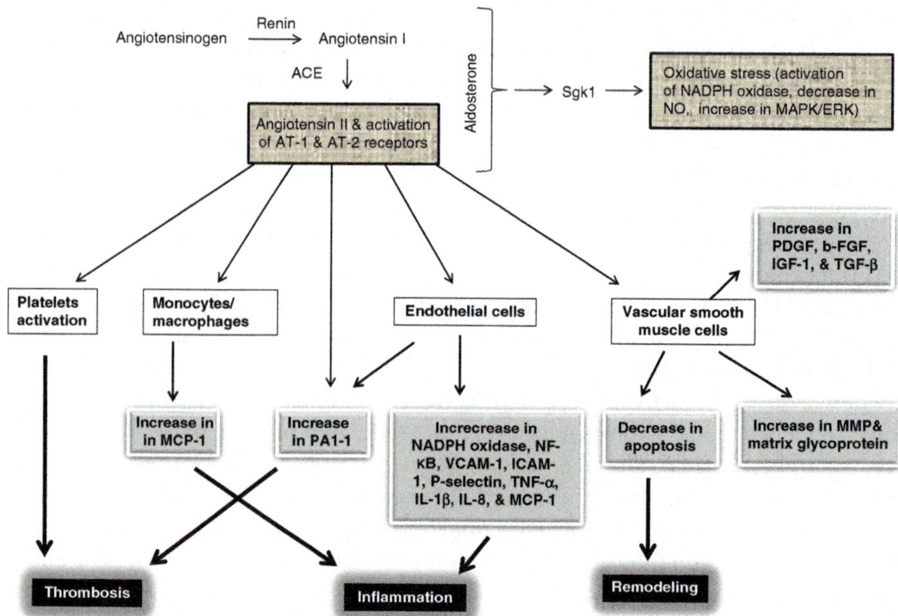

Fig. 6.4 Angiotensin II-mediated activation of platelets, endothelial cells, and vascular muscle cells and related biochemical processes. *AT1 receptor* Angiotensin II type-1 receptor, *AT2 receptor* angiotensin II type-2 receptor, *NF-κB* nuclear factor kappaB, *TNF-α* tumor necrosis factor-α, *IL-1β* Interleukin-1 beta, *VCAM-1* vascular cell adhesion molecule 1, *ICAM-1* intercellular Adhesion Molecule 1, *IL-6* Interleukin-6, *MMP* metalloproteinase, *PDGF* platelet-derived growth factor, *bFGF* Basic Fibroblast Growth Factor, *IGF-1* insulin-like growth factor 1, *TGFβ* transforming growth factor beta, *ACE* angiotensin converting enzyme, *NO* nitric oxide, *Sgk-1* serum and glucocorticoid-dependent kinase-1

effects on second messenger systems and signaling cascades of different tissues (Boldyreff and Wehling 2003; Dooley et al. 2012). An inappropriately high plasma level of Aldo contributes to progressive organ damage to the heart (Reil et al. 2012), vasculature, and kidney (Shibata and Fujita 2012), promoting myocardial (Nakamura et al. 2009) and vascular fibrosis (Tylicki et al. 2008), oxidative stress (Patni et al. 2007), and perivascular inflammation (Pu et al. 2003). RAAS is a key regulatory system of CVD, renal, and adrenal function, which maintains body fluid and electrolyte balance, as well as arterial pressure. The excessive activation of RAAS has been implicated in the development of metabolic syndrome (Massiera et al. 2001). Blockade of RAAS with angiotensin-converting enzyme inhibitors, angiotensin receptor blockers, and aldosterone receptor antagonists aldosterone antagonists has become the cornerstone of treatment in patients at various stages of heart and kidney disease, from early disease (hypertension and diabetes mellitus) to advanced severe end-organ cardiorenal failure (including nephropathy, heart failure, and combined cardiorenal failure). Thus, the activation of the sympathetic nervous system and RAAS, as well as increase in signaling through the mineralocorticoid receptor (MR)

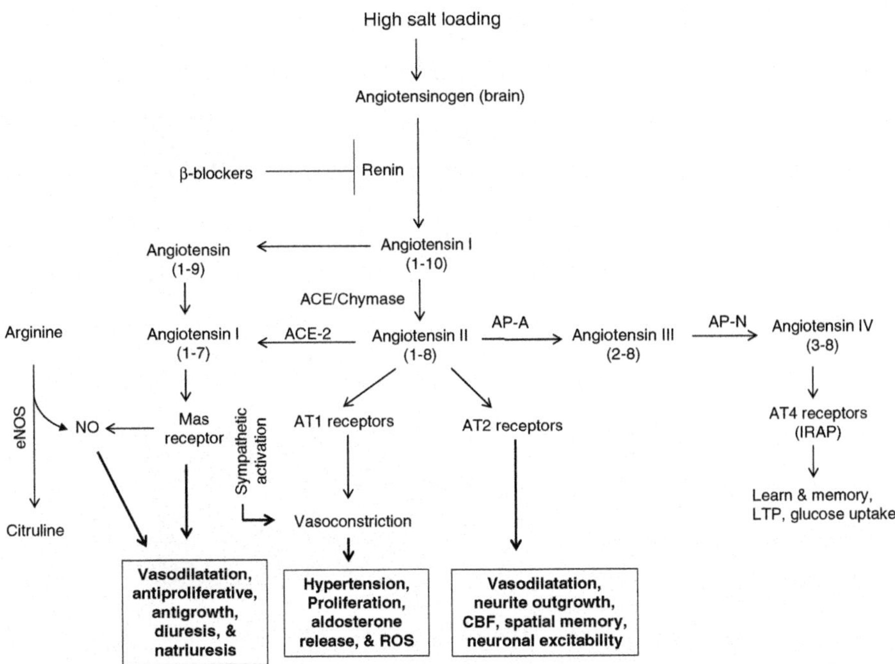

Fig. 6.5 Effect of salt loading on blood pressure in the brain. *ACE* Angiotensin converting enzyme, *AT1 receptor* angiotensin II type-1 receptor, *AT2 receptor* angiotensin II type-2 receptor, *ROS* reactive oxygen species, *AP-A* aminopeptidase A, *AP-N* aminopeptidase N, *IRAP* insulin-regulated aminopeptidase, *LTP* long term potentiation

result in increased production of reactive oxygen species (ROS), which are reactive chemical entities mainly generated by mitochondria, arachidonic acid cascade, and activation of NADPH oxidases, as intermediates in reduction–oxidation (redox) reactions (Farooqui 2013, 2014). Perturbations of the balance between ROS production and scavenging by antioxidant systems result in oxidative stress. In human subjects with obesity and type II diabetes, RAAS is involved in induction of oxidative stress and inflammation in the vascular tissue contributing to mineralocorticoid receptor-mediated insulin resistance (Aroor et al. 2013). Indeed, basic and clinical studies have demonstrated that elevated plasma aldosterone levels predict the development of insulin resistance by interfering with insulin signaling in vascular tissues. Aldosterone also suppresses insulin signaling via the downregulation of insulin receptor substrate-1 in vascular smooth muscle cells (Hitomi et al. 2007; Briet and Schiffrin 2011). Accumulating evidence suggests that oxidative stress is a common mediator of pathogenicity in cardiovascular diseases, type II diabetes, metabolic syndrome, neurodegenerative diseases, and various types of cancer (Lassègue et al. 2012). ROS also play an important role in the pathogenesis of inflammation through influencing platelet aggregation and migration of monocytes, hypertrophy, proliferation, fibrosis, and angiogenesis. These processes contribute to cardiovascular

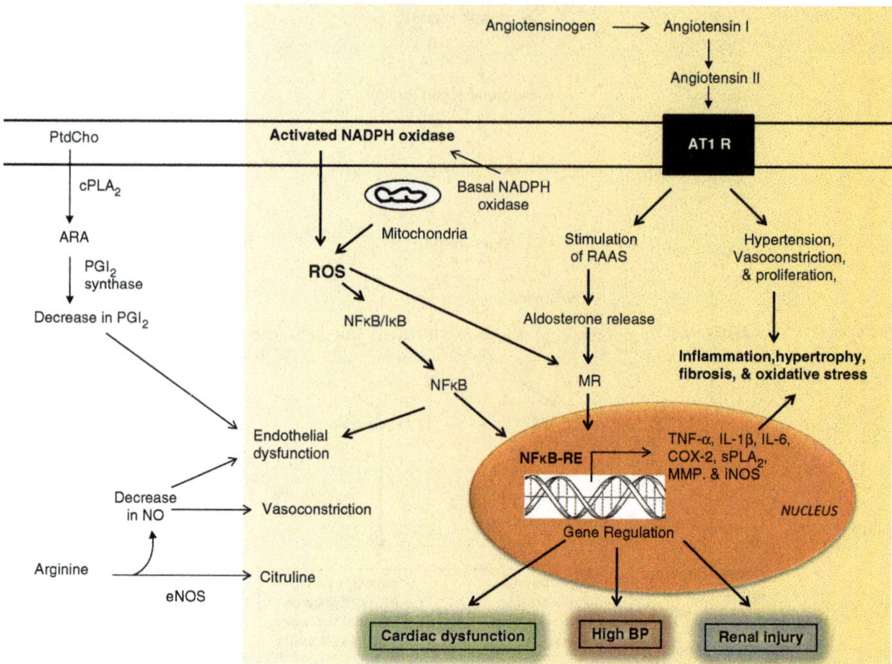

Fig. 6.6 Mineralocorticoid receptors-mediated signal transduction processes in cardiac and renal functions. *AT1 receptor* Angiotensin II type-1 receptor, *PM* Plasma membrane, *PtdCho* phosphatidylcholine, *cPLA₂* cytosolic phospholipase A₂, *COX-2* cyclooxygenase, *LOX* lipoxygenase, *NOS* inducible nitric oxide synthase, *NADPH oxidase* nicotinamide adenine dinucleotide phosphate-oxidase, *ARA* arachidonic acid, *ROS* reactive oxygen species, *NF-κB* nuclear factor kappaB, *NF-κB-RE* nuclear factor kappaB response element, *IκB* inhibitory subunit of NF-κB, *TNF-α* tumor necrosis factor-α, *IL-1β* Interleukin-1 beta, *IL-6* Interleukin-6, *sPLA₂* secretory phospholipase A₂, *MMP* matrix metalloproteinases, *NO* nitric oxide, *iNOS* inducible nitric oxide synthase

remodeling and endothelial dysfunction (Tian et al. 2008). The causal relationship between ROS production and hypertension following the long term consumption of high calorie diet, which is enriched in sodium, occurs at the vascular level where high calorie diet promotes oxidative stress, endothelial dysfunction, vascular inflammation, increased vascular reactivity and structural remodeling. Together, these responses lead to increased peripheral resistance and therefore to increased blood pressure (Bissonnette et al. 2008; Zhou et al. 2014). Production of ROS also contributes to insulin resistance and endothelial damage in persons predisposed to salt-sensitive hypertension and microalbuminuria (Zhou et al. 2014; Kim et al. 2006; Lastra et al. 2010; Laffer and Elijovich 2013). Sodium sensitive mechanism involves the abnormal activation of the aldosterone/MR pathway in the development of cardiovascular disease (CVD), metabolic syndrome, and chronic kidney disease (CKD). These pathological conditions lead to morbidity and mortality worldwide (Nagase et al. 2006, 2007; Matsui et al. 2008).

In CVD and CKD, endothelial dysfunction is also characterized by increase in vasoconstrictors, especially endothelin (ET)-1, which is a 21-amino-acid peptide with both mitogenic and vasoconstricting properties (Reriani et al. 2010). A growing body of evidence suggests that ET-1 plays a significant role in the regulation of vascular tone not only under normal conditions, but also in pathophysiological states such as atherosclerosis (Reriani et al. 2010). Administration of high concentrations ET1 results in vasoconstriction inducing physiological effects, such as alterations in arterial pressure. ET-1 produces its effects through two receptors, ET_A and ET_B. ET_A mediates contractions via activation of NOX, xanthine oxidase, lipoxygenase, uncoupled NO synthase, and mitochondrial respiratory chain enzymes. The ET_B induces relaxation on endothelial cells (Gomez-Alamillo et al. 2003). Many factors that normally stimulate ET-1 synthesis, (e.g., thrombin, AT-II) also cause the release of vasodilators such as prostacyclin (PGI_2) and/or NO, which oppose the vasoconstricting action of ET-1. Long-term administration of ET-1 receptor antagonist in humans improves coronary microvascular endothelial function, thus, supporting the role of endogenous ET system in the regulation of endothelial function in early atherosclerosis in humans (Reriani et al. 2010).

Consumption of fatty food in high calorie diet induces renal inflammation and aggravation of blood pressure in spontaneously hypertensive rats, via ROS (Chung et al. 2010).

ROS activate nuclear factor kappaB (NF-κB), which triggers a proinflammatory response characterized by increased levels of TNF-α, IL-1β, IL-6, and IL-8, increased expression of adhesion molecules, such as E-selectin, VCAM-1, and ICAM-1, and promotion of oxidative stress. NF-κB is one member of a ubiquitously expressed family of Rel-related transcription factors that serve as critical regulators of inflammatory related genes such as TNF-α and IL-1β (Baldwin 2001). NF-κB mainly exists in the cytosol as a pre-formed trimeric complex that consists of the inhibitory protein I-kB and the P50/P65 protein dimer. ROS induces redox changes that result in phosphorylation of the I-kB subunit, thereby activating its proteolytic digestion. When the inhibitor subunit is dislodged from the P60/P65 heterodimer, NF-κB can translocate to the nucleus, bind DNA, and initiate transcription (Gloire et al. 2006). Metabolic syndrome is a risk factor for CVD and CKD at least in part independent of type II diabetes and high BP per se, probably mediated by ROS. Moreover, the onset and maintenance of renal and cardiac damage may worsen metabolic syndrome features like high BP, leading to potential vicious cycles (Guarnieri et al. 2010).

Nitric oxide contributes to several other aspects of vascular health, including inhibition of platelet aggregation, monocyte adhesion to endothelial cells and abnormal smooth muscle cell proliferation, highlighting it as an important "antiatherogenic" moiety. Under physiologic conditions, endothelial cells synthesize and secrete a large spectrum of antiatherosclerotic factors, the most characterized of which is nitric oxide (NO), a gas synthesized from the metabolism of L-arginine by constitutive endothelial NO synthase (eNOS), an enzyme, which requires tetrahydrobiopterin as a cofactor (Fig. 6.6) (Forstermann and Munzel 2006). Once NO is produced, it diffuses to the vascular smooth muscle cells and activates guanylate cyclase, which leads to cGMP-mediated vasodilatation leading to cardioprotection, relaxation of smooth muscle cells, prevention of leukocyte adhesion and migration into the arterial wall, muscle cell proliferation, platelet adhesion and aggregation, and expression of adhesion molecule.

Under pathological conditions (CVD and CKD), the endothelium undergoes functional and structural changes losing its protective role, becoming a pro-atherosclerotic structure. The loss of the normal endothelial function results in impairment in NO bioavailability. This is followed by an increase in the production of ROS (Taddei et al. 2003). Following decrease in NO bioavailability, the endothelium activates various compensatory physiological pathways such as the release of prostacyclin, C-type natriuretic factor, 5-hydroxytryptamine serotonin, adenosine triphosphate, substance P, and acetylcholine, and other endothelium-derived hyperpolarizing factors (EDHFs) (Amiya et al. 2014). Another important point in relation to blood pressure is that there are pronounced sex differences in the release and activity of vasoactive mediators. Females have greater levels of the vasodilator NO compared to males (Thompson and Khalil 2003). In contrast, males tend to have greater Ang II (De Silva et al. 2009; Viegas et al. 2012), endothelin (ET) (Kittikulsuth et al. 2013; Tostes et al. 2008), and hydroxyeicosatetraenoic acids (HETE)-mediated vasoconstriction compared with females (Fava et al. 2012; Singh and Schwartzman 2008). There are also reports of sex differences in calcium handling (Devynck 2002).

The kallikrein-kinin system (KKS) is another intricate endogenous system associated with BP regulation, inflammation, cardiovascular homeostasis, analgesic responses, pain-transmitting mechanisms, release of cytokines and chemokines, generation of prostacyclin and nitric oxide, and cell proliferation (Pesquero and Bader 1998; Pereira et al. 2014). KKS comprises the serine proteases kallikreins, the protein substrates kininogens and the generation of two kinines peptides (bradykinin, a vasoactive nonapeptide (Arg^1-Pro^2-Pro^3-Gly^4-Phe^5-Ser^6-Pro^7-Phe^8-Arg^9) and kallidin, a vasoactive decapeptide (Lys^1-Arg^2-Pro^3-Pro^4-Gly^5-Phe^6-Ser^7-Pro^8-Phe^9-Arg^{10}), which act through activation of G-protein-coupled constitutive B_2 or inducible kinin B_1 receptors linked to signaling pathways involving increase in intracellular Ca^{2+}) concentrations and/or release of mediators including arachidonic acid metabolites, NO and endothelium-Derived Hyperpolarizing Factor (Fig. 6.7). Carboxypeptidase N, also known as kininase I, and carboxypeptidase M remove arginine from the carboxyl terminus of the kinins and generate their des-Arg derivatives, which are agonists mainly of the B_1 receptor. The kinins can also be inactivated by the action of kininase II (angiotensin-converting enzyme, ACE), neutral endopeptidase, and endothelin-converting enzyme which remove two amino acids from the carboxyl terminus (Kakoki and Smithies 2009). ACE has a 30 times lower K_m and 10 times higher k_{cat} for the kinins than for angiotensin I. Stimulation of bradykinin receptors by kinins elevates [Ca^{++}]i results in the activation of phosphatidylinositol (PtdIns)-specific phospholipase C (PI-PLC) in Gq-protein-dependent and Gq-protein-independent ways (Fig. 6.7). Allosteric activation of PI-PLCβ isoforms by Gq/11 proteins and direct tyrosine phosphorylation of PI-PLCγ isoforms by bradykinin receptors (Duchene et al. 2005) both play an important role in the bradykinin-induced changes in [Ca^{2+}]i.

Stimulation of either B_1R or B_2R results in activation of eNOS and phospholipase A_2 (PLA_2) activities along with the generation of nitric oxide and prostacyclin in endothelial cells, at least partly through increase in intracellular calcium levels ([Ca^{++}]i)

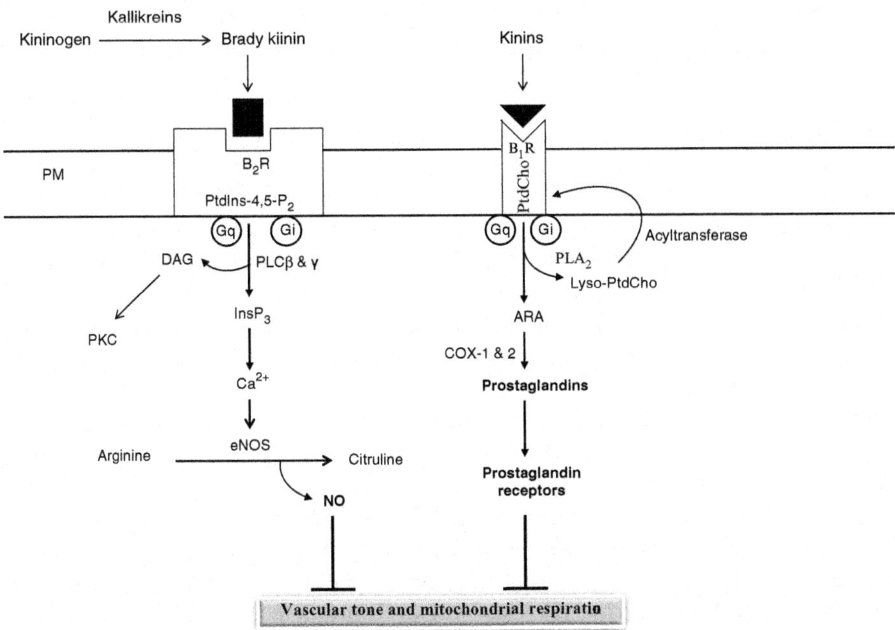

Fig. 6.7 Signal transduction mechanisms associated with stimulation of bradykinin receptors. *PM* Plasma membrane, *B$_1$R* and *B$_2$R* bradykinin receptors, *PLCβ* and *γ* phospholipase Cβ and γ, *PLA$_2$* phospholipase A$_2$, *PtdIns 4,5-P$_2$* phosphatidylinositol 4,5-bisphosphate, *PtdCho* phosphatidylcholine, *lyso-PtdCho* lyso phosphatidylcholine, *InsP$_3$* inositol 1,4,5-trisphosphate, *DAG* diacylglycerol, *ARA* arachidonic acid, *COX-1* and *2* cyclooxygenases 1 and 2, *Ca^{2+}* intracellular calcium, *PKC* protein kinase C, *eNOS* endothelial nitric oxide synthase, *NO* nitric oxide, *Gq* and *Gi* G protein

(Zubakova et al. 2008). KKS lowers blood pressure by promoting vasodilation, natriuresis and diuresis generally opposing the hypertensive effects of RAAS (Nolly et al. 1992). Moreover, activation of KKS reduces the generation of reactive oxygen species and plays a protective role against organ damage in the heart and kidney (Kayashima et al. 2012). Thus, in the cardiovascular system, the KKS exerts a fine control of vascular smooth muscle tone and arterial blood pressure, and plays a significant cardioprotective effect (Pesquero and Bader 1998; Pereira et al. 2014).

It is also reported that systolic BP and diastolic BP are correlated inversely with plasma antioxidant potential and directly with levels of F2-isoprostanes (prostaglandin-like mediators generated nonenzymically by free radical catalyzed peroxidation of arachidonic acid) (Rodrigo et al. 2007). In contrast, other plasma BP modulator levels, such as AT II, ET-1, renin, aldosterone, and homocysteine, among others, show no significant differences between normotensive and hypertensive subjects. These associations strongly support a role for oxidative stress in the modulation of BP.

6.3   Molecular Mechanisms Associated with Salt Sensitivity in the Brain

Like kidneys, brain also has specific and discrete short-term rapid-acting and long-term slow-acting pathways that modulate BP. The brain RAAS plays an essential role in the pathogenesis of hypertension (Clempus and Griendling 2006). Angiotensin II (Ang II) is the main effector peptide of the RAAS, and elicits its pressor response primarily through AT1R (D'Autreaux and Toledano 2007; Carlson and Wyss 2008). Although Ang-II is considered to be the primary effector peptide for the RAAS activation, it is becoming increasingly evident that angiotensin-1-7 (Ang (1-7)), which is generated by angiotensin converting enzyme 2 (ACE2) cleaving the carboxyl terminus phenylalanine residue from Ang II, plays an important role in controlling cardiovascular and cerebrovascular functions (Santos and Ferreira 2007). In the brain, Ang (1-7) increases nitric oxide levels via activation of the Mas receptor (MasR) and the angiotensin II type 2 receptor (AT2R) (Fig. 6.3) (Feng et al. 2010). In the brain, nitric oxide acts as a sympatho-inhibitory molecule (Gironacci et al. 2000; Patel 2000). It reacts with superoxide to form peroxynitrite. This reaction depends upon the diffusion of nitric oxide and is critical in the maintenance of sympathetic output. By acting through Mas, Ang (1-7) promotes vasodilation, antiproliferation and antihypertrophy (Ferrario et al. 2005; Santos et al. 2007). Accumulating evidence indicate that Ang (1-7) has beneficial effects in cardiovascular diseases. By cleaving Ang-II into Ang-(1-7), ACE2 may play a pivotal role in counter-regulating the actions of the well documented ACE/Ang II/AT1R axis and be beneficial for the cardiovascular system. Robust expression of ACE2 and Mas, and actions of Ang (1-7) has been reported to occur in the brain of both animals and humans. Central cardiovascular effects of Ang (1-7) include baroreflex facilitation, hypotension, bradykinin potentiation, vasopressin, bradykinin and NO release and prevention of norepinephrine release (Xia and Lazartigues 2008).

Ang II activates phospholipase A_2 (PLA_2) to generate ARA (Gorin et al. 2001), which has been implicated in certain cellular responses via activation of protein kinases and/or phosphatases (Gorin et al. 2001). ARA has been shown to activate JNK and Akt/PKB (protein kinase B) in response to interleukin-1 and Ang II respectively by a mechanism that does not require prostanoid production (Gorin et al. 2001; Droge 2002). Ang II-mediated generation of ROS (superoxide radical and H_2O_2) is closely associated with biological responses in cerebrovascular system involved in neural growth, migration and modification of the extracellular matrix. NAD(P)H oxidase systems of the Nox family are a major source of ROS implicated in Ang II-induced oxidative stress (Campbell 2014). The Nox oxidases are membrane-integrated proteins and their activation requires stimulus-induced membrane translocation of cytosolic proteins including the small GTPase Rac (Campbell 2014).

In the brain slow pathway involves the activation of central aldosterone (Aldo), mineralocorticoid receptors (MR), epithelial sodium channels (ENaCs), endogenous ouabain (EO), and AT1 receptors (Blaustein et al. 2012; Hamlyn and Blaustein 2013). High-salt diets raise plasma Na^+, which enters the brain via the fenestrated vasculature of the circumventricular organs raising cerebrospinal fluid (CSF) Na^+ levels.

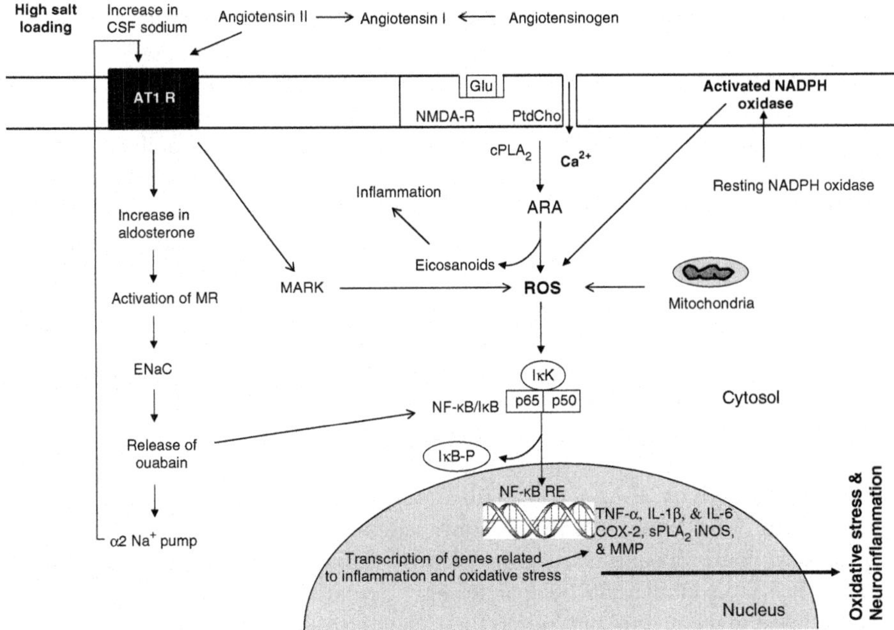

Fig. 6.8 Effect of high salt loading on signal transduction processes associated with oxidative stress and high blood pressure. *AT1 receptor* Angiotensin II type-1 receptor, *MR* mineralocorticoid receptors, *PM* Plasma membrane, *NMDA-R* N-methyl-D-aspartate receptor, *Glu* glutamate, *PtdCho* phosphatidylcholine, *cPLA₂* cytosolic phospholipase A₂, *COX-2* cyclooxygenase, *NOS* inducible nitric oxide synthase, *NADPH oxidase* nicotinamide adenine dinucleotide phosphate-oxidase, *ARA* arachidonic acid, *ROS* reactive oxygen species, *NF-κB* nuclear factor kappaB, *NF-κB-RE* nuclear factor kappaB response element, *IκB* inhibitory subunit of NF-κB, *TNF-α* tumor necrosis factor-α, *IL-1β* Interleukin-1 beta, *IL-6* Interleukin-6, *sPLA₂* secretory phospholipase A₂, *MMP* matrix metalloproteinases, *iNOS* inducible nitric oxide synthase

This process stimulates magnocellular neurons. Further, increasing brain Na^+ and Cl^- acutely by infusion has a well recognized short-term pressor effect that is associated with increased sympathetic outflow (Shah and Jandhyala 1991). Studies on Dahl salt-sensitive rats have indicated that in response to a high-salt intake endogenous ouabain (EO), a steroid hormone, is secreted and released by the brain (Blaustein et al. 2012) (Fig. 6.8). Secretion of EO not only promotes increase in BP, but also facilitates excretion of salt. In addition, EO also modulates Ca^{2+} signaling leading to cardiotonic and vasotonic effects. These observations support the view that EO regulates slow, modulatory pathways in the brain as well as in the peripheral tissues (Blaustein et al. 2012). EO has been reported to activate NF-κB in a time- and concentration-dependent manner in the hippocampus (de Sá et al. 2013; Kawamoto et al. 2012). In the hypothalamus, Na^+ activates the Aldo-EO-ANG II pathway that enhances sympathoexcitatory mechanisms and sustains increase in sympathetic nervous activity. In peripheral tissues, increase in plasma EO level produces the activation of a Src-mediated protein kinase cascade, which modulates Na^+ and Ca^{2+} transporter protein expression and protein phosphorylation leading not only to the upregulation of Ca^{2+}

signaling mechanisms and contractile responses in arterial smooth muscle, but also to the downregulation Ca^{2+} signaling and vasodilatory responses in endothelial cells. Because of the presence of RAAS in the brain and peripheral tissues (kidneys), there is a striking parallelism in the molecular changes. Thus, EO-mediated changes in α_2 Na^+ pumps and the protein kinase cascade play a key role in the modulation, even though the signals appear to alter the expression of different proteins in the brain and peripheral tissues (Huang et al. 2011; Zulian et al. 2010). These central and peripheral effects of EO, especially the slow, modulatory effects, are synergistic. These processes not only result in enhanced vasoconstriction and increased BP, but also initiate the vascular structural remodeling and total peripheral vascular resistance often observed in chronic hypertension (Blaustein et al. 2012).

All components of the RAAS and KKS are also found in various regions of the brain including cerebral cortex, brain stem, cerebellum, hypothalamus, hippocampus, and pineal gland, among others (Raidoo and Bhoola 1998). Both RAAS and KKS play an important role in the regulation of BP through the modulation of sympathetic nerve activity. RAAS has a stimulatory influence on the sympathetic nervous system. The brain RAAS not only augments presynaptic facilitation of sympathetic neurotransmitter release, but also enhances the central sympathetic outflow. It is well known that Ang II strongly potentiates sympathetic neurotransmitter release in the central nervous system. In contrast, the inhibitors of the RAAS (angiotensin converting enzyme inhibitors, angiotensin receptor blocker, and direct renin inhibitors) suppress sympathetic hyperactivity in the brain. The release of vasopressin, glutamate, and GABA is also modulated by the RAAS inhibition. The clinical significance of the modulation of central sympathetic neurotransmitter release by the RAAS inhibitors is not fully understood. However, above mentioned observations are consistent with the view that RAAS inhibitors regulate high BP by their neurosuppressive actions (Tsuda 2012). Similarly in the brain, KKS modulates BP through the generation and release of inflammatory mediators such as eicosanoids, cytokines, nitric oxide (NO) and free radicals, which contribute to oxidative stress and neuroinflammation. KKS also induces the release of excitatory amino acids, increasing intracellular $(Ca^{2+})i$ levels and inducing brain excitotoxicity. It is proposed that these peptides also modulate the disruption of the blood–brain-barrier (BBB) and dilation of the parenchyma of cerebral arteries causing edema (Sobey et al. 1997). The mitogen-activated protein kinase pathway, which culminates in the transcription of many genes involved in later responses (Leeb-Lundberg et al. 2005) is also activated by bradykinin receptors (B1R and B2R). The stimulation of both B1R and B2R lead to classical G-protein activation, leading to the generation of different second messengers.

6.4 High BP and Visceral Diseases

Metabolic syndrome is a multifactorial pathological condition, which is accompanied by obesity, insulin resistance, glucose intolerance, hyperlipidemia, high blood pressure, low grade inflammation, and oxidative stress. Over consumption of high

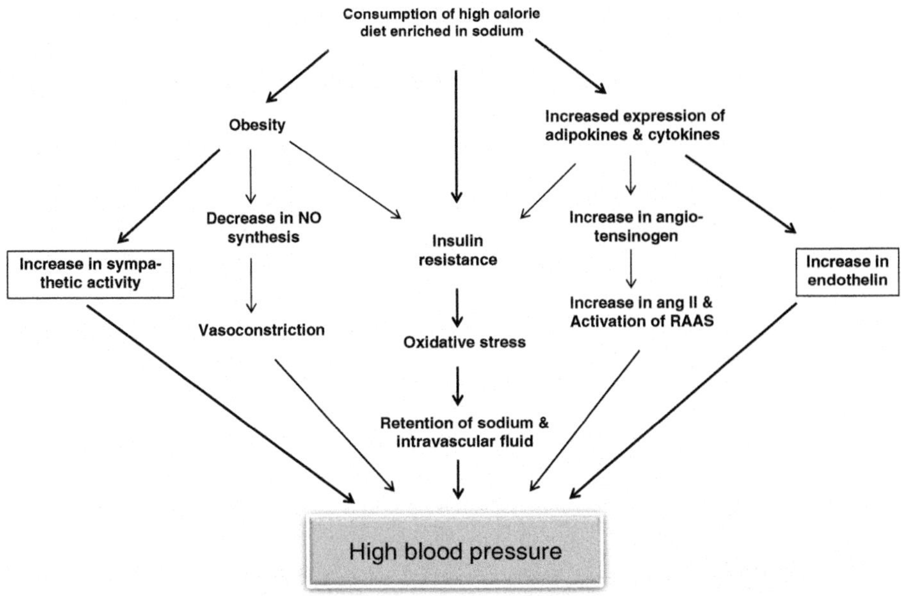

Fig. 6.9   Onset of high BP through the consumption of high calorie diet enriched in sodium

calorie diet (simple sugar, saturated and n-6 fatty acids, high salt, and low in fiber), stress, endocrine dysfunction, and low grade inflammation are prominent factors that contribute to the pathogenesis of metabolic syndrome (Cai 2013) (Fig. 6.9). Obesity, type II diabetes and hypertension are closely linked with insulin resistance and elevated sympathetic nervous activity. These factors are high risk factors for subsequent cardiovascular and renal complications.

High BP is a classical feature of the metabolic syndrome, and it has been reported that the metabolic syndrome is present in up to one third of hypertensive patients (Cuspidi et al. 2004).

The involvement of RAAS recently has been described in the metabolic syndrome. Ang-II induces adipogenesis (differentiation into adipocytes) (Clasen et al. 2005; Erbe et al. 2006) and lipogenesis (triglyceride storage in adipocytes) in vitro (Jones et al. 1997). However, the in vivo role of RAAS in the metabolic syndrome has been unclear. The effects of Ang-II on adipose tissue are mediated by Ang-II type 1 and type 2 receptors. Two mechanisms may be responsible for High BP in metabolic syndrome. One mechanism is associated with increase in endothelin and Ang-II and the other mechanism involves the increase in generation of uric acid. Contribution of endothelin and angiotensin-II in development of high blood pressure has been discussed above.

The endogenous synthesis of uric acid mainly occurs through the consumption of fructose corn syrup and sweets containing cane sugar in liver and intestines. Uric acid formation also takes place from purine containing foods (Fig. 6.10). The catabolism of purines occurs through a series of enzymic steps involving the enzyme xanthine oxidase. This enzyme transforms hypoxanthine to xanthine and subsequently to

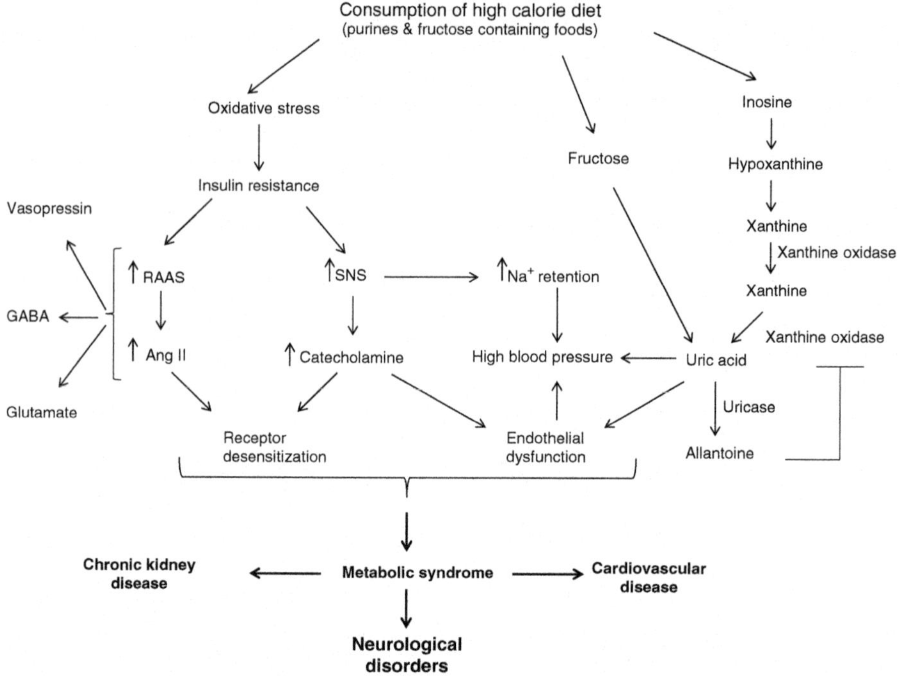

Fig. 6.10 Consumption of high calorie diet enriched in purine and fructose corn syrup containing food and development of high BP, endothelial cell dysfunction and metabolic syndrome. *RAAS* Renin–angiotensin–aldosterone system, *Ang II* angiotensine II, *SNS* sympathetic nervous system and upward arrow indicate increase in activity and levels

uric acid (Khitan and Kim 2013; Jalal et al. 2013). In the kidney, uric acid is readily filtered by the glomerulus and subsequently reabsorbed by the proximal tubular cells of the kidney; the normal fractional excretion of uric acid is approximately 10 %. In metabolic syndrome, reabsorption of uric acid is markedly increased. High levels of uric acid not only inhibit the generation of nitric oxide, but also contribute to worsening of insulin resistance in animal models of type II diabetes and metabolic syndrome (Khosla et al. 2005; Corry et al. 2008; Farooqui 2013). Uric acid contributes to kidney fibrosis mainly by inducing inflammation, endothelial dysfunction, oxidative stress, and activation of the renin-angiotensin system. In addition, elevated levels of uric acid in serum (hyperuricemia) are causative factor in gout and urolithiasis due to the formation and deposition of monosodium urate crystals. Hyperuricemia is a common finding in CKD due to decreased uric acid clearance.

Several mechanisms contribute to increase in BP. These mechanisms include increase in sympathetic nervous system activity (Farah et al. 2006; Verma et al. 1999), increase in oxidative stress, elevation in circulating catecholamines and increase in secretion of endothelin-1 (Tran et al. 2009; Juan et al. 1998). In addition, high levels of uric acid also produce deleterious effects on endothelial function, platelet adhesion and aggregation, and increase the risk of high blood pressure.

These processes are closely associated with cardiovascular diseases, kidney diseases, and gout (Alper et al. 2005). Chronic increase in uric acid levels is also linked to reduction in adiponectin and elevation in E-selectin, in parallel with positive effects such as reduction in nitrotyrosine and increase in total antioxidant capacity (Bo et al. 2008). Collective evidence suggests that long term over-consumption of high calorie diet is closely associated with the pathogenesis of metabolic syndrome, a disease, that may contribute to heart disease, type II diabetes, and chronic kidney diseases.

6.5 High BP and Neurological Disorders

Brain is the most perfused organ of the body. Blood flow to cerebral hemispheres is maintained through capillary beds connected to the pial vasculature by penetrating arterioles, and the pial vasculature stems from a network of arteries, which branch and supply blood to the anterior, middle, and posterior part of the brain. Maintenance of physiological function of brain depends on a constant blood supply through this network. To meet metabolic demands efficiently, the blood flow to the brain should be maintained to approximately 750 ml per minute or about 50 ml/100 ml of brain tissue per minute (Ito et al. 2005). Cerebral perfusion pressure is primarily maintained by an efficient cardiac output, patent elastic arterial walls modulated by the microvascular resistance of the brain, and adequate venous drainage. Other factors that modulate adequate cerebral blood flow include blood viscosity, blood flow velocity, and appropriate vascular resistance of the brain vasculature (Roher et al. 2012). High BP produces changes to the structure and function of these blood vessels, which impacts perfusion to various areas in the brain. Thus, hypoperfusion interferes with the delivery of oxygen and nutrients required to meet metabolic demands. Thus, hypoperfusion has been linked to white matter degradation, gray matter atrophy, and cognitive deficits. Several studies have indicated that prolonged reduction in blood flow due to aging and atherosclerosis not only results in high BP-mediated damage in occipito-temporal, prefrontal, and medial temporal lobe regions, but also impacts blood vessel function (Beason-Held et al. 2007). In fact, a mild chronic cerebrovascular hypoperfusion and hypometabolism caused by decrease in cerebral blood flow may lower metabolic rates of glucose, and oxygen. This may be one of the very early events in the pathogenesis of Alzheimer Disease (Iadecola 2004). The reason for brain hypometabolism may include defects in glucose transport at the blood–brain barrier, glycolysis, and/or mitochondrial function. Increase in high calorie-mediated obesity and BP in the middle age may also increase the risk of developing AD by decreasing the vascular integrity of the BBB, resulting in protein extravasation into brain tissue (Xu et al. 2011; Kalaria 2010). Furthermore, the decrease in cerebral blood flow due to atherosclerosis may negatively affect the synthesis of proteins required for learning and memory and eventually leading to neuritic injury and neuronal death. Thus, high BP not only contributes to cognitive dysfunction (Giordano et al. 2012), but in older adults with mild cognitive impairment and cardiovascular risk factors are twice as likely to develop dementia compared to those without such risk factors (Johnson et al. 2010). Moreover, cognitive declines may be faster in patients with mild cognitive impairment and high BP, compared to those without high BP (Goldstein et al. 2013). This makes high BP a major risk factor for vascular cognitive impairment in neurological disorders. High BP contributes to both the development and progression of cerebrovascular disease (MacMahon et al. 1990). There is growing evidence that high BP is the most powerful modifiable risk factor for cerebral vessel dysfunction and it is closely associated with cognitive decline (Nelson et al. 2011). The relationship among aging, high blood pressure and cognitive function is complex and not completely understood. Nevertheless, changes in BP are considered as a marker of cerebrovascular health (Barodka et al. 2011). In seniors, the consumption of high calorie diet containing high salt not only increases oxidative stress, insulin resistance, and elevates circulating catecholamines (Tran et al. 2009), decreases nitric oxide (NO) bioavailability, but also enhances RAAS, and Ang-II, levels (a potent vasoconstrictor) in a dose-dependent manner (Tran et al. 2009; Wright et al. 2013). As stated above, the brain RAAS system contains several functional components to produce the active ligands Ang-II, Ang-III, Ang-IV, Ang (1-7), and Ang-(3-7). These ligands interact with several receptor proteins including AT_1, AT_2, AT_4, and Mas, which are distributed within the central and peripheral nervous systems (Wright and Harding 2010) and modulate BP.

Long term consumption of high calorie diet causes metabolic syndrome. This pathological condition is characterized by hyperglycemia, insulin resistance, and high BP along with remarkable alterations in brain metabolism (Farooqui 2013; Farooqui et al. 2012). The exact mechanisms by which high BP in metabolic syndrome induces neurochemical changes in the brain remain elusive. However, cerebral insulin signaling, insulin resistance, glucose toxicity, oxidative stress, accumulation of advanced glycation end products, increase in cytokines, upregulation of amyloid precursor protein gene, promotion of amyloid precursor protein cleavage into Aβ peptide, or reduction in Aβ peptide clearance and the accumulation of Aβ peptide, and tau protein may contribute to a wide range of metabolic and vascular disturbances closely associated with the pathogenesis of Alzheimer disease (AD) (Fig. 6.11) (Shi et al. 2000; Saido et al. 1994; Weller et al. 2002; Farooqui 2013). In addition, increase in blood pressure may promote AD progression, rather than initiation, through mechanisms involving Aβ peptide accumulation. Studies in mouse models of AD with mutations in the amyloid precursor protein or presenilin genes have indicated that regulation of cerebral blood flow is impaired, and uncontrollable episodes of hypotension or hypertension result in undesired fluctuations in cerebral blood flow that may contribute to neuronal dysfunction (Niwa et al. 2000, 2002). Furthermore, in animal model of AD greater brain atrophy (specifically hippocampal and cortical atrophy) is accompanied with increased blood pressure in both non-demented and demented persons (Firbank et al. 2007; Jouvent et al. 2010). The mechanisms underlying this association are unclear. However, some studies have indicated that changes in BP can result in subcortical vascular pathology causing cortical neuronal apoptosis, which is thought to lead to brain atrophy (Viswanathan et al. 2006) along with salt induced changes in brain RAAS.

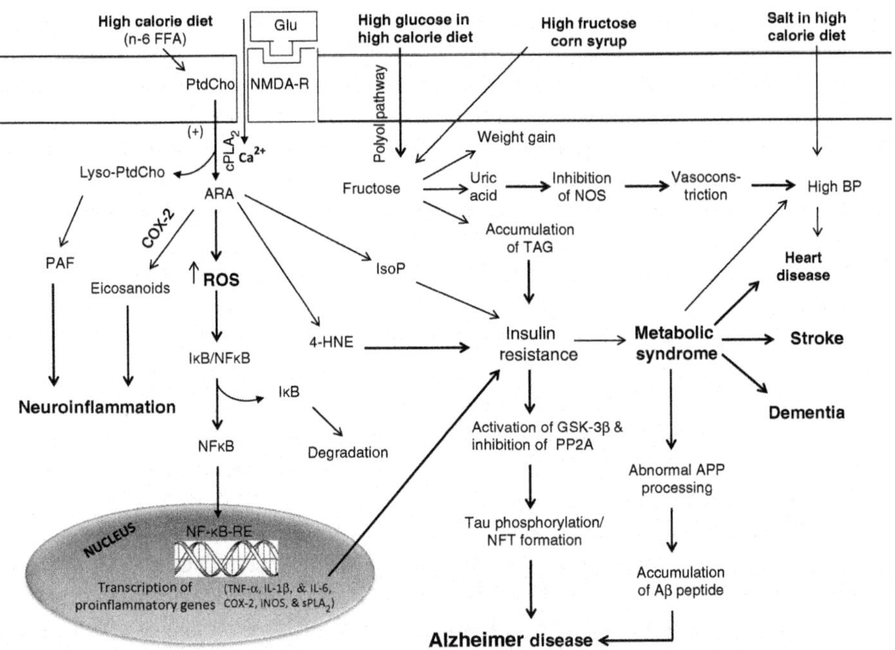

Fig. 6.11 Contribution of high BP and sodium rich high calorie diet in the pathogenesis of neurological disorders. *NMDA-R* N-methyl-D-aspartic acid receptor, *Glu* glutamate, *cPLA₂* cytosolic phospholipase A₂, *PtdCho* phosphatidylcholine, *COX-2* cyclooxygenase-2, *lyso-PtdCho* lyso-phosphatidylcholine, *ARA* Arachidonic acid, *PAF* platelet activating factor, *TNF-α* tumor necrosis factor-α, *IL-1β* interleukin-1β, *iNOS* inducible nitric oxide synthase, *sPLA₂* secretory phospholipase A₂, *ROS* reactive oxygen species, *TAG* triacylglycerol, *NF-κB* nuclear factor-kappa B, *4-HNE* 4-hydroxynonenal, *IsoP* isoprostane, *GSK-3β* Glycogen synthase kinase 3β, *PP2* Protein phosphatase 2, *NFT* neurofibrillary tangles

In RAAS system, ACE is a carboxy dipeptidase, which is fairly nonspecific. It not only hydrolyzes many peptides (as small as three amino acids), but also degrades Aβ1-42 peptide (Hemming and Selkoe 2005; Zou et al. 2009). Recent studies have indicated that decrease in ACE activity may increase the risk of AD. In addition, it is also reported that low levels of ACE in CSF are also associated with lower CSF tau and ptau (Jochemsen et al. 2014). The mechanisms explaining the relation between lower ACE and lower CSF tau levels in AD remain largely unknown. However, it is proposed that Ang-II may contribute to this effect. It is shown that central infusion of Ang-II in normal rat brains induced tau phosphorylation in a dose-dependent manner via activating glycogen synthase kinase 3β. Furthermore, tau phosphorylation can be reversed by co-administration of the Ang-II type 1 receptor antagonist, losartan, which is commonly used for the treatment of high BP (Tian et al. 2012). These results suggest that higher ACE levels which lead to higher Ang-II levels can induce tau pathology in the brain of AD patients, and this in turn result in higher tau levels in the CSF (Jochemsen et al. 2014). Evaluation of brain tissue from AD⁺ACE$^{10/10}$ mice, which are generated by crossing *ACE$^{10/10}$* mice with transgenic *APP$_{SWE}$/PS1$_{\Delta E9}$*

mice at 7 and 13 months indicates that levels of both soluble and insoluble brain $A\beta_{1-42}$ are significantly reduced compared with those in AD^+ mice (Bernstein et al. 2014). Furthermore, both plaque burden and astrogliosis are also drastically reduced in $AD^+ACE^{10/10}$ mice. Administration of the ACE inhibitor (ramipril) increases $A\beta$ levels in $AD^+ACE^{10/10}$ mice compared with the levels induced by the ACE-independent vasodilator hydralazine. Overall, $AD^+ACE^{10/10}$ mice have less brain-infiltrating cells, consistent with reduced AD-associated pathology. However, ACE-overexpressing macrophages are abundantly present around $A\beta$ plaques. At 11 and 12 months of age, the $AD^+ACE^{10/WT}$ and $AD^+ACE^{10/10}$ mice were virtually equivalent to non-AD mice in cognitive ability, as assessed by maze-based behavioral tests. Collectively these results demonstrate that an enhanced immune response, coupled with increased myelomonocytic expression of catalytically active ACE, prevents cognitive decline in a murine model of AD (Bernstein et al. 2014).

RAAS system has been identified in the nigrostriatal system (Yanagitani et al. 1999; Labandeira-Garcia et al. 2013). As mentioned above, Ang-II acts on the inflammatory cascade, via AT1R, producing high levels of ROS through the activation of the NADPH oxidase complex, a process closely associated with death of dopaminergic neurons in the nigrostriatal system, in Parkinson disease (Labandeira-Garcia et al. 2013; Ohshima et al. 2013) and AT1R blockage reduces dopaminergic neuron loss as well as lipid peroxidation in the 6-OHDA model of PD in rats. Collective evidence suggests that manipulation of RAAS using AT1R antagonists or ACE inhibitors may contribute to the treatment of PD (Ohshima et al. 2013; Saavedra 2012).

Alterations in brain RAAS have been reported to contribute to dementia and blockade of RAAS system may delay/retard the onset of dementia, a pathological condition that is characterized by a decrease in intelligence, memory, and perception and may be caused by various diseases. Logical and critical thinking, judgement, retentive memory, and short-term memory are impaired, while remote memory (long-term memory) can remain for a long time. In addition, personality may deteriorate (Iwanami et al. 2009; Wright and Harding 2010). The precise molecular mechanism of RAAS-mediated processes and their role are not fully understood. However, it is becoming increasingly evident that octapeptide Ang II disrupts learning and memory; while the hexapeptide angiotensin-IV (Ang-IV) facilitates memory acquisition and consolidation (Wright et al. 2013).

Following stroke-mediated injury, the production of inflammatory mediators promotes inflammatory responses by increasing the affinity of binding between immune cells and damaged neural cells. In cerebrovascular system, high levels of enzymic and non-enzymic lipid mediators also increase the build-up of atherosclerotic plaques in blood vessels and if a cerebral vessel becomes completely occluded, an ischemic event ensues leading to a cascade of inflammatory responses (Hjelstuen et al. 2006) triggering factors, which lead to microglial activation, resulting into more cytokine generation, induction of adhesion molecules, and selectin within the cerebral vasculature within 24 h of the ischemic insult (Farooqui 2013). Increase in chemokine stimulates inflammatory cell chemotaxis into ischemic brain, especially around the penumbra, or the infarct's border promoting brain damage.

Thus, high BP is the major modifiable risk factor for stroke and small vessel disease and is known to be the most-important factor for macrovascular cerebral complications such as atherothrombotic stroke and, consequently, vascular dementia. High BP also predisposes to more subtle cerebral processes based on arteriolar narrowing or microvascular pathological changes (Sierra 2012). It is proposed that cerebral microvascular disease is closely associated with vascular cognitive impairment. The mechanisms underlying high BP-mediated cognitive changes are complex and not yet fully understood. Both high and low BP (especially in the elderly) have been linked to progression of cognitive decline and dementia (Sierra 2012).

6.6 Dietary Sodium and Pathogenesis of Neurological Disorders

The consumption of processed foods containing high amounts of salt may contribute to increasing incidences of stroke, dementia, and autoimmune diseases along with alterations in cerebral blood flow (Rose et al. 1988; Ashby et al. 2012; Farrall and Wardlaw 2009; Farooqui and Farooqui 2015). Thus, marked increase in BP has been reported to occur in stroke patients (Lucke-Wold et al. 2012). High BP increases levels of cell adhesion molecules, vascular adhesion molecules, and selections. Subsequently, these molecules induce inflammatory responses by increasing the affinity of binding between immune cells and damaged tissue. The molecules also increase the build-up of atherosclerotic plaques in blood vessels through similar binding mechanisms. If a cerebral vessel becomes completely occluded, an ischemic event ensues leading to a cascade of inflammatory responses and still further immune cell binding (Hjelstuen et al. 2006). Similarly, decrease in cerebral flow and BBB disruption are also features of AD (Ashby et al. 2012).

The blockade of the RAAS can modulate immune responses and modulate symptoms of Experimental autoimmune encephalomyelitis (EAE) (Platten et al. 2009; Stegbauer et al. 2009). Excessive consumption of salt in high calorie diet affects the innate immune system through its effect on macrophage function (MacHnik et al. 2009). SGK1 has been reported to act as a mediator for sodium homeostasis. It can be induced by exogenous sodium chloride and is one of the major kinases that regulates Na^+ intake by phosphorylation of epithelial sodium channels (ENaCs) (Lang et al. 2006). Increase in sodium chloride concentrations found locally under physiological conditions in vivo dramatically boost the induction of murine and human Th17 cells. The molecular mechanism associated with induction of Th17 cell is not fully understood. However, it is suggested that high-salt conditions and activate the p38/MAPK pathway involving the tonicity-responsive enhancer binding protein (TonEBP/NFAT5) and the serum/glucocorticoid-regulated kinase 1 (SGK1) during cytokine-induced Th17 polarization. Gene silencing or chemical inhibition of p38/MAPK, NFAT5 or SGK1 abrogates the high-salt induced Th17 cell development. The Th17 cells generated under high-salt display a highly pathogenic and stable phenotype characterized by the up-regulation of the pro-inflammatory cytokines

GM-CSF, TNF-α and IL-2. It is also reported that mice fed with a high-salt diet develop a more severe form of experimental autoimmune encephalomyelitis (EAE), in line with augmented central nervous system infiltrating and peripherally induced antigen specific Th17 cells. Based on these observations, it is proposed that increase in dietary salt intake may represent an environmental risk factor for the development of autoimmune diseases through the induction of pathogenic Th17 cells (Kleinewietfeld et al. 2013; Wu et al. 2013). In addition, sodium also induces matrix metalloproteases (MMPs) that allow activated T cells to penetrate the extracellular matrix. Recent studies have also shown that IL-17 signals through a heteromeric receptor complex formed by IL-17R (IL-17RA) and IL-17RC, which are single-pass transmembrane proteins expressed by a variety of cells including astrocytes and microglia (Ruddy et al. 2004; Zepp et al. 2011). IL-17 plays a key role in the pathogenesis of a number of chronic inflammatory diseases, such as rheumatoid arthritis (RA), osteoarthritis, and multiple sclerosis.

6.7   Conclusion

Sodium in diet is a major contributor to high BP, but how much of an effect it has depends on the genetic makeup. This is because salt sensitivity is largely determined by several genes, one of which is the ACE gene. The RAAS plays a major role in the pathogenesis of high BP, which is a risk factor for coronary events, stroke, kidney failure, and heart failure. Within the RAAS, angiotensin converting enzyme (ACE) transforms angiotensin (Ang) I into the vasoconstrictor Ang II, which produces its effects via the angiotensin type 1 receptor (AT1R). Ang II induces increase in BP through vasoconstriction and salt and water retention. In experimental models, stimulation of angiotensin II type 2 receptors (AT2R) counteracts AT1R action either directly or by modulating AT1R signaling. In the RAAS, the ACE gene functions as a key modulator involved in BP regulation via control of sodium reabsorption in kidney. KKS also contribute to modulation of BP. Targeting the RAAS and KKS constitutes a major advance in the treatment of cardiovascular and cerebrovascular diseases. RAAS and KKS interact with each other at the level of ACE, which catalyzes the generation of angiotensin II and degradation of bradykinin. Moreover, angiotensin II enhances B1 and B2 receptor expression via transcriptional mechanisms. It is hypothesized that disruption of the RAAS/KKS balance may provoke increase in BP. Diet that lowers BP is rich in potassium, calcium and fiber. In addition, potassium causes increased excretion of sodium in the urine. Endothelial dysfunction is an important hallmark of the high BP. It reflects the premature aging of the intima exposed to the chronic increase in arterial BP. It is caused by complex changes in the balance between endothelium-dependent vasodilator and vasoconstrictor signals. Levels of released NO and the generation of vasoconstrictor eicosanoids contribute to endothelial dysfunction. RAAS-mediated changes in brain may contribute to the pathogenesis of neurological disorders. Thus, high BP may increase the risk of microvascular brain damage and impairment in mobility, cognition, and mood. Loss of cognitive function is one the most devastating manifestations of changes in BP and vascular disease. Cognitive decline is rapidly becoming an important cause of disability worldwide and contributes significantly to increased mortality. There is growing evidence that high BP is the most important modifiable vascular risk factor for development and progression of both cognitive decline and dementia. High BP contributes to cerebral small and large vessel disease resulting in brain damage and dementia. A decline in cerebrovascular reserve capacity and emerging degenerative vascular wall changes underlie complete and incomplete brain infarcts, hemorrhages, and white matter hyperintensities.

References

Allen AM, Zhuo J, Mendelsohn FA (2000) Localization and function of angiotensin AT1 receptors. Am J Hypertens 13:31S–38S

Alper AB Jr, Chen W, Yau L, Srinivasan SR, Berenson GS, Hamm LL (2005) Childhood uric acid predicts adult blood pressure: the Bogalusa Heart Study. Hypertension 45:34–38

Amiya E, Watanabe M, Komuro I (2014) The relationship between vascular function and the autonomic nervous system. Ann Vasc Dis 7:109–119

Aroor AR, Demarco VG, Jia G, Sun Z, Nistala R, Meininger GA, Sowers JR (2013) The role of tissue renin-angiotensin-aldosterone system in the development of endothelial dysfunction and arterial stiffness. Front Endocrinol (Lausanne) 4:161

Ashby EL, Love S, Kehoe PG (2012) Assessment of activation of the plasma kallikrein-kinin system in frontal and temporal cortex in Alzheimer's disease and vascular dementia. Neurobiol Aging 33:1345–1355

Audoly LP, Oliverio MI, Coffman TM (2000) Insights into the functions of type 1 (AT1) angiotensin II receptors provided by gene targeting. Trends Endocrinol Metab 11:263–269

Bader M, Ganten D (2008) Update on tissue renin-angiotensin systems. J Mol Med (Berl) 86:615–621

Baldwin AS Jr (2001) Series introduction: the transcription factor NF-κB and human disease. J Clin Invest 107:3–6

Barodka VM, Joshi BL, Berkowitz DE, Hogue CW, Nyhan D (2011) Review article: implications of vascular aging. Anesth Analg 112:1048–1060

Beason-Held LL, Moghekar A, Zonderman AB, Kraut MA, Resnick SM (2007) Longitudinal changes in cerebral blood flow in the older hypertensive brain. Stroke 38:1766–1773

Benigni A, Corna D, Zoja C, Sonzogni A, Latini R, Salio M, Conti S, Rottoli D, Longaretti L, Cassis P, Morigi M, Coffman TM, Remuzzi G (2009) Disruption of the Ang II type 1 receptor promotes longevity in mice. J Clin Invest 119:524–530

Bernstein KE, Koronyo Y, Salumbides BC, Sheyn J, Pelissier L, Lopes DH, Shah KH, Bernstein EA, Fuchs DT, Yu JJ, Pham M, Black KL, Shen XZ, Fuchs S, Koronyo-Hamaoui M (2014) Angiotensin-converting enzyme overexpression in myelomonocytes prevents Alzheimer's-like cognitive decline. J Clin Invest 124:1000–1012

Bissonnette SA, Glazier CM, Stewart MQ, Brown GE, Ellson CD, Yaffe MB (2008) Phosphatidylinositol 3-phosphate-dependent and -independent functions of p40phox in activation of the neutrophil NADPH oxidase. J Biol Chem 283:2108–2119

Blaustein MP, Leenen FH, Chen L, Golovina VA, Hamlyn JM, Pallone TL, Van Huysse JW, Zhang J, Wier WG (2012) How NaCl raises blood pressure: a new paradigm for the pathogenesis of salt-dependent hypertension. Am J Physiol Heart Circ Physiol 302:H1031–H1049

Blume A, Herdegen T, Unger T (1999) Angiotensin peptides and inducible transcription factors. J Mol Med 77:339–357

Bo S, Gambino R, Durazzo M, Ghione F, Musso G, Gentile L (2008) Associations between serum uric acid and adipokines, markers of inflammation, and endothelial dysfunction. J Endocrinol Invest 31:499–504

Boldyreff B, Wehling M (2003) Non-genomic actions of aldosterone: mechanisms and consequences in kidney cells. Nephrol Dial Transplant 18:1693–1695

Booth RE, Johnson JP, Stockand JD (2002) Aldosterone. Adv Physiol Educ 26:8–20

Brassard P, Amiri F, Schiffrin EL (2005) Combined angiotensin II type 1 and type 2 receptor blockade on vascular remodeling and matrix metalloproteinases in resistance arteries. Hypertension 46:598–606

Briet M, Schiffrin EL (2011) The role of aldosterone in the metabolic syndrome. Curr Hypertens Rep 13:163–172

Cai D (2013) Neuroinflammation and neurodegeneration in overnutrition-induced diseases. Trends Endocrinol Metab 24:40–47

Campbell DJ (2014) Clinical relevance of local renin angiotensin systems. Front Endocrinol (Lausanne) 5:113

Campbell NR, Johnson JA, Campbell TS (2012) Sodium consumption: an Individual's choice? Int J Hypertens 2012:860954

Carey RM, Padia SH (2008) Angiotensin AT2 receptors: control of renal sodium excretion and blood pressure. Trends Endocrinol Metab 19:84–87

Carlson SH, Wyss JM (2008) Neurohormonal regulation of the sympathetic nervous system: new insights into central mechanisms of action. Curr Hypertens Rep 10:233–240

CDC (2009) Application of lower sodium intake recommendations to adults–United States, 1999–2006. MMWR Morb Mortal Wkly Rep 58:281–283

Chung S, Park CW, Shin SJ, Lim JH, Chung HW, Youn DY, Kim HW, Kim BS, Lee JH, Kim GH, Chang YS (2010) Tempol or candesartan prevents high-fat diet-induced hypertension and renal damage in spontaneously hypertensive rats. Nephrol Dial Transplant 25:389–399

Clasen R, Schupp M, Foryst-Ludwig A, Sprang C, Clemenz M, Krikov M, Thöne-Reineke C, Unger T, Kintscher U (2005) PPARgamma-activating angiotensin type-1 receptor blockers induce adiponectin. Hypertension 46:137–143

Clempus RE, Griendling KK (2006) Reactive oxygen species signaling in vascular smooth muscle cells. Cardiovasc Res 71:216–225

Cook NR, Cutler JA, Obarzanek E, Buring JE, Rexrode KM, Kumanyika SK, Appel LJ, Whelton PK (2007) Long term effects of dietary sodium reduction on cardiovascular disease outcomes: observational follow-up of the trials of hypertension prevention (TOHP). BMJ 334:885–888

Corry DB, Eslami P, Yamamoto K, Nyby MD, Makino H, Tuck ML (2008) Uric acid stimulates vascular smooth muscle cell proliferation and oxidative stress via the vascular renin-angiotensin system. J Hypertens 26:269–275

Cuspidi C, Meani S, Fusi V, Severgnini B, Valerio C, Catini E, Leonetti G, Magrini F, Zanchetti A (2004) Metabolic syndrome and target organ damage in untreated essential hypertensives. J Hypertens 22:1991–1998

D'Autreaux B, Toledano MB (2007) ROS as signalling molecules: mechanisms that generate specificity in ROS homeostasis. Nat Rev Mol Cell Biol 8:813–824

de Sá LL, Kawamoto EM, Munhoz CD, Kinoshita PF, Orellana AM, Curi R, Rossoni LV, Avellar MC, Scavone C (2013) Ouabain activates NFκB through an NMDA signaling pathway in cultured cerebellar cells. Neuropharmacology 73:327–336

De Silva TM, Broughton BR, Drummond GR, Sobey CG, Miller AA (2009) Gender influences cerebral vascular responses to angiotensin II through Nox2-derived reactive oxygen species. Stroke 40:1091–1097

de Wardener HE, MacGregor GA (2002) Harmful effects of dietary salt in addition to hypertension. J Hum Hypertens 16:213–223

Devynck MA (2002) Gender and vascular smooth muscle cells: a direct influence on Ca^{2+} handling? J Hypertens 20:2139–2140

Dooley R, Harvey BJ, Thomas W (2012) Non-genomic actions of aldosterone: from receptors and signals to membrane targets. Mol Cell Endocrinol 350:223–234

Droge W (2002) Free radicals in the physiological control of cell function. Physiol Rev 82:47–95

Duchene J, Chauhan SD, Lopez F, Pecher C, Estève JP, Girolami JP, Bascands JL, Schanstra JP (2005) Direct protein-protein interaction between PLCgamma1 and the bradykinin B2 receptor–importance of growth conditions. Biochem Biophys Res Commun 326:894–900

Eap CB, Bochud M, Elston RC, Bovet P, Maillard MP, Nussberger J, Schild L, Shamlaye C, Burnier M (2007) CYP3A5 and ABCB1 genes influence blood pressure and response to treatment, and their effect is modified by salt. Hypertension 49:1007–1014

Erbe DV, Gartrell K, Zhang YL, Suri V, Kirincich SJ, Will S, Perreault M, Wang S, Tobin JF (2006) Molecular activation of PPARgamma by angiotensin II type 1-receptor antagonists. Vascul Pharmacol 45:154–162

Farah V, Elased KM, Chen Y, Key MP, Cunha TS, Ingoyen M, Morris M (2006) Nocturnal hypertension in mice consuming a high fructose diet. Auton Neurosci 130:41–50

Farooqui AA (2013) Metabolic syndrome: an important risk factor for stroke, Alzheimer disease, and depression. Springer, New York

Farooqui AA (2014) Inflammation and oxidative stress in neurological disorders. Springer, New York

Farooqui AA, Farooqui T (2015) Neurochemical effects of western diet consumption. In: Farooqui AA, Farooqui T (eds) Diet and exercise in cognitive function and neurological disease. Wiley-Blackwell/Wiley, Hoboken

Farooqui AA, Farooqui T, Panza F, Frisardi V (2012) Metabolic syndrome as a risk factor for neurological disorders. Cell Mol Life Sci 69:741–762

Farrall AJ, Wardlaw JM (2009) Blood-brain barrier: ageing and microvascular disease–systematic review and meta-analysis. Neurobiol Aging 30:337–352

Fava C, Ricci M, Melander O, Minuz P (2012) Hypertension, cardiovascular risk and polymorphisms in genes controlling the cytochrome P450 pathway of arachidonic acid: a sex-specific relation? Prostaglandins Other Lipid Mediat 98:75–85

Felder RA, Jose PA (2006) Mechanisms of disease: the role of GRK4 in the etiology of essential hypertension and salt sensitivity. Nat Clin Pract Nephrol 2:637–650

Feng Y, Xia H, Cai Y, Halabi CM, Becker LK, Santos RA, Speth RC, Sigmund CD, Lazartigues E (2010) Brain-selective overexpression of human Angiotensin-converting enzyme type 2 attenuates neurogenic hypertension. Circ Res 106:373–382

Ferrario CM, Trask AJ, Jessup JA (2005) Advances in the biochemical and functional roles of angiotensin converting enzyme 2 and angiotensin-(1-7) in the regulation of cardiovascular function. Am J Physiol Heart Circ Physiol 289:H2281–H2290

Fink GD, Arthur C (2009) Corcoran Memorial Lecture. Sympathetic activity, vascular capacitance, and long-term regulation of arterial pressure. Hypertension 53:307–312

Firbank MJ, Wiseman RM, Burton EJ, Saxby BK, O'Brien JT, Ford GA (2007) Brain atrophy and white matter hyperintensity change in older adults and relationship to blood pressure. J Neurol 254:713–721

Forstermann U, Munzel T (2006) Endothelial nitric oxide synthase in vascular disease: from marvel to menace. Circulation 113:1708–1714

Frohlich ED (2007) The salt conundrum: a hypothesis. Hypertension 50:161–166

Frohlich ED, Re RN (2009) Are local renin angiotensin systems the focal points for understanding salt sensitivity in hypertension? In: Frohlich ED, Re RN (eds) The local cardiac renin-angiotensin-aldosterone system. Springer, New York, pp 1–6

Fuenmayor N, Moreira E, Cubeddu LX (1998) Salt sensitivity is associated with insulin resistance in essential hypertension. Am J Hypertens 11:397–402

Fujita T (2010) Mineralocorticoid receptors, salt-sensitive hypertension, and metabolic syndrome. Hypertension 55:813–818

Fyhrquist F, Saijonmaa O (2008) Plasma renin activity: an assay with ongoing clinical relevance. Clin Chem 54:1400

Giordano N, Tikhonoff V, Palatini P, Bascelli A, Boschetti G, De Lazzari F, Grasselli C, Martini B, Caffi S, Piccoli A, Mazza A, Bisiacchi P, Casiglia E (2012) Cognitive functions and cognitive reserve in relation to blood pressure components in a population-based cohort aged 53 to 94 years. Int J Hypertens 2012:274851

Gironacci MM, Vatta M, Rodriguez-Fermepin M, Fernandez BE, Pena C (2000) Angiotensin-(1-7) reduces norepinephrine release through a nitric oxide mechanism in rat hypothalamus. Hypertension 35:1248–1252

Gloire G, Legrand-Poels S, Piette J (2006) NF-κB activation by reactive oxygen species: fifteen years later. Biochem Pharmacol 72:1493–1505

Goldstein FC, Levey AI, Steenland NK (2013) High blood pressure and cognitive decline in mild cognitive impairment. J Am Geriatr Soc 61:67–73

Gomez-Alamillo C, Juncos LA, Cases A, Haas JA, Romero JC (2003) Interactions between vaso-constrictors and vasodilators in regulating hemodynamics of distinct vascular beds. Hypertension 42:831–836

Gorin Y, Kim N-H, Feliers D, Bhandari B, Ghosh Choudhury G, Abboud HE (2001) Angiotensin II activates Akt/protein kinase B by an arachidonic acid/redox-dependent pathway and independent of phosphoinositide 3-kinase. FASEB J 15:1909–1920

Grassi G (2009) Assessment of sympathetic cardiovascular drive in human hypertension: achievements and perspectives. Hypertension 54:690–697

Guarnieri G, Zanetti M, Vinci P, Cattin MR, Pirulli A, Barazzoni R (2010) Metabolic syndrome and chronic kidney disease. J Ren Nutr 20:S19–S23

Guo DF, Sun YL, Hamet P, Inagami T (2001) The angiotensin II type 1 receptor and receptor-associated proteins. Cell Res 11:165–180

Hamlyn JM, Blaustein MP (2013) Salt sensitivity, endogenous ouabain and hypertension. Curr Opin Nephrol Hypertens 22:51–58

Harrison DG (2013) The mosaic theory revisited: common molecular mechanisms coordinating diverse organ and cellular events in hypertension. J Am Soc Hypertens 7:68–74

He FJ, MacGregor GA (2009) A comprehensive review on salt and health and current experience of worldwide salt reduction programmes. J Hum Hypertens 23:363–384

He FJ, MacGregor GA (2010) Reducing population salt intake worldwide: from evidence to implementation. Prog Cardiovasc Dis 52:363–382

Hemming ML, Selkoe DJ (2005) Amyloid β-protein is degraded by cellular angiotensin-converting enzyme (ACE) and elevated by an ACE inhibitor. J Biol Chem 280:37644–37650

Hitomi H, Kiyomoto H, Nishiyama A, Hara T, Moriwaki K, Kaifu K, Ihara G, Fujita Y, Ugawa T, Kohno M (2007) Aldosterone suppresses insulin signaling via the downregulation of insulin receptor substrate-1 in vascular smooth muscle cells. Hypertension 50:750–755

Hjelstuen A, Anderssen SA, Holme I, Seljeflot I, Klemsdal TO (2006) Markers of inflammation are inversely related to physical activity and fitness in sedentary men with treated hypertension. Am J Hypertens 19:669–675

Holbrook JT, Patterson KY, Bodner JE, Douglas LW, Veillon C, Kelsay JL, Mertz W, Smith JC Jr (1984) Sodium and potassium intake and balance in adults consuming self-selected diets. Am J Clin Nutr 40:786–793

Huang BS, Zheng H, Tan J, Patel KP, Leenen FH (2011) Regulation of hypothalamic renin-angiotensin system and oxidative stress by aldosterone. Exp Physiol 96:1028–1038

Iadecola C (2004) Neurovascular regulation in the normal brain and in Alzheimer's disease. Nat Rev Neurosci 5:347–360

Inagami T, Kambayashi Y, Ichiki T, Tsuzuki S, Eguchi S, Yamakawa T (1999) Angiotensin receptors: molecular biology and signaling. Clin Exp Pharmacol Physiol 26:544–549

Intengan HD, Schiffrin EL (2000) Structure and mechanical properties of resistance arteries in hypertension: role of adhesion molecules and extracellular matrix determinants. Hypertension 36:312–318

Ito H, Kanno I, Fukuda H (2005) Human cerebral circulation: positron emission tomography studies. Ann Nucl Med 19:65–74

Iwai N, Kajimoto K, Tomoike H, Takashima N (2007) Polymorphism of CYP11B2 determines salt sensitivity in Japanese. Hypertension 49:825–831

Iwanami U, Mogi M, Iwai M, Horiuchi M (2009) Inhibition of the renin-angiotensin system and target organ protection. Hypertens Res 32:229–237

Iwano M, Neilson EG (2004) Mechanisms of tubulointerstitial fibrosis. Curr Opin Nephrol Hypertens 13:279–284

Jalal DI, Chonchol M, Chen W, Targher G (2013) Uric acid as a target of therapy in CKD. Am J Kidney Dis 61:134–146

Jochemsen HM, Teunissen CE, Ashby EL, van der Flier WM, Jones RE, Geerlings MI, Scheltens P, Kehoe PG, Muller M (2014) The association of angiotensin-converting enzyme with biomarkers for Alzheimer's disease. Alzheimers Res Ther 6:27

Johnson JK, Pa J, Boxer AL, Kramer JH, Freeman K, Yaffe K (2010) Baseline predictors of clinical progression among patients with dysexecutive mild cognitive impairment. Dement Geriatr Cogn Disord 30:344–351

Jones BH, Standridge MK, Moustaid N (1997) Angiotensin II increases lipogenesis in 3T3-L1 and human adipose cells. Endocrinology 138:1512–1519

Jouvent E, Viswanathan A, Chabriat H (2010) Cerebral atrophy in cerebrovascular disorders. J Neuroimaging 20:213–218

Joyner MJ, Charkoudian N, Wallin BG (2008) A sympathetic view of the sympathetic nervous system and human blood pressure regulation. Exp Physiol 93:715–724

Juan CC, Fang VS, Hsu YP, Huang DB, Hsia DB, Yu PC, Kwok CF, Ho LT (1998) Overexpression of vascular endothelin-1 and endothelin-A receptors in a fructose-induced hypertensive rat model. J Hypertens 16:1775–1782

Kakoki M, Smithies O (2009) The kallikrein-kinin system in health and in diseases of the kidney. Kidney Int 75:1019–1030

Kalaria RN (2010) Vascular basis for brain degeneration: faltering controls and risk factors for dementia. Nutr Rev 68:S74–S87

Kaplan NM, Domino FJ (2013) Overview of hypertension in adults. 2013. Accessed 23 Jan 2014 http://www.uptodate.com/contents/overview-of-hypertension-in-adults?source=search_result &search=hypertension&selectedTitle=1~150

Kawamoto EM, Lima LS, Munhoz CD, Yshii LM, Kinoshita PF, Amara FG, Pestana RR, Orellana AM, Cipolla-Neto J, Britto LR, Avellar MC, Rossoni LV, Scavone C (2012) Influence of N-methyl-D-aspartate receptors on ouabain activation of nuclear factor-κB in the rat hippocampus. J Neurosci Res 90:213–228

Kayashima Y, Smithies O, Kakoki M (2012) The kallikrein-kinin system and oxidative stress. Curr Opin Nephrol Hypertens 21:92–96

Khitan Z, Kim DH (2013) Fructose: a key factor in the development of metabolic syndrome and hypertension. J Nutr Metab 2013:682673

Khosla UM, Zharikov S, Finch JL, Nakagawa T, Roncal C, Mu W, Krotova K, Block ER, Prabhakar S, Johnson RJ (2005) Hyperuricemia induces endothelial dysfunction. Kidney Int 67:1739–1742

Kim JA, Montagnani M, Koh KK, Quon MJ (2006) Reciprocal relationships between insulin resistance and endothelial dysfunction: molecular and pathophysiological mechanisms. Circulation 6:1888–1904

Kitiyakara C, Chabrashvili T, Chen Y, Blau J, Karber A, Aslam S, Welch WJ, Wilcox CS (2003) Salt intake, oxidative stress, and renal expression of NADPH oxidase and superoxide dismutase. J Am Soc Nephrol 14:2775–2782

Kittikulsuth W, Sullivan JC, Pollock DM (2013) ET-1 actions in the kidney: evidence for sex differences. Br J Pharmacol 168:318–326

Kleinewietfeld M, Manzel A, Titze J, Kvakan H, Yosef N, Linker RA, Muller DN, Hafler DA (2013) Sodium chloride drives autoimmune disease by the induction of pathogenic T_H17 cells. Nature 496:518–522

Kobori H, Nangaku M, Navar LG, Nishiyama A (2007) The intrarenal renin-angiotensin system: From physiology to the pathobiology of hypertension and kidney disease. Pharmacol Rev 59:251–287

Kurdi M, De Mello WC, Booz GW (2005) Working outside the system: an update on unconventional behavior of the renin angiotensin system components. Int J Biochem Cell Biol 37:1357–1367

Labandeira-Garcia JL, Rodriguez-Pallares J, Dominguez-Meijide A, Valenzuela R, Villar-Cheda B, Rodríguez-Perez AI (2013) Dopamine-angiotensin interactions in the basal ganglia and their relevance for Parkinson's disease. Mov Disord 28:1337–1342

Laffer CL, Elijovich F (2013) Differential predictors of insulin resistance in nondiabetic salt-resistant and salt-sensitive subjects. Hypertension 61:707–715

Lanaspa MA, Sanchez-Lozada LG, Cicerchi C, Li N, Roncal-Jimenez CA, Ishimoto T, Le M, Garcia GE, Thomas JB, Rivard CJ, Andres-Hernando A, Hunter B, Schreiner G, Rodriguez-Iturbe B, Sautin YY, Johnson RJ (2012) Uric acid stimulates fructokinase and accelerates fructose metabolism in the development of fatty liver. PLoS One 7:e47948

Lang F, Bohmer C, Palmada M, Seebohm G, Strutz-Seebohm N, Vallon V (2006) (Patho)physiological significance of the serum- and glucocorticoid-inducible kinase isoforms. Physiol Rev 86:1151–1178

Lassègue B, San Martín A, Griendling KK (2012) Biochemistry, physiology, and pathophysiology of NADPH oxidases in the cardiovascular system. Circ Res 110:1364–1390

Lastra G, Dhuper S, Johnson MS, Sowers JR (2010) Salt, aldosterone, and insulin resistance: impact on the cardiovascular system. Nat Rev Cardiol 7:577–584

Leeb-Lundberg LM, Marceau F, Müller-Esterl W, Pettibone DJ, Zuraw BL (2005) International union of pharmacology. XLV. Classification of the kinin receptor family: from molecular mechanisms to pathophysiological consequences. Pharmacol Rev 57:27–77

Lucke-Wold BP, Turner RC, Lucke-Wold AN, Rosen CL, Huber JD (2012) Age and the metabolic syndrome as risk factors for ischemic stroke: improving preclinical models of ischemic stroke. Yale J Biol Med 85:523–539

MacHnik A, Neuhofer W, Jantsch J, Dahlmann A, Tammela T, Machura K, Park JK, Beck FX, Müller DN, Derer W, Goss J, Ziomber A, Dietsch P, Wagner H, van Rooijen N, Kurtz A, Hilgers KF, Alitalo K, Eckardt KU, Luft FC, Kerjaschki D, Titze J (2009) Macrophages regulate salt-dependent volume and blood pressure by a vascular endothelial growth factor-C-dependent buffering mechanism. Nat Med 15:545–552

MacMahon S, Peto R, Cutler J, Collins R, Sorlie P, Neaton J, Abbott R, Godwin J, Dyer A, Stamler J (1990) Blood pressure, stroke, and coronary heart disease. Part 1, prolonged differences in blood pressure: prospective observational studies corrected for the regression dilution bias. Lancet 335:765–774

Massiera F, Seydoux J, Geloen A, Quignard-Boulange A, Turban S, Saint-Marc P, Fukamizu A, Negrel R, Ailhaud G, Teboul M (2001) Angiotensinogen-deficient mice exhibit impairment of diet-induced weight gain with alteration in adipose tissue development and increased locomotor activity. Endocrinology 142:5220–5225

Matsui H, Ando K, Kawarazaki H, Nagae A, Fujita M, Shimosawa T, Nagase M, Fujita T (2008) Salt excess causes left ventricular diastolic dysfunction in rats with metabolic disorder. Hypertension 52:287–294

Mezzano SA, Ruiz-Ortega M, Egido J (2001) Angiotensin II and renal fibrosis. Hypertension 38:635–638

Nagase M, Yoshida S, Shibata S, Nagase T, Gotoda T, Ando K, Fujita T (2006) Enhanced aldosterone signaling in the early nephropathy of rats with metabolic syndrome: possible contribution of fat-derived factors. J Am Soc Nephrol 17:3438–3446

Nagase M, Matsui H, Shibata S, Gotoda T, Fujita T (2007) Salt-induced nephropathy in obese spontaneously hypertensive rats via paradoxical activation of the mineralocorticoid receptor. Hypertension 50:877–883

Nakamura T, Kataoka K, Fukuda M, Nako H, Tokutomi Y, Dong YF (2009) Critical role of apoptosis signal-regulating kinase 1 in aldosterone/salt-induced cardiac inflammation and fibrosis. Hypertension 54:544–551

Nakamura T, Takahashi T, Fukul M, Eblhara I, Osada S, Tomino Y, Kolde H (1995) Enalapril attenuates increased gene expression of extracellular matrix components in diabetic rats. J. Am Soc Nephrol 5:1492–1497

Nelson PT, Head E, Schmitt FA, Davis PR, Neltner JH, Jicha GA, Abner EL, Smith CD, Van Eldik LJ, Kryscio RJ, Scheff SW (2011) Alzheimer's disease is not "brain aging": neuropathological, genetic, and epidemiological human studies. Acta Neuropathol 121: 571–587

Neves MF, Virdis A, Schiffrin EL (2003) Resistance artery mechanics and composition in angiotensin II-infused rats: effects of aldosterone antagonism. J Hypertens 21:189–198

Nguyen H, Odelola OA, Rangaswami J, Amanullah A (2013) A review of nutritional factors in hypertension management. Int J Hypertens 2013:698940

Niwa K, Younkin L, Ebeling C, Turner SK, Westaway D, Younkin S, Ashe KH, Carlson GA, Iadecola C (2000) Abeta 1-40-related reduction in functional hyperemia in mouse neocortex during somatosensory activation. Proc Natl Acad Sci U S A 97:9735–9740

Niwa K, Kazama K, Younkin L, Younkin SG, Carlson GA, Iadecola C (2002) Cerebrovascular autoregulation is profoundly impaired in mice overexpressing amyloid precursor protein. Am J Physiol Heart Circ Physiol 283:H315–H323

Nolly HL, Lama MC, Carretero OA, Scicli AG (1992) The kallikrein-kinin system in blood vessels. Agents Actions Suppl 38:1–9

Oberleithner H, Riethmüller C, Schillers H, MacGregor GA, de Wardener HE, Hausberg M (2007) Plasma sodium stiffens vascular endothelium and reduces nitric oxide release. Proc Natl Acad Sci U S A 104:16281–16286

Ogihara T, Asano T, Ando K, Sakoda H, Anai M, Shojima N, Ono H, Onishi Y, Fujishiro M, Abe M, Fukushima Y, Kikuchi M, Fujita T (2002) High-salt diet enhances insulin signaling and induces insulin resistance in Dahl salt-sensitive rats. Hypertension 40:83–89

Ohshima K, Mogi M, Horiuchi M (2013) Therapeutic approach for neuronal disease by regulating renin-angiotensin system. Curr Hypertens Rev 9:99–107

Padia SH, Kemp BA, Howell NL, Fournie-Zaluski MC, Roques BP, Carey RM (2008) Conversion of renal angiotensin II to angiotensin III is critical for AT2 receptor-mediated natriuresis in rats. Hypertension 51:460–465

Patel KP (2000) Role of paraventricular nucleus in mediating sympathetic outflow in heart failure. Heart Fail Rev 5:73–86

Patni H, Mathew JT, Luan L, Franki N, Chander PN, Singhal PC (2007) Aldosterone promotes proximal tubular cell apoptosis: role of oxidative stress. Am J Physiol Renal Physiol 293:F1065–F1071

Pereira RL, Felizardo RJ, Cenedeze MA, Hiyane MI, Bassi EJ, Amano MT, Origassa CS, Silva RC, Aguiar CF, Carneiro SM, Pesquero JB, Araújo RC, Keller Ade C, Monteiro RC, Moura IC, Pacheco-Silva A, Câmara NO (2014) Balance between the two kinin receptors in the progression of experimental focal and segmental glomerulosclerosis in mice. Dis Model Mech 7:701–710

Pesquero JB, Bader M (1998) Molecular biology of the kallikrein-kinin system: from structure to function. Braz J Med Biol Res 31:1197–1203

Platten M, Youssef S, Hur EM, Ho PP, Han MH, Lanz TV, Phillips LK, Goldstein MJ, Bhat R, Raine CS, Sobel RA, Steinman L (2009) Blocking angiotensin-converting enzyme induces potent regulatory T cells and modulates TH1- and TH17-mediated autoimmunity. Proc Natl Acad Sci U S A 106:14948–14953

Pu Q, Neves MF, Virdis A, Touyz RM, Schiffrin EL (2003) Endothelin antagonism on aldosterone-induced oxidative stress and vascular remodeling. Hypertension 42:49–55

Raidoo DM, Bhoola KD (1998) Pathophysiology of the kallikrein-kinin system in mammalian nervous tissue. Pharmacol Ther 79:105–127

Ray PE, Bruggeman LA, Horikoshi S, Aguilera G, Klotman PE (1994) Angiotensin II stimulates human fetal mesangial cell proliferation and fibronectin biosynthesis by binding to AT1 receptors. Kidney Int 45:177–184

Re RN (2009) Local renin angiotensin systems in the cardiovascular system. In: De Mello WC, Frohlich ED (eds) Renin angiotensin system and cardiovascular disease. Humana Press, New York, pp 25–34

Reil JC, Hohl M, Selejan S, Lipp P, Drautz F, Kazakow A, Münz BM, Müller P, Steendijk P, Reil GH, Allessie MA, Böhm M, Neuberger HR (2012) Aldosterone promotes atrial fibrillation. Eur Heart J 33:2098–2108

Reriani MK, Lerman LO, Lerman A (2010) Endothelial function as a functional expression of cardiovascular risk factors. Biomark Med 4:351–360

Rocha R, Chander PN, Khanna K, Zuckerman A, Stier CT Jr (1998) Mineralocorticoid blockade reduces vascular injury in stroke-prone hypertensive rats. Hypertension 31:451–458

Rodrigo R, Bächler JP, Araya J, Prat H, Passalacqua W (2007) Relationship between (Na + K)-ATPase activity, lipid peroxidation and fatty acid profile in erythrocytes of hypertensive and normotensive subjects. Mol Cell Biochem 303:73–81

Roher AE, Debbins JP, Malek-Ahmadi M, Chen K, Pipe JG, Maze S, Belden C, Maarouf CL, Thiyyagura P, Mo H, Hunter JM, Kokjohn TA, Walker DG, Kruchowsky JC, Belohlavek M, Sabbagh MN, Beach TG (2012) Cerebral blood flow in Alzheimer's disease. Vasc Health Risk Manag 8:599–611

Rose G, Stamler J, Stamler R (1988) Intersalt: an international study of electrolyte excretion and blood pressure. Results for 24 hour urinary sodium and potassium excretion. BMJ 297:319–328

Ruddy MJ, Wong GC, Liu XK, Yamamoto H, Kasayama S, Kirkwood KL, Gaffen SL (2004) Functional cooperation between interleukin-17 and tumor necrosis factor-alpha is mediated by CCAAT/enhancer-binding protein family members. J Biol Chem 279:2559–2567

Ruggenenti P, Cravedi P, Remuzzi G (2010) The RAAS in the pathogenesis and treatment of diabetic nephropathy. Nat Rev Nephrol 6:319–330

Ruiz-Ortega M, Lorenzo O, Rupérez M, Esteban V, Suzuki Y, Mezzano S, Plaza JJ, Egido J (2001) Role of the renin-angiotensin system in vascular diseases: expanding the field. Hypertension 38:1382–1387

Saavedra JM (2012) Angiotensin II AT(1) receptor blockers as treatments for inflammatory brain disorders. Clin Sci (Lond) 123:567–590

Saido TC, Yokota M, Maruyama K, Yamao-Harigaya W, Tani E, Ihara Y, Kawashima S (1994) Spatial resolution of the primary beta-amyloidogenic process induced in postischemic hippocampus. J Biol Chem 269:15253–15257

Santos RA, Ferreira AJ (2007) Angiotensin-(1-7) and the renin-angiotensin system. Curr Opin Nephrol Hypertens 16:122–128

Santos RAS, Silva ACS, Maric C, Silva DM, Machado RP, de Buhr I, Heringer-Walther S, Pinheiro SV, Lopes MT, Bader M, Mendes EP, Lemos VS, Campagnole-Santos MJ, Schultheiss HP, Speth R, Walther T (2007) Angiotensin-(1-7) is an endogenous ligand for the G protein-coupled receptor Mas. Proc Natl Acad Sci U S A 100:8258–8263

Schlaich MP, Krum H, Esler MD (2010) New therapeutic approaches to resistant hypertension. Curr Hypertens Rep 12:296–302

Shah J, Jandhyala BS (1991) Studies on the role(s) of cerebrospinal fluid osmolality and chloride ion in the centrally mediated pressor responses of sodium chloride. Clin Exp Hypertens A 13:297–312

Shi J, Yang SH, Stubley L, Day AL, Simpkins JW (2000) Hypoperfusion induces overexpression of beta-amyloid precursor protein mRNA in a focal ischemic rodent model. Brain Res 853:1–4

Shibata S, Fujita T (2012) Mineralocorticoid receptors in the pathophysiology of chronic kidney diseases and the metabolic syndrome. Mol Cell Endocrinol 350:273–280

Shibata S, Mu S, Kawarazaki H, Muraoka K, Ishizawa K, Yoshida S, Kawarazaki W, Takeuchi M, Ayuzawa N, Miyoshi J, Takai Y, Ishikawa A, Shimosawa T, Ando K, Nagase M, Fujita T (2011) Rac1 GTPase in rodent kidneys is essential for salt-sensitive hypertension via a mineralocorticoid receptor–dependent pathway. J Clin Invest 121:3233–3243

Sierra C (2012) Cerebral small vessel disease, cognitive impairment and vascular dementia. Panminerva Med 54:179–188

Singh H, Schwartzman ML (2008) Renal vascular cytochrome P450-derived eicosanoids in androgen-induced hypertension. Pharmacol Rep 60:29–37

Sobey CG, Heistad DD, Faraci FM (1997) Mechanisms of bradykinin-induced cerebral vasodilatation in rats. Evidence that reactive oxygen species activate K+ channels. Stroke 28:2290–2294, discussion 2295

Stegbauer J, Lee DH, Seubert S, Ellrichmann G, Manzel A, Kvakan H, Muller DN, Gaupp S, Rump LC, Gold R, Linker RA (2009) Role of the renin-angiotensin system in autoimmune inflammation of the central nervous system. Proc Natl Acad Sci U S A 106:14942–14947

Stolarz-Skrzypek K, Kuznetsova T, Thijs L, Tikhonoff V, Seidlerová J, Richart T, Jin Y, Olszanecka A, Malyutina S, Casiglia E, Filipovský J, Kawecka-Jaszcz K, Nikitin Y, Staessen JA, European Project on Genes in Hypertension (EPOGH) Investigators (2011) Fatal and nonfatal outcomes, incidence of hypertension, and blood pressure changes in relation to urinary sodium excretion. JAMA 305:1777–1785

Strazzullo P, D'Elia L, Kandala NB, Cappuccio FP (2009) Salt intake, stroke, and cardiovascular disease: meta-analysis of prospective studies. BMJ 339:b4567

Taddei S, Ghiadoni L, Virdis A, Versari D, Salvetti A (2003) Mechanisms of endothelial dysfunction: clinical significance and preventive non-pharmacological therapeutic strategies. Curr Pharm Des 9:2385–2402

Takahashi N, Sequeira Lopez MLS, Cowhig JE, Taylor MA, Hatada T, Riggs E, Lee G, Gomez RA, Kim H-S, Smithies O (2005) Ren1c homozygous null mice are hypotensive and polyuric, but heterozygotes are indistinguishable from wild type. J Am Soc Nephrol 16:125–132

Thamilselvan V, Menon M, Thamilselvan S (2012) Selective Rac1 inhibition protects renal tubular epithelial cells from oxalate-induced NADPH oxidase-mediated oxidative cell injury. Urol Res 40:415–423

Thompson J, Khalil RA (2003) Gender differences in the regulation of vascular tone. Clin Exp Pharmacol Physiol 30:1–15

Tian W, Li XJ, Stull ND, Ming W, Suh CI, Bissonnette SA, Yaffe MB, Grinstein S, Atkinson SJ, Dinauer MC (2008) Fc gamma r-stimulated activation of the NADPH oxidase: phosphoinositide-binding protein p40phox regulates NADPH oxidase activity after enzyme assembly on the phagosome. Blood 112:3867–3877

Tian M, Zhu D, Xie W, Shi J (2012) Central angiotensin II-induced Alzheimer-like tau phosphorylation in normal rat brains. FEBS Lett 586:3737–3745

Tostes RC, Fortes ZB, Callera GE, Montezano AC, Touyz RM, Webb RC, Carvalho MH (2008) Endothelin, sex and hypertension. Clin Sci (Lond) 114:85–97

Tran LT, MacLeod KM, McNeill JH (2009) Chronic etanercept treatment prevents the development of hypertension in fructose-fed rats. Mol Cell Biochem 330:219–228

Tsuda K (2012) Renin-angiotensin system and sympathetic neurotransmitter release in the central nervous system of hypertension. Int J Hypertens 2012:474870

Tylicki L, Rutkowski P, Renke M et al (2008) Triple pharmacological blockade of the renin-angiotensin-aldosterone system in nondiabetic CKD: an open-label crossover randomized controlled trial. Am J Kidney Dis 52:486–493

Unger T (2002) The role of the renin-angiotensin system in the development of cardiovascular disease. Am J Cardiol 89:3A–9A; discussion 10A

Unger T, Paulis L, Sica DA (2011) Therapeutic perspectives in hypertension: novel means for renin-angiotensin-aldosterone system modulation and emerging device-based approaches. Eur Heart J 32:2739–2747

Verma S, Bhanot S, McNeill JH (1999) Sympathectomy prevents fructose-induced hyperinsulinemia and hypertension. Eur J Pharmacol 373:R1–R4

Viegas VU, Liu ZZ, Nikitina T, Perlewitz A, Zavaritskaya O, Schlichting J, Persson PB, Regitz-Zagrosek V, Patzak A, Sendeski MM (2012) Angiotensin II type 2 receptor mediates sex differences in mice renal interlobar arteries response to angiotensin II. J Hypertens 30:1791–1798

Viswanathan A, Gray F, Bousser MG, Baudrimont M, Chabriat H (2006) Cortical neuronal apoptosis in CADASIL. Stroke 37:2690–2695

Weller RO, Yow HY, Preston SD, Mazanti I, Nicoll JA (2002) Cerebrovascular disease is a major factor in the failure of elimination of Abeta from the aging human brain: implications for therapy of Alzheimer's disease. Ann N Y Acad Sci 977:162–168

WHO (2010) Creating an enabling environment for population-based salt reduction strategies. Report of a joint technical meeting held by WHO and the Food Standards Agency, UK

Wright JW, Harding JW (2010) The brain RAS and Alzheimer's disease. Exp Neurol 223:326–333

Wright JW, Kawas LH, Harding JW (2013) A role for the brain RAS in Alzheimer's and Parkinson's diseases. Front Endocrinol (Lausanne) 4:158

Wu C, Yosef N, Thalhamer T, Zhu C, Xiao S, Kishi Y, Regev A, Kuchroo VK (2013) Induction of pathogenic T_H17 cells by inducible salt-sensing kinase SGK1. Nature 13:513–517

Xia H, Lazartigues E (2008) Angiotensin-converting enzyme 2 in the brain: properties and future directions. J Neurochem 107:1482–1494

Xu WL, Atti AR, Gatz M, Pedersen NL, Johansson B, Fratigleoni L (2011) Midlife overweight and obesity increase late-life dementia risk. Neurology 76:1568–1574

Yanagitani Y, Rakugi H, Okamura A, Moriguchi K, Takiuchi S, Ohishi M, Suzuki K, Higaki J, Ogihara T (1999) Angiotensin II type 1 receptor-mediated peroxide production in human macrophages. Hypertension 33:335–339

Yang J, Fuller PJ (2012) Interactions of the mineralocorticoid receptor—within and without. Mol Cell Endocrinol 350:196–205

Yang J, Chang CY, Safi R, Morgan J, McDonnell DP, Fuller PJ, Clyne CD, Young MJ (2011) Identification of ligand-selective peptide antagonists of the mineralocorticoid receptor using phage display. Mol Endocrinol 25:32–43

Zepp J, Wu L, Li X (2011) IL-17 receptor signaling and Th17-mediated autoimmune demyelinating disease. Trends Immunol 32:232–239

Zhao W, Chen SS, Chen Y, Ahokas RA, Sun Y (2008) Kidney fibrosis in hypertensive rats: role of oxidative stress. Am J Nephrol 28:548–554

Zhou MS, Wang A, Yu H (2014) Link between insulin resistance and hypertension: what is the evidence from evolutionary biology? Diabetol Metab Syndr 6:12

Zou K, Maeda T, Watanabe A, Liu J, Liu S, Oba R, Satoh Y, Komano H, Michikawa M (2009) $A\beta42$-to-$A\beta40$- and angiotensin-converting activities in different domains of angiotensin-converting enzyme. J Biol Chem 284:31914–31920

Zubakova R, Gille A, Faussner A, Hilgenfeldt U (2008) Ca^{2+} signalling of kinins in cells expressing rat, mouse and human B1/B2-receptor. Int Immunopharmacol 8:276–281

Zulian A, Baryshnikov SG, Linde CI, Hamlyn JM, Ferrari P, Golovina VA (2010) Upregulation of Na^+/Ca^{2+} exchanger and TRPC6 contributes to abnormal Ca^{2+} homeostasis in arterial smooth muscle cells from Milan hypertensive rats. Am J Physiol Heart Circ Physiol 299:H624–H633

Chapter 7
Importance and Roles of Fiber in the Diet

7.1 Introduction

Dietary fiber is a non-digestible food component of meal that includes non-starch polysaccharides, oligosaccharides, lignin, and analogous polysaccharides (Papathanasopoulos and Camilleri 2010; Raninen et al. 2011). Dietary fiber occurs in foods of plant origin. It not only helps in providing a feeling of fullness, but also promotes digestive health and laxation. Among Americans the mean intake of fiber ranges from 11 to 19 g/day for individuals aged 2 years and older in 2009–2010. However, women and men need 25 and 38 g fiber/day, respectively (Institute of Medicine 2005; U.S. Department of Agriculture 2012). Dietary fiber is classified on the basis of solubility in water, microbial fermentation in the large intestine, and viscosity. Soluble dietary fiber includes pectin, inulin, psyllium, some gums, and β-glucans, and polysaccharides, whereas insoluble dietary fiber includes cellulose, hemicelluloses, lignin, and resistance starch (Papathanasopoulos and Camilleri 2010; Slavin 2013) (Fig. 7.1). In recent years, the definition of dietary fiber has been expanded. It now includes oligosaccharides that have properties similar to soluble dietary fiber and resistant starch that escape enzymic digestion in the small intestine and act as dietary fiber in the large intestine. Soluble fiber is found in cucumbers, blueberries, beans, and nuts. It can be dissolved into a gel-like texture, helping to slow down the digestion. This helps one to feel full for longer time and is an important reason why fiber promotes weight control. Fiber stimulates the liver to produce more bile, thus helping digestion. Insoluble fiber also causes digested food to pass through the intestines more quickly, thus contributing to "regularity" and reducing the length of time the gut lining is in contact with any potentially harmful substances contained in the digested food. Soluble dietary fiber may induce diarrhea due to increase in osmotic pressure in the colon and simultaneously enlarges fecal output (Ishikawa et al. 1995). Insoluble fiber is found in foods like dark green leafy vegetables, green beans, celery, and carrot. Insoluble fiber does not dissolve at all

© Springer International Publishing Switzerland 2015

A.A. Farooqui, *High Calorie Diet and the Human Brain*,

DOI 10.1007/978-3-319-15254-7_7

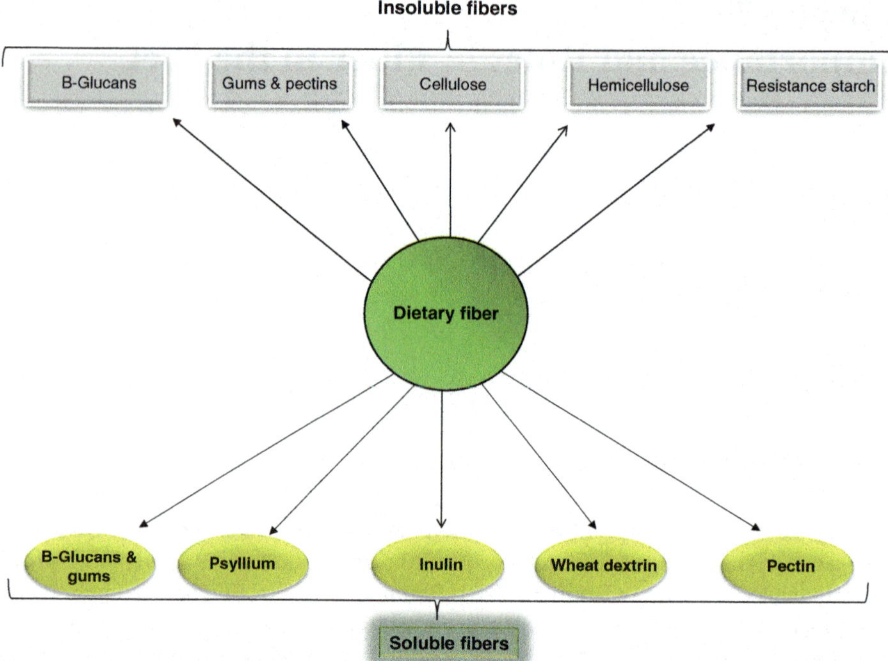

Fig. 7.1 Sources of soluble and insoluble fibers in the diet

and helps add bulk to feces. This helps food to move through the GI tract more quickly for healthy elimination. Insoluble dietary fiber may cause mucosal damage due to mechanical stimulation of the intestinal mucosa. A severe dietary problem with dietary fiber is that it interferes with mineral absorption because of its binding or bulking characteristics (Cassidy et al. 1986). Insoluble dietary fiber contains a relatively high level of phytate, which is known to bind calcium and magnesium, making them insoluble in the GI. Thus, level of phytate in fiber may be a key regulatory factor in mineral absorption, as well as the diluting effect on GI content due to fiber's bulking characteristics (Ishikawa et al. 1995; McGuire 2011). In small intestine, dietary fiber acts in three main physical forms: (a) soluble polymer chains in solution, (b) insoluble macromolecular assemblies, and (c) swollen, hydrated, sponge-like networks (Eastwood and Morris 1992). The principal physiological effect of dietary fiber in the small intestine is to reduce the rate (and in some cases the extent) of release of antioxidants (Brownlee 2001). Several factors modulate the influence of dietary fiber on antioxidant digestion including physical trapping of antioxidants within fiber assemblies, enhanced viscosity of gastric fluids restricting the peristaltic mixing process, and binding of bile salts to specific fiber components (Montagne et al. 2003; Eastwood and Morris 1992; Kaczmarczyk et al. 2012).

7.2   Biochemical Effects of Dietary Fiber Consumption

Little information is available on molecular aspects of beneficial effects of fiber consumption in human and animal health (Slavin 2001). Consumption of fiber retards excess dietary energy intake by a number of mechanisms (Heaton 1973) including: (a) displacement of energy-dense foods from the diet; (b) inhibition of absorption of macronutrients at the intestinal surface; (c) an altered luminal environment; and (d) an increased gastric distension (Heaton 1973). In general, soluble fiber is more completed fermented and has a higher viscosity than insoluble fibers. Prebiotic dietary fiber is not classified in terms of solubility or viscosity, but it is defined by the resistance to digestion and absorption in the small intestine, partial or complete fermentation by microbiota in the large intestine, and the ability to stimulate growth of selective bacteria. Only two such dietary fibers fit all of the above three criteria. They are considered as prebiotics: (a) inulin and (b) trans-galacto-oligosaccharides. Dietary fiber has ability to lower the glycemic response of the foods in which it is present. Foods, which are rich in dietary fiber release glucose more slowly into the blood, which is relevant to the prevention of disorders such as obesity, type II diabetes, and metabolic syndrome (Nugent 2005). Dietary fiber increases the gastrointestinal tract biomass, changes the composition of gut flora, and may decrease the risk of metabolic disorders like dyslipidemia, hypercholesterolemia, and hyperglycemia and also substantially influence immune-based actions. The consumption of dietary fiber is also known to decrease the oxidation of lipids, which in turn is associated with favorable effects on blood pressure (BP), serum lipid profile, insulin sensitivity, and decrease in inflammation (Ma et al. 2007; Ajani et al. 2004; Lattimer and Haub 2010; Turner and Lupton 2011). Onset of above mentioned processes may be due to dietary fiber-mediated modulation of gene expression and production of hormones, which modify lipid and carbohydrate metabolism. Acetyl-CoA carboxylase is the rate-limiting enzyme in lipogenesis, which is regulated by AMP-activated protein kinase (AMPK). In a 10-week study comparing obese and lean rats, inclusion of various fibers to a rat chow results in increased phosphorylation of AMPK, consequently inhibiting acetyl-CoA carboxylase. This inhibition of acetyl-CoA carboxylase in obese rats is comparable to the acetyl-CoA carboxylase activity in lean rats fed a control diet (Galisteo et al. 2010; Delzenne and Kok 2001). Collective evidence suggests that dietary fiber not only improves constipation, reduces serum cholesterol levels, and excretion of carcinogens into feces, but also increases satiety and reduces energy intake (Fig. 7.2). In contrast, the consumption of low-fiber diet with highly refined carbohydrate diet can contribute to hyperglycemia, which increases the proinflammatory cytokines, such as plasma interleukin-6 (IL-6), IL-18, and tumor necrosis factor α, (TNF-α) (Esposito et al. 2004). IL-6 is a primary determinant of C-reactive protein (CRP) production thus consistently elevated concentrations of IL-6 may result in elevated CRP concentrations.

It is well known that microorganisms reside in a symbiotic relationship in our GI tract. Some microbiota induces beneficial, while others cause harmful effects. The

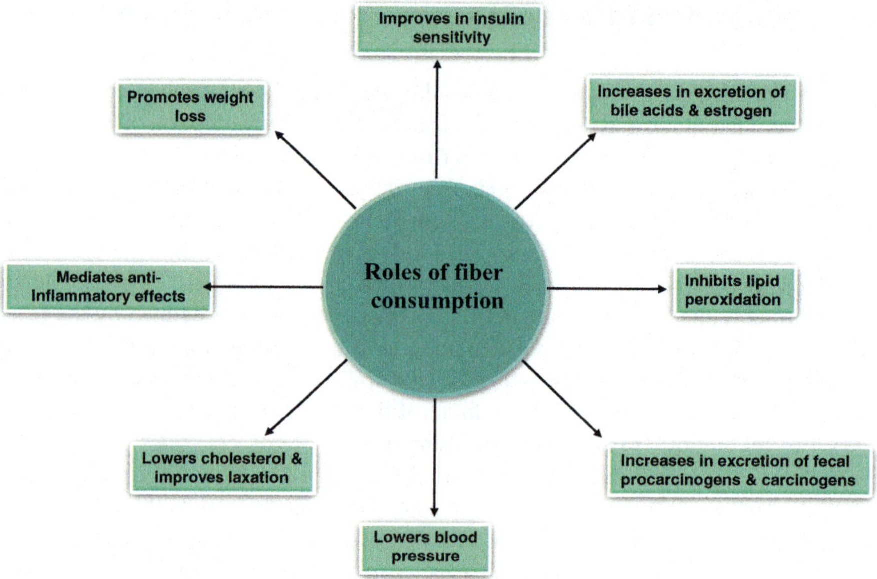

Fig. 7.2 Beneficial effects of dietary fibers on metabolic processes

mammalian microbiota is highly variable at lower taxonomic levels. However there are four dominant phyla: Firmicutes, Bacteroidetes, Actinobacteria and Proteobacteria (Dethlefsen et al. 2007). Firmicutes and Bacteroidetes account for >90 % of the bacterial population in the colon (Ley et al. 2008) while Actinobacteria and Proteobacteria (which includes Enterobacteriaceae) are regularly present but are scarce (<1–5 %) (Eckburg et al. 2005). Microbiota in the GI tract are essential not only for the breakdown of dietary fibers and production of short-chain fatty acids (SCFA), but also for the synthesis of vitamins (Tappenden and Deutsch 2007) supporting the view that the composition of the microbiota may contribute to the modulation of host metabolic functions. Accumulating evidence indicates that the gut microbiota play a significant role in the development of obesity, obesity-associated inflammation and insulin resistance (Kristiansen and Wang 2012). The biochemical mechanisms by which the microbiota participate in host functions impact the development and maintenance of the obese state, including host ingestive behavior, energy harvest, energy expenditure and fat storage are not fully understood. However, it is becoming increasingly evident that these processes are modulated by cross talk between microbiota and large intestinal tissue. The current challenge is to determine the relative importance of obesity-associated compositional and functional changes in the microbiota and to identify the relevant taxa and functional gene modules that promote leanness and obesity. The microbiota of the GI tract is also critical in determining the host's susceptibility to GI infections (Ghosh et al. 2011). Recent studies have indicated that there is a link between the gut microbiota and the immune system. Thus, innate immune cells, such as macrophages, neutrophils, and dendritic cells, and epithelial cells, which form the interface between the body and the external environment and are in close contact with the microbiota, express several membrane and intracellular proteins that sense microbial molecules (Vijay-Kumar et al. 2010). Examples of these sensors include pattern-recognition receptors such as Toll-like receptors (TLRs), C-type lectin, nucleotide oligomerization domain (NOD) receptors (NLRs), and retinoic acid inducible gene (RIG)-I-like receptors (RLRs), which are activated by microbial molecules including flagellin, lipopolysaccharide, lipoteichoic acid, peptidoglycans, N-acetylglucosamine, and double stranded-RNA. Above mentioned ligands and receptors not only participate in the microbiotic regulation of the immune system, but also serve as regulators themselves. These receptors also play a role in shaping the population of microbiota in the gut (Vijay-Kumar et al. 2010). Microbial fermentation of fiber in large intestine by microbiota results in the synthesis of SCFA according to the following equations:

$$59\,C_6H_{12}O_6 + 38\,H_2O \rightarrow 60\,\text{acetate} + 22\,\text{propionate} + 18\,\text{butyrate} + 96\,CO_2 + 256\,H^+$$

This production of SCFAs not only reduces circulating cholesterol levels, but produces anti-carcinogenic effects mediated by interference from numerous regulators of the cell cycle, proliferation, and apoptosis, such as β-catenin, p53, p21, Bax, and caspase three genes in chronic inflammation (Young et al. 2005; den Besten et al. 2013). During bacterial fermentation, different fibers produce different amounts and patterns of SCFAs. Thus, pectin forms high amounts of acetic acid, while guar gum yields propionic acid and β-glucan, fructo-oligosaccharides, some types of resistant starch and mixtures of dietary fibers forming high amounts of butyric acid (Henningsson et al. 2002; Nilsson and Nyman 2005). It is becoming increasingly evident that various SCFAs produce different physiological effects in GI. The bioactivities of butyrate are related to inhibition of class I and class II histone deacetylases (Davie 2003). Histone deacetylases regulate gene transcription through modification of chromatin structure by deacetylation of proteins, including histone proteins and transcription factors. In addition, butyrate is an important energy source for the colonic epithelial cells and has been shown to inhibit growth of cancer cells in vitro (Zoran et al. 1997). Butyrate also regulates and slows down fat transport from the intestine. The mechanism of butyrate action is not fully understood. However, it is proposed that promotion of energy expenditure and induction of mitochondrial function may be closely associated with butyrate-mediated processes. The stimulation of PGC-1α activity may be another molecular mechanism of butyrate action (Gao et al. 2009). Propionate has been shown to lower cholesterol levels (Fig. 7.3) (Hara et al. 1999) and produce a beneficial effect on glucose and lipid metabolism (Berggren et al. 1996). On the other hand, acetate stimulates hepatic de novo lipogenesis, and consequently increases adipocyte fatty acid storage. Furthermore, acetate and butyrate are also involved in epithelial barrier function maintenance as they stimulate the production and secretion of mucus by goblet cells, which protect the epithelium through increased expression of mucin (Barcelo et al. 2000). In vitro studies have indicated that butyrate increases MUC-2 expression by 23-fold in a goblet cell line demonstrating the importance of this function in

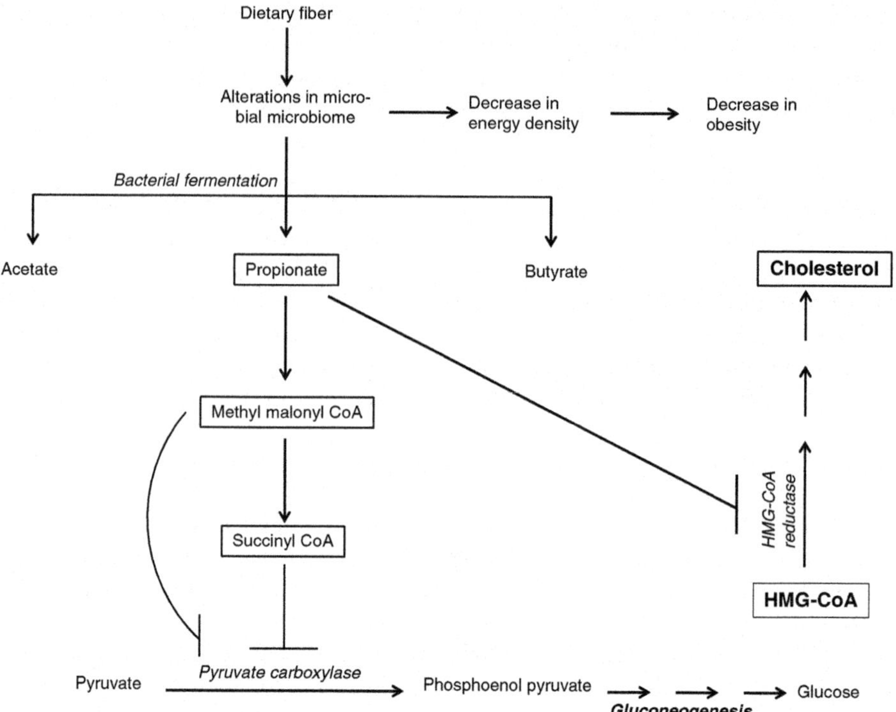

Fig. 7.3 Beneficial effects of short chain fatty acids in the metabolism. 3-Hydroxy-3-methylglutaryl-coenzyme (HMG-CoA; 3-Hydroxy-3-methylglutaryl-coenzyme reductase (HMG-CoA reductase)

modulating intestinal permeability (Gaudier et al. 2004). Butyrate also modulates the expression of tight junction protein expression (zonulin and occludin) and even low concentrations of this SCFA seem to reduce intestinal permeability (Bordin et al. 2004; Peng et al. 2007). The action of SCFA on leukocytes is mediated not only by the activation of G-protein coupled receptors (GPR41 and GPR43), but also by the inhibiton of histone deacetylase (HDAC). GPR receptors are expressed not only on adipose tissue, enteroendocrine cells, cells of the sympathetic nervous system, and gut epithelial cells, but also on immune cells (Brown et al. 2003; Briscoe et al. 2003). GPR41 has been reported to regulate host energy balance by modulating sympathetic activity and intestinal gluconeogenesis. By using in situ hybridization and quantitative RT-PCR analysis, it is shown that GPR41 mRNA is abundantly expressed in the mouse sympathetic ganglion (Kimura et al. 2011). GPR41 knockout mice exhibit the retardation of sympathetic nerve growth. However, more studies are required to elucidate the precise molecular mechanism by which GPR41 modulates sympathetic nerve differentiation and growth. SCFA (propionate) activates sympathetic nervous system, via GPR41 at the ganglionic level (Kimura et al. 2011). It is also shown that propionate increases the release of norepinephrine from sympathetic neurons through the GPR41–Gβγ–phospholipase C (PLC) β 3-ERK1/2-synapsin 2 (synaptic vesicle-associated phosphoprotein) pathway (Kimura et al. 2011; Inoue et al. 2012). β-Hydroxybutyrate (β-HB), a potent antagonist of GPR41 (Kimura et al. 2011) suppresses the propionate-induced sympathetic activation in both primary cultured sympathetic neurons and mice (Kimura et al. 2011; Inoue et al. 2012). Activation of GPR41 enhances production of peptide YY (PYY), an enteroendocrine cell hormone that normally inhibits gut motility, increases intestinal transit rate and reduces extraction of energy from the diet, thus affecting peripheral glucose utilization (Samuel et al. 2008). Collective evidence suggests that GPR receptors mediate their direct anti-inflammatory effects through interactions with SCFA. Insufficient intake of "healthy foodstuffs" adversely affects the production of bacterial metabolites. These metabolites and those derived directly from food drive beneficial downstream effects on immune pathways. It is proposed that insufficient exposure to dietary and bacterial metabolites may be closely associated with the development of inflammatory disorders in Western countries (Thorburn et al. 2014).

Inhibition of HDAC is associated with the inhibition of NFκ-B and IFNγ signaling pathways and an enhancement of peroxisome proliferator-activated receptor γ (PPARγ) expression, leading to the modulation of apoptosis and differentiation (Canani et al. 2011). Generation of SCFAs in the gut may also be responsible for the modulation of appetite, insulin signaling, and inflammation. It is proposed that SCFAs communicate with the immune system through G-protein-coupled free fatty acid receptors. Free fatty acid receptor 3 (FFA3) is expressed primarily by adipocytes and activated by propionate, butyrate, and acetate. The activation of adipocyte FFA3 by propionate leads to elevated leptin secretion, whereas activation by butyrate initiates adipogenesis (Vangaveti et al. 2010). FFA2 is expressed in leukocytes and colonic L-cells in addition to adipocytes. FFA2 is activated primarily by propionate (Vangaveti et al. 2010) and, to a lesser extent, acetate (Maslowski et al. 2009; Nilsson et al. 2003). Like FFA3, FFA2 may regulate the differentiation of adipocytes. However, unlike FFA3, it also lessens the fat cell triacylglycerol storage capacity with accumulation of body fat deposits (Sleeth et al. 2010). SCFAs also regulate several leukocyte functions including production of cytokines (TNF-α, IL-2, IL-6 and IL-10), eicosanoids and chemokines (e.g., MCP-1 and CINC-2). The ability of leukocytes to migrate to the foci of inflammation and to destroy microbial pathogens also seems to be affected by the SCFAs (Brown et al. 2003; Le Poul et al. 2003; Karaki et al. 2008).

Antiinflammatory effects of dietary fiber are mediated through the decrease in the production of proinflammatory cytokines (IL-6 and TNFα), inhibition of proinflammatory enzymes (cyclooxygenase 2, COX-2), and induction of nitric oxide synthase (iNOS) gene expression (Slavin 2005). The molecular mechanisms associated with beneficial effects of dietary fiber on metabolic health are not well established. However, it is proposed that fiber mediated changes in intestinal viscosity, nutrient absorption, rate of passage, production of short SCFAs and production of gut hormones are linked not only with decrease in body weight, insulin resistance, and type II diabetes, but also marked reduction in cardiovascular disease. Based on above beneficial effects, it is proposed that dietary fiber promotes longevity (Park et al. 2011). Collective evidence suggests that the consumption of fiber produces multiple beneficial effects on the body, such as, regulation of host gut bacterial community, hind gut fermentation and health, alleviation of weight gain, and increase in satiety.

7.3 Effects of Dietary Fiber on Insulin Resistance and Type II Diabetes

Insulin is an anabolic hormone, which regulates glucose homeostasis at several levels, including decreasing hepatic glucose synthesis and increasing peripheral glucose uptake, primarily in muscle and adipose tissue. Moreover, this hormone also stimulates lipogenesis and the synthesis of protein in hepatic and adipose tissues, while reducing lipolysis and proteolysis (Rorsman and Braun 2012). Insulin resistance is defined as the decrease in insulin capacity to stimulate glucose utilization, either due to insulin deficiency or due to impairment in its secretion and/or utilization. Systemic insulin resistance can result in impairment in the action of insulin in metabolically active organs and tissues, including skeletal muscle, the liver, and adipose tissue. In Western countries, the most common acquired factors that produce insulin resistance are obesity, sedentary lifestyle, and cerebral aging (Morino et al. 2006; Farooqui 2013, 2014). The molecular mechanism associated with insulin resistance is not fully understood. However, accumulation of lipids in the liver is considered to be one of the primary mechanisms involved in insulin resistance and type II diabetes. Thus, elevation in saturated fatty acids, triacylglycerol, diacylglycerol, acylcarnitines, and ceramide contribute to the molecular mechanism of insulin resistance (Adams et al. 2004, 2009; Holland et al. 2007) (Fig. 7.4). Ceramide has also been reported to promote insulin resistance by preventing insulin-stimulated Akt serine phosphorylation, activation, and translocation of Akt to its substrate (Summers et al. 1998). In addition, ceramide initiates inflammatory signaling pathways, leading to the activation of both c-jun NH_2-terminal kinase (JNK) and nuclear factor κB/inducer of κ kinase. All these processes have been reported to promote the development of insulin resistance (Cai et al. 2005; Chung et al. 2008; Ikonen and Vainio 2005). Collective evidence suggests that insulin resistance not only contributes to the defects in insulin receptor function, abnormalities in insulin signaling, alterations in glucose metabolism, induction of hyperinsulinemia, hyperglycemia and inflammation, but also increases BP (Wang and Jin 2009). Insulin resistance along with β-cell dysfunction, impaired with glucose tolerance and oxidative stress leads to type II diabetes (Ceriello and Motz 2004). This mechanism has been implicated as the underlying cause of both the macrovascular and microvascular complications associated with type II diabetes (Brownlee 2001).

Dietary fiber reduces the risk of type II diabetes through its effect on hepatic function and insulin sensitivity or by mediating the proinflammatory process. Dietary fiber decreases the oxidation of lipids, which in turn is associated with decreased inflammation (Ajani et al. 2004). It is postulated that a low-fiber diet with

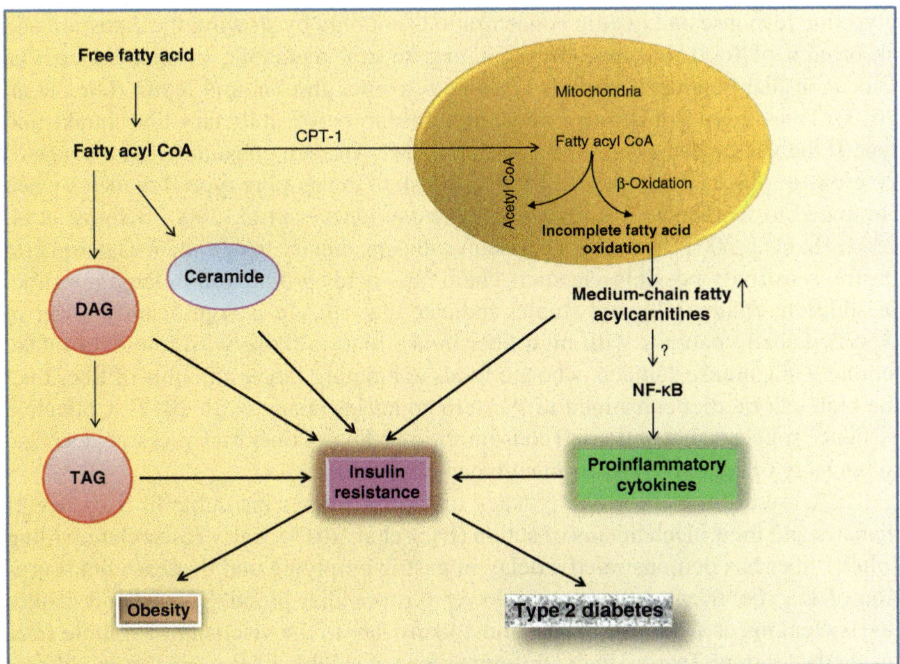

Fig. 7.4 Induction of insulin resistance by various lipids and their metabolites. *DAG* Diacylglycerol, *TAG* triacylglycerol, *CPT-1* carnitine palmitoyltransferase-1, *NF-κB* nuclear factor kappa-light-chain-enhancer of activated B cells

highly refined carbohydrates (high calorie diet) contributes to hyperglycemia, which increases the proinflammatory cytokines plasma interleukin-6 (IL-6), tumor necrosis factor α, and IL-18 (Esposito et al. 2004). IL-6 is a primary determinant of C-reactive protein (CRP) production; thus, consistently elevated concentrations of IL-6 might result in elevated CRP concentration. The consumption of high amounts of dietary fiber may lower concentration of CRP. The mechanism involved in lowering of CRP by fiber consumption remains elusive. However, it is proposed that dietary fiber decreases lipid oxidation, which in turn is associated with decreased inflammation (King 2005). Dietary fiber has the potential to attenuate glucose absorption rate, prevent weight gain, and increase the load of beneficial nutrients and antioxidants in the diet, all of which may help prevent type II diabetes. In addition, there is a positive correlation between consumption of high glycemic foods and pathogenesis of type II diabetes (Farooqui 2013). As stated above, dietary fiber not only resists digestion in the small intestine, thereby allowing it to enter the large intestine where it is fermented to produce SCFAs, which have anti-carcinogenic properties (Schatzkin et al. 2008; Lattimer and Haub 2010), but also increases the binding between bile acids and carcinogens and elevates the amount of estrogen excreted in the feces due to an inhibition of estrogen absorption in the intestines (Young et al. 2005; Lattimer and Haub 2010). Dietary fiber improves the postprandial glycemic response and insulin concentrations not only by slowing the digestion and absorption of food, but also by regulating several metabolic hormones, such as Glucagon-like peptide-1, peptide tyrosine tyrosine, ghrelin, and leptin (Díez et al. 2013). Thus, there is a strong inverse relationship between dietary fiber intake and type II diabetes when adjusted for age and BMI. Women consuming an average of 26 g/day of dietary fiber had a 22 % lower risk of developing type II diabetes when compared to women only consuming 13 g/day (Meyer et al. 2000; Schulze et al. 2004; Hu et al. 2001). In insulin-resistant subjects, dietary fiber enhances peripheral insulin sensitivity possibly via short-chain fatty acids production in large intestine. In addition, epidemiological studies indicate that HbA1c is significantly lower in type II diabetic patients with high fiber intake than in those with low fiber intake among 934 Chinese subjects who ate foods containing larger amounts of fiber than the high calorie diet consumed in Western countries (Jiang et al. 2012). Collective evidence suggests that increased consumption of dietary fiber may promote decrease in intensity of type II diabetes than glycemic index/load.

There are differences in the efficacy of soluble versus insoluble fiber in type II diabetes and their mechanisms of action (Díez et al. 2013). Early research regarding soluble fiber has demonstrated a delay in gastric emptying and decrease in absorption of macronutrients, resulting in lower postprandial blood glucose and insulin levels (Jenkins et al. 1978). This is most likely due to the viscosity of soluble fiber inside the GI tract. Interestingly, different types of soluble fiber exert varying effects on the viscosity and nutrient absorption. Guar has the highest viscosity as well as the greatest effect at decreasing postprandial blood glucose. Therefore, it is suggested that an increase in level of soluble fiber may be associated with a decrease in risk of type II diabetes. However, several recent studies have demonstrated the opposite effect, showing no correlation between soluble fiber and a reduced risk of type II diabetes (Meyer et al. 2000; Schulze et al. 2004; Montonen et al. 2003).

7.4   Effects of Dietary Fiber on Lipid Peroxidation

Lipid peroxidation is defined as an oxidative deterioration of membrane unsaturated fatty acids. It is a chain reaction initiated and propagated by any free radical that has sufficient reactivity to extract a hydrogen atom from the bis-allylic methylene group located between two adjacent double bonds of the hydrocarbon chain of an unsaturated fatty acid. The fatty acid is then transformed into a hydroperoxide. These hydroperoxides, in turn, may further breakdown into toxic aldehydes such as 4-hydroxynonenal, 4-hydroxyhexenal, acrolein, isoprostanes, and malondialdehyde (Farooqui and Horrocks 2007). The susceptibility of polyunsaturated fatty acid toward peroxidation increases with an increase in the number of unsaturated sites in the acyl chain. Lipid peroxidation in neural membranes has two broad outcomes. First is the structural damage to neural membranes and second is the generation of oxidized glycerophospholipids and their products. Some of these products are chemically reactive and are capable of covalently modifying neural membrane proteins. Oxidized glycerophospholipids and their breakdown products are major effectors of tissue damage following lipid peroxidation. Markers of lipid peroxidation include different molecules such as 4-HNE, acrolein, isoprostanes, and malondialdehyde. These markers are derived from arachidonic acid, which is released from neural membrane glycerophospholipids through the activation of phospholipases A_2 (Farooqui and Horrocks 2007).

The consumption of diet enriched in fiber decreases lipid peroxidation in several tissues (Thampi et al. 1991; Liu et al. 2003). Underlying mechanisms involved on the beneficial effects of fiber are not fully understood. However, it is proposed that constituents of fiber enriched diet upregulate endothelial nitric oxide synthase, decrease activities of carbohydrate digestive enzymes, decrease oxidative stress, and inhibit inflammatory gene expression in the liver and GI. In addition, dietary fiber has an impact on the expression of intestinal epithelial heat-shock proteins (HSP), which perform crucial housekeeping functions in order to maintain the mucosal barrier integrity (Liu et al. 2012).

7.5   Effects of Dietary Fiber on Satiation

Satiation is defined as the feeling of fullness during the consumption of a meal, which leads to meal termination (Schroeder et al. 2013). Satiation is controlled by numerous factors, including macronutrient composition, digestion, gastric emptying, and nutrient absorption, which together influence postprandial satiety responses. It is interesting to note that 100 cal from proteins produce different satiating effects than 100 cal from carbohydrate. The perception of fullness that comes from consuming protein may result in eating less carbohydrates and fats. Collective evidence suggests that a major function of satiation is to retard overconsumption during individual meals, thereby averting deleterious consequences from incomplete digestion as well as excessive disturbances in circulating levels of glucose and other nutrients.

Signals for satiation arise from multiple sites in the GI system, including the stomach, proximal small intestine, distal small intestine, colon, and pancreas. The consumption of food evokes satiation by two primary effects on the GI tract—gastric distention and release of peptides from enteroendocrine cells. The hindbrain is the principal central site receiving input from short-acting satiation signals, which are transmitted both by neural system (for example, by vagal afferents projecting to the nucleus of the solitary tract) as well as hormones (for example, by gut peptides acting directly on the area postrema (AP), which lies outside the blood–brain barrier. Although the perception of fullness clearly involves higher forebrain centers, conscious awareness of GI feedback signals is not required for satiation (Cummings and Overduin 2007). Satiation is also controlled by factors that begin when a food is consumed and continue as it enters the GI tract and is digested and absorbed (Benelam 2009; Schroeder et al. 2013). The consumption of fiber enriched food helps in controlling food intake by causing earlier termination of a meal or reducing inter-meal food intake. As food moves down the GI tract, signals are sent from brain to gut through neurotransmitter release to produce and modulate energy in a variety of ways, including slowing gastric emptying and reducing gastrointestinal secretions (gut hormones). These gut hormones modulate satiation. In addition, differences in particle size of dietary fiber may also modulate satiety, glycemic response, and other metabolic and biochemical (leptin, insulin) responses, which are related to insulin resistance and type II diabetes. Additionally, some dietary fibers produce prebiotic effects. For example, the presence of fermentable carbohydrates may increase the number of fecal bifidobacteria and lactobacilli (Costabile et al. 2008), thus potentially increasing the SCFAs production and thereby potentially altering the metabolic and physiological responses that affect body weight regulation. Collective evidence suggests that multiple mechanisms may be involved in fiber-mediated modulation of satiety (Slavin and Feirtag 2011). Greater satiation may be a product of the increased time required to chew certain fiber-rich foods. Increased time chewing is known to promote the release of saliva and gastric acid production, which may increase gastric distention leading optimal digestion. Some soluble/viscous fibers bind water, which also may increase distention. Stomach distension is known to trigger afferent vagal signals of fullness, which likely contribute to satiation during meals and satiety in the post-meal period.

Satiation is modulated by number of hormones. The glucagon-like peptides GLP-1 and GLP-2 are produced in enteroendocrine L cells of the small and large intestine and secreted in a nutrient-dependent manner. GLP-1 regulates nutrient assimilation via inhibition of gastric emptying and food intake. GLP-1 controls blood glucose following nutrient absorption via stimulation of glucose-dependent insulin secretion, insulin biosynthesis, islet proliferation, neogenesis, and inhibition of glucagon secretion. GLP-1 is an insulinotropic hormone. GLP-1 inhibits glucagon secretion. GLP-1 lowers blood glucose in normal subjects and in patients with type II diabetes. Collective evidence suggests that GLP-1 is secreted by the L cells of the ileum and proximal colon in a nutrient-dependent manner, enhances satiety and reduces food intake when administered to normal subjects (Bosch et al. 2009). Similarly, peptide tyrosine tyrosine (PYY), secreted from the same L cells as GLP-1. This hormone reduces food intake as well as delays gastric emptying (Plaisancie et al. 1996). The action of GLP-1 and PYY are opposed by ghrelin, a gut hormone, which induces hunger and stimulates food intake (Delmee et al. 2006). It is proposed that dietary fiber induces satiation not only by changing the secretion of satiety-related hormones, but also by interacting with leptin, a16-kDa protein, which is produced and secreted by adipocytes and is the product of the ob gene (Zhang et al. 1994; Frühbeck and Salvador 2000; Brubaker 2006; Plaisancie et al. 1996). Leptin acts as an afferent signal in a negative-feedback loop by inhibiting the appetite and regulating adiposity (Frühbeck and Gómez-Ambrosi 2003). Leptin also increases insulin sensitivity in normal and diabetic murine models (Frühbeck and Salvador 2000). Studies on feeding animals fermentable fiber (such as inulin, lactitol, resistant starch, and fructooligosaccharides) indicates that consumption of fiber not only results in an increase in concentration of plasma GLP-1, proximal colon GLP-1, and plasma PYY, but less weight gain and decrease in levels of ghrelin and leptin (Brubaker 2006; Plaisancie et al. 1996; Velázquez et al. 2006).

Several mechanisms have been proposed to explain fiber-induced ghrelin suppression, most importantly fermentation. These mechanisms include increase in circulating PYY levels and upregulation of somatostatin (Shimada et al. 2003; Sloth et al. 2007).

7.6   Effects of Dietary Fiber on Blood Pressure

High calorie diet-mediated obesity is the major cause of high BP. Obesity is related not only with increased activity of the renin-angiotensin-aldosterone and sympathetic nervous systems, and mineralocorticoid activity, but also with insulin resistance and reduction in kidney function. The consumption of high sodium chloride intake in high calorie diet strongly predisposes individual to high BP; whereas, the consumption of fiber is strongly associated with a reduced risk of BP, myocardial infarction, and ischemic stroke. At the molecular level fiber produces beneficial effects on gut hormones, such as cholecystokinin, glucagon-like peptide 1, and peptide YY and ghrelin, which not only control satiety but also modulate insulin sensitivity by providing potassium and magnesium (Weickert et al. 2005). In addition, soluble fibers also exert physiological effects on the stomach and small intestine by delaying the gastric emptying rate and small bowel transit time, increasing satiety, and slowing absorption of nutrients, such as glucose, triglycerides and cholesterol. Fermentation of insoluble fiber in the large intestine decreases cecal pH and increases bacterial biomass leading to an increase in fecal output and the production of gases (carbon dioxide, methane, hydrogen) and short-chain fatty acids (primarily acetate, propionate and butyrate), which produce antiinflammatory effects. These processes not only promote the reduction in weight, waist circumference, body mass index (BMI), percent body fat, and percent trunk fat mass; but also improve glucose metabolism and insulin sensitivity; and lower the risk of metabolic syndrome and type II diabetes (Satija and Hu 2012).

7.7   Effects of Dietary Fiber on Cholesterol Levels

Hypercholesterolemia is a major contributor for cardiovascular diseases (CVD) in both the developed and developing countries. Dietary fiber acts as low density lipoprotein cholesterol (LDL-C) lowering agents, offering an effective treatment against high blood cholesterol and CVD. Different types of fibers can vary in their viscosity, and studies in rats have indicated a positive relationship between viscosity and cholesterol-lowering ability. The molecular mechanism associated with cholesterol-lowering effect is not fully understood. Some investigators suggest that soluble fibers bind to bile acids or cholesterol during the intraluminal formation of micelles (Brown et al. 1999) resulting in reduction of cholesterol content of liver cells. However, increase in bile acid excretion may not be sufficient to account for the observed cholesterol reduction. Other investigators have suggested production of short-chain fatty acids (acetate, butyrate, and propionate) as possible cause of lower cholesterol (Nishina and Freedland 1990). It is also suggested that cholesterol lowering effects of dietary fibers are also supported by micronutrients and phytochemicals (Okarter and Liu 2010).

7.8 Antiinflammatory Effects of Dietary Fiber

Inflammation is an essential and protective component of mammalian physiology. It facilitates restoration of tissue homeostasis. The molecular mechanisms associated with beneficial effects of dietary fiber on metabolic health are not well established. However, it is proposed that fiber mediated changes in intestinal viscosity, nutrient absorption, rate of passage, production of SCFA and production of gut hormones may be linked not only with decrease in body weight, insulin resistance, and type II diabetes, but also marked reduction in inflammation and cardiovascular diseases.

It is well known that human intestinal microbiome contains up to 10^{14} bacterial species (mostly in the colon) that belong to approximately 1,000 different species (Musso et al. 2010). However, only about 40 of these species constitute 99 % of the total composition. Gut microbiota has an essential role in energy homeostasis, immune regulation, and protection against pathogens. The consumption of low or high fiber diet may have profound effects on gut microbiota and it is suggested that interactions of dietary components with microbiota may regulate inflammatory conditions and responses. Thus, germ-free mice have 40 % less body fat than conventionally raised mice, which can be reversed following colonization of their gut (Backhed et al. 2004). Furthermore, the microbial composition of mice given a high calorie diet containing high fat and sugar content can be modified after only 1 day, with lower numbers of Bacteroidetes species (Backhed et al. 2007), a group of Gram negative bacteria, which harbor lipopolysaccharides (LPS) in their cell wall. LPS is a large molecule formed by a lipid and a polysaccharide that elicits a strong immune response, promoting inflammation to protect the organism from bacterial infection (Akira and Takeda 2004). LPS is a potent activator of pathogen-associated molecular pattern (PAMP) responses, primarily via toll-like receptor 4 (TLR4), which activates an extensive cell signaling pathway that induces the inflammatory response and cytokine expression and secretion (Medzhitov and Horng 2009). Circulating levels of LPS are increased high calories diet consuming mice, rats, and humans. In rodents, increased levels of LPS are directly related to increase in intestinal permeability (Cani et al. 2008). Increase in intestinal permeability is due to decrease in expression and activity of tight junction proteins (zonula occludens-1 (ZO-1) and occludin), which together with gut epithelial cells form a barrier that separates the intestinal lumen and its bacterial population and products from peritoneal tissues. This degradation of tight junction function leads to the leakage of bacterial products, such as LPS, and bacterial translocation, which have recently been described as key factors in both human and mice insulin resistance and inflammation (Fig. 7.5) (Burcelin et al. 2012; Caricilli et al. 2011; Amar et al. 2011). In contrast, germ-free mice also

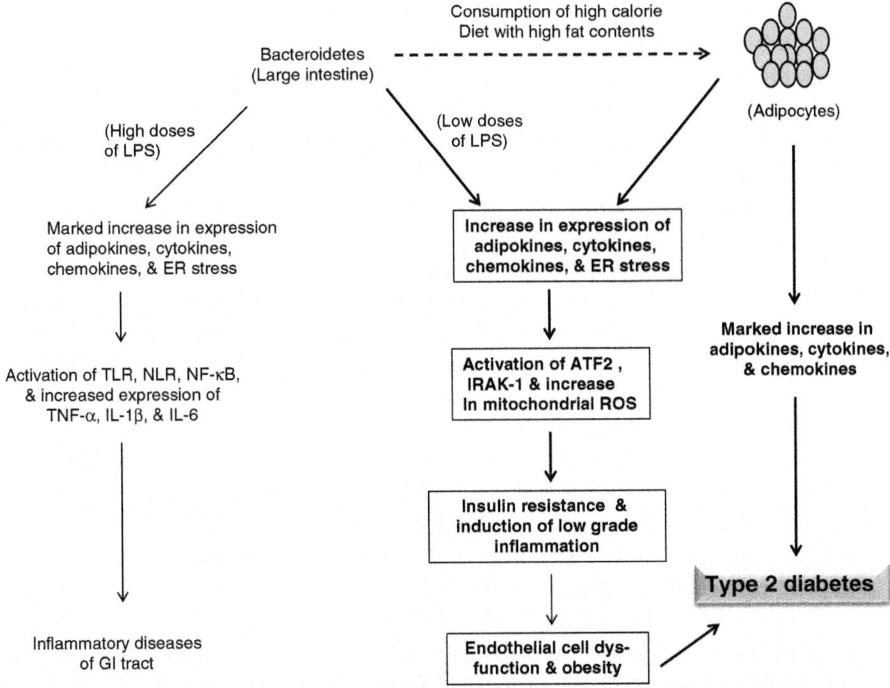

Fig. 7.5 Contribution of microbiota (Bacteroidetes) in pathogenesis of obesity and endothelial dysfunction. *LPS* Lipopolysaccharide; *ER* endoplasmic reticulum; *TLR* toll-like receptor; *NLR* Nod-like receptor; *NF-κB* nuclear factor kappa-light-chain-enhancer of activated B cells; *TNF-α* tumor necrosis factor-alpha; *IL-1β* interleukin-1 beta; *GI* gastrointestinal tract; *ATF* transcription factor 2; *IRAK* IL-1R-associated kinases-1; *ROS* reactive oxygen species

have an impaired ability to develop oral tolerance (Tang 2009) to food or aeroallergens, demonstrating an important connection between a healthy microbiome and immune regulation. Changes in the gut microbiota composition are classically considered as one of the many factors involved in the pathogenesis of either inflammatory bowel disease or irritable bowel syndrome. The use of particular food products with a prebiotic effect has thus been tested in clinical trials with the objective to improve the clinical activity and well-being of patients with such disorders. Promising beneficial effects have been demonstrated in some preliminary studies, including changes in gut microbiota composition (especially increase in bifidobacteria concentration).

Gut microbiota interacts with immune system to allow the host to tolerate the large amount of antigens present in the gut. It is becoming increasingly evident that gut evidence has highlighted the role of the microbiota in health and disease. Thus, microbiota have been implicated in the pathogenesis of diseases such as nonalcoholic steatohepatitis (Dumas et al. 2006), allergy (Noverr et al. 2004), the formation of gallstones (Ridlon et al. 2006), and inflammatory bowel disease (Alverdy and Chang 2008). The composition of the gut microbiota is highly variable

(Turnbaugh et al. 2007), and its diversity can be significantly affected by alterations in diet (Turnbaugh et al. 2006) or antibiotic use (Jernberg et al. 2007). Detailed investigations on gut microbial biodiversity have provided information on how does different microorganisms influence host function and promote beneficial effects and how does alterations in gut microbiota (dysbiosis) promote gastrointestinal disorders (Vieira et al. 2013).

Production of SCFA (butyrate) by the bacterial fermentation of dietary fiber colonic lumen induces changes in gene expression influencing colonic function related to an anti-inflammatory effect (Daly and Shirazi-Beechey 2006; Peng et al. 2009; Hamer et al. 2008). Thus, butyrate has been shown to modulate inflammation not only through the inhibition of NF-κB, but also through the up-regulation of PPAR-γ (Segain et al. 2000). The supplementation of butyrate at 5 % wt/wt in high-fat diet prevents the development of dietary obesity and insulin resistance. It also reduced obesity and insulin resistance in obese mice (Gao et al. 2009). The mechanism of butyrate action may be related to promotion of energy expenditure and induction of mitochondria function. Several in vivo studies have also indicated a decrease in inflammation after rectal administration of butyrate or mixtures of SCFA in patients with active ulcerative colitis (Lührs et al. 2002), but detailed information on butyrate-mediated decrease in inflammation activity remains unclear. It is also likely that the consumption of high fiber diet causes reduction in the expression of proinflammatory cytokines through the butyrate-mediated inhibition of NF-κB in the intestinal mucosa (Segain et al. 2000). Collective evidence suggests that the generation of butyrate is associated with maintenance of gut homeostasis and epithelial integrity, since it not only serves as the main energy source for colonocytes, directly influences host gene expression by inhibiting histone deacetylases, but also by interfering with proinflammatory signals, such as NF-κB (Fig. 7.6) (Segain et al. 2000; Hamer et al. 2008). A breakdown of epithelial integrity is associated with emerging diseases such as inflammatory bowel diseases and type II diabetes (Cani et al. 2007; Macia et al. 2012), and butyrate-producing members specifically are reduced in such patients (Clemente et al. 2012; Qin et al. 2012).

7.9 Consumption of Dietary Fiber and Weight Control

Dietary fiber facilitates body-weight control through different physiological mechanisms (Pereira and Ludwig 2001). Thus, fiber-rich foods promote satiating effect due to their relatively low energy density and palatability as compared with low-fiber foods. Furthermore, dietary fiber, especially soluble fiber, increases the viscosity of diets and slows down the digestion, which stimulates satiety regulating gut hormones, such as cholecystokinin and glucagon-like peptide 1, which modulate satiety. In addition, dietary fiber provides a mechanical barrier to the enzymic digestion of other macronutrients such as fat and starch in the small intestine. Moreover, the slower digestion and absorption rate of carbohydrate may reduce the postprandial blood glucose response, leading to improvement in insulin sensitivity, fatty acid oxidation, and weight control (Pereira and Ludwig 2001).

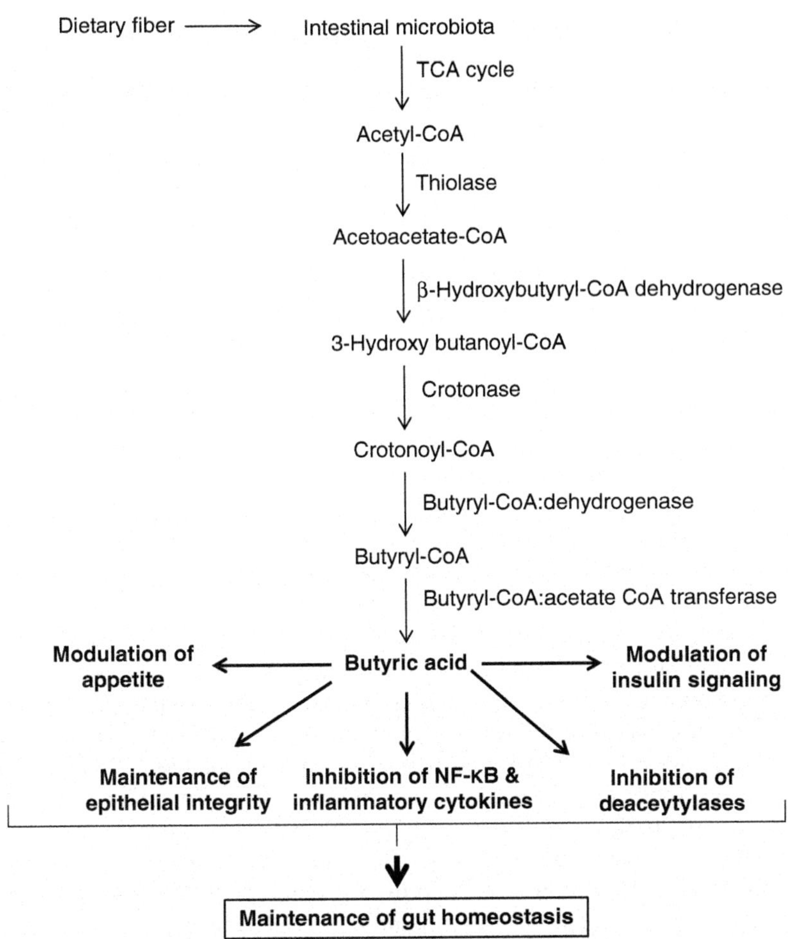

Fig. 7.6 Generation of butyric acid from dietary fiber by large intestinal microbiota and roles of butyric acid in large intestine

7.10   Effects of Dietary Fibers on the Excretion of Estrogen, Pro-Carcinogens and Carcinogens

Epidemiological studies support the view that dietary fiber reduces the risk of colorectal cancer. Several plausible mechanisms have been proposed to explain these observations, including increased stool bulk and dilution of carcinogens in the colonic lumen, reduced transit time, and bacterial fermentation of fiber to SCFAs (Lipkin et al. 1999; Howe et al. 1992).

In addition, dietary fiber is a major source of several vitamins (especially B-vitamins), minerals (magnesium and zinc), phytochemicals (phytic acid, tannins, and enzyme inhibitors), and phyto-oestrogens, which has anticancer properties and may decrease the risk of colorectal cancer by several potential mechanisms (Slavin et al. 1999) including interference in cell cycle, uncontrolled cell growth, and induction of chronic inflammation (Young et al. 2005; den Besten et al. 2013). Over the past decade several cohort studies have published results suggesting the ability of dietary fiber to the risk of colorectal cancer, with mixed results (Sanjoaquin et al. 2004; Schatzkin et al. 2007; Larsson et al. 2005).

7.11 Effect of Dietary Fiber on Diseases

It is well known that fiber is not absorbed into the systemic blood circulation. However, dietary fiber is fermented in the GI and modulates systemic immune response along with inflammation (Kau et al. 2011). Supplementation of 2.5 % fiber (plant-based fructooligosaccharide) in diet suppresses allergic airway inflammation caused by mite allergen in mice (Yasuda et al. 2010). Similarly, supplementation of plant-based fructooligosaccharide also reduces the ovalbumin-mediated allergic peritonitis (Yasuda et al. 2012) and allergic asthma (Vos et al. 2007) in mice. Mice consuming 5 % plant-based fructooligosaccharide in diet show less 2,4-dinitrofluorobenzene-induced contact sensitivity and less increase in plasma inflammatory markers (Watanabe et al. 2008; Fujiwara et al. 2010a). Furthermore, the same amount of plant-based fructooligosaccharide supplementation also reduces spontaneous skin lesions and inflammation-related cytokines in NC/Nga mice (Fujiwara et al. 2010a) supporting the view that supplementation of plant-based fructooligosaccharide induces anti-inflammatory protection for offspring (Fujiwara et al. 2010b).

Dietary fiber has also been reported to decrease the risk of developing metabolic syndrome. Thus, diet rich in fiber improves glycemic control in type II diabetes (Brennan 2005), reduces low-density lipoprotein (LDL) cholesterol in hypercholesterolemia (Brown et al. 1999), and contributes positively to long-term weight management (Slavin 2005). Collective evidence suggests that adequate consumption of fiber is associated with reduced risk for cardiovascular disease, cancer, type II diabetes, metabolic syndrome, certain gastrointestinal disorders and obesity. Despite ongoing efforts to promote adequate fiber through increased vegetable, fruit and whole-grain consumption, average fiber consumption has remained flat at approximately half of the recommended daily amounts.

7.12 Conclusion

Dietary fiber is the non-digestible form of carbohydrates and lignin. Dietary fiber occurs in plants. The consumption of high fiber diet produces beneficial effects on human health through many potential mechanisms. Soluble, viscous fiber modulates the absorption food from the small intestine because of the formation of gels that attenuate postprandial blood glucose and lipid rises. The formation of gels also slows gastric emptying, maintaining levels of satiety and contributing towards less weight gain. Dietary fiber also increases fecal bulking and viscosity and there is less contact time between potential carcinogens and mucosal cells. In addition, dietary fiber increases the binding between bile acids and carcinogens. The consumption of dietary fiber not only increases the amount of estrogen excreted in the feces due to an inhibition of estrogen absorption in the intestine, but also promotes healthy lipid profiles, glucose tolerance, and ensures normal gastrointestinal function. In addition, dietary fiber is a substrate for fermentation by microbiota found in the rectum. Microbiota produce SCFAs (acetate, propionate, and butyrate). The rate and amount of SCFA production depends on the microbiota species and amounts of microbiota present in the colon, the substrate source and gut transit time. SCFAs are readily absorbed. Butyrate is the major energy source for colonocytes. Propionate is largely taken up by the liver. Acetate enters the peripheral circulation to be metabolized by peripheral tissues. It is the principal SCFA in the colon, and after absorption it has been shown to increase cholesterol synthesis. However, propionate, a gluconeogenerator, has been shown to inhibit cholesterol synthesis. Therefore, substrates that can decrease the acetate: propionate ratio may reduce serum lipids and possibly cardiovascular disease risk. In colonic mucosa, butyrate not only promotes colon cancer prevention of colon by inducing cell differentiation, cell-cycle arrest and apoptosis of transformed colonocytes, but also by inhibiting the enzyme histone deacetylase and decreasing the transformation of primary to secondary bile acids as a result of colonic acidification. Therefore, a greater increase in SCFA production and potentially a greater delivery of SCFA, specifically butyrate, to the distal colon may result in a protective effect. Collective evidence suggests that dietary fiber lowers the risk of coronary heart disease, type II diabetes, some cancers, obesity, and premature death. Dietary fiber acts in different ways: (a) improves laxation by increasing bulk and reducing transit time of feces through the bowel; (b) increases excretion of bile acid, estrogen, and fecal procarcinogens and carcinogens by interacting with them; (c) lowers serum cholesterol; (d) slows glucose absorption and improves insulin sensitivity; (e) lowers blood pressure; (f) promotes weight loss;(g) inhibits lipid peroxidation; and (h) produces anti-inflammatory effects.

References

Adams JM II, Pratipanawatr T, Berria R, Wang E, DeFronzo RA, Sullards MC, Mandarino LJ (2004) Ceramide content is increased in skeletal muscle from obese insulin-resistant humans. Diabetes 53:25–31

Adams SH, Hoppel CL, Lok KH, Zhao L, Wong SW, Minkler PE, Hwang DH, Newman JW, Garvey WT (2009) Plasma acylcarnitine profiles suggest incomplete long-chain fatty acid beta-oxidation and altered tricarboxylic acid cycle activity in type 2 diabetic African-American women. J Nutr 139:1073–1081

Ajani UA, Ford ES, Mokdad AH (2004) Dietary fiber and C-reactive protein: findings from national health and nutrition examination survey data. J Nutr 134:1181–1185

Akira S, Takeda K (2004) Toll-like receptor signalling. Nat Rev Immunol 4:499–511

Alverdy JC, Chang EB (2008) The re-emerging role of the intestinal microflora in critical illness and inflammation: why the gut hypothesis of sepsis syndrome will not go away. J Leukoc Biol 83:461–466

Amar J, Chabo C, Waget A, Klopp P, Vachoux C, Bermúdez-Humarán LG, Smirnova N, Bergé M, Sulpice T, Lahtinen S, Ouwehand A, Langella P, Rautonen N, Sansonetti PJ, Burcelin R (2011) Intestinal mucosal adherence and translocation of commensal bacteria at the early onset of type 2 diabetes: molecular mechanisms and probiotic treatment. EMBO Mol Med 3:559–572

Backhed F, Ding H, Wang T, Hooper LV, Koh GY, Nagy A, Semenkovich CF, Gordon JI (2004) The gut microbiota as an environmental factor that regulates fat storage. Proc Natl Acad Sci U S A 101:15718–15723

Backhed F, Manchester JK, Semenkovich CF, Gordon JI (2007) Mechanisms underlying the resistance to diet-induced obesity in germ-free mice. Proc Natl Acad Sci U S A 104:979–984

Barcelo A, Claustre J, Moro F, Chayvialle JA, Cuber JC, Plaisancié P (2000) Mucin secretion is modulated by luminal factors in the isolated vascularly perfused rat colon. Gut 46:218–224

Benelam B (2009) Satiation, satiety and their effects on eating behaviour. Nutr Bulletin 34:126–173

Berggren AM, Nyman EM, Lundquist I, Bjorck IM (1996) Influence of orally and rectally administered propionate on cholesterol and glucose metabolism in obese rats. Br J Nutr 76:287–294

Bordin M, D'Atri F, Guillemot L, Citi S (2004) Histone deacetylase inhibitors up-regulate the expression of tight junction proteins. Mol Cancer Res 2:692–701

Bosch G, Verbrugghe A, Hesta M, Holst JJ, van der Poel AF, Janssens GP, Hendriks WH (2009) The effects of dietary fibre type on satiety-related hormones and voluntary food intake in dogs. Br J Nutr 102:318–325

Brennan CS (2005) Dietary fibre, glycaemic response, and diabetes. Mol Nutr Food Res 49:560–570

Briscoe CP, Tadayyon M, Andrews JL, Benson WG, Chambers JK, Eilert MM, Ellis C, Elshourbagy NA, Goetz AS, Minnick DT, Murdock PR, Sauls HR Jr, Shabon U, Spinage LD, Strum JC, Szekeres PG, Tan KB, Way JM, Ignar DM, Wilson S, Muir AI (2003) The orphan G protein-coupled receptor GPR40 is activated by medium and long chain fatty acids. J Biol Chem 278:11303–11311

Brown L, Rosner B, Willett WW, Sacks FM (1999) Cholesterol-lowering effects of dietary fiber: a meta-analysis. Am J Clin Nutr 69:30–42

Brown AJ, Goldsworthy SM, Barnes AA, Eilert MM, Tcheang L, Daniels D et al (2003) The orphan G protein-coupled receptors GPR41 and GPR43 are activated by propionate and other short chain carboxylic acids. J Biol Chem 278:11312–11319

Brownlee M (2001) Biochemistry and molecular cell biology of diabetic complications. Nature 12:813–820

Brubaker PL (2006) The glucagon-like peptides: pleiotropic regulators of nutrient homeostasis. Ann N Y Acad Sci 1070:10–26

Burcelin R, Garidou L, Pomie C (2012) Immuno-microbiota cross and talk: the new paradigm of metabolic diseases. Semin Immunol 24:67–74

Cai D, Yuan M, Frantz DF, Melendez PA, Hansen L, Lee J, Shoelson SE (2005) Local and systemic insulin resistance resulting from hepatic activation of IKK-beta and NF-kappaB. Nat Med 11:183–190

Canani RB, Costanzo MD, Leone L, Pedata M, Meli R, Calignano A (2011) Potential beneficial effects of butyrate in intestinal and extraintestinal diseases. World J Gastroenterol 17: 1519–1528

Cani PD, Amar J, Iglesias MA, Poggi M, Knauf C, Bastelica D, Neyrinck AM, Fava F, Tuohy KM, Chabo C, Waget A, Delmée E, Cousin B, Sulpice T, Chamontin B, Ferrières J, Tanti JF, Gibson GR, Casteilla L, Delzenne NM, Alessi MC, Burcelin R (2007) Metabolic endotoxemia initiates obesity and insulin resistance. Diabetes 56:1761–1772

Cani PD, Bibiloni R, Knauf C, Waget A, Neyrinck AM, Delzenne NM, Burcelin R (2008) Changes in gut microbiota control metabolic endotoxemia-induced inflammation in high-fat diet-induced obesity and diabetes in mice. Diabetes 57:1470–1481

Caricilli AM, Picardi PK, de Abreu LL, Ueno M, Prada PO, Ropelle ER, Hirabara SM, Castoldi Â, Vieira P, Camara NO, Curi R, Carvalheira JB, Saad MJ (2011) Gut microbiota is a key modulator of insulin resistance in TLR 2 knockout mice. PLoS Biol 9:e1001212

Cassidy MM, Fitzpatrick LR, Vahouny GV (1986) The effect of fiber in the postweaning diet on nutritional and intestinal morphological indices in the rats. In: Vahouny CV, Kritchevsky D (eds) Dietary fiber, basic and clinical aspects. Plenum Press, New York, pp 229–252

Ceriello A, Motz E (2004) Is oxidative stress the pathogenic mechanism underlying insulin resistance, diabetes, and cardiovascular disease? The common soil hypothesis revisited. Arterioscler Thromb Vasc Biol 12:816–823

Chung J, Nguyen AK, Henstridge DC, Holmes AG, Chan MH, Mesa JL, Lancaster GI, Southgate RJ, Bruce CR, Duffy SJ, Horvath I, Mestril R, Watt MJ, Hooper PL, Kingwell BA, Vigh L, Hevener A, Febbraio MA (2008) HSP72 protects against obesity-induced insulin resistance. Proc Natl Acad Sci U S A 105:1739–1744

Clemente JC, Ursell LK, Parfrey LW, Knight R (2012) The impact of the gut microbiota on human health: an integrative view. Cell 148:1258–1270

Costabile A, Klinder A, Fava F, Napolitano A, Fogliano V, Leonard C, Gibson GR, Tuohy KM (2008) Whole-grain wheat breakfast cereal has a prebiotic effect on the human gut microbiota: a double-blind, placebo-controlled, crossover study. Br J Nutr 99:110–120

Cummings DE, Overduin J (2007) Gastrointestinal regulation of food intake. J Clin Invest 117:13–23

Daly K, Shirazi-Beechey SP (2006) Microarray analysis of butyrate regulated genes in colonic epithelial cells. DNA Cell Biol 25:49–62

Davie JR (2003) Inhibition of histone deacetylase activity by butyrate. J Nutr 133:2485S–2493S

Delmee E, Cani PD, Gual G, Knauf C, Burcelin R, Maton N, Delzenne NM (2006) Relation between colonic proglucagon expression and metabolic response to oligofructose in high fat diet-fed mice. Life Sci 79:1007–1013

Delzenne NM, Kok N (2001) Effects of fructans-type prebiotics on lipid metabolism. Am J Clin Nutr 73:456S–458S

den Besten G, van Eunen K, Groen AK, Venema K, Reijngoud DJ, Bakker BM (2013) The role of short-chain fatty acids in the interplay between diet, gut microbiota, and host energy metabolism. J Lipid Res 54:2325–2340

Dethlefsen L, McFall-Ngai M, Relman DA (2007) An ecological and evolutionary perspective on human-microbe mutualism and disease. Nature 449:811–818

Díez R, García JJ, Diez MJ, Sierra M, Sahagún AM, Calle ÁP, Fernández N (2013) Hypoglycemic and hypolipidemic potential of a high fiber diet in healthy versus diabetic rabbits. Biomed Res Int 2013:960568

Dumas ME, Barton RH, Toye A, Cloarec O, Blancher C, Rothwell A, Fearnside J, Tatoud R, Blanc V, Lindon JC, Mitchell SC, Holmes E, McCarthy MI, Scott J, Gauguier D, Nicholson JK (2006) Metabolic profiling reveals a contribution of gut microbiota to fatty liver phenotype in insulin-resistant mice. Proc Natl Acad Sci U S A 103:12511–12516

Eastwood MA, Morris ER (1992) Physical properties of dietary fiber that influence physiological function: a model for polymers along the gastrointestinal tract. Am J Clin Nutr 55:436–442

Eckburg PB, Bik EM, Bernstein CN, Purdom E, Dethlefsen L, Sargent M, Gill SR, Nelson KE, Relman DA (2005) Diversity of the human intestinal microbial flora. Science 308:1635–1638

Esposito K, Marfella R, Ciotola M, Di Palo C, Giugliano F, Giugliano G, D'Armiento M, D'Andrea F, Giugliano D (2004) Effect of a Mediterranean-style diet on endothelial dysfunction and markers of vascular inflammation in the metabolic syndrome: a randomized trial. JAMA 292:1440–1446

Farooqui AA (2013) Metabolic syndrome: an important risk factor for stroke, Alzheimer disease, and depression. Springer, New York

Farooqui AA (2014) Inflammation and oxidative stress in neurological disorders: effect of lifestyle, genes, and age. Springer, New York

Farooqui AA, Horrocks LA (2007) Glycerophospholipids in Brain. Springer, New York

Frühbeck G, Gómez-Ambrosi J (2003) Control of body weight: a physiologic and transgenic perspective. Diabetologia 46:143–172

Frühbeck G, Salvador J (2000) Relation between leptin and the regulation of glucose metabolism. Diabetologia 43:3–12

Fujiwara R, Sasajima N, Takemura N, Ozawa K, Nagasaka Y, Okubo T, Sahasakul Y, Watanabe J, Sonoyama K (2010a) 2,4-Dinitrofluorobenzene-induced contact hypersensitivity response in NC/Nga mice fed fructo-oligosaccharide. J Nutr Sci Vitaminol (Tokyo) 56:260–265

Fujiwara R, Takemura N, Watanabe J, Sonoyama K (2010b) Maternal consumption of fructo-oligosaccharide diminishes the severity of skin inflammation in offspring of NC/Nga mice. Br J Nutr 103:530–538

Galisteo M, Moron R, Rivera L, Romero R, Anguera A, Zarzuelo A (2010) Plantago ovata huskssupplemented diet ameliorates metabolic alterations in obese Zucker rats through activation of AMP-activated protein kinase. Comparative study with other dietary fibers. Clin Nutr 29:261–267

Gao Z, Yin J, Zhang J, Ward RE, Martin RJ, Lefevre M, Cefalu WT, Ye J (2009) Butyrate Improves Insulin Sensitivity and Increases Energy Expenditure in Mice. Diabetes 58:1509–1517

Gaudier E, Jarry A, Blottière HM, de Coppet P, Buisine MP, Aubert JP, Laboisse C, Cherbut C, Hoebler C (2004) Butyrate specifically modulates MUC gene expression in intestinal epithelial goblet cells deprived of glucose. Am J Physiol 287:G1168–G1174

Ghosh S, Dai C, Brown K, Rajendiran E, Makarenko S, Baker J, Ma C, Halder S, Montero M, Ionescu VA, Klegeris A, Vallance BA, Gibson DL (2011) Colonic microbiota alters host susceptibility to infectious colitis by modulating inflammation, redox status, and ion transporter gene expression. Am J Physiol Gastrointest Liver Physiol 301:G39–G49

Hamer HM, Jonkers D, Venema K, Vanhoutvin S, Troost FJ, Brummer RJ (2008) Review article: the role of butyrate on colonic function. Aliment Pharmacol Ther 27:104–119

Hara H, Haga S, Aoyama Y, Kiriyama S (1999) Short-chain fatty acids suppress cholesterol synthesis in rat liver and intestine. J Nutr 129:942–948

Heaton KW (1973) Food fibre as an obstacle to energy intake. Lancet 22(2):1418–1421

Henningsson AM, Bjorck IM, Nyman EM (2002) Combinations of indigestible carbohydrates affect short-chain fatty acid formation in the hindgut of rats. J Nutr 132:3098–3104

Holland WL, Brozinick JT, Wang LP, Hawkins ED, Sargent KM, Liu Y, Narra K, Hoehn KL, Knotts TA, Siesky A, Nelson DH, Karathanasis SK, Fontenot GK, Birnbaum MJ, Summers SA (2007) Inhibition of ceramide synthesis ameliorates glucocorticoid-, saturated-fat- and obesity-induced insulin resistance. Cell Metab 5:167–179

Howe GR, Benito E, Castelleto R, Cornee J, Esteve J, Gallagher RP, Iscovich JM, Deng-ao J, Kaaks R, Kune GA (1992) Dietary intake of fiber and decreased risk of cancers of the colon and rectum: evidence from the combined analysis of 13 case-control studies. J Natl Cancer Inst 84:1887–1896

Hu FB, Manson JE, Stampfer MJ, Colditz G, Liu S, Solomon CG, Willett WC (2001) Diet, lifestyle, and the risk of type 2 diabetes mellitus in women. N Engl J Med 345:790–797

Ikonen E, Vainio S (2005) Lipid microdomains and insulin resistance: is there a connection? Sci STKE 2005: pe3

Inoue D, Kimura I, Wakabayashi M, Tsumoto H, Ozawa K, Hara T, Takei Y, Hirasawa A, Ishihama Y, Tsujimoto G (2012) Short-chain fatty acid receptor GPR41-mediated activation of sympathetic neurons involves synapsin 2b phosphorylation. FEBS Lett 586:1547–1554

Institute of Medicine (U.S.) Panel on Macronutrients, Institute of Medicine (U.S.) (2005) Standing committee on the scientific evaluation of dietary reference intakes. Dietary reference intakes for energy, carbohydrate, fiber, fat, fatty acids, cholesterol, protein, and amino acids. National Academies Press, Washington, p 339

Ishikawa F, Takayama H, Matsumoto K (1995) Effect of β 1-4 linked galacto oligosaccharides on human fecal microflora. Bifidus 9:5–18

Jenkins DJ, Wolever TM, Leeds AR, Gassull MA, Haisman P, Dilawari J, Goff DV, Metz GL, Alberti KG (1978) Dietary fibres, fibre analogues, and glucose tolerance: importance of viscosity. Br Med J 1:1392–1394

Jernberg C, Lofmark S, Edlund C, Jansson JK (2007) Long-term ecological impacts of antibiotic administration on the human intestinal microbiota. ISME J 1:56–66

Jiang J, Qiu H, Zhao G, Zhou Y, Zhang Z, Zhang H, Jiang Q, Sun Q, Wu H, Yang L, Ruan X, Xu WH (2012) Dietary fiber intake is associated with HbA1c level among prevalent patients with type 2 diabetes in pudong new area of Shanghai, China. PLoS One 12:e46552

Kaczmarczyk MM, Miller MJ, Freund GG (2012) The health benefits of dietary fiber: beyond the usual suspects of type 2 diabetes mellitus, cardiovascular disease and colon cancer. Metabolism 61:1058–1066

Karaki S, Tazoe H, Hayashi H, Kashiwabara H, Tooyama K, Suzuki Y et al (2008) Expression of the short-chain fatty acid receptor, GPR43, in the human colon. J Mol Histol 39:135–142

Kau AL, Ahern P, Griffin N, Goodman A, Gordon J (2011) Human nutrition, the gut microbiome and the immune system. Nature 474:327–336

Kimura I, Inoue D, Maeda T, Hara T, Ichimura A, Miyauchi S, Kobayashi M, Hirasawa A, Tsujimoto G (2011) Short-chain fatty acids and ketones directly regulate sympathetic nervous system via G protein-coupled receptor 41 (GPR41). Proc Natl Acad Sci U S A 108: 8030–8035

King DE (2005) Dietary fiber, inflammation, and cardiovascular disease. Mol Nutr Food Res 49:594–600

Kristiansen K, Wang J (2012) A metagenome-wide association study of gut microbiota in type 2 diabetes. Nature 490:55–60

Larsson SC, Giovannucci E, Bergkvist L, Wolk A (2005) Whole grain consumption and risk of colorectal cancer: a population-based cohort of 60,000 women. Br J Cancer 92:1803–1807

Lattimer JM, Haub MD (2010) Effects of dietary fiber and its components on metabolic health. Nutrients 2:1266–1289

Le Poul E, Loison C, Struyf S, Springael JY, Lannoy V, Decobecq ME, Brezillon S, Dupriez V, Vassart G, Van Damme J, Parmentier M, Detheux M (2003) Functional characterization of human receptors for short chain fatty acids and their role in polymorphonuclear cell activation. J Biol Chem 278:25481–25489

Ley RE, Hamady M, Lozupone C, Turnbaugh PJ, Ramey RR, Bircher JS, Schlegel ML, Tucker TA, Schrenzel MD, Knight R, Gordon JI (2008) Evolution of mammals and their gut microbes. Science 320:1647–1651

Lipkin M, Reddy B, Newmark H, Lamprecht SA (1999) Dietary factors in human colorectal cancer. Annu Rev Nutr 19:545–586

Liu S, Willett WC, Manson JE, Hu FB, Rosner B, Colditz G (2003) Relation between changes in intakes of dietary fiber and grain products and changes in weight and development of obesity among middle-aged women. Am J Clin Nutr 78:920

Liu HY, Lundh T, Dicksved J, Lindberg JE (2012) Expression of heat shock protein 27 in gut tissue of growing pigs fed diets without and with inclusion of chicory fiber. J Anim Sci 5(Suppl 4):25–27

Lührs H, Gerke T, Müller JG, Melcher R, Schauber J, Boxberge F, Scheppach W, Menzel T (2002) Butyrate inhibits NF-kappa B activation in lamina propria macrophages of patients with ulcerative colitis. Scand J Gastroenterol 37:458–466

Ma Y, Griffith J, Chasan-Tuber L (2007) Association between dietary fiber and serum C-reactive protein. Am J Clin Nutr 83:760–766

Macia L, Thorburn AN, Binge LC, Marino E, Rogers KE, Maslowski KM, Vieira AT, Kranich J, Mackay CR (2012) Microbial influences on epithelial integrity and immune function as a basis for inflammatory diseases. Immunol Rev 245:164–176

Maslowski KM, Vieira AT, Ng A, Kranich J, Sierro F, Yu D, Schilter HC, Rolph MS, Mackay F, Artis D, Xavier RJ, Teixeira MM, Mackay CR (2009) Regulation of inflammatory responses by gut microbiota and chemoattractant receptor GPR43. Nature 461:1282–1286

McGuire S (2011) U.S. Department of Agriculture and U.S. Department of Health and Human Services, Dietary Guidelines for Americans, 2010. Adv Nutr 2:293–294

Medzhitov R, Horng T (2009) Transcriptional control of the inflammatory response. Nat Rev Immunol 9:692–703

Meyer KA, Kushi LH, Jacobs DR, Slavin J, Sellers TA, Folsom AR (2000) Carbohydrates, dietary fiber, and incident type 2 diabetes in older women. Am J Clin Nutr 71:921–930

Montagne L, Pluske JR, Hampson DJ (2003) A review of interactions between dietary fibre and the intestinal mucosa, and their consequences on digestive health in young non-ruminant animals. Anim Feed Sci Technol 108:95–117

Montonen J, Knekt P, Jarvinen R, Aromaa A, Reunanen A (2003) Whole-grain and fiber intake and the incidence of type 2 diabetes. Am J Clin Nutr 77:622–629

Morino K, Petersen KF, Shulman GI (2006) Molecular mechanisms of insulin resistance in humans and their potential links with mitochondrial dysfunction. Diabetes 55(Suppl 2):S9–S15

Musso G, Gambino R, Cassader M (2010) Obesity, diabetes, and gut microbiota: the hygiene hypothesis expanded? Diabetes Care 33:2277–2284

Nilsson U, Nyman M (2005) Short-chain fatty acid formation in the hindgut of rats fed oligosaccharides varying in monomeric composition, degree of polymerisation and solubility. Br J Nutr 94:705–713

Nilsson NE, Kotarsky K, Owman C, Olde B (2003) Identification of a free fatty acid receptor, FFA_2R, expressed on leukocytes and activated by short-chain fatty acids. Biochem Biophys Res Commun 303:1047–1052

Nishina PM, Freedland RA (1990) The effects of dietary fiber feeding on cholesterol metabolism in rats. J Nutr 120:800–805

Noverr MC, Noggle RM, Toews GB, Huffnagle GB (2004) Role of antibiotics and fungal microbiota in driving pulmonary allergic responses. Infect Immun 72:4996–5003

Nugent AP (2005) Health properties of resistant starch. Nutr Bulletin 30:27–54

Okarter N, Liu RH (2010) Health benefits of whole grain phytochemicals. Crit Rev Food Sci Nutr 50:193–208

Papathanasopoulos A, Camilleri M (2010) Dietary fiber supplements: effects in obesity and metabolic syndrome and relationship to gastrointestinal functions. Gastroenterology 138:65–72, e1–2

Park Y, Subar AF, Hollenbeck A, Schatzkin A (2011) Dietary fiber intake and mortality in the NIH-AARP diet and health study. Arch Intern Med 171:1061–1068

Peng L, He Z, Chen W, Holzman IR, Lin J (2007) Effects of butyrate on intestinal barrier function in a Caco-2 cell monolayer model of intestinal barrier. Pediatr Res 61:37–41

Peng L, Li ZR, Green RS, Holzman IR, Lin J (2009) Butyrate enhances the intestinal barrier by facilitating tight junction assembly via activation of AMP-activated protein kinase in Caco-2 cell monolayers. J Nutr 139:1619–1625

Pereira MA, Ludwig DS (2001) Dietary fiber and body-weight regulation: observations and mechanisms. Pediatr Clin North Am 48:969–980

Plaisancie P, Dumoulin V, Chayvialle JA, Cuber JC (1996) Luminal peptide YY-releasing factors in the isolated vascularly perfused rat colon. J Endocrinol 151:421–429

Qin J, Li Y, Cai Z, Li S, Zhu J, Zhang F, Liang S, Zhang W, Guan Y, Shen D, Peng Y, Zhang D, Jie Z, Wu W, Qin Y, Xue W, Li J, Han L, Lu D, Wu P, Dai Y, Sun X, Li Z, Tang A, Zhong S, Li X, Chen W, Xu R, Wang M, Feng Q, Gong M, Yu J, Zhang Y, Zhang M, Hansen T, Sanchez G, Raes J, Falony G, Okuda S, Almeida M, LeChatelier E, Renault P, Pons N, Batto JM, Zhang Z, Chen H, Yang R, Zheng W, Li S, Yang H, Wang J, Ehrlich SD, Nielsen R, Pedersen O, Kristiansen K, Wang J (2012) A metagenome-wide association study of gut microbiota in type 2 diabetes. Nature 490:55–60

Raninen K, Lappi J, Mykkänen H, Poutanen K (2011) Dietary fiber type reflects physiological functionality: comparison of grain fiber, inulin, and polydextrose. Nutr Rev 69:9–21

Ridlon JM, Kang DJ, Hylemon PB (2006) Bile salt biotransformations by human intestinal bacteria. J Lipid Res 47:241–259

Rorsman P, Braun M (2012) Regulation of insulin secretion in human pancreatic islets. Ann Rev Physiol 75:21–25

Samuel BS, Shaito A, Motoike T, Rey FE, Backhed F, Manchester JK, Hammer RE, Williams SC, Crowley J, Yanagisawa M, Gordon JI (2008) Effects of the gut microbiota on host adiposity are modulated by the short-chain fatty-acid binding G protein-coupled receptor, Gpr41. Proc Natl Acad Sci U S A 105:16767–16772

Sanjoaquin MA, Appleby PN, Thorogood M, Mann JI, Key TJ (2004) Nutrition, lifestyle and colorectal cancer incidence: a prospective investigation of 10998 vegetarians and non-vegetarians in the United Kingdom. Br J Cancer 90:118–121

Satija A, Hu FB (2012) Cardiovascular benefits of dietary fiber. Curr Atheroscler Rep 14: 505–514

Schatzkin A, Mouw T, Park Y, Subar AF, Kipnis V, Hollenbeck A, Leitzmann MF, Thompson FE (2007) Dietary fiber and whole-grain consumption in relation to colorectal cancer in the NIH-AARP Diet and Health Study. Am J Clin Nutr 85:1353–1360

Schatzkin A, Park Y, Leitzmann MF, Hollenbeck AR, Cross AJ (2008) Prospective study of dietary fiber, whole grain foods, and small intestinal cancer. Gastroenterology 135:1163–1167

Schroeder N, Marquart LF, Gallaher DD (2013) The role of viscosity and fermentability of dietary fibers on satiety- and adiposity-related hormones in rats. Nutrients 5:2093–2113

Schulze MB, Liu S, Rimm EB, Manson JE, Willett WC, Hu FB (2004) Glycemic index, glycemic load, and dietary fiber intake and incidence of type 2 diabetes in younger and middle–aged women. Am J Clin Nutr 80:348–356

Segain JP, Raingeard de la Blétière D, Bourreille A, Leray V, Gervois N, Rosales C, Ferrier L, Bonnet C, Blottière HM, Galmiche JP (2000) Butyrate inhibits inflammatory responses through NFkappaB inhibition: implications for Crohn's disease. Gut 47:397–403

Shimada M, Date Y, Mondal MS, Toshinai K, Shimbara T, Fukunaga K, Murakami N, Miyazato M, Kangawa K, Yoshimatsu H, Matsuo H, Nakazato M (2003) Somatostatin suppresses ghrelin secretion from the rat stomach. Biochem Biophys Res Commun 302:520–525

Slavin JL (2001) Dietary fiber and colon cancer. In: Sungsoo Cho S, Dreher ML (eds) Handbook of dietary fiber. Marcel Dekker, New York, pp 31–46

Slavin JL (2005) Dietary fiber and body weight. Nutrition 21:411–418

Slavin JL (2013) Carbohydrates, dietary fiber, and resistant starch in white vegetables: links to health outcomes. Adv Nutr 4:351S–315S

Slavin J, Feirtag J (2011) Chicory inulin does not increase stool weight or speed up intestinal transit time in healthy male subjects. Food Funct 2:72–77

Slavin JL, Martini MC, Jacobs DR Jr, Marquart L (1999) Plausible mechanisms for the protectiveness of whole grains. Am J Clin Nutr 70:S459–S463

Sleeth ML, Thompson EL, Ford HE, Zac-Varghese SE, Frost G (2010) Free fatty acid receptor 2 and nutrient sensing: a proposed role for fibre, fermentable carbohydrates and short-chain fatty acids in appetite regulation. Nutr Res Rev 23:135–145

Sloth B, Davidsen L, Holst JJ, Flint A, Astrup A (2007) Effect of subcutaneous injections of PYY1-36 and PYY 3-36 on appetite, ad libitum energy intake, and plasma free fatty acid concentration in obese males. Am J Physiol 293:E604–E609

Summers SA, Garza L, Zhou H, Birnbaum MJ (1998) Regulation of insulin-stimulated glucose transporter GLUT4 translocation and Akt kinase activity by ceramide. Mol Cell Biol 18:5457–5464

Tang MLK (2009) Probiotics and prebiotics: immunological and clinical effects in allergic disease. Nestle Nutrition Workshop Series. Pediatr Prog 64:219–238

Tappenden KA, Deutsch AS (2007) The physiological relevance of the intestinal microbiota—contributions to human health. J Am Coll Nutr 26:679S–683S

Thampi BS, Manoj G, Leelamma S, Menon VP (1991) Dietary fiber and lipid peroxidation: effect of dietary fiber on levels of lipids and lipid peroxides in high fat diet. Indian J Exp Biol 29:563–567

Thorburn AN, Macia L, Mackay CR (2014) Diet, metabolites, and "western-lifestyle" inflammatory diseases. Immunity 40:833–842

Turnbaugh PJ, Ley RE, Mahowald MA, Magrini V, Mardis ER, Gordon JI (2006) An obesity-associated gut microbiome with increased capacity for energy harvest. Nature 444: 1027–1031

Turnbaugh PJ, Ley RE, Hamady M, Fraser-Liggett CM, Knight R, Gordon JI (2007) The human microbiome project. Nature 449:804–810

Turner ND, Lupton JR (2011) Dietary fiber. Adv Nutr 2:151–152

U.S. Department of Agriculture, Agricultural Research Service (2012) Nutrient intakes from food: mean amounts consumed per individual, by gender and age, what we eat in America, NHANES 2009–2010. Accessed 15 Feb 2014 from http://www.ars.usda.gov/ba/bhnrc/fsrg

Vangaveti V, Shashidhar V, Jarrod G, Baune BT, Kennedy RL (2010) Free fatty acid receptors – emerging targets for treatment of diabetes and its complications. Ther Adv Endocrinol Metab 1:165–175

Velázquez M, Davies C, Marett R, Slavin JL, Feirtag JM (2006) Effect of oligosaccharides and fibre substitutes on short-chain fatty acid production by human faecal microflora. Anaerobe 6:87–92

Vieira AT, Teixeira MM, Martins FS (2013) The role of probiotics and prebiotics in inducing gut immunity. Front Immunol 4:445

Vijay-Kumar M, Aitken JD, Carvalho FA, Cullender TC, Mwangi S, Srinivasan S, Sitaraman SV, Knight R, Ley RE, Gewirtz AT (2010) Metabolie syndrome and altered gut microbiota in mice lacking toll-like receptor 5. Science 328:228–231

Vos AP, van Esch B, Stahl B, M'Rabet L, Folkerts G, Nijkamp F, Garssen J (2007) Dietary supplementation with specific oligosaccharide mixtures decreases parameters of allergic asthma in mice. Int Immunopharmacol 7:1582–1587

Wang Q, Jin T (2009) The role of insulin signaling in the development of beta-cell dysfunction and diabetes. Islets 1:95–101

Watanabe J, Sasajima N, Aramaki A, Sonoyama K (2008) Consumption of fructo-oligosaccharide reduces 2,4-dinitrofluorobenzene-induced contact hypersensitivity in mice. Br J Nutr 100:339–346

Weickert MO, Mohlig M, Koebnick C, Holst JJ, Namsolleck P, Ristow M, Osterhoff M, Rochlitz H, Rudovich N, Spranger J et al (2005) Impact of cereal fibre on glucose-regulating factors. Diabetologia 48:2343–2353

Yasuda A, Inoue K, Sanbongi C, Yanagisawa R, Ichinose T, Yoshikawa T, Takano H (2010) Dietary supplementation with fructooligosaccharides attenuates airway inflammation related to house dust mite allergen in mice. Int J Immunopathol Pharmacol 23:727–735

Yasuda A, Inoue K, Sanbongi C, Yanagisawa R, Ichinose T, Tanaka M, Yoshikawa T, Takano H (2012) Dietary supplementation with fructooligosaccharides attenuates allergic peritonitis in mice. Biochem Biophys Res Commun 422:546–550

Young GP, Hu Y, Le Leu RK, Nyskohus L (2005) Dietary fibre and colorectal cancer: a model for environment—gene interactions. Mol. Nutr. Food Res 49:571–584

Zhang Y, Proenca R, Maffei M, Barone M, Leopold L, Friedman JM (1994) Positional cloning of the mouse obese gene and its human homologue. Nature 372:425–432

Zoran DL, Turner ND, Taddeo SS, Chapkin RS, Lupton JR (1997) Wheat bran diet reduces tumor incidence in a rat model of colon cancer independent of effects on distal luminal butyrate concentrations. J Nutr 127:2217–2225

Chapter 8
Effects of the High Calorie Diet on the Development of Chronic Visceral Disease

8.1 Introduction

In past 100 years human diet has changed dramatically in developed countries. As stated in Chap. 1, high calorie diet provides about 50 % of total daily calories from refined carbohydrates, 35 % calories from fat and refined oils, and 15 % from proteins of animal origin. In addition in high calorie diet, the ratio between n-6 (arachidonic acid, ARA) to n-3 fatty acid (docosahexaenoic acid, DHA) is about 20:1 (Farooqui 2009). In contrast, Paleolithic diet on which our forefathers lived and survived thousands of years contained carbohydrates (40 %), fats (21 %), and proteins (39 %) (Cordain et al. 2005; Simopoulos 2009, 2013). Changes in the composition of macronutrients and emergence of fast food industry have been largely driven by technological advances in food production and processing to provide high calorie and taste appealing (high levels of simple sugars, refined grains and high intake of saturated and n-6 fatty acids) foods for large urban populations (Cordain et al. 2005; Simopoulos 2009, 2013; Rook 2010). The consumption of high-calorie diets produces serious metabolic imbalances in signaling processes, resulting in obesity. The molecular mechanism involved in high calorie induced obesity remains elusive. However, recent studies have indicated that in mice deficient in Sirtuin1 (SIRT1 Knock-out), a NAD^+ dependent metabolic-sensor deacetylase in pro-opiomelanocortin (POMC) neurons cause hypersensitivity to diet inducing reduction in energy expenditure resulting in the obesity (Ramadori et al. 2010). When placed on high calorie diet, these mice become more prone to develop hepatic steatosis and metabolic damage (Ramadori et al. 2010; Xu et al. 2010; Purushotham et al. 2012). These observations suggest that deficiency of SIRT1 increases the risk of fatty liver in response to dietary fat. The liver steatosis is probably caused by increase in lipogenesis and reduction in liver fat export. The inflammation may contribute to the pathogenesis of hepatic steatosis as well. A reduction in lipid mobilization may be associated with hepatic steatosis and low energy expenditure (Ramadori et al. 2010; Xu et al. 2010). The stimulation of SIRT1 results in increased

© Springer International Publishing Switzerland 2015
A.A. Farooqui, *High Calorie Diet and the Human Brain*,
DOI 10.1007/978-3-319-15254-7_8

glucose-dependent insulin secretion from pancreatic beta cells, and directly stimulates the insulin signaling pathways in insulin-sensitive organs. Furthermore, SIRT1 regulates adiponectin secretion, inflammatory responses, gluconeogenesis, and levels of reactive oxygen species, which together contribute to the development of insulin resistance. Moreover, overexpression of SIRT1 and several SIRT1 activators has beneficial effects on glucose homeostasis and insulin sensitivity in obese mice models (Liang et al. 2009). Collectively, these studies suggest that high calorie diet, if not consumed in moderation, can not only damage the heart, kidneys, waistlines, and immune system, but also leads to increase in obesity, metabolic syndrome, cardiovascular disease, asthma, allergies, chronic joint disease, skin and digestive disorders, and cancer. The consumption of high calorie diet also reduces exposure to microorganisms, increases exposure to pollutions, increases levels of stress and contributes to immune dysfunction (Rook 2010; Cordain et al. 2005; Simopoulos 2009, 2013). In contrast, diet consisting of low caloric intake not only alleviate symptoms of the metabolic syndrome in humans and mice, but also slows primary aging resulting in an increase in maximal longevity (Holloszy and Fontana 2007; Schenk et al. 2011). In addition, low calorie diet also has a protective effect against incidence of malignancies. The molecular mechanism associated with above processes is not fully understood. However, low calorie diet not only increases SIRT1 deacetylase activity in skeletal muscle in mice, but also enhances insulin-stimulated PtdIns 3K) signaling and glucose uptake. These adaptations in skeletal muscle insulin action can be completely abrogated in mice lacking SIRT1 deacetylase activity (Schenk et al. 2011). At the molecular level, SIRT1 is required for the deacetylation and inactivation of the transcription factor STAT3 during low energy diet consumption leading to decrease in gene and protein expression of the p55α/p50α subunits of PtdIns 3K, thereby promoting more efficient PtdIns 3K signaling during insulin stimulation. These data indicate that SIRT1 is an integral signaling node in skeletal muscle linking low calorie diet with improved insulin action, primarily via modulation of PtdIns 3K signaling (Schenk et al. 2011). Alternatively, SIRT1 has also been shown to be a repressor for the protein tyrosine phosphatase 1B (PTP1b), a major tyrosine phosphatase for the insulin receptor and the insulin receptor substrate proteins, IRS1 and IRS2 (Sun et al. 2007). Therefore, it is likely that SIRT1 deficient muscles also display higher PTP1b activity, which may also antagonize the augmentation of insulin signaling in response to low calorie diet. These results support the view that involvement of SIRT1 may contribute to metabolic adaptations triggered by nutrient deprivation in skeletal muscle (Boutant and Cantó 2013).

8.2 Molecular Mechanisms Associated with Harmful Effects of High Calorie Diet in Visceral Tissues

Long term consumption of high calorie diet comprising excess energy from fat and refined carbohydrates along with low fiber, Ca^{2+}, and Mg^{2+} increases glucose levels in blood and visceral tissues. Glucose is transformed into fructose through polyol pathway. High levels of fructose produce increase in the synthesis of ceramide, di- and triacylglycerols (DAG and TAG), and uric acid along with increase in the expression of transcription factor SREBP-1c, the principal inducer of hepatic lipogenesis inducing insulin resistance (Matsuzaka et al. 2004; Vila et al. 2008; Seneff et al. 2011). This effect of fructose on SREBP-1c requires peroxisome proliferator-activated receptor γcoactivator1β (PCG-1β). Fructose also activates the hepatic transcription factors carbohydrate-responsive element binding protein (ChREBP), which upregulates the expression of hepatic fatty acid synthase and acetyl-CoA carboxylase (Koo et al. 2008). At high concentrations, fructose is converted into AGEs leading to the glycation of various proteins. The glycation of proteins not only impairs their function, but also makes proteins highly susceptible to oxidative damage. Glycated proteins are also resistant to the degradation by lysosomal enzymes. For example, glycated haemoglobin (haemoglobin damaged by glucose exposure) is used as marker for type II diabetes and glycated LDL are poorly recognized by lipoprotein receptors and scavenger receptors (Zimmermann et al. 2001). Over time, these proteins and their debris accumulate in the blood serum and along arterial walls. These damaged proteins are collectively known as AGEs (Brownlee 2005; Giacco and Brownlee 2010).

The fructose-mediated generation of AGEs also increases the production of ROS, which activate redox-sensitive transcription factor (NF-κB). This transcription factor migrates to the nucleus and induces transcription of genes associated with the production of inflammatory cytokines, chemokines, and adhesion molecules (TNF-α, IL-1β, IL-6, IL-8, MCP-1, ICAM-1, and VCAM-1) (Fig. 8.1). The production of cytokines, chemokines, and adhesion molecules along with the generation prostaglandins, leukotrienes, and thromboxanes may contribute to inflammation. Cytokines and chemokines also stimulate activities of phospholipases A_2 and cyclooxygenase-2. These enzymes hydrolyze arachidonic acid (ARA) from membrane phospholipid and oxidize it to produce more eicosanoids and ROS. Furthermore, nonenzymic oxidation of ARA also produces 4-hydroxynonenal (4-HNE), isoprotanes, isofurans, and isoketals. These metabolites contribute to oxidative stress and insulin resistance by impairing the ability of muscle and liver cells to respond to insulin (Mattson 2009; Pillon et al. 2011; Farooqui 2013). 4-HNE also promotes atherosclerosis by modifying lipoproteins and inducing cardiac cell damage by impairing metabolic enzymes.

Proteins are the major source of dietary nutrients. When proteins are digested, amino acids are released to the body for biosynthetic purposes or for generating cellular energy. High calorie diet contains high levels of proteins derived from red meat and milk. These proteins are enriched in leucine. Studies on the effect of leucine, a component of protein have been controversial. However, some studies indicate that high levels of leucine contribute to insulin resistance by over-stimulating mammalian target of rapamycin complex 1 (mTORC1). mTORC1, a pivotal nutrient-sensitive kinase, promotes growth and cell proliferation in response to glucose, energy, growth factors and amino acids (Melnik 2012). The downstream target of mTORC1, the kinase S6K1, induces insulin resistance by phosphorylation of insulin receptor substrate-1, thereby increasing the metabolic burden of β-cells. Moreover, leucine-mediated mTORC1-S6K1-signaling plays an important role in

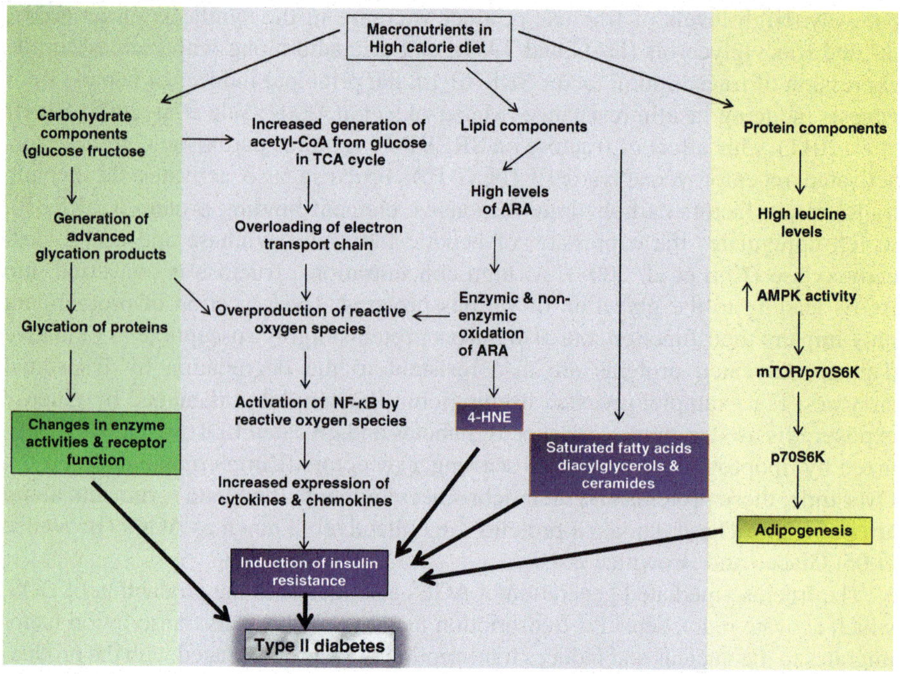

Fig. 8.1 Metabolic consequences of long term consumption of high calorie diet. *TCA* Tricarboxylic acid cycle; *ARA* arachidonic acid; *NF-κB* nuclear factor kappaB; *4-HNE* 4-hydroxynonenal; *AMPK* 5' adenosine monophosphate-activated protein kinase; and *PtdIns 3K* phosphatidylinositol 3-kinase/Akt; *mTOR* mammalian target of rapamycin/p70S6K) pathways

adipogenesis, thus increasing the risk of obesity-mediated insulin resistance (Fig. 8.1) (Melnik 2012). Based on the above information, it is suggested that high calorie diet produces insulin resistance, chronic oxidative stress and inflammation in visceral tissues (Farooqui 2013). In contrast, other investigators have shown that leucine supplementation retards high-fat diet-mediated obesity, hyperglycemia, and dyslipidemia in animal models. The molecular mechanisms associated with this process are not fully understood. However, it is reported that leucine-mediated activation of SIRT1 may contribute to the maintenance of energy, metabolic homeostasis by preventing insulin resistance and obesity (Li et al. 2012). Thus, in male C57BL/6 J mice, supplementation of leucine with high fat diet increases the expression of SIRT1 leading to reduced acetylation of PGC1α and FOXO1. The factors are associated with attenuation of high fat-induced mitochondrial dysfunction, insulin resistance, and obesity (Li et al. 2012). Collective evidence suggests that long term consumption of high calorie diet not only produces insulin resistance by significantly reducing the protein expression of insulin receptor, insulin receptor substrate-1, and dysregulated secretion of adipokines/cytokines, but also initiating and maintaining chronic oxidative stress and inflammation due to the increased generation of ARA-derived lipid mediators and eicosanoids along with increased expression of proinflammatory cytokines, respectively (Farooqui 2013). It is also reported that the consumption of red and processed meats may increase risk of type II diabetes by mechanisms that increase circulating proinflammatory markers. Positive associations have been observed between red meat or processed meat and the proinflammatory blood marker C-reactive protein (CRP), which in turn has been associated with higher risk of type II diabetes (Lee et al. 2009; Azadbakht and Esmaillzadeh 2009). The positive association between intake of meat and CRP can be explained by the presence of high levels of free heme iron (hemoglobin and myoglobin) originating from red meat as well as by the presence of AGEs which occur naturally in meat and are formed through heat processing (Fernández-Real et al. 2002; Uribarri et al. 2010). Based on these studies, it is proposed that high levels of CRP are considered as a risk of type II diabetes. In addition, long term consumption of high calorie diet also influences the gut microbiome, metabolic responses, and immune function - all of which may contribute to the rising propensity for chronic low-grade inflammation and altering homeostatic mechanisms which are common risk factors for virtually all non-communicable diseases (NCDs) (Tulic et al. 2011).

8.2.1   Effects of High Calorie Diet on Immune Function

Obesity is associated with immune dysfunction (Smith et al. 2007). In high calorie diet consuming humans, high levels of circulating proinflammatory cytokines have been detected, suggesting a state of dysregulated inflammatory response. In both genetic and diet-induced animal models, obesity has been reported to be associated with immune dysfunction (Lamas et al. 2004). The mechanistic link among high calorie diet, obesity, and low grade inflammation involves the endoplasmic reticulum stress, activation of the transcription factor NF-κB, and expression of adipokines (adiponectin and resistin), proinflammatory cytokines (TNF-α, IL-1β, and IL-6) from adipose tissue leading to the development of insulin resistance and type II diabetes (Solinas and Karin 2010). Recent studies have indicated that gut bacteria can initiate the inflammatory state of obesity and insulin resistance through lipopolysaccharide (LPS), a component of the gram-negative bacterial cell walls, which can trigger the inflammatory process by binding to the CD14 toll-like receptor-4 (TLR-4) complex at the surface of innate immune cells. The involvement of the TLR-4 signaling pathways for metabolic diseases is supported by the finding that the deletion of TLR-4 prevents the high-fat diet–induced insulin resistance (Shi et al. 2006).

8.3   Effects of Long Term Consumption of High Calorie Diet on Chronic Visceral Diseases

Long term consumption of high calorie diet results in obesity, a chronic and multifactorial pathologic condition caused by an imbalance in energy intake and energy expenditure leading to the pathological growth of adipocytes and deposition of fat (Farooqui 2013). Other factors that initiate, support, and maintain obesity-mediated visceral diseases are lack of exercise, accessibility to energy dense foods, consumption of refined carbohydrates, and trans fats (Farooqui 2014). Chronic inflammation in obesity is accompanied by increased plasma levels of C-reactive protein, tumor necrosis factor (TNF)-α, interleukin (IL)-6, monocyte chemotactic protein-1, IL-8, leptin and osteopontin (Zeyda and Stulnig 2009). Furthermore, positive correlation between body mass index Z-score and C-reactive protein has been reported to be associated with obesity-related complications (Spagnuolo et al. 2010), such as increase in inflammatory, oxidative/nitric oxide and endoplasmic reticulum stress gene expression, which may lead to inflammatory reaction in the gut (Duparc et al. 2011). Long term existence of obesity in the body is linked to a number of chronic pathologies, including hypertension, poor glycemic control, dyslipidemia, and the development of type II diabetes and metabolic syndrome, heart disease, nonalcoholic fatty liver disease (NAFLD), nonalcoholic steatohepatitis (NASH), COPD, arthritis, osteoporosis, autoimmune diseases, and cancer (Bayol et al. 2010; Elahi et al. 2009; Chechi et al. 2009).

8.3.1 Effects of High Calorie Diet on Type II Diabetes and Metabolic Syndrome

Type II diabetes is a complex endocrine and metabolic disorder characterized by impaired insulin secretion, insulin resistance, and β-cell dysfunction (Stumvoll et al. 2005). It is the leading cause of premature deaths. Improperly managed, type II diabetes can lead to a number of health issues, including heart diseases, stroke, kidney disease, blindness, nerve damage, leg and foot amputations, and death. There are two types of diabetes, namely type 1 and type II diabetes. The molecular mechanisms associated with insulin resistance in type 1 and type II diabetes are not fully understood. However, as stated earlier, long term over consumption of high calorie diet and accumulation of lipids in the liver is considered to be one of the primary mechanisms involved in insulin resistance and type II diabetes (Fig. 8.2). Elevations in saturated fatty acids (palmitic and stearic acids), triacylglycerol, diacylglycerol, acylcarnitines, and ceramide are closely associated with the molecular mechanism of insulin resistance and type II diabetes (Itani et al. 2002; Adams et al. 2004, 2009; Holland et al. 2007). In type 1 diabetes insulin resistance is caused by the inability of insulin-secreting pancreatic β cells to produce insulin (Stumvoll et al. 2005). As stated in Chaps. 2 and 3, insulin resistance is broadly defined as the reduction in insulin ability to stimulate glucose uptake from body peripheral tissues. At the molecular level, insulin activates glucose uptake by stimulating the canonical IRS-PtdIns 3K-Akt pathway and by phosphorylating and inactivating Akt substrate 160 (AS160) for a protein which activates and retards, retards the translocation of glucose transporter (GLUT) 4 to the membrane. Thus, by inhibiting AS160, insulin facilitates the GLUT4 translocation from inner vesicules, promoting fusion to the plasma membrane and consequently glucose uptake (Sakamoto and Holman 2008). Several mechanisms have been proposed to explain the metabolic consequences of

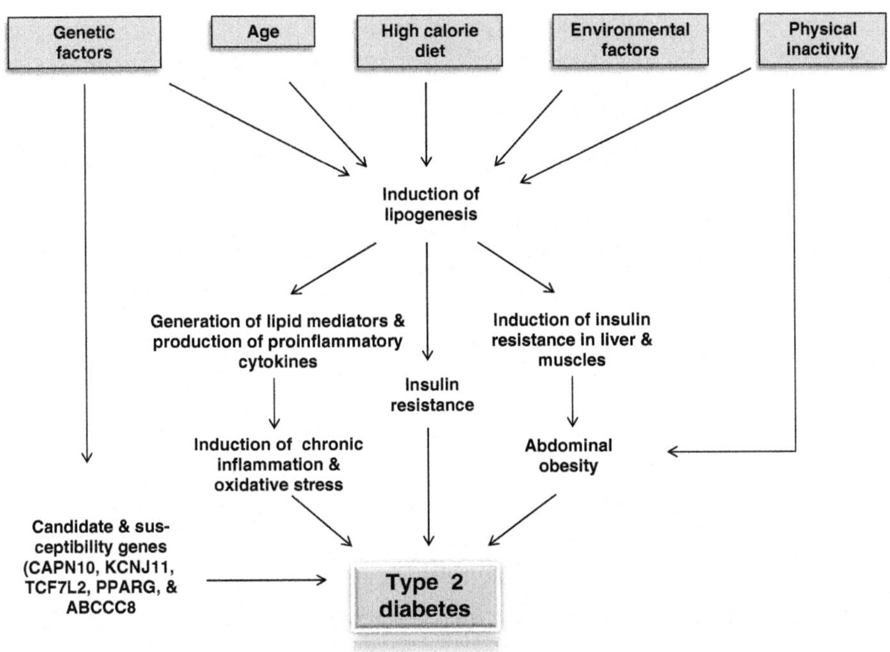

Fig. 8.2 Factors modulating the induction of type II diabetes

insulin resistance. These mechanisms include increase in high levels of saturated non-esterified fatty acids, induction of proinflammatory cytokines, adipokines, mitochondrial dysfunction, and deficiency of magnesium. In addition, the development of insulin resistance in type II diabetes correlates with immune cell infiltration and inflammation in white adipose tissue. Collectively, these studies indicate that in insulin resistance, glucotoxicity, lipotoxicity, and amyloid formation cause beta-cell dysfunction. These factors along with interactions between multiple genetic and environmental factors contribute to pathogenesis of type 2 diabetes (Stumvoll et al. 2005; Dedoussis et al. 2007).

Type II diabetes is also accompanied by endothelial dysfunction, which is linked with increased circulating levels of endothelium-derived adhesion molecules and plasminogen activator inhibitor-1 (Meigs et al. 2004), reflecting a pro-inflammatory and pro-thrombotic endothelial activity (Fig. 8.3). Recent studies have indicated abnormalities of mitochondrial function as a proximate mechanism for increased oxidative stress and PKC activation in the diabetic vasculature. In addition to the production of ATP, mitochondria also generate 1–2 % ROS under physiological conditions, which not only modulate cell growth, differentiation, and apoptosis (Duchen 2004; Tang et al. 2014) but also play a role in the insulin resistance and activation of AMP kinase, a central regulator of cellular energy status (Fig. 8.4) (Quintero et al. 2006). Type II diabetes also has a strong genetic component, but only a handful of genes have been identified. These genes include genes for calpain 10,

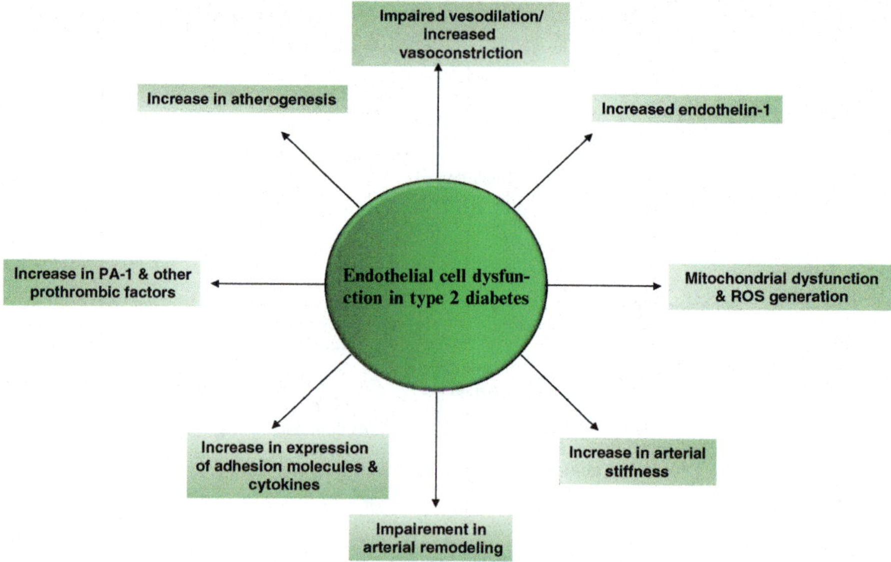

Fig. 8.3 Consequences of endothelial cell dysfunction in type II diabetes

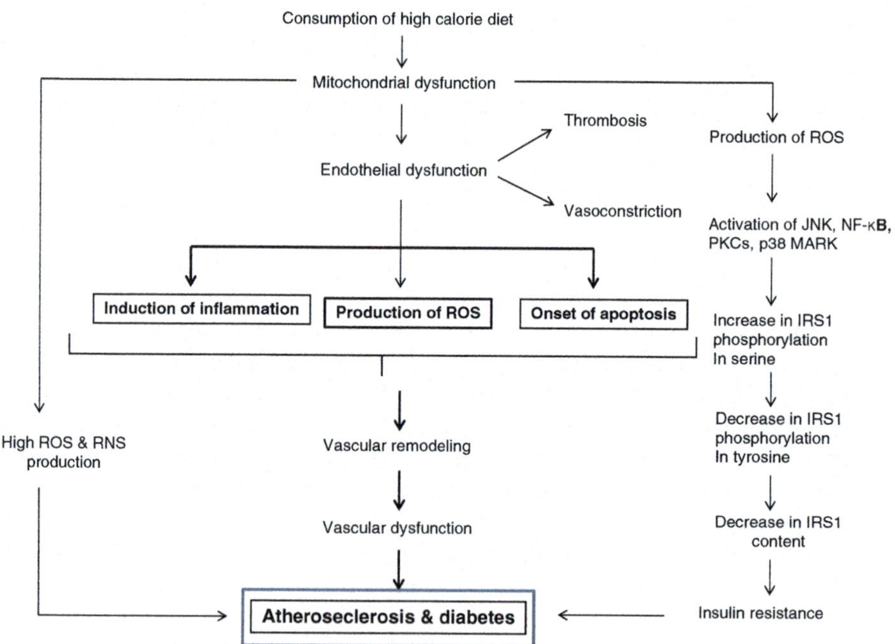

Fig. 8.4 Processes associated with the pathogenesis of atherosclerosis and type II diabetes

potassium inward-rectifier 6.2, peroxisome proliferator-activated receptor gamma, insulin receptor substrate-1, KCNJ11 gene, TCF7L2 gene, ABCCC8 gene and gene for adiponectin (Fig. 8.2) (Stumvoll et al. 2005; Dedoussis et al. 2007). Management includes not only diet and exercise, but also combinations of anti-hyperglycemic drug treatment with lipid-lowering, antihypertensive, and antiplatelet therapy for type II diabetes.

The MetS is a complex pathological condition, which involves abdominal obesity, dyslipidemia, hypertension, and insulin resistance. Biochemically, MetS is accompanied by hyperglycemia, elevation in triglycerides, low high-density lipo-protein cholesterol (HDL-C), impaired mitochondrial oxidative phosphorylation and mitochondrial biogenesis, dampened insulin metabolic signaling, and endothe-lial dysfunction, These factors predispose the individual to increased risk of devel-oping type II diabetes mellitus, cardiovascular disease (CVD), fatty liver disease, and some forms of cancer (Lakka and Laaksonen 2007; Ren et al. 2010; Farooqui et al. 2012; Song et al. 2004; Villegas et al. 2006; Muoio and Newgard 2008).

Collective evidence suggests that the consumption of high calorie diet and lack of exercise are the main drivers of the epidemic rates of obesity, type II diabetes, and MetS, which are characterized by chronic low grade inflammation and oxida-tive stress that affect insulin activity in metabolically sensitive tissues, notably the liver, muscle, and adipose tissue, and drives a metabolic disorder that culminates in the deregulation of glucose homeostasis (Hotamisligil 2006; Farooqui 2013).

8.3.2   High Calorie Diet and Heart Disease

The pathophysiology of heart disease is very complex due to the involvement of multiple factors, such as activation of endothelial activation, development of ath-erosclerosis, and high blood pressure (Peluso et al. 2012; Touyz 2004; Rodrigo et al. 2011). In the pathogenesis of heart disease related to ischemia/reperfusion injury, redox imbalance triggers the activity of a number of signaling pathways mediated by reactive oxygen species (ROS) and reactive nitrogen species (RNSs) (Elahi et al. 2009). Consequently, in cardiac surgery with extracorporeal circulation, electrical and structural myocardial remodeling due to the excessive production of these ROS may lead to the development of arrhythmias such as atrial fibrillation (Rodrigo et al. 2008). Furthermore, percutaneous transluminal coronary angioplasty following acute myocardial infarction results in heart reperfusion damage, thus enhancing the infarct size (Rodrigo et al. 2013).

The consumption of high calorie diet induces coronary atherosclerosis, a condi-tion, which is responsible for the vast majority of the cardiovascular events associ-ated with heart disease. High calorie-mediated inflammation and oxidative stress have been reported to impair beta-cell function and exacerbate insulin resistance in type II diabetes (Wellen and Hotamisligil 2005), which often coexists in patients with atherosclerosis and cardiovascular disease (Stocker and Keaney 2004). In addition, endothelial dysfunction plays a central role in the pathogenesis of heart disease.

As stated above, high-calorie dietary intake is a major risk factor for the development of obesity. Genetic factors and family history are also important risk factors for the cardiovascular disease. Among modifiable risk factors, obesity is accompanied by type II diabetes mellitus, insulin resistance, dyslipidemia, hypercholesterolemia, and hypertension, which cumulatively damage endothelial function (Meyers and Gokce 2007; Poirier et al. 2006). The structural and functional integrity of the vascular endothelium has been reported to play a crucial role in cardiovascular homeostasis. Under normal conditions, the vascular endothelium releases various vasodilator and vasoconstrictor substances (prostacyclin, platelet activating factor, endothelium-derived hyperpolarizing factor, nitric oxide, endothelin-1, and angiotensin II) that regulate local vascular tone to ensure adequate blood flow as well as regulate platelet aggregation and leukocyte adhesion to the endothelium (Widlansky et al. 2003). Furthermore, under physiological conditions, NO prevents leukocyte adhesion and maintains the endothelium in a quiescent, anti-inflammatory state (Widlansky et al. 2003). These factors effectively retard or delay the onset of atherosclerosis (Davignon and Ganz 2004). However, long term occurrence of high calorie-mediated insulin resistance results in endothelial and mitochondrial dysfunctions (Stepp 2006; Tabit et al. 2010; Bakker et al. 2009; Kizhakekuttu et al. 2012). During high calorie diet-mediated insulin resistance, high levels of ROS and RNS species, which are produced by cellular disturbances in glucose and lipid metabolism decrease levels of NO contributing to vasoconstriction. Furthermore, in the presence of ROS, endothelium activates NF-κB, which migrates to the nucleus where it upregulates the expression of proinflammatory cytokines and chemokines, such as tumor necrosis factor-beta (TNF-β), IL-1β, IL-6, and MCP-1 and adhesion molecules, such as vascular cell adhesion molecule-1 (VCAM-1) and intercellular adhesion molecule-1 (ICAM-1), which are required for the adhesion of leukocytes to the endothelial surface (Fig. 8.5) (Libby et al. 2002; Dessì et al. 2013). The endothelial expression of these factors contributes to the development of inflammation within the arterial wall and promotes the formation of atherosclerotic plaques, which are composed of vascular smooth muscle cells (VSMCs), monocyte/macrophages, T lymphocytes, and other inflammatory cells, in addition to intra- and extracellular lipid and cellular debris (Li et al. 1993). Furthermore, endothelial dysfunction contributes to impairment in insulin action by altering the transcapillary passage of insulin to target tissues (Cersosimo and DeFronzo 2006). Reduced expansion of the capillary network, with attenuation of microcirculatory blood flow to metabolically active tissues, contributes to the impairment of insulin-stimulated glucose and lipid metabolism. Vascular injury caused by increased levels of proinflammatory eicosanoids, and oxidative stress to the vessel wall triggers inflammatory reactions and responses by releasing more chemoattractants, cytokines, and chemokines, which worsen the insulin resistance and endothelial dysfunction (Cersosimo and DeFronzo 2006). Elevated levels of circulating lipids and lipid mediators in low-density lipoproteins (LDL) are a significant risk factor for the development of plaques (Libby et al. 2011). The migration of LDL into the vessel wall with subsequent oxidation and subsequent endothelial dysfunction are key processes initiating atherogenesis. In patients with type II diabetes, mitochondrial

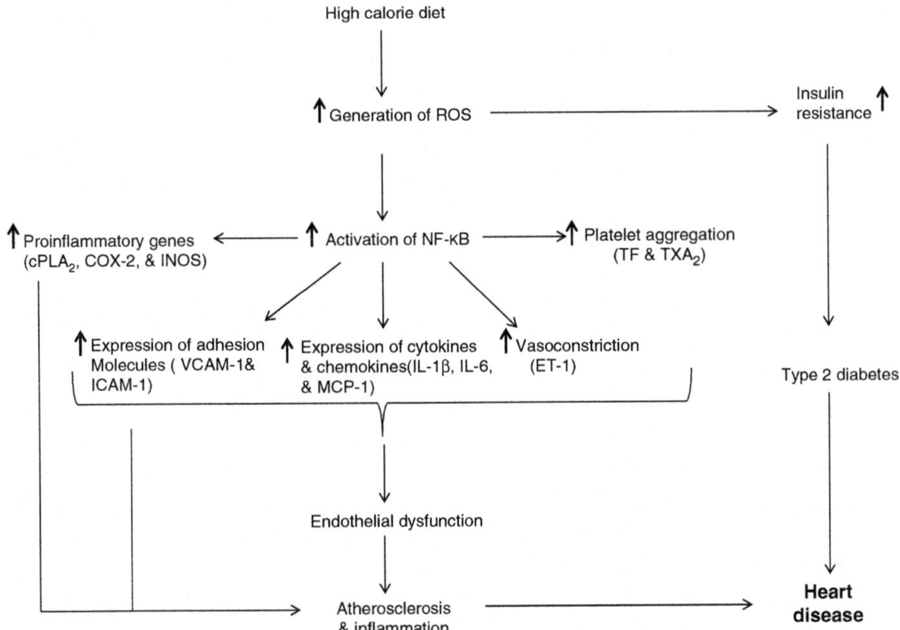

Fig. 8.5 Contribution of high calorie diet-mediated endothelial dysfunction and insulin resistance in the pathogenesis of heart disease. *NF-κB* Nuclear factor kappa light-chain enhancer of activated B cells; *ROS* reactive oxygen species; *cPLA2* cytosolic phospholipase A₂; *COX-2* cyclooxygenase-2; *i-NOS* inducible nitric oxide synthase; *VCAM-1* vascular cell adhesion protein 1; *ICAM-1* intercellular adhesion molecule-1; *IL-8* interlukin-8; *MCP-1* monocyte chemoattractant protein-1; *TF* tissue factor; *ET-1* endothelin-1; and *TX A₂* thromboxane A. *Dark upward arrows* indicate increase and *dark downward arrow* indicate decrease

dysfunction is accompanied by lower mitochondrial O_2 consumption, $\Delta\psi_m$, polymorphonuclear cell rolling velocity, and GSH/GSSG ratio, and higher mROS production and rolling flux (Hernandez-Mijares et al. 2013). These factors are closely associated with the pathogenesis of heart disease.

High calorie diet is deficient in magnesium. Hypomagnesemia contributes to the development of hypertension and heart disease. In experimental studies, magnesium has been shown to participate in the regulation of vascular tone (Altura et al. 1984; Bernardini et al. 2005), endothelial function (Bernardini et al. 2005), vascular inflammation (Blache et al. 2006), and alterations in glucose and lipid metabolism (Barbagallo et al. 2003). Supplementation with magnesium exerts beneficial effects on the cardiovascular system by enhancing endothelium-dependent vasodilation, improving lipid metabolism, reducing inflammation, and inhibiting platelet function (Shechter 2010). Magnesium is involved in regulation of cation flux across cardiomyocytes through direct binding and allosteric effect on potassium and calcium channels. Magnesium is also required for normal cardiac electrophysiology (Mubagwa et al. 2007). Abnormally low circulating magnesium (hypomagnesemia,

<0.65 mmol/L) is a known risk factor for cardiac arrest (AHA-ECC guidelines, 2000). Severe dietary magnesium restriction also adversely affects oxidative metabolism, glucose homeostasis, and retention and excretion of other electrolytes (Nielsen et al. 2007a, b).

8.3.3   High Calorie Diet and Osteoporosis

Osteoporosis is a multifactorial progressive degenerative skeletal disorder characterized by compromised bone strength predisposing a person to an increased risk of fractures. Bones are composed of 30 % protein and because the amino acids in bone collagen that are necessary for bone metabolism cannot be reused, dietary protein intake is essential for bone metabolism (Cao and Nielsen 2010). Bone strength involves the integration of two main features: (a) bone mineral density (BMD) that accounts for almost 70 % of bone strength, and (b) bone quality that accounts for the remaining 30 % of bone strength. Risk factors for bone fractures vary among individuals, and include presence or absence of fragility fractures, family history, lifestyle factors, BMD, postural instability and/or quadriceps weakness, a history of falls, and prior fracture (Orimo et al. 2012). Osteoporosis also has a heritable component, but this is reduced with age as environmental factors such as risk of falling come into play. Susceptibility to osteoporosis is not only governed by many different genetic variants, but also by nutritional factors, including calcium, vitamin D, magnesium, and protein intake, which play an important role in both the acquisition of peak bone mass and its maintenance in later life.

It is known that about 60 % of total magnesium is stored in the bone. One third of skeletal magnesium resides on cortical bone either on the surface of hydroxyapatite or in the hydration shell around the crystal (Alfrey and Miller 1973; Castiglioni et al. 2013). Magnesium serves as a reservoir of exchangeable magnesium useful to maintain physiological extracellular concentrations of the cations (Jahnen-Dechent and Ketteler 2012). Deficiency of magnesium has also been reported to promote osteoporosis (Rude et al. 2009; Castiglioni et al. 2013). Thus, bones of magnesium deficient animals are brittle and fragile (Boskey et al. 1992). Several mechanisms have been proposed to explain the role of magnesium in increasing the bone density. Magnesium deficiency rapidly leads to hypomagnesemia, which is in part buffered through the mobilization of surface magnesium from the bone. In addition, the newly formed crystals of apatite are larger and better structured in magnesium deficient animals than controls, and this affects bone stiffness (Creedon et al. 1999). In addition, magnesium deficiency not only retards cartilage and bone differentiation as well as matrix calcification (Schwartz and Reddi 1979), but also inhibits osteoblast growth by increasing the release of nitric oxide through the upregulation of inducible nitric oxide synthase (Leidi et al. 2012). Other consequences of magnesium deficiency include decrease in secretion of parathyroid hormone (PTH), $1,25(OH)_2$-vitamin D levels, and calcium absorption. Reduction in secretion of PTH results not only in impaired peripheral response to this hormone, but also reduces

25-hydroxycholecalciferol-1-hydroxylase activity. This enzyme is involved in the synthesis of $1,25(OH)_2$-vitamin D and requires magnesium for its activity (Gray et al. 1972; Rude et al. 2003, 2009). Other important risk factors for osteoporosis and reduced BMD include physical inactivity (which may increase fracture risk in part by reducing muscle mass), cigarette smoking, alcohol use, low sun exposure (for production of vitamin D), and use of some medications including glucocorticoids and anticonvulsants (Nordin 2010). Inadequately supply of vitamin D and calcium is known to decrease in the intestinal calcium and phosphate absorption (Rizzoli and Bonjour 2006). Insufficient vitamin D supply resulting in inadequately low intestinal calcium absorption leads to the overproduction of PTH, which, in turn, increases bone resorption (Lips 2001). Osteoclastogenesis plays a pivotal role in bone homeostasis; bone is maintained by active remodeling through the balance between bone resorption by osteoclasts and bone synthesis by osteoblasts. Osteoclasts are bone resorbing cells and are essential for bone remodeling. Osteoclasts are formed from hematopoietic progenitors in the monocyte/macrophage lineage (Courtial et al. 2012). Thus, bone homeostasis depends on the balance between osteoblastic bone formation and osteoclastic bone resorption, but an imbalance caused by an increased number of osteoclasts or overactivation can lead to impaired bone structure and low bone mass, which are common characteristics in patients with bone disorders (Boyle et al. 2003). Osteoclasts originate from haematopoietic stem cells and are closely related to monocytes and macrophages. On the other hand, osteoblasts represent a unique bone-forming cell derived from mesenchymal stem cells. At the cellular and molecular level, osteoclast-mediated bone resorption commences by osteoblasts initiating proliferation of osteoclast precursors and their differentiation into mature osteoclasts by secreting a cytokine called macrophage colony stimulating factor (MCSF) (Teitelbaum 2000; Matsuo and Irie 2008). Osteoblasts also release the key mediator for osteoclastogenesis, receptor activator of nuclear factor-κB (RANK) ligand (RANKL), which interacts with its receptor on the plasma membrane of osteoclast precursors and stimulates pre-osteoclasts differentiation into mature osteoclasts (Fig. 8.6) (Khosla 2001). Synthesis and expression of RANKL and MCSF are stimulated by various osteoclastogenic factors, such as parathyroid hormone (PTH) and PTH-related peptide and prolactin (Seriwatanachai et al. 2008; Zaidi 2007). In addition to counterbalance RANKL action, osteoblasts synthesize and secrete osteoprotegerin (OPG), a soluble decoy receptor capable of inhibiting RANK-RANKL interaction and osteoclastogenesis (Simonet et al. 1997; Matsuo and Irie 2008; Asagiri and Takayanagi 2007). In the presence of activated osteoclasts, the bone resorption begins with dissolution of inorganic and organic components by hydrochloric acid, cathepsin K and lysosomal protease from osteoclasts (Matsuo and Irie 2008; Supanchart and Kornak 2008; Sobacchi et al. 2007). OPG retards RANKL and RANK receptor interactions. The rate of bone formation depends on the presence of mature osteoblasts, calcium, phosphate, and other minerals. This process is enhanced by vitamin D as well as by intermittent pulses of parathyroid hormone. Collective evidence suggests that deficiencies of vitamin D, calcium, phosphate, along with changes in mediators of inflammatory immune-mediated responses

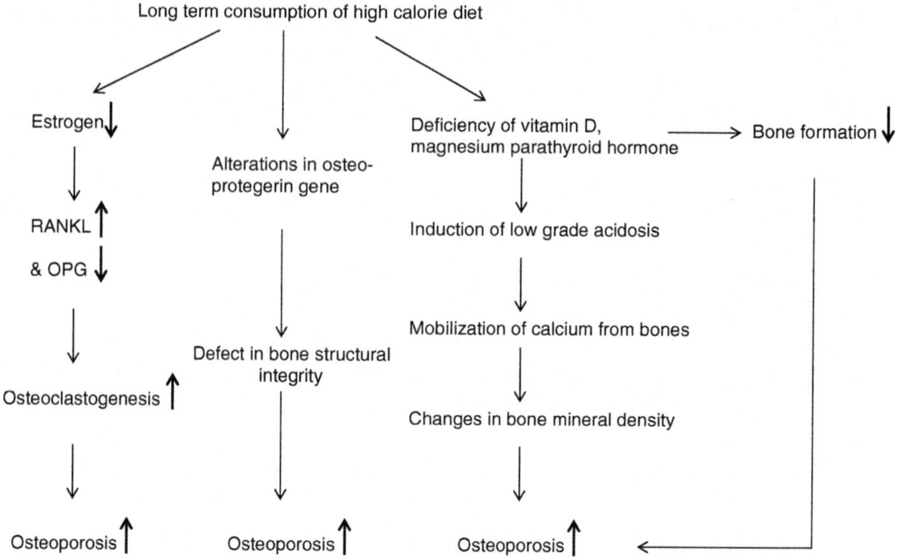

Fig. 8.6 Biochemical processes involved in the induction of osteoporosis. OPG Osteoprotegerin and *RANK* receptor activator of nuclear factor kappa B; *RANKL* ligand for receptor activator of nuclear factor kappa B. *Dark upward arrows* indicate increase and *dark downward arrow* indicate decrease

(e.g., cytokines), parathyroid hormone, estrogens, androgens, corticosteroids and alterations in osteoprotegerin gene are associated with defective bone structural integrity, as expressed by either a prominent defect in mineral deposition that characterizes the pathological condition of osteomalacia or by a loss of the entire mineralized organic matrix, an important feature of osteoporosis (Bischoff-Ferrari and Staehelin 2008).

After resorption of bone, osteoblast-induced bone production starts to fill the resorption pits with newly mineralized bone. The type I collagen fibrils, which are secreted by osteoblasts, are arranged into the organic matrix osteoid. Organic matrix osteoid is then mineralized by calcium and phosphate in the presence of alkaline phosphatase, osteocalcin and osteopontin. Eventually, hydroxide ions are gradually added and mature hydroxyapatite crystals $[Ca_{10}(PO_4)_6(OH)_2]$ are formed (Sims and Gooi 2008). Bone formation is stimulated by Insulin-like growth factor (IGF)-1, insulin, bone morphogenetic proteins, and OPG. These mediators serve as anabolic signals for bone formation (Zaidi 2007; Mohan and Baylink 2005). Among these anabolic mediators, liver-derived IGF-1 is of particular interest because IGF-I production is a major regulator of bone mass and deficiency of IGF-1 and its receptor results in osteoporosis (Zaidi 2007; Mohan and Baylink 2005).

Recent studies have indicated that osteoporosis and hip fracture are common complications observed in AD patients. The underlying mechanisms between osteoporosis are not poorly understood (Cornelius et al. 2014). However, the generation of ROS has emerged as intracellular redox signaling molecules involved in the regulation of bone metabolism, including receptor activator of nuclear factor-κB ligand-dependent osteoclast differentiation. ROS also produce cytotoxic effects such as lipid peroxidation and oxidative damage to proteins and DNA (Farooqui 2010; Cornelius et al. 2014). The production of ROS, which has been implicated in the regulation of cellular stress response mechanisms, is an integrated, highly regulated, process under control of redox sensitive genes coding for redox proteins called vitagenes. Vitagenes, encode for proteins such as heat shock proteins (Hsps) Hsp32, Hsp70, the thioredoxin, and the sirtuin protein, represent a systems controlling a complex network of intracellular signaling pathways relevant to life span and involved in the preservation of cellular homeostasis under stress conditions (Cornelius et al. 2014).

High calorie diet is rich in energy (high amounts of meat proteins and refined carbohydrates) but poor in nutrients (vitamin D and magnesium) (Drewnowski 2010). Thus, long term consumption of high calorie diet not only results in nutrients deficiency (vitamin D and magnesium) in elderly (Wang and Beydoun 2007), but produces low-grade acidosis (approximately 50–100 mEq acid/day) which is intensified by declining kidney function and aging (Fig. 8.6). Recent studies have also indicated that the consumption of high calorie diet enriched in animal origin results in acid load and net acid excretion, which may contribute to the pathophysiology of osteoporosis. According to acid–base hypothesis of osteoporosis "diets high in acid-forming nutrients promotes the release of calcium from bone resulting in bone resorption (Remer and Manz 1995). It is therefore proposed that the chronic exposure to high dietary acid load may contribute to low bone mass. Dietary acid load in human diets can be estimated by calculating the net endogenous acid production or the potential renal acid load of the diet using information on dietary intakes. Many studies indicate that metabolic acidosis leads to calcium loss from bone, inhibits osteoblast function, stimulates osteoclast activity, and impairs bone mineralization (Arnett 2008; Vormann and Remer 2008). In contrast, a neutralizing diet improves bone micro-architecture and bone mineral density (Jehle et al. 2013). There is no cure for osteoporosis at the present time, but there is evidence that some lifestyle factors, notably dietary calcium, vitamin D and exercise may reduce this age-related reduction in bone mass.

8.3.4 High Calorie Diet and Rheumatoid Arthritis

Arthritis is a group of diseases characterized by inflammation in joints. There are different forms of arthritis including osteoarthritis, rheumatoid arthritis (RA), psoriatic arthritis, and arthritis associated with autoimmune diseases. Each form of arthritis has its own biochemical mechanism and causes. The most common form of arthritis, osteoarthritis is caused by trauma to the joint, infection of the joint, or age. RA is a chronic autoimmune inflammatory disease effecting joints of hands, wrists, feet, knees, cubitus, ankles, shoulder, and is manifested by swelling and pain producing functional impairment. Psoriatic arthritis is a chronic autoimmune disease caused by malfunctioning immune system and characterized by inflammation of the skin psoriasis (a serious skin condition) and joints. Arthritis associated with autoimmune diseases is caused by the production of anti-nuclear antibodies (ANA), which instead of attacking the legitimate infectious, viral invaders, and mutated cells of the body, turns and attacks aspects of the healthy body cells and tissues itself. The immune system of patient becomes confused, poorly defending against outside invading agents, while attacking healthy tissues and organs. The consumption of high calorie diet, which is enriched in meat, dairy, fat, and refined grains and simple sugars produces low-grade systemic inflammation and oxidative tissue stress and irritation, placing the immune system in an overactive state initiating the pathogenesis of many chronic diseases including arthritis (Carrera-Bastos et al. 2012). Macronutrients components of high calorie diet have been reported to produce inflammation and oxidative stress even in healthy people (Gregersen et al. 2012).

RA is a chronic systemic autoimmune inflammatory disease of the joint characterized by inflammatory cell infiltration, synovial lining hyperplasia, and destruction of cartilage and bone. Clinically, RA manifests as joint pain, stiffness, and swelling. If left untreated, persistent synovial inflammation can progress to cartilage and bone destruction ultimately leading to major long-term disability and mortality. The molecular mechanism of RA is not yet been fully understood. However, it is becoming increasingly evident that the production of autoantibody, immune complex formation, inflammatory cell infiltration, and tumor-like proliferation of synovium are closely associated with the pathogenesis of RA (Steiner and Smolen 2002; Pap et al. 2000). Autoantibodies, such as rheumatoid factor (anti-IgG antibody) and anti–type II collagen (CII) antibody, are detected in RA patients with high probability. These autoantibodies contribute to the formation of immune complexes within the joint, producing the activation of the complement cascade and inflammatory cell infiltration into the joint. At the molecular level, RA is accompanied by increased production of cytokines, chemokines, and growth factors by infiltrated macrophages and neutrophils (Naka et al. 2002; Dayer 2003). Macrophages and neutrophils, which infiltrate joint release IL-1β. This cytokine activates synovial cells leading to the secretion of other chemokines, cytokines, and growth factors such as CCL2 (monocyte chemoattractant protein 1 [MCP-1]), CCL3 (macrophage inflammatory protein 1α), IL-6, TNF-α, interleukin-1, interleukin-17 interferon-γ, cell adhesion molecules (intercellular adhesion molecule-1, vascular cell adhesion molecule-1) and matrix metalloproteinases, and fibroblast growth factors (FGFs), which contribute to joint inflammation through increased production of proinflammatory eicosanoids (PGE$_2$ PGI$_2$, and prostacyclin) to the leading amplification of inflammation and slow destruction of joints (Koch 2005; Inada and Krane 2002; Pulichino et al. 2006). PGI receptor-deficient (IP$^{-/-}$) mice in collagen-induced arthritis (CIA model show a significant reduction in arthritic scores and reduction in IL-1β and IL-6 levels in the arthritic paws (Honda et al. 2006). Inhibition of both PGE receptors (EP2 and EP4) lowers inflammatory events and arthritis in CIA supporting the view that both PGE$_2$ and PGI$_2$ participate in the rheumatoid arthritis. Supplementation with n-3 PUFA retards the activity of inflammatory factors and delays the destruction of cartilage during arthritis (Calder 2008; Curtis et al. 2000). Moreover, decreasing the consumption of ARA to 90 mg/day results in lowering levels of inflammatory eicosanoids and improving the clinical symptoms of RA (Adam et al. 2003; Kobayashi et al. 2007). n-3 fatty acids act both directly (e.g., by replacing ARA in cellular membranes and decreasing precursors for eicosanoids and competing for cyclooxygenases and lipoxygenases) and indirectly (e.g., by downregulating the expression of inflammatory genes through effects on transcription factor activation). Moreover, n-3 fatty acids also give rise to a family of antiinflammatory mediators termed as resolvins, protectins, and maresins (Serhan et al. 2011). Thus, n-3 fatty acids are potentially potent antiinflammatory drugs that can be used for the treatment of variety of acute and chronic inflammatory diseases including rheumatoid arthritis, inflammatory bowel diseases, and asthma (Calder 2008). Collective evidence suggests that high intake of n-3 fatty acids decreases the generation of eicosanoids, inhibits the expression of proinflammatory cytokines, retards the formation of ROS, and blocks the expression of adhesion molecules (Calder 2008).

8.3.5 High Calorie Diet and Cancer

It is well known that genetic mutations (changes in gene expression), epigenetic changes (acetylation, methylation, or phosphorylation), chronic inflammation, high fat diet and sedentary lifestyle are risk factors for cancer (Vogelstein and Kinzler 2004; Ting et al. 2006; Woutersen et al. 1999). Epidemiological and animal studies have indicated that the consumption of a high fat diet also increases the risk of developing cancer, in particular colorectal, breast, pancreatic and prostate cancer (Woutersen et al. 1999). Arachidonic acid (ARA), a fatty acid belonging to n-6 fatty acid family, is not only a major component of animal fats but also a precursor for proinflammatory lipid mediators, which contribute to the chronic inflammation and cancer. The metabolism of arachidonic acid by cyclooxygenase (COX), lipoxygenase (LOX) and P450 epoxygenase (EPOX) pathways generates eicosanoids, including prostaglandins, leukotrienes, thromboxanes, hydroxyeicosatetraenoic acids (HETEs), epoxyeicosatrienoic acids (EETs) and hydroperoxyeicosatetraenoic acids (HPETEs) (Fig. 8.7). These mediators are responsible for inflammation, a process that is now recognized to be a critical component for tumor progression and one of the recently added "hallmarks of cancer" (Coussens and Werb 2002; Hanahan and Weinberg 2011; Greene et al. 2011). In several types of cancers (breast, lung, and pancreas) expression of COX-2 and 5-LOX is markedly increased (Wang and Dubois 2010). In addition, cancer patients not only have increased circulating markers of inflammation such as C-reactive protein (CRP), but they also show increased levels of proinflammatory cytokines (interleukin-6 and soluble tumor necrosis factor-α) and their receptors (Wallace 2002; McMillan et al. 2000; Barber et al. 1999). Many cancers arise from sites of infection, chronic irritation and inflammation. It is now becoming evident that the tumor microenvironment is largely supported by inflammatory cells, which are indispensable for the neoplastic process, fostering proliferation, survival and migration. In addition, tumor cells also share

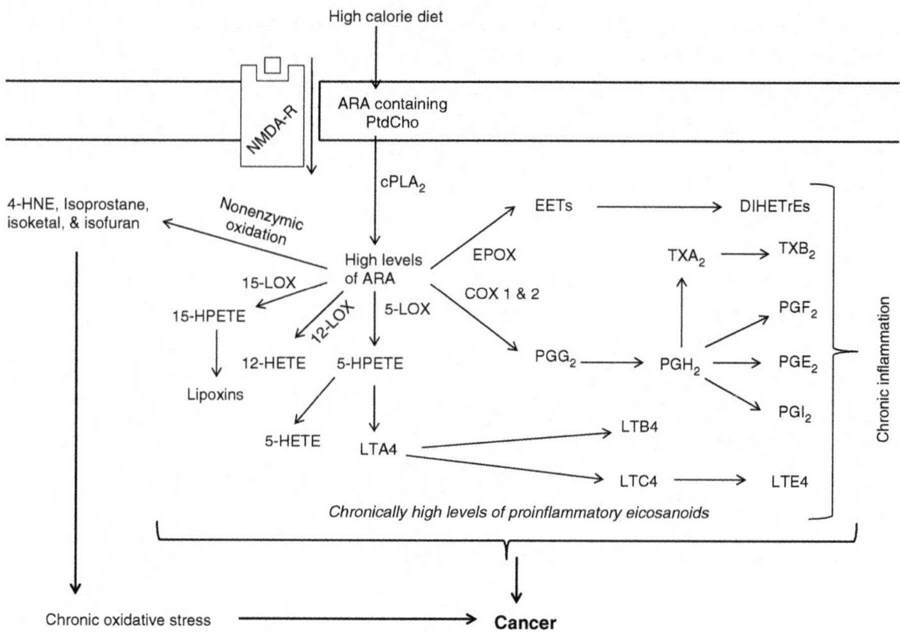

Fig. 8.7 Consumption of high calorie diet and generation and contribution of ARA-derived lipid mediators in the pathogenesis of cancer. *PtsCho* Phosphatidylcholine; cPLA$_2$ cytosolic phospholipase A$_2$; *ARA* arachidonic acid; *COX-1* and *2* cyclooxygenase-1 and 2; *5-LOX* 5-lipoxygenase; *EPOX* epoxygenase; and *4-HNE* 4-hydroxynonenal

some signaling molecules with innate immune system, such as selectins, chemokines and their receptors for invasion, migration and metastasis (Coussens and Werb 2002). Based on these findings, it is suggested that increased consumption of animal fat in high calorie diet may be one mechanism for the contribution of dietary fats to carcinogenesis, a process, which is also supported by metabolic defects involving increased production of ROS in mitochondria and alterations in redox state. These studies support the view that 30–40 % of all kinds of cancers can be delayed or prevented with a healthy diet and lifestyle. High calorie diet not only has high amounts of red or processed meat and low levels of calcium, magnesium, vitamin D and methyl-donor nutrients, but also contains an imbalance of n-3 and n-6 fatty acids ratio with low in fiber contents. High levels of n-6 fatty acids may contribute to prostaglandin-mediated inflammation, which may increase the risk of cancer (Bingham and Riboli 2004). On the other hand, the consumption of fruit, vegetables, and n-3 fatty acids and n-3 fatty acid-derived lipid mediators (resolvins, protectins, and maresins) may decrease the risk of cancer. Protective components in a cancer-preventive diet include selenium, folic acid, vitamin B12, vitamin D, chlorophyll and antioxidants such as carotenoids (alpha-carotene, beta-carotene, lycopene, lutein, cryptoxanthin). In addition, the consumption of dietary fiber in vegetable, fruits, and probiotics may also produce anticancer effects.

The relationship between the consumption of high calorie diet-mediated obesity and cancer has is becoming increasingly evident. For example, many obese patients have greater chance of developing breast cancer, colorectal, and prostate cancer (Pérez-Hernández et al. 2014). It is estimated that 15–20 % of all cancer deaths can be attributed to the obesity. As stated above, tumor growth is regulated by interactions between tumor cells and their tissue microenvironment. In this sense, obesity may lead to cancer development through dysfunctional adipose tissue and altered signaling pathways. Pathways relating to the obesity and cancer may include: (a) inflammatory changes leading to macrophage polarization and abnormal adipokine/cytokine profiles, which support low grade inflammation through the production of ARA-derived eicosanoids; and (b) insulin resistance that is caused by high levels of saturated fatty acids.

Studies on the analysis of cancer-mediated changes in the lipid profile indicate important clues linking obesity with cancer. It is reported that changes in lipid metabolism may promote cancer development due to increase in free fatty acid, which contributes to oncogenic signals through changes in lipid signaling, inflammatory responses, insulin resistance, and alterations in adipokines (Louie et al. 2013). Furthermore, marked increase in fatty acid synthase (FASN) activity has been observed in breast cancer cell lines (Hunt et al. 2007), ovarian tumors (Gansler et al. 1997), or cancer precursor lesions in different locations (colon, stomach, esophagus, and oral cavity) (Alli et al. 2005). The inhibition of the FASN activity reduces the cancer cells' proliferative capacity, suggesting that FFA act as an energy source for cancer cells. Since obesity and cancer present a high prevalence, more studies on the molecular mechanism linking obesity with cancers are urgently needed (Pérez-Hernández et al. 2014).

8.4 Conclusion

The long term consumption of high calorie diet induces the synthesis of proinflammatory lipid mediators (eicosanoids and platelet activating factor), proinflammatory cytokines (TNF-α, IL-1β, and IL-6), upregulates gp91(phox) subunit of NADPH oxidase, and downregulates superoxide dismutase (SOD) isoforms, glutathione peroxidase (GPX), and heme oxygenase-2 (HO-2) in various body tissues including brain. These biochemical changes produce oxidative stress, and low grade inflammation. These processes promote weight gain, obesity and insulin resistance leading to type II diabetes and metabolic syndrome. These pathological conditions are risk factors for cardiovascular disease, osteoporosis, arthritis, and various types of cancers.

References

Adam O, Beringer C, Kless T, Lemmen C, Adam A, Wiseman M, Adam P, Klimmek R, Forth W (2003) Anti-inflammatory effects of a low arachidonic acid diet and fish oil in patients with rheumatoid arthritis. Rheumatol Int 23:27–36

Adams JM II, Pratipanawatr T, Berria R, Wang E, DeFronzo RA, Sullards MC, Mandarino LJ (2004) Ceramide content is increased in skeletal muscle from obese insulin-resistant humans. Diabetes 53:25–31

Adams SH, Hoppel CL, Lok KH, Zhao L, Wong SW, Minkler PE, Hwang DH, Newman JW, Garvey WT (2009) Plasma acylcarnitine profiles suggest incomplete long-chain fatty acid beta-oxidation and altered tricarboxylic acid cycle activity in type 2 diabetic African-American women. J Nutr 139:1073–1081

AHA (ECC Guidelines) (2000) Guidelines for cardiopulmonary resuscitation and emergency cardiovascular care, part 8: advanced challenges in resuscitation: section 1: life-threatening electrolyte abnormalities. Circulation 102:I217–I222

Alfrey AC, Miller NL (1973) Bone magnesium pools in uremia. J Clin Invest 52:3019–3027

Alli PM, Pinn ML, Jaffee EM, McFadden JM, Kuhajda FP (2005) Fatty acid synthase inhibitors are chemopreventive for mammary cancer in neu-N transgenic mice. Oncogene 24:39–46

Altura BM, Altura BT, Gebrewold A, Ising H, Günther T (1984) Magnesium deficiency and hypertension: correlation between magnesium-deficient diets and microcirculatory changes in situ. Science 223:1315–1317

Arnett TR (2008) Extracellular pH regulates bone cell function. J Nutr 138:415S–418S

Asagiri M, Takayanagi H (2007) The molecular understanding of osteoclast differentiation. Bone 40:251–264

Azadbakht L, Esmaillzadeh A (2009) Red meat intake is associated with metabolic syndrome and plasma c-reactive protein concentrations in women. J Nutr 139:335–339

Bakker W, Eringa EC, Sipkema P, van Hinsbergh VW (2009) Endothelial dysfunction and diabetes: roles of hyperglycemia, impaired insulin signaling and obesity. Cell Tissue Res 335:165–189

Barbagallo M, Dominguez LJ, Galioto A, Ferlisi A, Cani C, Malfa L, Pineo A, Busardo' A, Paolisso G (2003) Role of magnesium in insulin action, diabetes and cardio-metabolic syndrome X. Mol Aspects Med 24:39–52

Barber MD, Fearon KC, Ross JA (1999) Relationship of serum levels of interleukin-6, soluble interleukin-6 receptor and tumour necrosis factor receptors to the acute-phase protein response in advanced pancreatic cancer. Clin Sci (Lond) 96:83–87

Bayol SA, Simbi BH, Fowkes RC, Stickland NC (2010) A maternal "Junk Food" diet in pregnancy and lactation promotes nonalcoholic fatty liver disease in rat offspring. Endocrinology 151:1451–1461

Bernardini D, Nasulewic A, Mazur A, Maier JA (2005) Magnesium and microvascular endothelial cells: a role in inflammation and angiogenesis. Front Biosci 10:1177–1182

Bingham S, Riboli E (2004) Diet and cancer–the European prospective investigation into cancer and nutrition. Nat Rev Cancer 4:206–215

Bischoff-Ferrari HA, Staehelin HB (2008) Importance of vitamin D and calcium at older age. Int J Vitam Nutr Res 78:286–292

Blache D, Devaux S, Joubert O, Loreau N, Schneider M, Durand P, Prost M, Gaume V, Adrian M, Laurant P, Berthelot A (2006) Long-term moderate magnesium-deficient diet shows relationships between blood pressure, inflammation and oxidant stress defense in aging rats. Free Radic Biol Med 41:277–284

Boskey AL, Rimnac CM, Bansal M, Federman M, Lian J, Boyan BD (1992) Effect of short-term hypomagnesemia on the chemical and mechanical properties of rat bone. J Orthop Res 10:774–783

Boutant M, Cantó C (2013) SIRT1 metabolic actions: integrating recent advances from mouse models. Mol Metab 3:5–18

Boyle WJ, Simonet WS, Lacey DL (2003) Osteoclast differentiation and activation. Nature 423:337–342

Brownlee M (2005) The pathobiology of diabetic complications: a unifying mechanism. Diabetes 54:1615–1625

Calder PC (2008) PUFA, inflammatory processes and rheumatoid arthritis. Proc Nutr Soc 67:409–418

Cao JJ, Nielsen FH (2010) Acid diet (high-meat protein) effects on calcium metabolism and bone health. Curr Opin Clin Nutr Metab Care 13:698–702

Carrera-Bastos P, Fontes-Villalba M, Keefe JHO, Lindeberg S, Cordain L (2012) The western diet and lifestyle and diseases of civilization. Res Rep Clin Cardiol 2:15–35

Castiglioni S, Cazzaniga A, Albisetti W, Maier JA (2013) Magnesium and osteoporosis: current state of knowledge and future research directions. Nutrients 5:3022–3033

Cersosimo E, DeFronzo RA (2006) Insulin resistance and endothelial dysfunction: the road map to cardiovascular diseases. Diabetes Metab Res Rev 22:423–436

Chechi K, McGuire JJ, Cheema SK (2009) Developmental programming of lipid metabolism and aortic vascular function in C57BL/6 mice: a novel study suggesting an involvement of LDL-receptor. Am J Physiol Regul Integr Comp Physiol 296:R1029–R1040

Cordain L, Eaton SB, Sebastian A, Mann N, Lindeberg S, Watkins BA, O'Keefe JH, Brand-Miller J (2005) Origins and evolution of the Western diet: health implications for the 21st century. Am J Clin Nutr 81:341–354

Cornelius C, Koverech G, Crupi R, Di Paola R, Koverech A, Lodato F, Scuto M, Salinaro AT, Cuzzocrea S, Calabrese EJ, Calabrese V (2014) Osteoporosis and Alzheimer pathology: role of cellular stress response and hormetic redox signaling in aging and bone remodeling. Front Pharmacol 5:120

Courtial N, Smink JJ, Kuvardina ON, Leutz A, Göthert JR, Lausen J (2012) Tal1 regulates osteoclast differentiation through suppression of the master regulator of cell fusion dc-stamp. FASEB J 26:523–532

Coussens LM, Werb Z (2002) Inflammation and cancer. Nature 420:860–867

Creedon A, Flynn A, Cashman K (1999) The effect of moderately and severely restricted dietary magnesium intakes on bone composition and bone metabolism in the rat. Br J Nutr 82:63–71

Curtis CL, Hughes CE, Flannery CR, Little CB, Harwood JL, Caterson B (2000) n-3 Fatty acids specifically modulate catabolic factors involved in articular cartilage degradation. J Biol Chem 275:721–724

Davignon J, Ganz P (2004) Role of endothelial dysfunction in atherosclerosis. Circulation 109(23 Suppl 1):III27–III32

Dayer JM (2003) The pivotal role of interleukin-1 in the clinical manifestations of rheumatoid arthritis. Rheumatology (Oxford) 42:3–10

Dedoussis GVZ, Kaliora AC, Panagiotakos DB (2007) Genes, diet and type 2 diabetes mellitus: a review. Rev Diabet Stud 4:13–24

Dessì M, Noce A, Bertucci P, Manca di Villahermosa S, Zenobi R, Castagnola V, Addessi E, Di Daniele N (2013) Atherosclerosis, dyslipidemia, and inflammation: the significant role of poly-unsaturated fatty acids. ISRN Inflamm 2013:191823

Drewnowski A (2010) The cost of US foods as related to their nutritive value. Am J Clin Nutr 92:1181–1188

Duchen MR (2004) Roles of mitochondria in health and disease. Diabetes 53:S96–S102

Duparc T, Naslain D, Colom A, Muccioli GG, Massaly N, Delzenne NM, Valet P, Cani PD, Knauf C (2011) Jejunum inflammation in obese and diabetic mice impairs enteric glucose detection and modifies nitric oxide release in the hypothalamus. Antioxid Redox Signal 14:415

Elahi MM, Cagampang FR, Mukhtar D, Anthony FW, Ohri SK, Hanson MA (2009) Long-term maternal high-fat feeding from weaning through pregnancy and lactation predisposes offspring to hypertension, raised plasma lipids and fatty liver in mice. Br J Nutr 102:514–519

Farooqui AA (2009) Beneficial effects of fish oil on human brain. Springer, New York

Farooqui AA (2010) Neurochemical aspects of neurotraumatic and neurodegenerative diseases. Springer, New York

Farooqui AA (2013) Metabolic syndrome: an important risk factor for stroke, Alzheimer, and depression. Spinger, New York

Farooqui AA (2014) Inflammation and oxidative stress in neurological disorders. Springer, New York

Farooqui AA, Farooqui T, Panza F, Frisardi V (2012) Metabolic syndrome as a risk factor for neurological disorders. Cell Mol Life Sci 69:741–762

Fernández-Real JM, López-Bermejo A, Ricart W (2002) Cross-talk between iron metabolism and diabetes. Diabetes 51:2348–2354

Gansler TS, Hardman W III, Hunt DA, Schaffel S, Hennigar RA (1997) Increased expression of fatty acid synthase (OA-519) in ovarian neoplasms predicts shorter survival. Hum Pathol 28:686–692

Giacco F, Brownlee M (2010) Oxidative stress and diabetic complications. Cir Res 107:1058–1070

Gray RW, Omdahl JL, Ghazarian JG, DeLuca HF (1972) 25-Hydroxycholecalciferol-1 hydroxylase. Subcellular location and properties. J Biol Chem 247:7528–7532

Greene ER, Huang S, Serhan CN, Panigrahy D (2011) Regulation of inflammation in cancer by eicosanoids. Prostaglandins Other Lipid Mediat 96:27–36

Gregersen S, Samocha-Bonet D, Heilbronn LK, Campbell LV (2012) Inflammatory and oxidative stress responses to high-carbohydrate and high-fat meals in healthy humans. J Nutr Metab 8:238056

Hanahan D, Weinberg RA (2011) Hallmarks of cancer: the next generation. Cell 144:646–674

Hernandez-Mijares A, Rocha M, Rovira-Llopis S, Bañuls C, Bellod L, de Pablo C, Alvarez A, Roldan-Torres I, Sola-Izquierdo E, Victor VM (2013) Human leukocyte/endothelial cell interactions and mitochondrial dysfunction in type 2 diabetic patients and their association with silent myocardial ischemia. Diabetes Care 36:1695–1702

Holland WL, Brozinick JT, Wang LP, Hawkins ED, Sargent KM, Liu Y, Narra K, Hoehn KL, Knotts TA, Siesky A, Nelson DH, Karathanasis SK, Fontenot GK, Birnbaum MJ, Summers SA (2007) Inhibition of ceramide synthesis ameliorates glucocorticoid-, saturated-fat- and obesity-induced insulin resistance. Cell Metab 5:167–179

Holloszy JO, Fontana L (2007) Caloric restriction in humans. Exp Gerontol 42:709–712

Honda T, Segi-Nishida E, Miyachi Y, Narumiya S (2006) Prostacyclin-IP signaling and prostaglandin E2-EP2/EP4 signaling both mediate joint inflammation in mouse collagen-induced arthritis. J Exp Med 203:325–335

Hotamisligil GS (2006) Inflammation and metabolic disorders. Nature 444:860–867

Hunt DA, Lane HM, Zygmont ME, Dervan PA, Hennigar RA (2007) MRNA stability and overexpression of fatty acid synthase in human breast cancer cell lines. Anticancer Res 27:27–34

Inada M, Krane SM (2002) Targeting rheumatoid inflammation and joint destruction in the mouse. J Clin Invest 110:611–612

Itani SI, Ruderman NB, Schmieder F, Boden G (2002) Lipid-induced insulin resistance in human muscle is associated with changes in diacylglycerol, protein kinase C, and IkappaB-alpha. Diabetes 51:2005–2011

Jahnen-Dechent W, Ketteler M (2012) Magnesium basics. Clin Kidney J 5:i3–i14

Jehle S, Hulter HN, Krapf R (2013) Effect of potassium citrate on bone density, microarchitecture, and fracture risk in healthy older adults without osteoporosis: a randomized controlled trial. J Clin Endocrinol Metab 98:207–217

Khosla S (2001) Minireview: the OPG/RANKL/RANK system. Endocrinology 142:5050–5055

Kizhakekuttu TJ, Wang J, Dharmashankar K, Ying R, Gutterman DD, Vita JA, Wildlansky ME (2012) Adverse alterations in mitochondrial function contribute to type 2 diabetes mellitus–related endothelial dysfunction in humans. Arterioscler Thromb Vasc Biol 32:2531–2539

Kobayashi T, Ito S, Kuroda T, Yamamoto K, Sugita N, Narita I, Sumida T, Fumitake Gejyo F, Yoshie H (2007) The interleukin-1 and Fcgamma receptor gene polymorphisms in Japanese patients with rheumatoid arthritis and periodontitis. J Periodontol 78:2311–2318

Koch AE (2005) Chemokines and their receptors in rheumatoid arthritis: future targets? Arthritis Rheum 52:710–721

Koo HY, Wallig MA, Chung BH, Nara TY, Cho BH, Nakamura MT (2008) Dietary fructose induces a wide range of genes with distinct shift in carbohydrate and lipid metabolism in fed and fasted rat liver. Biochim Biophys Acta 1782:341–348

Lakka TA, Laaksonen DE (2007) Physical activity in prevention and treatment of the metabolic syndrome. Appl Physiol Nutr Metab 32:76–88

Lamas O, Martinez JA, Marti A (2004) Energy restriction restores the impaired immune response in overweight (cafeteria) rats. J Nutr Biochem 15:418–425

Lee CC, Adler AI, Sandhu MS, Sharp SJ, Forouhi NG, Erqou S, Luben R, Bingham S, Khaw KT, Wareham NJ (2009) Association of C-reactive protein with type 2 diabetes: prospective analysis and meta-analysis. Diabetologia 52:1040–1047

Leidi M, Dellera F, Mariotti M, Banfi G, Crapanzano C, Albisetti W, Maier JA (2012) Nitric oxide mediates low magnesium inhibition of osteoblast-like cell proliferation. J Nutr Biochem 23:1224–1229

Li H, Cybulsky MI, Gimbrone MA, Libby P (1993) An atherogenic diet rapidly induces VCAM-1, a cytokine-regulatable mononuclear leukocyte adhesion molecule, in rabbit aortic endothelium. Arterioscler Thromb 13:197–204

Li H, Xu M, Lee J, He C, Xie Z (2012) Leucine supplementation increases SIRT1 expression and prevents mitochondrial dysfunction and metabolic disorders in high-fat diet-induced obese mice. Am J Physiol Endocrinol Metab 303:E1234–E1244

Liang F, Kume S, Koya D (2009) SIRT1 and insulin resistance. Nat Rev Endocrinol 5:367–373

Libby P, Ridker PM, Maseri A (2002) Inflammation and atherosclerosis. Circulation 105:1135–1143

Libby P, Ridker PM, Hansson GK (2011) Progress and challenges in translating the biology of atherosclerosis. Nature 473:317–325

Lips P (2001) Vitamin D deficiency and secondary hyperparathyroidism in the elderly: consequences for bone loss and fractures and therapeutic implications. Endocr Rev 22:477–501

Louie SM, Roberts LS, Nomura DK (2013) Mechanisms linking obesity and cancer. Biochim Biophys Acta 1831:1499–1508

Matsuo K, Irie N (2008) Osteoclast-osteoblast communication. Arch Biochem Biophys 473:201–209

Matsuzaka T, Shimano H, Yahagi N, Amemiya-Kudo M, Okazaki H, Tamura Y, Iizuka Y, Ohashi K, Tomita S, Sekiya M, Hasty A, Nakagawa Y, Sone H, Toyoshima H, Ishibashi S, Osuga J, Yamada N (2004) Insulin-independent induction of sterol regulatory element-binding protein-1c expression in the livers of streptozotocin-treated mice. Diabetes 53:560–569

Mattson MP (2009) Roles of the lipid peroxidation product 4-hydroxynonenal in obesity, the metabolic syndrome, and associated vascular and neurodegenerative disorders. Exp Gerontol 44:625–633

McMillan DC, Sattar N, Talwar D, O'Reilly DS, McArdle CS (2000) Changes in micronutrient concentrations following anti-inflammatory treatment in patients with gastrointestinal cancer. Nutrition 16:425–428

Meigs JB, Hu FB, Rifai N, Manson JE (2004) Biomarkers of endothelial dysfunction and risk of type 2 diabetes mellitus. JAMA 291:1978–1986

Melnik BC (2012) Leucine signaling in the pathogenesis of type 2 diabetes and obesity. World J Diabetes 3:38–53

Meyers MR, Gokce N (2007) Endothelial dysfunction in obesity: etiological role in atherosclerosis. Curr Opin Endocrinol Diabetes Obes 14:365–369

Mohan S, Baylink DJ (2005) Impaired skeletal growth in mice with haploinsufficiency of IGF-I: genetic evidence that differences in IGF-I expression could contribute to peak bone mineral density differences. J Endocrinol 185:415–420

Mubagwa K, Gwanyanya A, Zakharov S, Macianskiene R (2007) Regulation of cation channels in cardiac and smooth muscle cells by intracellular magnesium. Arch Biochem Biophys 458:73–89

Muoio DM, Newgard CB (2008) Molecular and metabolic mechanisms of insulin resistance and β-cell failure in type 2 diabetes. Nat Rev Mol Cell Biol 9:193–205

Naka T, Nishimoto N, Kishimoto T (2002) The paradigm of IL-6: from basic science to medicine. Arthritis Res 4:S233–S242

Nielsen FH, Milne DB, Klevay LM, Gallagher S, Johnson L (2007a) Dietary magnesium deficiency induces heart rhythm changes, impairs glucose tolerance, and decreases serum cholesterol in postmenopausal women. J Am Coll Nutr 26:121–132

Nielsen FH, Milne DB, Gallagher S, Johnson L, Hoverson B (2007b) Moderate magnesium deprivation results in calcium retention and altered potassium and phosphorus excretion by postmenopausal women. Magnes Res 20:19–31

Nordin BEC (2010) Evolution of the calcium paradigm: the relation between vitamin D, serum calcium and calcium absorption. Nutrients 2:997–1004

Orimo H, Nakamura T, Hosoi T, Iki M, Uenishi K, Endo N, Ohta H, Shiraki M, Sugimoto T, Suzuki T, Soen S, Nishizawa Y, Hagino H, Fukunaga M, Fujiwara S (2012) Japanese 2011 guidelines for prevention and treatment of osteoporosis—executive summary. Arch Osteoporos 7:3–20

Pap T, Muller-Ladner U, Gay RE, Gay S (2000) Fibroblast biology. Role of synovial fibroblasts in the pathogenesis of rheumatoid arthritis. Arthritis Res 2:361–367

Peluso I, Morabito G, Urban L, Ioannone F, Serafini M (2012) Oxidative stress in atherosclerosis development: the central role of LDL and oxidative burst. Endocr Metab Immune Disord Drug Targets 12:351–360

Pérez-Hernández AI, Catalán V, Gómez-Ambrosi J, Rodríguez A, Frühbeck G (2014) Mechanisms linking excess adiposity and carcinogenesis promotion. Front Endocrinol (Lausanne) 5:65

Pillon NJ, Vella RE, Souleere L, Becchi M, Lagarde M, Soulage CO (2011) Structural and functional changes in human insulin induced by the lipid peroxidation byproducts 4-hydroxy-2-nonenal and 4-hydroxy-2-hexenal. Chem Res Toxicol 24:752–762

Poirier P, Giles TD, Bray GA, Hong Y, Stern JS, Pi-Sunyer FX, Eckel RH, American Heart Association, Obesity Committee of the Council on Nutrition, Physical Activity, and Metabolism (2006) Obesity and cardiovascular disease: pathophysiology, evaluation, and effect of weight loss: an update of the 1997 American Heart Association scientific statement on obesity and heart disease from the Obesity Committee of the Council On Nutrition, Physical Activity, And Metabolism. Circulation 113:898–918

Pulichino AM, Rowland S, Wu T, Clark P, Xu D, Mathieu MC, Riendeau D, Audoly LP (2006) Prostacyclin antagonism reduces pain and inflammation in rodent models of hyperalgesia and chronic arthritis. J Pharmacol Exp Therap 319:1043–1050

Purushotham A, Xu Q, Li X (2012) Systemic SIRT1 insufficiency results in disruption of energy homeostasis and steroid hormone metabolism upon high-fat-diet feeding. FASEB J 26: 656–667

Quintero M, Colombo SL, Godfrey A, Moncada S (2006) Mitochondria as signaling organelles in the vascular endothelium. Proc Natl Acad Sci U S A 103:5379–5384

Ramadori G, Fujikawa T, Fukuda M, Anderson J, Morgan DA, Mostoslavsky R, Stuart RC, Perello M, Vianna CR, Nillni EA, Rahmouni K, Coppari R (2010) SIRT1 deacetylase in POMC neurons is required for homeostatic defenses against diet-induced obesity. Cell Metab 12:78–87

Remer T, Manz F (1995) Potential renal acid load of foods and its influence on urine pH. J Am Diet Assoc 95:791–797

Ren J, Pulakat L, Whaley-Connell A, Sowers JR (2010) Mitochondrial biogenesis in the metabolic syndrome and cardiovascular disease. J Mol Med (Berl) 88:993–1001

Rizzoli R, Bonjour JP (2006) Physiology of calcium and phosphate homeostasis. In: Seibel MJ, Robins SP, Bilezikian JP (eds) Dynamics of bone and cartilage metabolism: principles and clinical applications. Academic, San Diego, CA, pp 345–360

Rodrigo R, Cereceda M, Castillo R, Asenjo R, Zamorano J, Araya J, Castillo-Koch R, Espinoza J, Larraín E (2008) Prevention of atrial fibrillation following cardiac surgery: basis for a novel therapeutic strategy based on non-hypoxic myocardial preconditioning. Pharmacol Ther 118:104–127

Rodrigo R, González J, Paoletto F (2011) The role of oxidative stress in the pathophysiology of hypertension. Hypertens Res 34:431–440

Rodrigo R, Prieto JC, Castillo R (2013) Cardioprotection against ischaemia/reperfusion by vitamins C and E plus n-3 fatty acids: molecular mechanisms and potential clinical applications. Clin Sci 124:1–115

Rook GA (2010) 99th Dahlem conference on infection, inflammation and chronic inflammatory disorders: darwinian medicine and the 'hygiene' or 'old friends' hypothesis. Clin Exp Immunol 13:70–79

Rude RK, Gruber HE, Wie LY, Frausto A, Mills BG (2003) Magnesium deficiency: effect on bone and mineral metabolism in the mouse. Calcif Tissue Int 72:32–41

Rude RK, Singer FR, Gruber HE (2009) Skeletal and hormonal effects of magnesium deficiency. J Am Coll Nutr 28:131–141

Sakamoto K, Holman GD (2008) Emerging role for AS160/TBC1D4 and TBC1D1 in the regulation of GLUT4 traffic. Am J Physiol Endocrinol Metab 295:29–37

Schenk S, McCurdy CE, Philp A, Chen MZ, Holliday MJ, Bandyopadhyay GK, Osborn O, Baar K, Olefsky JM (2011) Sirt1 enhances skeletal muscle insulin sensitivity in mice during caloric restriction. J Clin Invest 121:4281–4288

Schwartz R, Reddi AH (1979) Influence of magnesium depletion on matrix-induced endochondral bone formation. Calcif Tissue Int 29:15–20

Seneff S, Wainwright G, Mascitelli L (2011) Is the metabolic syndrome caused by a high fructose, and relatively low fat, low cholesterol diet? Arch Med Sci 7:8–20

Serhan CN, Fredman G, Yang R, Karamnov S, Belayev LS, Bazan NG, Zhu M, Winkler JW, Petasis NA (2011) Novel proresolving aspirin-triggered DHA pathway. Chem Biol 18:976–987

Seriwatanachai D, Thongchote K, Charoenphandhu N, Pandaranandaka J, Tudpor K, Teerapornpuntakit J, Suthiphongchai T, Krishnamra N (2008) Prolactin directly enhances bone turnover by raising osteoblast-expressed receptor activator of nuclear factor κB ligand/osteoprotegerin ratio. Bone 42:535–546

Shechter M (2010) Magnesium and cardiovascular system. Magnes Res 23:60–72

Shi H, Kokoeva MV, Inouye K, Tzameli I, Yin H, Flier JS (2006) TLR4 links innate immunity and fatty acid–induced insulin resistance. J Clin Invest 116:3015–3025

Simonet WS, Lacey DL, Dunstan CR, Kelley M, Chang MS, Lüthy R, Nguyen HQ, Wooden S, Bennett L, Boone T, Shimamoto G, DeRose M, Elliott R, Colombero A, Tan HL, Trail G, Sullivan J, Davy E, Bucay N, Renshaw-Gegg L, Hughes TM, Hill D, Pattison W, Campbell P, Sander S, Van G, Tarpley J, Derby P, Lee R, Boyle WJ (1997) Osteoprotegerin: a novel secreted protein involved in the regulation of bone density. Cell 89:309–319

Simopoulos AP (2009) Evolutionary aspects of the dietary omega-6:omega-3 fatty acid ratio: medical implications. World Rev Nutr Diet 100:1–21

Simopoulos AP (2013) Dietary omega-3 fatty acid deficiency and high fructose intake in the development of metabolic syndrome, brain metabolic abnormalities, and non-alcoholic fatty liver disease. Nutrients 5:2901–2923

Sims NA, Gooi JH (2008) Bone remodeling: multiple cellular interactions required for coupling of bone formation and resorption. Semin Cell Dev Biol 19:444–451

Smith AG, Sheridan PA, Harp JB, Beck MA (2007) Diet-induced obese mice have increased mortality and altered immune responses when infected with influenza virus. J Nutr 137:1236–1243

Sobacchi C, Frattini A, Guerrini MM, Abinun M, Pangrazio A, Susani L, Bredius R, Mancini G, Cant A, Bishop N, Grabowski P, Del Fattore A, Messina C, Errigo G, Coxon FP, Scott DI, Teti A, Rogers MJ, Vezzoni P, Villa A, Helfrich MH (2007) Osteoclast-poor human osteopetrosis due to mutations in the gene encoding RANKL. Nat Genet 39:960–962

Solinas G, Karin M (2010) JNK1 and IKKbeta: molecular links between obesity and metabolic dysfunction. FASEB J 24:2596–2611

Song Y, Manson JE, Buring JE, Liu S (2004) A prospective study of red meat consumption and type 2 diabetes in middle-aged and elderly women: the women's health study. Diabetes Care 27:2108–2115

Spagnuolo MI, Cicalese MP, Caiazzo MA, Franzese A, Squeglia V, Assante LR, Valerio G, Merone R, Guarino A (2010) Relationship between severe obesity and gut inflammation in children: what's next? Ital J Pediatr 36:66–71

Steiner G, Smolen J (2002) Autoantibodies in rheumatoid arthritis and their clinical significance. Arthritis Res 4:S1–S5

Stepp DW (2006) Impact of obesity and insulin resistance on vasomotor tone: nitric oxide and beyond. Clin Exp Pharmacol Physiol 33:407–414

Stocker R, Keaney JF Jr (2004) Role of oxidative modifications in atherosclerosis. Physiol Rev 84:1381–1478

Stumvoll M, Goldstein BJ, van Haeften TW (2005) Type 2 diabetes: principles of pathogenesis and therapy. Lancet 365:1333–1346

Sun C, Zhang F, Ge X, Yan T, Chen X, Shi X (2007) SIRT1 improves insulin sensitivity under insulin-resistant conditions by repressing PTP1B. Cell Metab 6:307–319

Supanchart C, Kornak U (2008) Ion channels and transporters in osteoclasts. Arch Biochem Biophys 473:161–165

Tabit CE, Chung WB, Hamburg NM, Vita JA (2010) Endothelial dysfunction in diabetes mellitus: molecular mechanisms and clinical implications. Rev Endocr Metab Disord 11:61–74

Tang X, Luo Y-X, Chen H-Z, Liu DP (2014) Mitochondria, endothelial cell function, and vascular diseases. Front Physiol 5:175

Teitelbaum SL (2000) Bone resorption by osteoclasts. Science 289:1504–1508

Ting AH, McGarvey KM, Baylin SB (2006) The cancer epigenome—components and functional correlates. Genes Dev 20:3215–3231

Touyz RM (2004) Reactive oxygen species, vascular oxidative stress, and redox signaling in hypertension: what is the clinical significance? Hypertension 44:248–252

Tulic MK, Hodder M, Forsberg A, McCarthy S, Richman T, D'Vaz N, van den Biggelaar AH, Thornton CA, Prescott SL (2011) Differences in innate immune function between allergic and nonallergic children: new insights into immune ontogeny. J Allergy Clin Immunol 127: 470–478

Uribarri J, Woodruff S, Goodman S, Cai W, Chen X, Pyzik R, Yong A, Striker GE, Vlassara H (2010) Advanced glycation end products in foods and a practical guide to their reduction in the diet. J Am Diet Assoc 110:911–916, e12

Vila L, Roglans N, Alegret M, Sanchez RM, Vazquez-Carrera M, Laguna JC (2008) Suppressor of cytokine signaling-3 (SOCS-3) and a deficit of serine/threonine (Ser/Thr) phosphoproteins involved in leptin transduction mediate the effect of fructose on rat liver lipid metabolism. Hepatology 48:1506–1516

Villegas R, Shu XO, Gao YT, Yang G, Cai H, Li H, Zheng W (2006) The association of meat intake and the risk of type 2 diabetes may be modified by body weight. Int J Med Sci 3:152–159

Vogelstein B, Kinzler KW (2004) Cancer genes and the pathways they control. Nat Med 10:789–799

Vormann J, Remer T (2008) Dietary, metabolic, physiologic, and disease-related aspects of acid-base balance: foreword to the contributions of the second International Acid-Base Symposium. J Nutr 138:413S–414S

Wallace JM (2002) Nutritional and botanical modulation of the inflammatory cascade–eicosanoids, cyclooxygenases, and lipoxygenases–as an adjunct in cancer therapy. Integr Cancer Ther 1:7–37; discussion 37

Wang D, Dubois RN (2010) Eicosanoids and cancer. Nat Rev Cancer 10:181–193

Wang Y, Beydoun MA (2007) The obesity epidemic in the United States—gender, age, socioeconomic, racial/ethnic, and geographic characteristics: a systematic review and meta-regression analysis. Epidemiol Rev 29:6–28

Wellen KE, Hotamisligil GS (2005) Inflammation, stress, and diabetes. J Clin Invest 115: 1111–1119

Widlansky ME, Gokce N, Keaney JF Jr, Vita JA (2003) The clinical implications of endothelial dysfunction. J Am Coll Cardiol 42:1149–1160

Woutersen RA, Appel MJ, van Garderen-Hoetmer A, Wijnands MV (1999) Dietary fat and carcinogenesis. Mutat Res 443:111–127

Xu F, Gao Z, Zhang J, Rivera CA, Yin J, Weng J, Ye J (2010) Lack of SIRT1 (Mammalian Sirtuin 1) activity leads to liver steatosis in the SIRT1+/− mice: a role of lipid mobilization and inflammation. Endocrinology 151:2504–2514

Zaidi M (2007) Skeletal remodeling in health and disease. Nat Med 13:791–801

Zeyda M, Stulnig TM (2009) Obesity, inflammation, and insulin resistance—a mini-review. Gerontology 55:379–386

Zimmermann R, Panzenböck U, Wintersperger A, Levak-Frank S, Graier W, Glatter O, Fritz G, Kostner GM, Zechner R (2001) Lipoprotein lipase mediates the uptake of glycated LDL in fibroblasts, endothelial cells, and macrophages. Diabetes 50:1643–1653

Chapter 9
Effects of Long Term Consumption of High Calorie Diet on Neurological Disorders

9.1 Introduction

As stated in Chap. 1, high calorie diet not only contains high amounts of processed macronutrients (fats, cholesterol, proteins and simple sugars), but also has high salt (sodium chloride). In addition, high calorie diet is low in fiber and deficient in minerals. High calorie diet is characterized by the consumption of high fat sandwich spreads, red meat, sandwich meat, potatoes, butter, and lard, eggs, sauces, pizza, soda, and desserts (Villegas et al. 2004, 2010; Esmaillzadeh et al. 2007). It is estimated that present day high calorie diet provides about 50 % of total daily calories from refined carbohydrates, 35 % calories from fat and refined oils, and 15 % calories from proteins of animal origin. In high calorie diet the ratio between n-6 (arachidonic acid, ARA) to n-3 fatty acid (docosahexaenoic acid, DHA) is about 20:1 (Farooqui 2014). In contrast, Paleolithic diet on which our forefathers lived and survived thousands of years contained unprocessed carbohydrates (40 %), fats (21 %), and proteins (39 %) (Simopoulos 2009, 2013; Bengmark 2013). In general, long term consumption of high calorie diet leads to alterations in the metabolic homeostasis, a metabolic state where energy intake exceeds energy expenditure (obesity). This condition leads to cellular oxidative stress and inflammation, which are closely associated with psychiatric and cardiovascular diseases and neurodevelopmental deficits (O'Keefe et al. 2008; Harris 2008). The generation of oxidative stress is initiated in mitochondria, where overloading of metabolites of fats and carbohydrates (free fatty acids and glucose) produces an increase in the production of acetyl CoA. Elevated levels of acetyl CoA cause an increase in NADH generation from the TCA cycle. Elevation in availability of NADH at complex I of mitochondrial electron transport chain increases membrane potential to the extent that complex III is stalled resulting in a longer half-life for coenzyme Q. Increased availability of coenzyme Q leads to an increased reduction of oxygen to superoxide ($O_2^{\cdot}$). Thus, the main impact of the consumption of high calorie diet is higher levels of superoxide in the mitochondria (Ceriello and Motz 2004). Superoxide is a relatively unstable

© Springer International Publishing Switzerland 2015
A.A. Farooqui, *High Calorie Diet and the Human Brain*,
DOI 10.1007/978-3-319-15254-7_9

intermediate, which is transformed into hydrogen peroxide in the mitochondria by superoxide dismutase. The newly formed hydrogen peroxide can then undergo a Haber-Weiss or Fenton reaction, yielding a highly reactive hydroxyl radical, which can oxidize mitochondrial proteins, DNA, and lipids amplifying the effects of the superoxide-initiated oxidative stress (Yin et al. 2012).

9.2 Effects of High Calorie Diet on the Brain

The long term consumption of high calorie diet negatively affects cognitive performance in animal models of cognition (Wu et al. 2004; Stranahan et al. 2008; Kanoski et al. 2007; Kanoski and Davidson 2011). Much of the experimental evidence linking excessive consumption of high calorie diet to cognitive decline has been obtained from studies on rodent hippocampus, a brain region that plays major roles in learning and memory processes. The large pyramidal hippocampal neurons are especially sensitive to damage from a variety of environmental and biological insults (Michaelis 2012). These neurons rely heavily both on oxidative phosphorylation and mitochondria for energy, and abnormalities in either of these mechanisms may compromise hippocampal integrity (Michaelis 2012). Dietary composition is one such environmental factor that can negatively impact hippocampal metabolism and function. There is a putative causal relationship between high calorie diet consumption and hippocampal abnormalities related to learning and memory (Stranahan et al. 2008; Kanoski et al. 2007; Kanoski and Davidson 2011). While spatial reference memory impairments arise and remain relatively stable only 72 h post consumption, nonspatial reference impairments are not robustly observed until after 30 days of high calorie diet consumption. This suggests that hippocampal dependent spatial memory is particularly sensitive to the disruption by high calorie diet intake. These impairments can arise prior to the development of high calorie diet-induced metabolic derangements and increased adiposity (Kanoski and Davidson 2011; Kanoski 2012).

The molecular mechanisms by which high calorie diet impairs cognitive function are not fully understood. However, it is suggested that high energy diet may impair hippocampal plasticity by reducing the expression of brain-derived neurotrophic factor (BDNF) (Fig. 9.1), a neurotrophin which is synthesized by neurons in an activity-dependent manner (Stranahan et al. 2008). This neurotrophin plays an important role in the survival, maintenance, and growth of many types of neurons. BDNF is abundantly expressed in the hippocampus, hypothalamus, and cerebral cortex (McNay et al. 2010). BDNF activates a high-affinity receptor tyrosine kinase called trkB. This receptor is located in the plasma membrane, dendrites and presynaptic terminals. Activation of trkB is coupled with a signaling pathway, which involves PtdIns 3 kinase, Akt kinase and FOXO transcription factors. This pathway modulates the expression of genes that enhance synaptic plasticity (glutamate receptor subunits and growth-associated protein 43) and cell survival (due to antioxidant enzyme superoxide dismutase 2, and the anti-apoptotic protein Bcl-2)

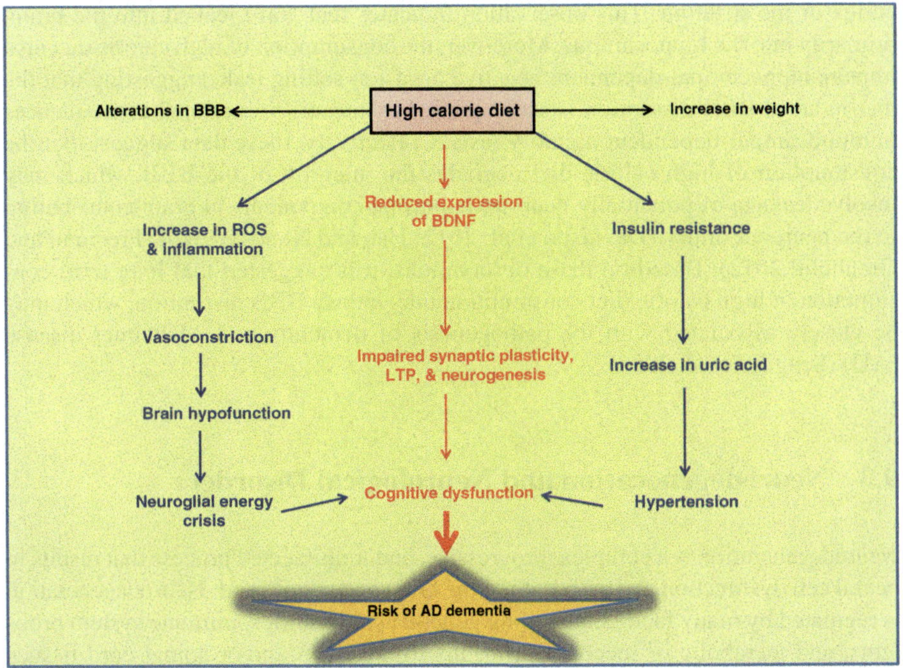

Fig. 9.1 Neurochemical consequences of high calorie diet consumption

(Koponen et al. 2004; Mattson et al. 2004; Pang and Lu 2004; Bramham and Messaoudi 2005). In addition to enhancing synaptic plasticity and neuronal survival, BDNF also stimulates the neurogenesis (the differentiation of neurons from neural stem cells) (Schmidt and Duman 2007), which may be associated with the beneficial effects of BDNF on cognition. High calorie diet also increases fasting blood glucose, serum cholesterol, and triacylglycerols leading to endothelial cell dysfunction and atherosclerotic changes in the cerebrovascular system. It is also reported that the consumption of high calorie diet upregulates mechanisms involved in glutamate clearance and simultaneously impairs glutamate metabolism. Collective evidence suggests that high calorie diet triggers neurochemical changes, leading to a desensitization of NMDA receptors within the hippocampus, which may contribute to cognitive deficits (Valladolid-Acebes et al. 2012).

Molecular mechanisms involved in the pathogenesis of high fat diet-mediated neurological disorders remain elusive (Kanoski and Davidson 2011). Recent studies have indicated that the long term consumption of high calorie diet decreases blood–brain barrier (BBB) permeability (Davidson et al. 2012). Thus, rats fed with high calorie diet for 90+ days show decrease in the expression of tight junction proteins, particularly claudin 5 and claudin 12, within the BBB and choroid plexus. Moreover, within the hippocampus of high calorie diet fed rats, there is an increase in NaFl fluorescence compared to control groups, an effect, which is not observed in the cortex or the striatum. This observation indicates that NaFl leaked into the brain, primarily into the hippocampus. Moreover, the consumption of high calorie diet also impairs hippocampal-dependent negative occasion setting task, suggesting that the diet-induced BBB disruption is accompanied by negative functional consequences in hippocampal-dependent memory tests. Collectively, these data suggest that the consumption of high calorie diet degrades the integrity of the BBB, which may involve leakage of potentially neurotoxic plasma components in brain contributing to the neuronal injury (Davidson et al. 2012; Hsu and Kanoski 2014; Freeman and Granholm 2012). Based on these observations, it is suggested that long term consumption of high calorie diet consumption may induce BBB disruption, which may be closely associated with the pathogenesis of dementia and Alzheimer disease (AD) (Sengillo et al. 2013).

9.3 Neurodegeneration and Neurological Disorders

Neurodegeneration is a complex, progressive, and multifaceted process that results in neural cell dysfunction and cell death in the brain and spinal cord. Neurodegeneration is regulated by many factors, including genetic abnormalities, immune system problems, and metabolic or mechanical insults to the brain and/or spinal cord tissues (Farooqui 2010). Neurodegeneration occurs in neurotraumatic diseases, which are accompanied by metabolic trauma (ischemia or stroke), traumatic brain injury (TBI), and spinal cord trauma (SCI), neurodegenerative diseases (Alzheimer disease, AD; Parkinson disease, PD; Huntington disease, HD; and amyotrophic lateral sclerosis, ALS) and neuropsychiatric disorders (schizophrenia and depression) (Farooqui 2010). Neurotraumatic, neurodegenerative, and neuropsychiatric diseases fall under the umbrella term neurological disorders. It is becoming increasingly evident that neurodegeneration is accompanied by the upregulation of interplay among excitotoxicity, oxidative stress, and neuroinflammation (Farooqui et al. 2008; Farooqui 2010). In neurotraumatic diseases, neurodegeneration occurs rapidly (in a matter of hours to days) because of sudden lack of oxygen, rapid decrease in ATP, disturbance in transmembrane potential and sudden collapse of ion gradients at very early stage (Farooqui and Horrocks 2007; Farooqui 2010). In addition, in neurotraumatic diseases, acute neuroinflammation develops rapidly because of rapid generation and accumulation of eicosanoids, platelet-activating factor, and the release of proinflammatory cytokines (Farooqui and Horrocks 2007; Farooqui 2010). In contrast, in neurodegenerative diseases, oxygen, nutrients, and reduced levels of ATP continue to be available to the nerve cells, and ionic homeostasis is maintained to a limited extent. The interplay among excitotoxicity, oxidative stress, and neuroinflammation occurs at a slow rate along the accumulation of post-translationally modified, especially misfolded, proteins, leading to a neurodegenerative process that takes many years to develop (Farooqui and Horrocks 2007; Farooqui et al. 2010). Furthermore, in neurodegenerative diseases due to abnormalities in immune system, chronic inflammation lingers for years, causing continued insult to the brain tissue and ultimately reaching the threshold of detection many years after the onset of the neurodegenerative diseases (Williams 2002; Farooqui et al. 2010; Farooqui 2010). Examples of neurodegenerative diseases are Alzheimer disease (AD), Parkinson disease (PD), Huntington disease (HD), amyotrophic lateral sclerosis (ALS), and prion diseases. In AD, neurodegeneration occurs in the nucleus basalis, hippocampus, and entorhinal cortex; in PD, neurodegeneration takes place in the substantia nigra; degeneration of striatal medium spiny neurons occurs in HD; and ALS is characterized by damage to motor neurons in the brain and spinal cord. Among these neurodegenerative conditions, it is not clear when does the onset of disease start and how long does it take for neuropathological changes to appear. Neuropsychiatric disorders include both neurodevelopmental disorders and behavioral or psychological difficulties associated with some neurological disorders. An important characteristic of neuropsychiatric disorders is the impairment of cognitive processing. This includes not only ability to learn and store the memory but also to retrieve stored memory for further use and to apply the stored memory to efficiently solve problems (Gallagher 2004). The impairment of cognitive process may be caused by the overexpression or underexpression of certain genes or other unknown factors that result in behavioral symptoms, such as thoughts or actions, delusions, and hallucinations, which are the hallmarks of many neuropsychiatric disorders including schizophrenia, depression, and bipolar disorders.

9.4 Molecular Mechanisms Associated with Neurological Disorders

The molecular mechanisms associated with neurological disorders are not fully understood. However, it is suggested that the pathogenesis of neurological disorders may involve aging, genetic and environmental factors, insulin resistance, reduction in cellular antioxidant defenses (activities of superoxide dismutase, glutathione peroxidase, catalase, and glutathione reductase), mitochondrial dysfunction, increase in the production of ROS, elevation in calcium influx, stimulation of calcium-dependent enzymes, stimulation and translocation of NF-κB to the nucleus leading to increase in expression of proinflammatory cytokines, and accumulation of peroxidized lipids, proteins and DNA oxidative products, supporting the view that neurodegeneration is a multifactorial process involving genetic, environmental, and endogenous factors (Fig. 9.2) (Farooqui 2010). Endogenous factors that contribute to neurological disorders also include excitotoxicity, oxidative stress, neuroinflammation, abnormal protein dynamics with defective protein degradation and aggregation related to the ubiquitin-proteasomal system, resulting in the generation and accumulation of misfolded proteins, autoimmunity, and abnormal mitochondrial function leading to increase in Ca^{2+} levels, and impairment in energy metabolism (Fig. 9.2) (Williams 2002; Farooqui and Horrocks 2007; Farooqui 2010; Jellinger 2009).

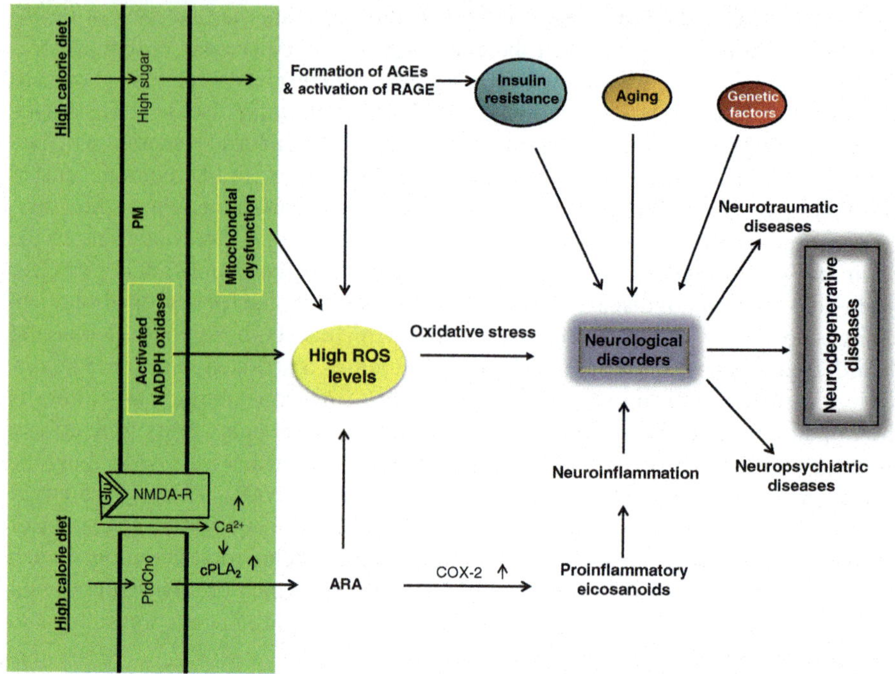

Fig. 9.2 Risk factors contributing to the pathogenesis of neurological disorders

9.4.1   Molecular Mechanisms Associated with Neurotraumatic Disease

Among neurotraumatic diseases, stroke (ischemia) is a metabolic insult, which is induced by severe reduction or blockade in cerebral blood flow due to cerebrovascular disease. This blockade not only decreases oxygen and glucose delivery to brain tissue but also results in the breakdown of blood–brain barrier (BBB) and buildup of potentially toxic products in the brain. Breakdown of BBB integrity in stroke-mediated injury results in the transmigration of numerous immune system cells including monocytes and lymphocytes and also causes hyperpermeability induced by enhanced transcytosis and gap formation between endothelial cells (Farooqui 2010). Stroke is classified into two major categories: ischemic and hemorrhagic. Ischemic stroke is due to interruption of the blood supply due to blood clot, while hemorrhagic stroke is due to the rupture of a blood vessel or an abnormal vascular structure. Long-term outcome of stroke depends on the region of the brain that is affected and the severity and generally paralysis, cognitive deficits, and speech and language problems may occur in stroke patients. Risk factors for stroke include age, obesity, type II diabetes, hypertension, high cholesterol, and poor diet (Farooqui 2010). Neurochemical events in ischemic injury are the release of glutamate, overstimulation of glutamate receptor, rapid calcium influx, and activation of calcium-dependent enzymes (phospholipase A_2, phospholipase C, cyclooxygenase, and nitric oxide synthases) (Farooqui 2010), induction of oxidative stress, and neuroinflammation. Age is a prominent risk factor for stroke. Thus, at the age of 55–64 years the prevalence of stroke is 11 %. The risk increases to 43 % in subjects that are older than 85 years. The reason for age-mediated vulnerability for stroke is not fully understood. However, it is becoming increasingly evident that cerebral aging is a complex and heterogenous process related to a large variety of molecular changes involving multiple neuronal networks, due to alterations of neurons (synapses, axons, dendrites, etc), particularly affecting strategically important regions, such as hippocampus and prefrontal areas. Potential mechanisms of age-mediated vulnerability may include changes in brain plasticity-promoting factors, unregulated expression of neurotoxic factors, or differences in the generation of scar tissue that impedes the formation of new axons and blood vessels in the infarcted region (Popa-Wagner et al. 2007). Other risk factors include the long term consumption of high calorie diet, smoking, lack of exercise, and insulin resistance (Farooqui 2013). During the stroke, lack of oxygen impairs oxidative phosphorylation and maintains electron transport chain proteins in a reduced state. Upon reperfusion, oxygen is restored. The interactions between oxygen and reduced proteins promote a burst of ROS production, which mediates injury. In addition, ROS are also generated in the cytoplasm and the plasma membrane by means of xanthine oxidase, NOS and NADPH oxidase (Manzanero et al. 2013).

TBI and SCI are accompanied by acute trauma to the brain and spinal cord tissues. Motor cycle and car accidents are major causes of TBI and SCI among young people. TBI and SCI consist of two broadly defined components: a primary component, attributable to the mechanical insult itself, and a secondary component, attributable to the series of systemic and local neurochemical and pathophysiological changes that occur in the brain and spinal cord after the initial insult (Arundine and Tymianski 2004; Klussmann and Martin-Villalba 2005; Raghupathi 2004; Farooqui 2010). The primary injury rapidly causes rapid deformation of brain or spinal cord tissue, hemorrhage, and rupture of neural cell membranes at the impact site leading to the release of intracellular contents, disruption of blood flow, and breakdown of BBB. In contrast, secondary injury to the brain or spinal cord induces neurochemical alterations, activation of microglial cells and astrocytes, and demyelination involving oligodendroglia leading to the increase in calcium influx, activation of calcium-dependent enzymes, increased expression of proinflammatory cytokines, and delayed cell death (Farooqui 2010). The generation of ROS, neuroinflammation, and apoptotic cell death are critical components of secondary injury evolution.

9.4.2 Molecular Mechanisms Associated with Neurodegenerative Diseases

Neurodegenerative diseases are a heterogeneous group of illnesses with distinct clinical phenotypes and genetic etiologies. Most neurodegenerative diseases (95 %) are sporadic with unknown mechanism of pathogenesis. Toxic environmental factors and unhealthy lifestyle have been suggested to contribute to the pathogenesis of neurodegenerative processes (BenMoyal-Segal and Soreq 2006; Seidl et al. 2014). This view is based on the observation that some neurodegenerative diseases arise in geographic or temporal clusters. For example, the consumption of β-methylaminoalanine (BMAA) in *Cycas circinalis*, a plant commonly ingested as a food or medicine by the Chamorros of Guam has been reported to cause Guam-type amyotrophic lateral sclerosis/parkinsonism dementia (ALS/PDC) (Ince and Codd 2005). Intoxication with 1-methyl-4-phenyl-1,2,3,6-tetrahydropyridine (MPTP) associated with severe and irreversible parkinsonian like syndrome that is similar but not identical to PD pathology and progression. In addition, exposure to certain insecticides and herbicides, such as paraquat and rotenone, also produces a Parkinson-like syndrome (Brown et al. 2006; Kamel and Hoppin 2004; Keifer and Firestone 2007). Dairy product consumption and drinking milk may increase one's risk of PD independently of calcium intake (Park et al. 2005; Kyrozis et al. 2013), particularly in men (Chen et al. 2007). Preliminary studies have indicated that individuals who consume large amounts of dairy products may often have low serum uric acid levels (Choi et al. 2005). Serum uric acid is inversely correlated with the risk of PD and disease duration (Weisskopf et al. 2007; Shen et al. 2013). Only 5 % neurodegenerative diseases can be traced to specific genetic mutations for example in AD patients presenilin-1 and 2, APP and APOE4 genes , in PD patients genes for α-synuclein, Parkin, DJ-1, PINK 1, UCH-L1, and LRRK2; SOD-1 gene in ALS. The most important risk factors for neurodegenerative diseases are old age, positive family history, unhealthy lifestyle, and exposure to toxic environment (Farooqui 2010). Like neurotraumatic diseases, neurodegenerative are also accompanied by onset of excitotoxicity, oxidative stress, and neuroinflammation as common mechanisms of neurodegeneration. Thus, neurodegenerative diseases not only lead to enhanced neural membrane phospholipid degradation, increased generation of lipid mediators, and abnormal protein aggregation along with initiation of vicious cycles of aberrant neuronal activity and compensatory alterations in neurotransmitter receptor signaling leading to loss of synapse, but also disintegration of neural networks and ultimate failure of neurological functions (Farooqui and Horrocks 2007; Farooqui 2010; Jellinger 2009). In addition to above mentioned processes, neurodegenerative diseases are accompanied by the accumulation of misfolded proteins, mitochondrial and proteasomal dysfunction, and premature and slow death of certain neuronal populations in brain tissue (Graeber and Moran 2002). Neurodegenerative diseases result in a progressive loss of cognitive function and motor disabilities with devastating consequences to their patients. It is proposed that an upregulation of interplay among excitotoxicity, oxidative stress, and neuroinflammation may be a common mechanism of brain damage in neurodegenerative diseases (Farooqui and Horrocks 2007; Farooqui 2010). Importantly, neurogenesis, a process associated with birth and maturation of functional new hippocampal neurons, is impaired by interplay among excitotoxicity, oxidative stress, and neuroinflammation accounting for brain atrophy in patients with neurodegenerative diseases.

9.4.3 Molecular Mechanisms Associated with Neuropsychiatric Diseases

Neuropsychiatric disorders not only involve abnormalities in cerebral cortex and limbic system (thalamus, hypothalamus, hippocampus, and amygdale), but also alterations in cognitive processing, which are mediated by changes in signal transduction processes associated with everyday problem-solving behavior. This includes the ability to learn and store the memory to retrieve stored memory for further use, and to apply the stored memory to efficiently solve problems (Gallagher 2004). At the cellular level, abnormalities in neuropsychiatric disorders are linked to gray matter atrophy caused by decreased neuronal and glial size, increased cellular packing density suggesting a disruption in neuronal connectivity, particularly in the dorsolateral prefrontal cortex, and distortions in neuronal orientation (Arnold and Trojanowski 1996; Blitzer et al. 2005). These observations are supported by neuroimaging studies that indicate a number of anatomical and neurochemical abnormalities in neurocircuits in specific brain area of neuropsychiatric patients. At the molecular level cognitive abnormalities are not only accompanied by alterations in several neurotransmitters, but also overexpression or underexpression of genes related to signal transduction processes involved in the modulation of behavioral symptoms such as thoughts or actions, delusions, and hallucinations. Neurochemical and neuroimaging studies have also indicated alterations in cerebral blood flow and glucose utilization in the limbic system and prefrontal cortex of patients with major depression and other neuropsychiatric diseases (Ito et al. 1996; Kimbrell et al. 2002). Collective evidence suggests that genetic factors, alterations in blood flow, disruption of cellular connectivity, decrease in neurogenesis, alterations in microcircuitry, decrease in neuroplasticity along with mild oxidative stress, and mild neuroinflammation are major risk factors for neuropsychiatric diseases. In addition, both AD, PD, and HD are accompanied by neuropsychiatric symptoms due to age-related changes in neurotransmission, neuroplasticity, and signal transduction processes (Becker et al. 1997; Blitzer et al. 2005; Perlis et al. 2010), supporting the view that there is an overlap among some neurochemical mechanisms associated with neurodegenerative and neuropsychiatric diseases.

9.5   High Calorie Diet and Stroke

Brain accounts for only 2 % body weight, but it utilizes 20 % of the oxygen and 25 % of glucose for the production of energy, which is required for the maintenance of ionic balance across neural membranes, production of action potential and initiation and maintenance of post-synaptic currents, creation of proper cellular redox potentials, and recycling of neurotransmitters (Rolfe and Brown 1997). Because of high oxygen consumption, brain is highly susceptible to oxidative and nitrosative damage. As stated above, stroke (ischemia) is a metabolic insult caused by severe reduction or blockade in cerebral blood flow due to cerebrovascular disease. This blockade not only decreases oxygen and glucose delivery to the brain tissue but also results in the breakdown of BBB and buildup of potentially toxic products in the brain. Breakdown of BBB integrity in ischemic injury not only results in transmigration of numerous immune system cells including monocytes and lymphocytes but also causes hyperpermeability induced by enhanced transcytosis and gap formation between endothelial cells (Bousser 2012). Interruption in glucose and oxygen supply to the brain during stroke may result in a decrease in ATP generation, loss of ion homeostasis, mitochondrial dysfunction, production of ROS, such as superoxide, and hydroxyl anion and reactive nitrogen species (RNS), such as NO and $ONOO^-$, and changes in the redox status of neural cells. These processes also contribute to cerebral edema, which is the primary cause of patient mortality after stroke. The initial response to a transient insufficiency of energy is depolarization, which causes Na^+ influx into axons. Prolonged depletion of ATP produces a massive influx of Ca^{2+} that facilitates neural cell death (Fig. 9.3). In addition, stroke-induced injury is accompanied by excessive release of glutamate in the extracellular space and over-stimulation of glutamate receptors, which releases ARA and eicosanoids through the stimulation of phospholipases A_2 (PLA_2) and cyclooxygenase-2 (COX-2). This process not only generates more oxidative stress, but also increases the expression and release of cytokines from activated microglial cells and astrocytes (Farooqui and Horrocks 1994).

High calorie diet is enriched in "red meat (high amounts of n-6 fatty acids and amino acid leucine), refined carbohydrate (sugar, white flour and white rice), and soft drinks containing fructose corn syrup". High calorie diet is associated with a higher risk for stroke, (Fig. 9.3). It is becoming increasingly evident that the consumption of high calorie diet causes behavioral impairment due to decrease in levels of BDNF, a growth factor crucial for synaptic plasticity and learning and memory (Molteni et al. 2002; Pistell et al. 2010; Bhat 2010; Dhungana et al. 2013). Long term consumption of high calorie diet results in hyperglycemia, insulin resistance, hypertriglyceridemia, and low HDL-cholesterol leading to type II diabetes and metabolic syndrome (MetS). The onset of MetS increases the chances of stroke in patients of all ages, ranging from neonates through to the elderly. The risk of stroke is high within 5 years of diagnosis for type II diabetes is 9 % (mortality 21 %), that is more than doubles the rate for the general population. The molecular mechanisms associated with increased risk of stroke in type II diabetes and MetS is

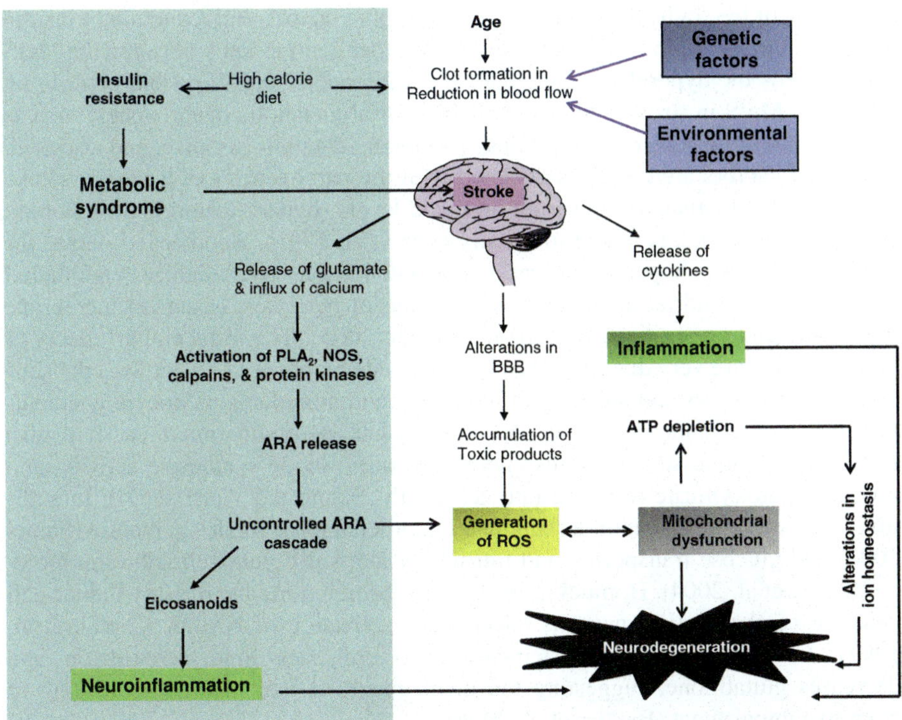

Fig. 9.3 Neurochemical effects of high calorie diet on neurodegeneration in stroke. *PLA₂* Phospholipase A₂, *NOS* nitric oxide synthase, *ARA* arachidonic acid, *BBB* blood–brain barrier, *ROS* reactive oxygen species

not fully understood. However, long term consumption of high calorie diet produces prolonged hyperglycemia, which may cause brain damage not only through the generation of high levels of advanced glycation end-products (AGEs) by the polyol pathway, but also through increase in the expression of the receptor for advanced glycation end products (RAGE) (Brownlee 1995; Farooqui 2013). Long term consumption of high calorie diet also induces the synthesis of proinflammatory lipid mediators (eicosanoids and platelet activating factor), proinflammatory cytokines (TNF-α, IL-1β, and IL-6), upregulates of gp91(phox) subunit of NADPH oxidase, and downregulates superoxide dismutase (SOD) isoforms, glutathione peroxidase (GPX), and heme oxygenase-2 (HO-2) in various body tissues including brain. These neurochemical changes produce oxidative stress, and inflammation (Farooqui et al. 2012; Farooqui 2013). In addition, MetS increases the risk of vascular disease and stroke through the involvement of hypertension, insulin insensitivity and dyslipidemia (Li et al. 2010; Abarquez 2003). These processes lead to atherogenesis and the prothrombotic state, which increases the risk of stroke (Uchino et al. 2010). The association between MetS and stroke is not attenuated after the adjustment for the presence of established cardiovascular and cerebrovascular risk factors. Like stroke patients, MetS patients show the inflammatory activity, increase in the generation of free radicals, changes of neurotrophic factors, and reduction of insulin transport into the brain (Lusis et al. 2008). Another commonality between the MetS and stroke is the hyperglycemia, which is detrimental to cognition and other brain functions. MetS in stroke patients involves several metabolic disturbances, such as high levels of low density lipoproteins, fibrinogen, insulin resistance, and uric acid. These risk factors are closely associated with the pathogenesis of ischemic stroke (Farooqui 2013). Increase in serum uric acid levels (hyperuricemia), an important risk factor of stroke, is linked to obesity (Lee et al. 1995). Another risk factor for stroke and MetS is hyperhomocysteinemia, a condition, which induces endothelial damage, mitochondrial disintegration, swelling of pericytes, basement membrane thickening and perivascular detachment (Troen 2005). The intracellular effects of homocysteine are very divergent. Hyperhomocysteinemia not only induces the activation of caspase-8 and subsequent apoptosis and stimulates monocyte chemoattractant protein-1/interleukin-8 and subsequent neuroinflammation, but also increases oxidative stress, retards endothelial nitric oxide synthetase activity, and produces peroxynitrite resulting into cell death (Skurk and Walsh 2004). In addition, homocysteine also inhibits capillary endothelial nitric oxide synthetase (Faraci 2003) and glucose transporter, and transiently alters different cell adhesion molecules (Lee et al. 2004). Homocysteine has also been reported to directly induce cell death of cerebrocortical neurons through the activation of NMDA (Lipton et al. 1997). Chronic hyperhomocysteinemia also enhances lipid peroxidaton and decreases glutathione, suggesting the involvement of oxidative stress leading to cognitive impairment (Baydas et al. 2005).

Collective evidence suggests that high energy diet-mediated increase in the release of free fatty acids and triglycerides not only produces abnormalities in insulin signaling pathway and induces endothelial dysfunction due to increase in the production of ROS, but also lowers concentrations of HDLs contributing to dyslipidemia and proinflammatory state. Inflammation, the key pathogenic component of atherosclerosis, facilitates thrombosis, a process, which is closely associated with stroke. In brain stroke is accompanied by increase in free fatty acid release, elevation in ROS, increase in the expression of proinflammatory cytokines, and activation of microglial cells and astrocytes. Interplay among proinflammatory cytokines, insulin signaling, endothelial cell dysfunction, alterations in the expression of transcription factors, and platelet aggregation may increase the risk of stroke in MetS patients.

9.6 High Calorie Diet and Alzheimer Disease

Alzheimer's disease (AD), the most common cause of dementia, is a progressive neurodegenerative disorder affecting seniors. The prevalence of AD rises exponentially with age, increasing drastically after 65 years. Thus, approximately 0.6 % in persons ages 65–69 years, 1.0 % in ages 70–74, 2.0 % in ages 75–79, 3.3 % in ages 80–84, and 8.4 % in persons 85 years and older. Sporadic AD, affecting more than

15 million people worldwide, is more common than familial AD (Richard and Brayne 2014). Mild cognitive impairment (MCI) is regarded as a transition state between normal aging and dementia (Swomley et al. 2014). Importantly, almost one half of these individuals evolves to late onset AD (LOAD), accounting for about 60 % of the total cases of dementia in USA and Western countries (Swomley et al. 2014). AD is a multifactorial disease of unknown pathogenesis. Clinically, AD is characterized by a decline in many functions such as memory, speech, personality and judgment, vision, association sensory-motor function, and culminates in the death of the individual typically within 3–9 years after diagnosis (Castellani et al. 2010). The hallmarks of AD include extracellular amyloid beta (senile) plaques (Aβ), intracellular neurofibrillary tangles, chronic oxidative stress, neuroinflammation, mitochondrial dysfunction, brain atrophy, and vascular pathology. Vascular defects include cerebrovascular dysfunction, decreased cerebral blood flow, and blood–brain barrier (BBB) disruption.

Many investigators suggest that poor blood flow may also contribute to the pathogenesis of AD. Based on many studies, it is suggested that decrease in energy production and oxygen may be caused by insufficient blood flow potentially initiating the signaling pathways modulating the production of Aβ and the brain's ability to remove this toxic protein. It is also shown that β and γ-secretases are regulated in response to stress caused by energy deprivation at the translational level (O'Connor et al. (2008). An insufficient supply of glucose induces phosphorylation of the translation initiation factor eIF2alpha (eIF2α), which consequently increases β and γ-secretases, resulting in the overproduction of Aβ through amyloidogenic pathway (O'Connor et al. 2008) (Fig. 9.4). Furthermore, the lack of oxygen may also cause the impairment of brain clearance of Aβ through stimulation of serum response factor (SRF) and myocardin (MYOCD) expressions (Bell et al. 2009), which are much more active in the blood vessels of AD patients than in normal subjects (Chow et al. 2007). Overexpression of SRF and MYOCD in cerebrovascular smooth muscle cells (CVSMCs) negatively regulates the expression of low-density lipoprotein receptor-related protein-1 (LRP1), which is the major Aβ clearance receptor in the BBB. The stimulation of LRP1 along with the transactivation properties of the sterol regulatory element binding transcription factor 2 (SREBF2), ultimately leads to toxic Aβ accumulation. Based on these studies, it is suggested that the components of this signaling pathway may be involved in reduced blood flow, implicating as a potential target for AD treatment. Improving blood flow by exercise, healthy eating, or using dietary supplements may also be effective for preventing AD. Substantial evidence demonstrates an association between physical activity and improvement of cognitive decline in AD (Winchester et al. 2012; Maesako et al. 2012; Farooqui 2014). Decrease in cerebral blood flow may negatively affect the synthesis of proteins required for learning and memory and eventually promote neuritic injury and neuronal death leading to cognitive decline and loss of memory. The cognitive decline observed in AD has its roots at the synapse, the space between neurons, through which they communicate (Fig. 9.4). The synapse is also a site at which Aβ peptide, the characteristic amyloid protein associated with AD, is believed to first deposit (Terry et al. 1991; Farooqui 2010). Aβ peptides are generated by successive

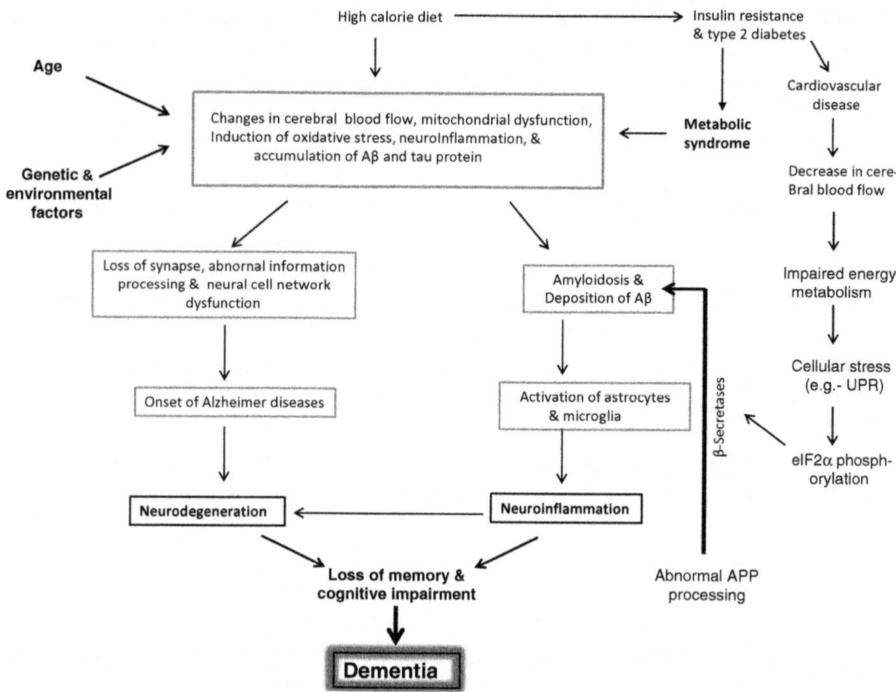

Fig. 9.4 Neurochemical effects of high calorie diet on the pathogenesis of Alzheimer disease

proteolysis of amyloid β precursor protein (APP), a large transmembrane glycoprotein, which is initially cleaved by the β-site APP-cleaving enzyme 1 and subsequently by γ-secretase in the transmembrane domain (Findeis 2007). Aβ peptide is a 40–43 amino acid peptide, which forms soluble oligomeric species (two to six peptides) and insoluble fibrils (β-pleated sheets). The most toxic peptide specie is soluble/oligomeric Aβ (Glabe and Kayed 2006; Haass and Selkoe 2007). Hyperphosphorylated tau is insoluble and aggregates in paired helical filamentous structures causing an impaired axonal transport (Castellani et al. 2010). The oligomeric tau is considered to be the most toxic form, which has been reported to be correlated with memory impairment and neuronal loss (Oddo et al. 2006; Yoshiyama et al. 2007). Activated microglia and astrocytes are usually found to be associated with the amyloid plaques. Activation of microglia leads to uptake and clearance of Aβ (Farooqui 2010). Impaired clearance of Aβ from the brain by the cells of the neurovascular unit may lead to the accumulation of Aβ on blood vessels and in brain parenchyma. The accumulation of Aβ on the cerebral blood vessels, known as cerebral amyloid angiopathy (CAA), is also associated with cognitive decline in AD.

Neurochemically, AD is characterized by the induction of neuroinflammation, oxidative stress due to enhancement in metabolism of neural membrane phospholipids, sphingoloipids, and cholesterol (Farooqui 2011). This enhancement is due to the activation of phospholipases A_2 (PLA$_2$), sphingomyelinases (SMase), and cholesterol

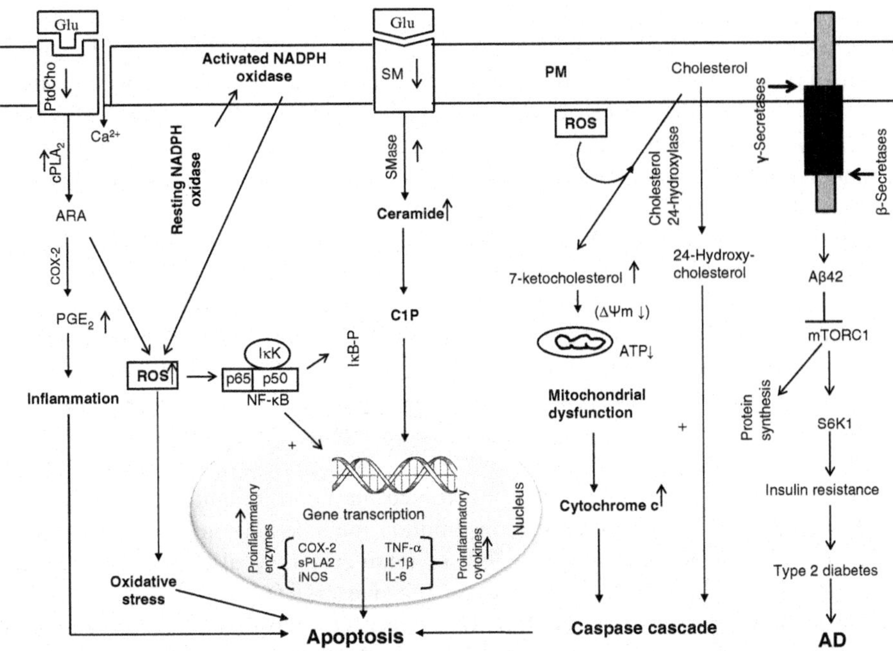

Fig. 9.5 Hypothetical diagram showing enrichment of phospholipid, sphingolipid , and cholesterol metabolism due to the consumption of high calorie diet and interactions among phospholipid-, sphingolipid-, and cholesterol-derived lipid mediators leading to apoptotic cell death. *PtdCho* Phosphatidylcholine; *SM* sphingomyelin, *cPLA2* cytosolic phospholipase A₂, *SMase* sphingomyelinase, *ARA* arachidonic acid; *ROS* reactive oxygen species, *COX-2* cyclooxygenase-2, *ROS* reactive oxygen species, *ceramide 1 P* ceramide-1-phosophate, *PM* plasma membrane. Proinflammatory genes include TNF-α; IL-1β; nitric oxide synthase; *COX-2* cyclooxygenase-2, *sPLA₂* secretory phospholipase A₂. Positive sign (+) indicates stimulation, *Upward arrow* indicate increase, and *downward arrow* indicates decrease in levels of precursors and lipid mediators

hydroxylases (CYP46), increase in levels of phospholipid, sphingolipid, and cholesterol-derived lipid mediators, energy metabolism/mitochondrial dysfunction, activation of caspases, stimulation of protein phosphorylation, loss of synapses, progressive impairment of memory, severe dementia possibly because of advancing age and decline in cellular protein quality control processes, vascular dysfunction, enhancement in oxidative stress and neuroinflammation, neurotrophic factor dysregulation, and disrupted leptin signaling (Fig. 9.5) (Barnham et al. 2004; Powers et al. 2009; Price et al. 2009; Farooqui 2010; Farooqui et al. 2010, 2012). Significant progressive atrophy has been reported to occur in seniors (Resnick et al. 2003), but atrophy occurs with much accelerated rate in subjects suffering from AD (Jack et al. 2004). An intermediate rate of atrophy is found in patients with mild cognitive impairment (MCI) (Ries et al. 2008). Many cross-sectional and prospective studies have indicated that increased levels of plasma total homocysteine (tHcy) are an important factor that contributes to the pathogenesis atrophy, dementia, notably AD (Seshadri 2006; Zylberstein et al. 2009). Raised tHcy is also associated with both regional and whole brain atrophy, not only in AD (Clarke et al. 1998) but also in healthy seniors (Seshadri et al. 2008; den Heijer et al. 2003). More recently a wide range of potentially modifiable risk factors for AD and dementia have been reported. These factors include cardiovascular risk factors (hypertension, diabetes, and obesity), psychosocial factors (depression) and health behaviors (low level of physical or mental activity, and smoking) (Daviglus et al. 2010). Impairment of memory, abstract thinking and judgment are the hallmarks of dementia.

Epidemiological studies have indicated that the consumption of high calorie diet is closely associated with the pathogenesis of AD (Kalmijn 2000; Grant et al. 2002; Bhat 2010). The consumption of high calorie diet significantly increases ROS generation and expression of gp91phox, p22phox, p47phox, and p67phox NADPH oxidase subunits in the cerebral cortex (Zhang et al. 2005). In addition, levels of PGE_2 levels and PLA_2 and COX-2 activities are also increased in brain animals fed with high calorie diet. Activation of these enzymes produces oxidative stress and neuroinflammation. ROS stimulates NF-κB, a transcription factor responsible for regulating the expression of genes involved in neuroinflammation (Fig. 9.4) (Farooqui 2010). High calorie diet consumption-mediated chronic neuroinflammation and age-related decrease in length of telomeres may also contribute to the pathogenesis of AD and thus shorten the lifespan of AD patients. Telomeres are DNA–protein complexes that protect the ends of chromosomes, and are essential for maintaining the chromosomal integrity during replication (Lin et al. 2012; Zanni and Wick 2011). The consumption of large portion of high calorie diet and impaired sleep are now emerging as powerful predictors of accelerated telomere shortening and reduced telomerase activity (Lin et al. 2012; Barcelo et al. 2010). It is proposed that telomere degradation can be accelerated by cytokine induced chronic inflammation (Bekaert et al. 2007; Kaszubowska 2008; Carrero et al. 2008), which is linked with high fat consumption, increase in inflammatory cytokines (IL-6, TNF-α, and C-reactive protein, CRP), decrease in cerebral volume, mood disturbance, sleep loss and cognitive dysfunction (Gorelick 2010). All these biomarkers are increased in patients with AD.

The consumption of high calorie diet is also an important risk factor for MetS. Adverse effects of high calorie diet on metabolic homeostasis are linked to adipose tissue physiology. It is reported that the imbalance between caloric intake and energy expenditure may cause hyperplasia and hypertrophy of adipocytes (Funaki 2009). Caloric intake and energy expenditure involves mammalian target of rapamycin complexl (mTORC1), a serine/threonine kinase member of the PtdIns 3K-related kinase family and the mammalian PtdIns. High calorie diet consumption up-regulates three major pathways involved in mTORC1 activation: (a) increased supply of glucose and fat; (b) stimulation of high calorie diet-induced insulin/IGF-signaling; and (c) abundance of leucine which is supplied by meat and dairy proteins. High levels of glucose and fat increase cellular energy (ATP) levels and thus suppress AMP kinase (AMPK) activity resulting in mTORC1 activation, which represents an essential intracellular target for the actions of hormones and nutrients on food intake and body weight regulation. By being at the crossroads of a nutrient-hormonal signaling network, hypothalamic mTORC1 controls important functions in peripheral organs, such as muscle oxidative metabolism, white adipose tissue differentiation and β-cell-dependent insulin secretion. Notably, dysregulation of the mTORC1 pathway by high protein and high fat diet has been implicated in the development of obesity and obesity-related conditions, such as type II diabetes (Melnik 2012). This factor is also associated with the basic molecular mechanisms of the aging brain, neuronal development and plasticity (Jaworski and Sheng 2006; Bishop et al. 2010).

The consumption of high calorie diet produces insulin resistance and reduces insulin transport across the BBB resulting in lowering insulin levels and activity in brain. These effects may be responsible not only for the reduction in CSF insulin and decrease in brain insulin receptor activity, but also in the progressive reduction of brain glucose metabolism in AD (Craft et al. 1998). Reduction in brain insulin signaling not only leads to increase in tau protein phosphorylation and increase in Aβ levels in a streptozotocin mouse model of type II diabetes (Jolivalt et al. 2008). Insulin also facilitates the release of intracellular Aβ in neuronal cultures and stimulates Aβ trafficking to the plasma membrane (Gasparini et al. 2001). Furthermore, intravenous injections of insulin increase plasma levels of Aβ42 in AD patients, but not in normal subjects, an effect that can be exaggerated in AD patients with higher body mass index (Kulstad et al. 2006). 4-HNE, a metabolite of arachidonic acid metabolism not only accumulates in visceral tissues in MetS and in brains from AD patients, but also stimulates the production of Aβ through the upregulation of β-secretases (Mattson 2009). Accumulated Aβ oligomer induces neuronal insulin resistance in the AD brain not only by inhibiting the insulin network by targeting the insulin/Akt pathway (Townsend et al. 2007), but also by removing insulin receptors after binding at the dendrites of synaptic sites (Zhao et al. 2008). Impaired insulin signaling cannot efficiently block GSK3β and therefore, the activated GSK3β not only triggers APP γ-secretase activity, but also increases tau phosphorylation (Hooper et al. 2008) simultaneously aggravating the two major pathological substrates of AD. Insulin resistance in type II diabetes and AD also contributes to the development of endothelial dysfunction, which is closely associated with IKKβ/NF-κB signaling (Shoelson and Goldfine 2009; Cai 2009; Madonna and De Caterina 2011) in the hypothalamus, a brain region, which modulates energy, body weight, and glucose imbalance (Kleinridders et al. 2009; Posey et al. 2009; Purkayastha et al. 2011; Ewing 2010).

Type II diabetes (an important component of MetS) increases the risk of AD. The molecular mechanisms associated with risk remains unclear. It is proposed that aging, insulin resistance, inflammatory cytokines, and oxidative stress may contribute to the risk for developing AD (Haan 2006; Bhat 2010; de la Monte 2009; de la Monte and Tong 2013). Insulin promotes the accumulation of Aβ by limiting Aβ degradation via direct competition for the IDE, a 110-kDa thiol zinc-metalloendopeptidase that cleaves small proteins of diverse sequence, many of which share a propensity to form β-pleated sheet-rich amyloid fibrils. Furthermore, as stated above, insulin itself promotes Aβ accumulation by accelerating amyloid precursor protein (APP)/Aβ trafficking from the trans-Golgi network, a major cellular site for Aβ generation, to the plasma membrane (Gasparini et al. 2001, 2002;

Farris et al. 2003). High calorie diet mediated chronic hyperglycemia produces AGEs via Maillard reactions (Brownlee 1995). AGEs not only produce complications of type II diabetes and promote the pathogenesis of AD by interacting with the receptors for AGE (RAGE). The activation of RAGE promotes oxidative stress and inflammatory responses, which play a crucial role in the development of cardiovascular and neurodegenerative diseases (Farooqui 2010).

Recent studies on the contribution of high fat diet in a mouse model of genetically induced AD-like neuropathology (3xTg-AD) have indicated that Defects in insulin production and signaling are closely associated with the pathogenesis of type II diabetes and AD two age-related pathologies (Vandal et al. 2014). In 3xTg-AD the cerebral expression of human AD transgenes leads to peripheral glucose intolerance, which is associated with pancreatic human Aβ accumulation. High-fat diet not only enhances glucose intolerance, but also promotes soluble Aβ-mediated memory impairment in 3xTg-AD mice. Importantly, a single insulin injection reverses the deleterious effects of high fat diet on memory and soluble Aβ levels, partly through changes in Aβ production and/or clearance. The molecular mechanisms through which insulin functions in the brain are the same as those operating in the periphery. However, certain insulin actions are different in the central nervous system, such as hormone-induced glucose uptake due to a low insulin-sensitive GLUT-4 activity, and because of the predominant presence of GLUT-1 and GLUT-3. In addition, insulin in the brain contributes to the control of nutrient homeostasis, cognitive function, and memory, as well as to neurotrophic, neuromodulatory, and neuroprotective effects. Alterations of these functional activities may contribute to the manifestation of several clinical entities, such as central insulin resistance, type II diabetes mellitus, and AD. Collective evidence suggests that several potential mechanisms link type II diabetes and MetS with AD: (a) hyperglycemia-mediated increase in oxidative stress with accumulation of advanced glycation end-products results in progressive functional and structural abnormalities in the brain (Farooqui 2013; de la Monte 2009); (b) Type II diabetes is associated with insulin resistance and hyperinsulinaemia, which interferes with β-amyloid peptides metabolism through the reduction of insulin-degrading enzyme activity (Craft 2007). In addition, mitochondrial dysfunction, ER stress, abnormal protein processing, abnormalities in insulin signaling, dysregulated glucose metabolism, hypercholesterolemia, and the activation of inflammatory pathways are features common to type II diabetes and AD (Sims-Robinson et al. 2010; Farooqui 2013). Another common factor for type II diabetes and AD is the desensitization of insulin receptors in the brain. Insulin acts as a growth factor in the brain and is neuroprotective leading to the activation of dendritic sprouting, regeneration and stem cell proliferation (Duarte et al. 2013). The impairment of insulin signaling may promote the development of AD. Based on these studies, it is suggested that high calorie diet-mediated neurochemical changes are associated with a higher risk for developing type II diabetes, metabolic syndrome and AD as well as other neurodegenerative diseases (Seneff et al. 2011; Farooqui 2013; Haan 2006; Farooqui et al. 2012; Craft 2007; Talbot et al. 2012). This suggestion is supported by the development of drugs that re-sensitize insulin receptors and treat type II diabetes. Same drugs can be utilized to neurodegenerative processes in the brain. In particular, the incretins GLP-1 (glucagon-like peptide-1) and GIP (glucose-dependent insolinotropic polypeptide) are hormones that re-sensitize insulin signaling. Incretins also have similar growth-factor-like properties as insulin and are neuroprotective. In mouse models of AD, GLP-1 receptor agonists reduce amyloid plaque formation, reduce the inflammation response in the brain, protect neurons from oxidative stress, induce neurite outgrowth, and protect synaptic plasticity and memory formation from the detrimental effects caused by β-amyloid production and inflammation (Duarte et al. 2013).

According to published research the consumption of fish, the primary dietary source of n-3 fatty acids, is associated with a reduced risk of cognitive decline or dementia (Kalmijn 2000; Morris et al. 2003; Fotuhi et al. 2009). Some studies have found that the consumption of DHA, but not other n-3 fatty acids, is associated with a reduced risk of type II diabetes, metabolic syndrome, and AD (Morris et al. 2003). DHA, the principle omega-3 fatty acid in the brain and heart, plays an important role in neural and cardiac function. DHA is the most abundant long-chain polyunsaturated fatty acid in the brain. It is enriched in the synaptic plasma membrane fractions, and it is reduced in the brains of patients with AD (Söderberg et al. 1991; Prasad et al. 1998). Other major n-3 fatty acid found in fish, eicosapentaenoic acid, is virtually absent from the brain. Decrease in plasma DHA contributes to cognitive decline in healthy elderly and AD patients. Higher DHA intake and plasma levels are inversely correlated with increased relative risk of AD and coronary heart disease. It is well known that DHA provides cardiovascular benefits (lower triglycerides, increased HDL cholesterol, reduction in resting heart rate) in older adults. Preclinically, DHA supplementation restores brain DHA levels, enhances learning and memory tasks in aged animals, and significantly reduces beta amyloid, plaques, and tau in transgenic AD models.

9.7 High Calorie Diet and Dementia

As stated in Chap. 2, dementia is a major cause of disability, which is clinically-defined not only by memory deficits, and disturbances of other higher cortical functions, but also by deterioration in emotional control and social behavior (Sonnen et al. 2009). Two major types of dementia have been identified. Generalized atrophy in the cortical area of the brain results in dementia associated with AD and that due to vascular dementia mainly due to stroke (Farooqui 2010). It is becoming increasingly evident that long term consumption of high calorie diet results in deficits in hippocampal-dependent memory formation. In a series of elegant studies, Kanoski, Davidson, and colleagues demonstrated that high calorie diet, which is enriched in saturated fatty acids, n-6 fatty acids, and simple carbohydrates, produces deficits in hippocampal-dependent memory processes before the onset of obesity (Kanoski and Davidson 2011; Kanoski 2012). The molecular mechanisms associated with high calorie-mediated alterations in memory formations are not fully understood. However, it is suggested that high calorie diet not only reduces levels of BDNF in the hippocampus (Kanoski et al. 2007; Stranahan et al. 2008) and in the prefrontal cortex (Kanoski et al. 2007), but also alters dendritic morphology (Granholm et al. 2008), impairs synaptic plasticity (Stranahan et al. 2008), causes alterations in blood vessel structure (Freeman et al. 2011), and increases neuroinflammation (White et al. 2009; Pistell et al. 2010) in the hippocampus following long term consumption of high calorie diet. Furthermore, high calorie diet produces changes in glutamatergic signaling in hippocampal neurons, specifically via upregulation of synaptic clearance mechanisms and alterations in glutamate metabolism leading to NMDA receptor desensitization (Freeman et al. 2011; Kanoski and Davidson 2011). These neurochemicals changes contribute to the pathogenesis of dementia specifically in elder subjects, who have less blood flow due to development of atherosclerosis in their blood vessels.

9.8   High Calorie Diet and Depression

Depression is a multifactorial disorder characterized by poor mood, anorexia, changes in weight (decrease or increase), fatigue, lethargy, sleep disturbances (insomnia or hypersomnia), psychomotor retardation or agitation, feelings of worthlessness or guilt, diminished cognitive functioning, and recurrent thoughts of death (American Psychiatric Association). Depression is linked with multiple biological abnormalities, including vascular pathological changes, autonomic functional changes, psychomotor changes, hypercoagulability, abnormality in the sympatho-adrenal system, and hypothalamic-pituitary-adrenal axis hyperactivity leading to low self esteem (Davidson et al. 2002; Evans et al. 2005). These abnormalities may produce profound effects on memory and behavior in persons both with and without cognitive impairment, which can not only lead to increased risk for incident MCI and dementia in older adults, but can also be linked with accelerated cognitive decline (Kim and Diamond 2002; Wilson et al. 2007). In addition, multiple biological abnormalities may also lead to deleterious neuroendocrine and associated inflammatory changes including suppression of IGF-1 and other neuroprotective factors, impaired synaptic plasticity, suppressed neurogenesis, reduced neuronal survival, and other adverse morphological and functional changes in the hippocampus, prefrontal cortex and other brain structures. All these changes can profoundly affect mood, sleep, learning, and memory (Lucassen et al. 2010). Not all individuals show all neurochemical changes and symptoms of depression (Anisman and Matheson 2005). Depression common occurs in AD, with prevalence estimates ranging from 25 to 74.9 % in a group of recent studies (Benoit et al. 2012). Neuroimaging studies have indicated that above mentioned abnormalities in depressed subjects show an increase in ventricle/brain ratio, localized atrophy of the prefrontal cortex, cingulated gyrus, ventral striatum, amygdale, cerebellum, and hippocampus (aan het Rot et al. 2009) along with increase in IFN-γ, IFN-α, and TNF-α, and ROS-mediated activation of the enzyme, indolamine 2, 3 dioxygenase (IDO) in microglia, which metabolizes tryptophan via the kyneurenine pathway

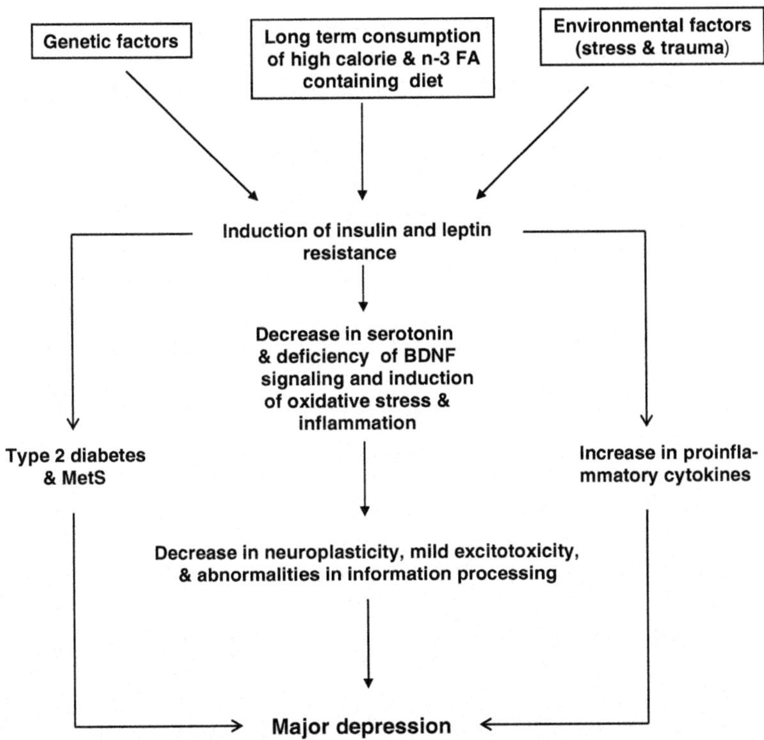

Fig. 9.6 Hypothetical diagram showing the induction of major depression by the consumption of high calorie diet. *MetS* Metabolic syndrome, *BDNF* brain-derived neurotrophic factor; and proinflammatory cytokines (TNF-α, IL-1β, IL-6)

(Dantzer et al. 2008; Maes et al. 2012). Neurochemically, depression is not only characterized by the reduction in plasma levels of BDNF, a growth factor, which regulates synaptic plasticity in neuronal networks (Sen et al. 2008), elevated blood levels of proinflammatory cytokines (IL-6 and TNF-α) (Dowlati et al. 2010), low levels of magnesium, and omega-3 fatty acids (Farooqui 2013). Furthermore, depression is also accompanied by changes in neurotransmitters, neuropeptides (vasopressin), and cytokine levels (Fig. 9.6). In AD, depression is associated with specific neurochemical changes contributing to alterations in Geriatric Depression scale (GDS) scores, but not in agitation scores. These changes are correlated with choline/creatine ratio in left dorsolateral prefrontal cortex (Tsai et al. 2013). Cortical atrophy associated with depression is observed in wide regions of the prefrontal cortex and temporal cortex (Lebedev et al. 2014) and decreased gray matter volume in the left inferior temporal gyrus (Son et al. 2013). Depressed AD patients also exhibited greater white matter atrophy in frontal, temporal, and parietal lobes than AD patients without depressive symptoms (Lee et al. 2012). Collective evidence suggests that depression is not only associated with chronic low-grade inflammatory response, activation of cell-mediated immunity, and activation of the compensatory anti-inflammatory reflex system, but also with increased oxidative and nitrosative stress and autoimmune responses directed against oxidative and nitrosative stress modified neoepitopes (Maes et al. 2012; Moylan et al. 2012). In addition, genetic and environmental factors can also predispose individuals to major depressive disorder (Sullivan et al. 2000; Wurtman 2005; Levinson 2006).

Besides above mentioned factors, dietary factors also play an important role in the susceptibility and pathogenesis of depression. Long term consumption of high calorie diet (energy-dense food) increases the risk of developing obesity and depression, especially among women. High calorie diet induces anabolic/catabolic imbalance due to changes in cortisol, glucose, insulin, and decrease in BDNF levels leading to oxidative stress and systemic chronic inflammation, which is coupled with induction of insulin resistance, generation of ROS, formation of AGEs, and deposition of abdominal fat tissue (Yamagishi et al. 2008; Rustad et al. 2011; Wang et al. 2009; Farooqui et al. 2012). It is also reported that humans with poorly controlled diabetes show hyperglycemia accompanied by an accelerated rate of ROS, AGEs formation and accumulation, impairment in hippocampal neurogenesis, and induction of depressive behavior (Yamagishi et al. 2008; Wang et al. 2009). Thus, it is becoming increasingly evident that high calorie diet-mediated metabolic abnormalities (obesity, dyslipidemia, glucose intolerance, hypertension, and chronic inflammation), contribute to the pathogenesis of MetS, a pathological condition which is an important risk factor for the pathogenesis of depression (Farooqui et al. 2012; Farooqui 2013).

Major depression, type II diabetes, and MetS share overlapping abnormalities in metabolic networks defined as alterations in glucocorticoid signaling, changes in glucose-insulin homeostasis, induction of oxidative stress and inflammatory processes, and development of insulin and leptin resistance (Mommersteeg et al. 2013; Farooqui 2013, 2014). Meta-analysis has indicated that patients with depression have an elevated risk of developing type two diabetes (Knol et al. 2006), and conversely that patients with diabetes have significantly increased risk of developing depression (Rotella and Mannucci 2013). This bi-directional relationship is suggestive of convergent pathological processes rather than a simplistic cause and effect relationship. Furthermore, it has also been hypothesized that the doubled rates of depression in female diabetic patients may help in explaining the high prevalence of coronary heart disease in women with diabetes (Clouse et al. 2003). However, many of the above finding are contradictory (Holt et al. 2014; Champaneri et al. 2010) and more double blind and randomize studies are needed on the relationship between type II diabetes and depression in larger human populations.

9.9 Conclusion

Long-term consumption of high calorie diet, which is enriched in saturated fats, cholesterol, and n-6 fatty acids produces obesity, insulin resistance, oxidative stress, low grade inflammation, and cognitive dysfunction due to abnormalities in mitochondrial function and marked alterations in signal transduction processes. Onset of chronic inflammation and oxidative stress promotes type II diabetes, and metabolic syndrome, which are risk factors for stroke, AD, and depression. The long term consumption of high calorie diet along with lack of exercise, suppresses adaptive cellular response signaling by inhibiting expression of neurotrophic factors, protein chaperons, DNA-repair proteins, autophagy, and mitochondrial biogenesis In addition, high calorie diet also alters hippocampal morphology/plasticity leading to the impairment of cognitive function in rodents. This brain region is involved in learning and memory formation. Accumulating evidence suggests that long term consumption of high calorie diet not only causes oxidative stress through multiple biochemical mechanisms, but also promotes low grade chronic inflammation through increased expression of proinflammatory cytokines.

References

aan het Rot M, Mathew SJ, Charney DS (2009) Neurobiological mechanisms in major depressive disorder. CMAJ 180:305–313

Abarquez RF Jr (2003) Microvascular disease relevance in the hypertension syndrome. Clin Hemorheol Microcirc 29:295–300

Anisman H, Matheson K (2005) Stress, depression, and anhedonia: caveats concerning animal models. Neurosci Biobehav Rev 29:525–546

Arnold SE, Trojanowski JQ (1996) Recent advances in defining the neuropathology of schizophrenia. Acta Neuropathol (Berl) 92:217–231

Arundine M, Tymianski M (2004) Molecular mechanisms of glutamate-dependent neurodegeneration in ischemia and traumatic brain injury. Cell Mol Life Sci 61:657–668

Barcelo A, Pierola J, Lopez-Escribano H, de la Pena M, Soriano JB, Alonso-Fernandez A, Ladaria A, Agustí A (2010) Telomere shortening in sleep apnea syndrome. Respir Med 104:1225

Barnham KJ, Masters CL, Bush AI (2004) Neurodegenerative diseases and oxidatives stress. Nat Rev Drug Discov 3:205–214

Baydas G, Ozer M, Yasar A, Tuzcu M, Koz ST (2005) Melatonin improves learning and memory performances impaired by hyperhomocysteinemia in rats. Brain Res 1046:187–194

Becker T, Becker G, Seufert J, Hofmann E, Lange KW, Naumann M (1997) Parkinson's disease and depression: evidence for an alteration of the basal limbic system detected by transcranial sonography. J Neurol Neurosurg Psychiatry 63:590–596

Bekaert S, De Meyer T, Rietzschel ER, De Buyzere ML, De Bacquer D, Langlois M, Segers P, Cooman L, Van Damme P, Cassiman P, Criekinge V (2007) Telomere length and cardiovascular risk factors in a middle-aged population free of overt cardiovascular disease. Aging Cell 6:639–647

Bell RD, Deane R, Chow N, Long X, Sagare A, Singh I, Streb JW, Guo H, Rubio A, Van Nostrand W, Miano JM, Zlokovic BV (2009) SRF and myocardin regulate LRP-mediated amyloid-β clearance in brain vascular cells. Nat Cell Biol 11:143–153

Bengmark S (2013) Processed foods, dysbiosis, systemic inflammation, and poor health. Curr Nutr Food Sci 9:113–143

BenMoyal-Segal L, Soreq H (2006) Gene-environment interactions in sporadic Parkinson's disease. J Neurochem 97:1740–1755

Benoit M, Berrut G, Doussaint J, Bakchine S, Bonin-Guillaume S, Frémont P, Gallarda T, Krolak-Salmon P, Marquet T, Mékiès C, Sellal F, Schuck S, David R, Robert P (2012) Apathy and depression in mild Alzheimer's disease: a cross-sectional study using diagnostic criteria. J Alzheimers Dis 31:325–334

Bhat NR (2010) Linking cardiometabolic disorders to sporadic Alzheimer's disease: a perspective on potential mechanisms and mediators. J Neurochem 115:551–562

Bishop NA, Lu T, Yankner BA (2010) Neural mechanisms of ageing and cognitive decline. Nature 464:529–535

Blitzer RD, Iyengar R, Landau EM (2005) Postsynaptic signaling networks: cellular cogwheels underlying long-term plasticity. Biol Psychiatry 57:113–119

Bousser MG (2012) Stroke prevention: an update. Front Med 6:22–34

Bramham CR, Messaoudi E (2005) BDNF function in adult synaptic plasticity: the synaptic consolidation hypothesis. Prog Neurobiol 76:99–12510

Brown TP, Rumsby PC, Capleton AC, Rushton L, Levy LS (2006) Pesticides and Parkinson's disease—is there a link? Environ Health Perspect 114:156–164

Brownlee M (1995) Advanced protein glycosylation in diabetes and aging. Annu Rev Med 46:223–234

Cai D (2009) NFkappaB-mediated metabolic inflammation in peripheral tissues versus central nervous system. Cell Cycle 8:2542–2548

Carrero J, Stenvinkel P, Fellstrom B, Qureshi AR, Lamb K, Heimburger O, Bárány P, Radhakrishnan K, Lindholm B, Soveri I, Nordfors L, Shiels PG (2008) Telomere attrition is associated with inflammation, low fetuin-A levels and high mortality in prevalent haemodialysis patients. J Intern Med 263:302–312

Castellani RJ, Rolston RK, Smith MA (2010) Alzheimer disease. Dis Mon 56:484–546

Ceriello A, Motz E (2004) Is oxidative stress the pathogenic mechanism underlying insulin resistance, diabetes, and cardiovascular disease? The common soil hypothesis revisited. Arterioscler Thromb Vasc Biol 24:816–823

Champaneri S, Wand GS, Malhotra SS, Casagrande SS, Golden SH (2010) Biological basis of depression in adults with diabetes. Curr Diab Rep 10:396–405

Chen H, O'Reilly E, McCullough ML, Rodriguez C, Schwarzschild MA, Calle EE, Thun MJ, Ascherio A (2007) Consumption of dairy products and risk of Parkinson's disease. Am J Epidemiol 165:998–1006

Choi HK, Liu S, Curhan G (2005) Intake of purine-rich foods, protein, and dairy products and relationship to serum levels of uric acid: the Third National Health and Nutrition Examination Survey. Arthritis Rheum 52:283–289

Chow N, Bell RD, Deane R, Streb JW, Chen J, Brooks A, Van Nostrand W, Miano JM, Zlokovic BV (2007) Serum response factor and myocardin mediate arterial hypercontractility and cerebral blood flow dysregulation in Alzheimer's phenotype. Proc Natl Acad Sci U S A 104: 823–828

Clarke R, Smith AD, Jobst KA, Refsum H, Sutton L, Ueland PM (1998) Folate, vitamin B12, and serum total homocysteine levels in confirmed Alzheimer disease. Arch Neurol 55:1449–1455

Clouse RE, Lustman PJ, Freedland KE, Griffith LS, McGill JB, Carney RM (2003) Depression and coronary heart disease in women with diabetes. Psychosom Med 65:376–383

Craft S (2007) Insulin resistance and Alzheimer's disease pathogenesis: potential mechanisms and implications for treatment. Curr Alzheimer Res 4:147–152

Craft S, Peskind E, Schwartz MW, Schellenberg GD, Raskind M, Porte D Jr (1998) Cerebrospinal fluid and plasma insulin levels in Alzheimer's disease: relationship to severity of dementia and apolipoprotein E genotype. Neurology 50:164–168

Dantzer R, O'Connor JC, Freund GG, Johnson RW, Kelley KW (2008) From inflammation to sickness and depression: when the immune system subjugates the brain. Nat Rev Neurosci 9:46–56

Davidson RJ, Pizzagalli D, Nitschike JB, Putnam K (2002) Depression: perspectives from affective neuroscience. Annu Rev Psychol 53:545–574

Davidson TL, Monnot A, Neal AU, Martin AA, Horton JJ, Zheng W (2012) The effects of a high-energy diet on hippocampal-dependent discrimination performance and blood-brain barrier integrity differ for diet-induced obese and diet-resistant rats. Physiol Behav 107:26–33

Daviglus ML, Bell CC, Berrettini W, Bowen PE, Connolly ES Jr, Cox NJ, Dunbar-Jacob JM, Granieri EC, Hunt G, McGarry K, Patel D, Potosky AL, Sanders-Bush E, Silberberg D,

Trevisan M (2010) National Institutes of Health State-of-the-Science Conference statement: preventing Alzheimer disease and cognitive decline. Ann Intern Med 153:176–181

de la Monte SM (2009) Insulin resistance and Alzheimer's disease. BMB Rep 42:475–481

de la Monte SM, Tong M (2013) Insulin resistance and metabolic failure underlie Alzheimer disease. In: Farooqui T, Farooqui AA (eds) Metabolic syndrome and neurological disorders. Wiley, Oxford, pp 1–30

den Heijer T, Vermeer SE, Clarke R, Oudkerk M, Koudstaal PJ, Hofman A, Breteler MM (2003) Homocysteine and brain atrophy on MRI of non-demented elderly. Brain 126:170–175

Dhungana H, Rolova T, Savchenko E, Wojciechowski S, Savolainen K, Ruotsalainen AK, Sullivan PM, Koistinaho J, Malm T (2013) Western-type diet modulates inflammatory responses and impairs functional outcome following permanent middle cerebral artery occlusion in aged mice expressing the human apolipoprotein E4 allele. J Neuroinflammation 10:102

Dowlati Y, Herrmann N, Swardfager W, Thomson S, Oh PI, Van Uum S, Koren G, Lanctôt KL (2010) Relationship between hair cortisol concentrations and depressive symptoms in patients with coronary artery disease. Neuropsychiatr Dis Treat 6:393–400

Duarte AI, Candeias E, Correia SC, Santos RX, Carvalho C, Cardoso S, Plácido A, Santos MS, Oliveira CR, Moreira PI (2013) Crosstalk between diabetes and brain: glucagon-like peptide-1 mimetics as a promising therapy against neurodegeneration. Biochim Biophys Acta 1832:527–541

Esmaillzadeh A, Kimiagar M, Mehrabi Y, Azadbakht L, Hu FB, Willett WC (2007) Dietary patterns, insulin resistance, and prevalence of the metabolic syndrome in women. Am J Clin Nutr 85:910–918

Evans DL, Charney DS, Lewis L, Golden RN, Gorman JM, Krishnan KR, Nemeroff CB, Bremner JD, Carney RM, Coyne JC, Delong MR, Frasure-Smith N, Glassman AH, Gold PW, Grant I, Gwyther L, Ironson G, Johnson RL, Kanner AM, Katon WJ, Kaufmann PG, Keefe FJ, Ketter T, Laughren TP, Leserman J, Lyketsos CG, McDonald WM, McEwen BS, Miller AH, Musselman D, O'Connor C, Petitto JM, Pollock BG, Robinson RG, Roose SP, Rowland J, Sheline Y, Sheps DS, Simon G, Spiegel D, Stunkard A, Sunderland T, Tibbits P Jr, Valvo WJ (2005) Mood disorders in the medically ill: scientific review and recommendations. Biol Psychiatry 58:175–189

Ewing GW (2010) Mathematical modeling the neuroregulation of blood pressure using a cognitive top-down approach. N Am J Med Sci 2:341–352

Faraci FM (2003) Hyperhomocysteinemia, a million ways to lose control. Arterioscler Thromb Vasc Biol 23:371–373

Farooqui AA (2010) Neurochemical aspects of neurotraumatic and neurodegenerative diseases. Springer, New York

Farooqui AA (2011) Lipid mediators and their metabolism in the brain. Springer, New York

Farooqui AA (2013) Metabolic syndrome: an important risk factor for stroke, Alzheimer, and depression. Springer, New York

Farooqui AA (2014) Inflammation and oxidative stress in neurological disorders. Springer, New York

Farooqui AA, Horrocks LA (1994) Excitotoxicity and neurological disorders: involvement of membrane phospholipids. Int Rev Neurobiol 36:267–323

Farooqui AA, Horrocks LA (2007) Glycerophospholipids in the brain: phospholipases A_2 in neurological disorders. Springer, New York

Farooqui AA, Ong WY, Horrocks LA (2008) Neurochemical aspects of excitotoxicity. Springer, New York

Farooqui AA, Ong WY, Farooqui T (2010) Lipid mediators in the nucleus: their potential contribution to Alzheimer's disease. Biochim Biophys Acta 1801:906–916

Farooqui AA, Farooqui T, Panza F, Frisardi V (2012) Metabolic syndrome as a risk factor for neurological disorders. Cell Mol Life Sci 69:741–762

Farris W, Mansourian S, Chang Y, Lindsley L, Eckman EA, Frosch MP, Eckman CB, Tanzi RE, Selkoe DJ, Guenette S (2003) Insulin-degrading enzyme regulates the levels of insulin, amyloid beta-protein, and the beta-amyloid precursor protein intracellular domain in vivo. Proc Natl Acad Sci U S A 100:4162–4167

Findeis MA (2007) The role of amyloid beta peptide 42 in Alzheimer's disease. Pharmacol Ther 116:266–286

Fotuhi M, Mohassel P, Yaffe K (2009) Fish consumption, long-chain omega-3 fatty acids and risk of cognitive decline or Alzheimer disease: a complex association. Nat Clin Pract Neurol 5:140–152

Freeman LR, Granholm AC (2012) Vascular changes in rat hippocampus following a high saturated fat and cholesterol diet. J Cereb Blood Flow Metab 32:643–653

Freeman LR, Haley-Zitlin V, Stevens C, Granholm AC (2011) Diet-induced effects on neuronal and glial elements in the middle-aged rat hippocampus. Nutr Neurosci 14:32–44

Funaki M (2009) Saturated fatty acids and insulin resistance. J Med Invest 56:88–92

Gallagher S (2004) Neurocognitive models of schizophrenia: a neurophenomenological critique. Psychopathology 37:8–19

Gasparini L, Gouras GK, Wang R, Gross RS, Beal MF, Greengard P, Xu H (2001) Stimulation of beta-amyloid precursor protein trafficking by insulin reduces intraneuronal beta-amyloid and requires mitogen-activated protein kinase signaling. J Neurosci 21:2561–2570

Gasparini L, Netzer WJ, Greengard P, Xu H (2002) Does insulin dysfunction play a role in Alzheimer's disease? Trends Pharmacol Sci 23:288–293

Glabe CG, Kayed R (2006) Common structure and toxic function of amyloid oligomers implies a common mechanism of pathogenesis. Neurology 66:S74–S78

Gorelick PB (2010) Role of inflammation in cognitive impairment: results of observational epidemiological studies and clinical trials. Innate Inflam Stroke 1207:155–162

Graeber MB, Moran LB (2002) Mechanisms of cell death in neurodegenerative diseases: fashion, fiction, and facts. Brain Pathol 12:385–390

Granholm AC, Bimonte-Nelson HA, Moore AB, Nelson ME, Freeman LR, Sambamurti K (2008) Effects of a saturated fat and high cholesterol diet on memory and hippocampal morphology in the middle-aged rat. J Alzheimers Dis 14:133–145

Grant WB, Campbell A, Itzhaki RF, Savory J (2002) The significance of environmental factors in the etiology of Alzheimer's disease. J Alzheimers Dis 4:179–189

Haan MN (2006) Therapy Insight: type 2 diabetes mellitus and the risk of late-onset alzheimer's disease. Nat Clin Pract Neurol 2:159–166

Haass C, Selkoe DJ (2007) Soluble protein oligomers in neurodegeneration: lessons from the Alzheimer's amyloid beta-peptide. Nat Rev Mol Cell Biol 8:101–112

Harris WS (2008) The omega-3 index as a risk factor for coronary heart disease. Am J Clin Nutr 87:1997S–2002S

Holt RI, de Groot M, Golden SH (2014) Diabetes and depression. Curr Diab Rep 14:491

Hooper C, Killick R, Lovestone S (2008) The GSK3 hypothesis of Alzheimer's disease. J Neurochem 104:1433–1439

Hsu TM, Kanoski SE (2014) Blood-brain barrier disruption: mechanistic links between Western diet consumption and dementia. Front Aging Neurosci 6:88

Ince PG, Codd GA (2005) Return of the cycad hypothesis – does the amyotrophic lateral sclerosis/parkinsonism dementia complex (ALS/PDC) of Guam have new implications for global health? Neuropathol Appl Neurobiol 31:345–353

Ito H, Kawashima R, Awata S, Ono S, Sato K, Goto R, Koyama M, Sato M, Fukuda H (1996) Hypoperfusion in the limbic system and prefrontal cortex in depression: SPECT with anatomic standardization technique. J Nucl Med 37:410–414

Jack CR Jr, Shiung MM, Gunter JL, O'Brien PC, Weigand SD, Knopman DS, Boeve BF, Ivnik RJ, Smith GE, Cha RH, Tangalos EG, Petersen RC (2004) Comparison of different MRI brain atrophy rate measures with clinical disease progression in AD. Neurology 62:591–600

Jaworski J, Sheng M (2006) The growing role of mTOR in neuronal development and plasticity. Mol Neurobiol 34:205–219

Jellinger KA (2009) Recent advances in our understanding of neurodegeneration. J Neural Transm 116:1111–1162

Jolivalt CG, Lee CA, Beiswenger KK, Smith JL, Orlov M, Torrance MA, Masliah E (2008) Defective insulin signaling pathway and increased glycogen synthase kinase-3 activity in the brain of diabetic mice: parallels with Alzheimer's disease and correction by insulin. J Neurosci Res 86:3265–3274

Kalmijn S (2000) Fatty acid intake and the risk of dementia and cognitive decline: a review of clinical and epidemiological studies. J Nutr Health Aging 4:202–207

Kamel F, Hoppin JA (2004) Association of pesticide exposure with neurologic dysfunction and disease. Environ Health Perspect 112:950–958

Kanoski SE (2012) Cognitive and neuronal systems underlying obesity. Physiol Behav 106:337–344

Kanoski SE, Davidson TL (2011) Western diet consumption and cognitive impairment: links to hippocampal dysfunction and obesity. Physiol Behav 103:59–68

Kanoski SE, Meisel RL, Mullins AJ, Davidson TL (2007) The effects of energy-rich diets on discrimination reversal learning and on BDNF in the hippocampus and prefrontal cortex of the rat. Behav Brain Res 182:57–66

Kaszubowska L (2008) Telomere shortening and ageing of the immune system. J Physiol Pharmacol 59:169–186

Keifer MC, Firestone J (2007) Neurotoxicity of pesticides. J Agromedicine 12:17–25

Kim JJ, Diamond DM (2002) The stressed hippocampus, synaptic plasticity and lost memories. Nat Rev Neurosci 3(6):453–462

Kimbrell TA, Ketter TA, George MS, Little JT, Benson BE, Willis MW, Herscovitch P, Post RM (2002) Regional cerebral glucose utilization in patients with a range of severities of unipolar depression. Biol Psychiatry 51:237–252

Kleinridders A, Schenten D, Könner AC, Belgardt BF, Mauer J, Okamura T, Wunderlich FT, Medzhitov R, Brüning JC (2009) MyD88 signaling in the CNS is required for development of fatty acid-induced leptin resistance and diet-induced obesity. Cell Metab 10:249–259

Klussmann S, Martin-Villalba A (2005) Molecular targets in spinal cord injury. J Mol Med 83:657–671

Knol MJ, Twisk JW, Beekman AT, Heine RJ, Snoek FJ, Pouwer F (2006) Depression as a risk factor for the onset of type 2 diabetes mellitus. A meta-analysis. Diabetologia 49:837–845

Koponen E, Lakso M, Castrén E (2004) Overexpression of the full-length neurotrophin receptor trkB regulates the expression of plasticity-related genes in mouse brain. Brain Res Mol Brain Res 130:81–94

Kulstad JJ, Green PS, Cook DG, Watson GS, Reger MA, Baker LD, Plymate SR, Asthana S, Rhoads K, Mehta PD, Craft S (2006) Differential modulation of plasma beta-amyloid by insulin in patients with Alzheimer disease. Neurology 66:1506–1510

Kyrozis A, Ghika A, Stathopoulos P, Vassilopoulos D, Trichopoulos D, Trichopoulou A (2013) Dietary and lifestyle variables in relation to incidence of Parkinson's disease in Greece. Eur J Epidemiol 28:67–77

Lebedev AV, Beyer MK, Fritze F, Westman E, Ballard C, Aarsland D (2014) Cortical changes associated with depression and antidepressant use in Alzheimer and Lewy body dementia: an MRI surface-based morphometric study. Am J Geriatric Psychiatry 22:4–13

Lee J, Sparrow D, Vokonas PS, Landsberg L, Weiss ST (1995) Uric acid and coronary heart disease risk: evidence for a role of uric acid in the obesity-insulin resistance syndrome. The Normative Aging Study. Am J Epidemiol 142:288–294

Lee H, Kim H-J, Kim J-M, Chang N (2004) Effects of dietary folic acid supplementation on cerebrovascular endothelial dysfunction in rats with induced hyperhomocysteinemia. Brain Res 996:139–147

Lee GJ, Lu PH, Hua X, Lee S, Wu S, Nguyen K, Teng E, Leow AD, Jack CR Jr, Toga AW, Weiner MW, Bartzokis G, Thompson PM, Alzheimer's Disease Neuroimaging Initiative (2012) Depressive symptoms in mild cognitive impairment predict greater atrophy in alzheimer's disease-related regions. Biol Psychiatry 71:814–821

Levinson DF (2006) The genetics of depression: a review. Biol Psychiatry 60:84–92

Li R, Zhang H, Wang W, Wang X, Huang Y, Huang C (2010) Vascular insulin resistance in prehypertensive rats: role of PI3-kinase/Akt/eNOS signaling. Eur J Pharmacol 628:140–147

Lin J, Epel E, Blackburn E (2012) Telomeres and lifestyle factors: roles in cellular aging. Mutat Res 730:85–910

Lipton SA, Kim W-K, Choi Y-B, Kumar S, D'Emilia DM, Rayudu PV, Arnelle DR, Stamler JS (1997) Neurotoxicity associated with dual actions of homocysteine at the N-methyl-D-aspartate receptor. Proc Natl Acad Sci U S A 94:5923–5928

Lucassen PJ, Meerlo P, Naylor AS, van Dam AM, Dayer AG, Fuchs E, Oomen CA, Czéh B (2010) Regulation of adult neurogenesis by stress, sleep disruption, exercise and inflammation: implications for depression and antidepressant action. Eur Neuropsychopharmacol 20:1–1710

Lusis AJ, Attie AD, Reue K (2008) Metabolic syndrome: from epidemiology to systems biology. Nat Rev Genet 9:819–830

Madonna R, De Caterina R (2011) Cellular and molecular mechanisms of vascular injury in diabetes–part I: pathways of vascular disease in diabetes. Vascul Pharmacol 54:68–74

Maes M, Berk M, Goehler L, Song C, Anderson G, Galecki P, Lepnard B (2012) Depression and sickness behavior are Janus-faced responses to shared inflammatory pathways. BMC Med 10:66

Maesako M, Uemura K, Kubota M, Kuzuya A, Sasaki K, Hayashida N, Asada-Utsugi M, Watanabe K, Uemura M, Kihara T, Takahashi R, Shimohama S, Kinoshita A (2012) Exercise is more effective than diet control in preventing high fat diet-induced beta-amyloid deposition and memory deficit in amyloid precursor protein transgenic mice. J Biol Chem 287:23024–23033

Manzanero S, Santro T, Arumugam TV (2013) Neuronal oxidative stress in acute ischemic stroke: sources and contribution to cell injury. Neurochem Int 62:712–718

Mattson MP (2009) Roles of the lipid peroxidation product 4-hydroxynonenal in obesity, the metabolic syndrome, and associated vascular and neurodegenerative disorders. Exp Gerontol 44:625–633

Mattson MP, Maudsley S, Martin B (2004) BDNF and 5-HT: a dynamic duo in age-related neuronal plasticity and neurodegenerative disorders. Trends Neurosci 27:589–594

McNay EC, Ong CT, McCrimmon RJ, Cresswell J, Bogan JS, Sherwin RS (2010) Hippocampal memory processes are modulated by insulin and high-fat-induced insulin resistance. Neurobiol Learn Mem 93:546–553

Melnik BC (2012) Leucine signaling in the pathogenesis of type 2 diabetes and obesity. World J Diabetes 3:38–53

Michaelis EK (2012) Selective neuronal vulnerability in the hippocampus: relationship to neurological diseases and mechanisms for differential sensitivity of neurons to stress. In: Bartsch (ed) The clinical neurobiology of the hippocampus: an integrative review. Oxford University Press, Oxford, pp 54–76

Molteni R, Barnard RJ, Ying Z, Roberts CK, Gomez-Pinilla F (2002) A high-fat, refined sugar diet reduces hippocampal brain-derived neurotrophic factor, neuronal plasticity, and learning. Neuroscience 10:803–814

Mommersteeg PM, Herr R, Pouwer F, Holt RI, Loerbroks A (2013) The association between diabetes and an episode of depressive symptoms in the 2002 World Health Survey: an analysis of 231, 797 individuals from 47 countries. Diabetic Med 30:e208–e214

Morris MC, Evans DA, Bienias JL, Tangney CC, Bennett DA, Wilson RS, Aggarwal N, Schneider J (2003) Consumption of fish and n-3 fatty acids and risk of incident Alzheimer disease. Arch Neurol 60:940–946

Moylan S, Maes M, Wray NR, Berk M (2012) The neuroprogressive nature of major depressive disorder: pathways to disease evolution and resistance, and therapeutic implications. Mol Psychiatry 18:595–606

O'Connor T, Sadleir KR, Maus E, Velliquette RA, Zhao J, Cole SL, Eimer WA, Hitt B, Bembinster LA, Lammich S, Lichtenthaler SF, Hébert SS, De Strooper B, Haass C, Bennett DA, Vassar R (2008) Phosphorylation of the translation initiation factor eIF2α increases BACE1 levels and promotes amyloidogenesis. Neuron 60:988–1009

Oddo S, Vasilevko V, Caccamo A, Kitazawa M, Cribbs DH, LaFerla FM (2006) Reduction of soluble Abeta and tau, but not soluble Abeta alone, ameliorates cognitive decline in transgenic mice with plaques and tangles. J Biol Chem 281:39413–39423

O'Keefe JH, Gheewala NM, O'Keefe JO (2008) Dietary strategies for improving post-prandial glucose, lipids, inflammation, and cardiovascular health. J Am Coll Cardiol 51:249–255

Pang PT, Lu B (2004) Mechanisms of late-phase LTP and long-term memory in normal and aging hippocampus. Aging Res Rev 4:407–430

Park M, Ross GW, Petrovitch H, White LR, Masaki KH, Nelson JS, Tanner CM, Curb JD, Blanchette PL, Abbott RD (2005) Consumption of milk and calcium in midlife and the future risk of Parkinson disease. Neurology 64:1047–1051

Perlis RH, Smoller JW, Mysore J, Sun M, Gillis T, Purcell S, Rietschel M, Nothen MM, Witt S, Maier W, Iosifescu DV, Sullivan P, Rush AJ, Fava M, Breiter H, Macdonald M, Gusella J (2010) Prevalence of incompletely penetrant Huntington's disease alleles among individuals with major depressive disorder. Am J Psychiatry 167:574–579

Pistell PJ, Morrison CD, Gupta S, Knight AG, Keller JN, Ingram DK, Bruce-Keller AJ (2010) Cognitive impairment following high fat diet consumption is associated with brain inflammation. J Neuroimmunol 10:25–32

Popa-Wagner A, Carmichael ST, Kokaja Z, Kessler C, Walker LC (2007) The response of the aged brain to stroke: too much, too soon? Curr Neurovasc Res 4:216–227

Posey KA, Clegg DJ, Printz DJ, Byun J, Morton GJ, Vivekanandan-Giri A, Pennathur S, Baskin DG, Heinecke JW, Woods SC, Schwartz MW, Niswender KD (2009) Hypothalamic proinflammatory lipid accumulation, inflammation, and insulin resistance in rats fed a high-fat diet. Am J Physiol Endocrinol Metab 296:E1003–E1012

Powers ET, Morimoto RI, Dillin A, Kelly JW, Balch WE (2009) Biological and chemical approaches to diseases of proteostasis deficiency. Annu Rev Biochem 78:959–991

Prasad MR, Lovell MA, Yatin M, Dhillon H, Markesbery WR (1998) Regional membrane phospholipid alterations in Alzheimer's disease. Neurochem Res 23:81–88

Price JL, McKeel DW Jr, Buckles VD, Roe CM, Xiong C, Grundman M, Hansen LA, Petersen RC, Parisi JE, Dickson DW, Smith CD, Davis DG, Schmitt FA, Markesbery WR, Kaye J, Kurlan R, Hulette C, Kurland BF, Higdon R, Kukull W, Morris JC (2009) Neuropathology of nondemented aging: presumptive evidence for preclinical Alzheimer disease. Neurobiol Aging 30:1026–1036

Purkayastha S, Zhang G, Cai D (2011) Uncoupling the mechanisms of obesity and hypertension by targeting hypothalamic IKK-β and NF-κB. Nat Med 17:883–887

Raghupathi R (2004) Cell death mechanisms following traumatic brain injury. Brain Pathol 14:215–222

Resnick SM, Pham DL, Kraut MA, Zonderman AB, Davatzikos C (2003) Longitudinal magnetic resonance imaging studies of older adults: a shrinking brain. J Neurosci 23:3295–3301

Richard E, Brayne C (2014) Dementia: mild cognitive impairment – not always what it seems. Nat. Rev. Neurol 10:130–131

Ries ML, Carlsson CM, Rowley HA, Sager MA, Gleason CE, Asthana S, Johnson SC (2008) Magnetic resonance imaging characterization of brain structure and function in mild cognitive impairment: a review. J Am Geriatr Soc 56:920–934

Rolfe DFS, Brown GC (1997) Cellular energy utilization and molecular origin of standard metabolic rate in mammals. Physiol Rev 77:731–758

Rotella F, Mannucci E (2013) Diabetes mellitus as a risk factor for depression. A meta-analysis of longitudinal studies. Diabetes Res Clin Pract 99:98–104

Rustad JK, Musselman DL, Nemeroff CB (2011) The relationship of depression and diabetes: pathophysiological and treatment implications. Psychoneuroendocrinology 36:1276–1286

Schmidt HD, Duman RS (2007) The role of neurotrophic factors in adult hippocampal neurogenesis, antidepressant treatments and animal models of depressive-like behavior. Behav Pharmacol 18:391–418

Seidl SE, Santiago JA, Bilyk H, Potashkin JA (2014) The emerging role of nutrition in Parkinson's disease. Front Aging Neurosci 6:36

Sen S, Duman R, Sanacora G (2008) Serum brain-derived neurotrophic factor, depression, and antidepressant medications: meta-analyses and implications. Biol Psychiatry 64:527–532

Seneff S, Wainwright G, Mascitelli L (2011) Nutrition and Alzheimer's disease: the detrimental role of a high carbohydrate diet. Eur J Intern Med 22:134–140

Sengillo JD, Winkler EA, Walker CT, Sullivan JS, Johnson M, Zlokovic BV (2013) Deficiency in mural vascular cells coincides with blood-brain barrier disruption in Alzheimer's disease. Brain Pathol 23:303–310

Seshadri S (2006) Elevated plasma homocysteine levels: Risk factor or risk marker for the development of dementia and Alzheimer's disease? J Alzheimers Dis 9:393–398

Seshadri S, Wolf PA, Beiser AS, Selhub J, Au R, Jacques PF, Yoshita M, Rosenberg IH, D'Agostino RB, DeCarli C (2008) Association of plasma total homocysteine levels with subclinical brain injury: cerebral volumes, white matter hyperintensity, and silent brain infarcts at volumetric magnetic resonance imaging in the Framingham Offspring Study. Arch Neurol 65:642–649

Shen C, Guo Y, Luo W, Lin C, Ding M (2013) Serum urate and the risk of Parkinson's disease: results from a meta-analysis. Can J Neurol Sci 40:73–79

Shoelson SE, Goldfine AB (2009) Getting away from glucose: fanning the flames of obesity-induced inflammation. Nat Med 15:373–374

Simopoulos AP (2009) Evolutionary aspects of the dietary omega-6:omega-3 fatty acid ratio: medical implications. World Rev Nutr Diet 100:1–21

Simopoulos AP (2013) Dietary omega-3 fatty acid deficiency and high fructose intake in the development of metabolic syndrome, brain metabolic abnormalities, and non-alcoholic fatty liver disease. Nutrients 5:2901–2923

Sims-Robinson C, Kim B, Rosko A, Feldman EL (2010) How does diabetes accelerate Alzheimer disease pathology? Nat Rev Neurol 6:551–559

Skurk C, Walsh K (2004) Death receptor induced apoptosis, a new mechanism of homocysteine-mediated endothelial cell cytotoxicity. Hypertension 43:1168–1170

Söderberg M, Edlund C, Kristensson K, Dallner G (1991) Fatty acid composition of brain phospholipids in aging and in Alzheimer's disease. Lipids 26:421–425

Son JH, Han DH, Min KJ, Kee BS (2013) Correlation between gray matter volume in the temporal lobe and depressive symptoms in patients with Alzheimer's disease. Neurosci Lett 548:15–20

Sonnen JA, Larson EB, Haneuse S, Woltjer R, Li G, Crane PK, Craft S, Montine TJ (2009) Neuropathology in the adult changes in thought study: a review. J Alzheimers Dis 18:703–711

Stranahan AM, Norman ED, Lee K, Cutler RG, Telljohann RS, Egan JM, Mattson MP (2008) Diet-induced insulin resistance impairs hippocampal synaptic plasticity and cognition in middle-aged rats. Hippocampus 18:1085–1088

Sullivan PF, Neale MC, Kendler KS (2000) Genetic epidemiology of major depression: review and meta-analysis. Am J Psychiatry 157:1552–1562

Swomley AM, Förster S, Keeney JT, Triplett J, Zhang Z, Sultana R, Butterfield DA (2014) Abeta, oxidative stress in Alzheimer disease: evidence based on proteomics studies. Biochim Biophys Acta 1842:1248–1257

Talbot K, Wang HY, Kazi H, Han LY, Bakshi KP, Stucky A, Fuino RL, Kawaguchi KR, Samoyedny AJ, Wilson RS, Arvanitakis Z, Schneider JA, Wolf BA, Bennett DA, Trojanowski JQ, Arnold SE (2012) Demonstrated brain insulin resistance in Alzheimer's disease patients is associated with IGF-1 resistance, IRS-1 dysregulation, and cognitive decline. J Clin Invest 122:1316–1338

Terry RD, Masliah E, Salmon DP, Butters N, DeTeresa R, Hill R, Hansen LA, Katzman R (1991) Physical basis of cognitive alterations in Alzheimer's disease: synapse loss is the major correlate of cognitive impairment. Ann Neurol 30:572–580

Townsend M, Mehta T, Selkoe DJ (2007) Soluble Abeta inhibits specific signal transduction cascades common to the insulin receptor pathway. J Biol Chem 282:33305–33312

Troen AM (2005) The central nervous system in animal models of hyperhomocysteinemia. Prog Neuropsychopharmacol Biol Psychiatry 29:1140–1151

Tsai CF, Hung CW, Lirng JF, Wang SJ, Fuh JL (2013) Differences in brain metabolism associated with agitation and depression in Alzheimer's disease. East Asian Arch Psychiatry 23:86–90

Uchino K, Lin R, Zaidi SF, Kuwabara H, Sashin D, Bircher N, Chang YF, Hammer MD, Reddy V, Jovin TG, Vora N, Jumaa M, Massaro L, Billigen J, Boada F, Yonas H, Nemoto EM (2010)

Increased cerebral oxygen metabolism and ischemic stress in subjects with metabolic syndrome-associated risk factors: preliminary observations. Transl Stroke Res 1:178–183

Valladolid-Acebes I, Merino B, Principato A, Fole A, Barbas C, Lorenzo MP, García A, Del Olmo N, Ruiz-Gayo M, Cano V (2012) High-fat diets induce changes in hippocampal glutamate metabolism and neurotransmission. Am J Physiol Endocrinol Metab 302:E396–E402

Vandal M, White PJ, Tremblay C, St-Amour I, Chevrier G, Emond V, Lefrançois D, Virgili J, Planel E, Giguere Y, Marette A, Calon F (2014) Insulin reverses the high-fat diet-induced increase in brain Aβ and improves memory in an animal model of Alzheimer disease. Diabetes. pii: DB_140375

Villegas R, Salim A, Flynn A, Perry IJ (2004) Prudent diet and the risk of insulin resistance. Nutr Metab Cardiovasc Dis 14:334–343

Villegas R, Yang G, Gao YT, Cai H, Li H, Zheng W, Shu XO (2010) Dietary patterns are associated with lower incidence of type 2 diabetes in middle-aged women: the Shanghai Women's Health Study. Int J Epidemiol 39:889–899

Wang S, Sun Z, Guo Y, Yuan Y, Yang B (2009) Diabetes impairs hippocampal function via advanced glycation end product mediated new neuron generation in animals with diabetes-related depression. Toxicol Sci 111:72–79

Weisskopf MG, O'Reilly E, Chen H, Schwarzschild MA, Ascherio A (2007) Plasma urate and risk of Parkinson's disease. Am J Epidemiol 166:561–567

White CL, Pistell PJ, Purpera MN, Gupta S, Fernandez-Kim SO, Hise TL, Keller JN, Ingram DK, Morrison CD, Bruce-Keller AJ (2009) Effects of high fat diet on Morris maze performance, oxidative stress and inflammation in rats: contributions of maternal diet. Neurobiol Dis 35:3–13

Williams A (2002) Defining neurodegenerative diseases. BMJ 324:1465–1466

Wilson RS, Arnold SE, Schneider JA, Li Y, Bennett DA (2007) Chronic distress, age-related neuropathology, and late-life dementia. Psychosom Med 69:47–53

Winchester J, Dick MB, Gillen D, Reed B, Miller B, Tinklenberg J, Mungas D, Chui H, Galasko D, Hewett L, Cotman CW (2012) Walking stabilizes cognitive functioning in Alzheimer's disease (AD) across one year. Arch Gerontol Geriatr 56:96–103

Wu A, Ying Z, Gomez-Pinilla F (2004) The interplay between oxidative stress and brain-derived neurotrophic factor modulates the outcome of a saturated fat diet on synaptic plasticity and cognition. Eur J Neurosci 19:1699–1707

Wurtman RJ (2005) Genes, stress, and depression. Metabolism 54(Suppl 1):16–19

Yamagishi S, Matsui T, Nakamura K, Takeuchi M, Inoue H (2008) Telmisartan inhibits advanced glycation end products (AGEs)-elicited endothelial cell injury by suppressing AGE receptor (RAGE) expression via peroxisome proliferator-activated receptor-gammaactivation. Protein Pept Lett 15:850–853

Yin F, Sancheti YH, Cadenas E (2012) Mitochondrial thiols in the regulation of cell death pathways. Antioxid Redox Signal 17:1714–1727

Yoshiyama Y, Higuchi M, Zhang B et al (2007) Synapse loss and microglial activation precede tangles in a P301S tauopathy mouse model. Neuron 53:337–351

Zanni GR, Wick JY (2011) Telomeres: unlocking the mystery of cell division and aging. Consult Pharm 26:78–90

Zhang X, Dong F, Ren J, Driscoll MJ, Culver B (2005) High dietary fat induces NADPH oxidase-associated oxidative stress and inflammation in rat cerebral cortex. Exp Neurol 191:318–325

Zhao WQ, De Felice FG, Fernandez S, Chen H, Lambert MP, Quon MJ, Krafft GA, Klein WL (2008) Amyloid beta oligomers induce impairment of neuronal insulin receptors. FASEB J 22:246–260

Zylberstein DE, Lissner L, Bjorkelund C, Mehlig K, Thelle DS, Gustafson D, Ostling S, Waern M, Guo X, Skoog I (2009) Midlife homocysteine and late-life dementia in women, 2009. A prospective population study. Neurobiol Aging 32:380–386

Chapter 10
Perspective and Direction for Future Research: Modification of High Calorie Diet Needed for Optimal Health of Human Visceral and Brain Tissues

10.1 Introduction

The consumption of diet is closely linked with health state of an individual. Healthy diet, moderate exercise, and 6–7 h of sleep contribute to the maintenance of optimal health. Conversely, poor unhealthy diet, lack of exercise, and lack of adequate sleep favor the occurrence of various disease states such as cancer, obesity, type II diabetes, dyslipidemia and hypertension. While the exact mechanisms regarding how obesity detrimentally affects health remain unclear, increased chronic inflammation and oxidative stress are key physiologic features of obesity (Hotamisligil 2006). Chronic inflammation is characterized by abnormal cytokine secretion, increase in acute-phase reactants and other mediators, and the activation of a network of inflammatory signaling pathways (Chandalia and Abate 2007). Indeed, inflammatory markers are not only correlated tightly with the degree of obesity and insulin resistance (Pickup and Crook 1998), but also with vascular disease risk factors (Rader 2000). The inflammatory response that emerges in the presence of obesity is probably triggered predominantly by adipose tissue, although other sites such as liver may also contribute (Shoelson et al. 2007). Furthermore, inflammatory and innate immune responses are also activated by increased levels of serum lipids, such as cholesterol and saturated long-chain fatty acids (Kennedy et al. 2009; Averill and Bornfeldt 2009).

Long term overconsumption of high calorie diet, which is enriched in refined grains (refining of grains specifically wheat reduces levels of betaine, a component that lowers levels of homocysteine), saturated and omega-6 fats, proteins of animal origins, high salt, and low in fiber, increases obesity, hyperglycemia, insulin resistance, dyslipidemia, and hypertension (Fig. 10.1a). The contribution of refined sugars and lipids in above mentioned conditions is hard to distinguish. However, prolonged hyperglycemia, accumulation of reactive oxygen species (ROS) and advanced glycation products (AGE), and activation of protein kinase C along with endothelial dysfunction are often suggested to be the primary cause of obesity, insulin resistance, dyslipidemia, hypertension, and diabetic complications

© Springer International Publishing Switzerland 2015
A.A. Farooqui, *High Calorie Diet and the Human Brain*,
DOI 10.1007/978-3-319-15254-7_10

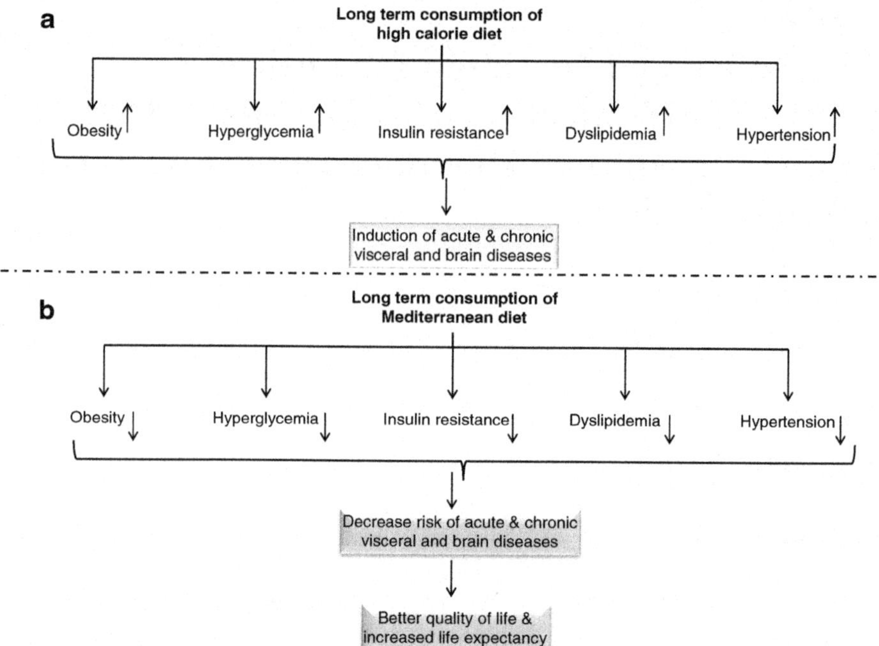

Fig. 10.1 Effects of long term consumption of high calorie diet and Mediterranean diet on obesity, hyperglycemia, insulin resistance, dyslipidemia, and hypertension

(Aronson 2008; Farooqui 2013). The combination of obesity, type II diabetes, dyslipidemia and hypertension may lead to metabolic syndrome that in turn increases the risk of death from cardiovascular and other chronic age-related neurodegenerative diseases (Farooqui 2012). Metabolic syndrome not only increases the risk of heart disease, stroke, and age-related chronic neurodegenerative diseases, but also induces immune system dysfunction throughout the body leading to adverse effects on cognitive function. Collective evidence suggests that effective interventions and public health policies are thus required to address the burden of above mentioned chronic disease with nutrition. The development and repair of neural tissue depend on the proper intake of essential structural nutrients, minerals, phytochemicals, and vitamins. Therefore, what one consumes, or avoid, may have an important impact on our cognitive ability and mental performance. There are two of the key areas in which nutrition is thought to play an important role include optimizing neurodevelopment in children and in preventing neurodegeneration and cognitive decline during aging (Sizonenko et al. 2013). Thus, most chronic diseases are associated with underlying preventable risk factors, such as elevated blood pressure, high blood glucose or glucose intolerance, hyperlipidemia, physical inactivity, excessive sedentary behaviors, overweight and obesity, and tobacco usage. The development of chronic diseases can be prevented or delayed if these risk factors are addressed before they progress to overt disease.

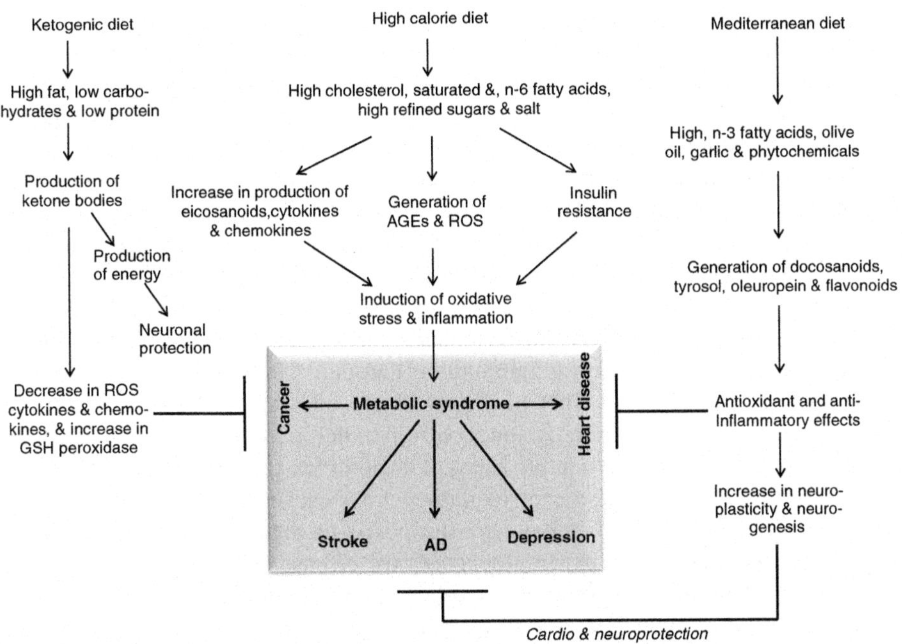

Fig. 10.2 Effects of various types of diet patterns on metabolic syndrome and neurological disorders. *AGEs* Advanced glycation products, *ROS* reactive oxygen species, *AD* Alzheimer disease

Vascular disorders, such as hypercholesterolemia and atherosclerosis, are important risk factors for Alzheimer disease (AD) (Iadecola 2004; Skoog and Gustafson 2006). Furthermore, cardiovascular disease risk factors, such as a sedentary lifestyle, high saturated fatty acid (SFA) intake, type II diabetes, smoking and obesity, are associated with a higher risk of developing neurological disorders (Whitmer et al. 2008; Solomon et al. 2009). As stated in Chap. 1 that long term consumption of high calorie diet is accompanied by the development of chronic oxidative stress, which characterized by production of high levels of ROS and AGE along with decline in antioxidant capacity in visceral tissues and the brain (Fig. 10.2). Overproduction of ROS arising either from mitochondrial electron-transport chain or excessive stimulation of NADPH results in oxidative stress, a deleterious process that can be an important mediator of damage to cell structures, including lipids, proteins, and DNA. High levels of ROS and AGE not only neutralize reducing capacity of cells and produce changes in signal transduction pathways, but also produce changes in telomeres, the 'biological clock' of the cellular aging. Telomeres are protective structures at the end of chromosomes that consist of six recurring nucleotide bases (TTAGGG). Small amounts of these terminal sequences are lost with each cell division. The enzyme telomerase compensates for this loss by rebuilding telomeres. The negative correlation between telomere length and the ability to replicate in somatic cells coupled with reduction in telomere length with age has led to the suggestion that telomeres play a central role in the aging process.

Telomeres are highly susceptible to oxidative stress because of their high content of guanines. Increase in intracellular ROS levels is associated with acceleration in the rate of telomere shortening. Oxidative stress-mediated progressive shortening of telomeres leads to senescence, apoptotic cell death, or the oncogenic transformation of somatic cells in various tissues (Farooqui 2013).

Chronic inflammation in the brain, which is caused by long term consumption of high calorie diet, is characterized not only by the activation of microglia and astrocytes, but also due to sustained activation of NF-κB along with generation of inflammatory mediators such as tumor necrosis factor-α (TNF-α), interleukin 1β (IL-1β), and interleukin 6 (IL 6). The sustained expression and release of these inflammatory mediators cause an imbalance in the inflammatory cycle homeostasis promoting further intensification of neuroinflammation (Farooqui 2010). Oxidative stress and inflammation caused by distinct biochemical cascades are processes, which are closely intertwined and generally function in parallel, particularly in the brain, an organ especially prone to oxidative stress. The complex interplay between inflammatory mediators and markers for oxidative stress caused by the long term consumption of high calorie diet has been proposed to regulate the progression of chronic neurodegeneration in neurodegenerative diseases (Farooqui 2010, 2013). Thus, the most common hypotheses to explain the pathogenesis of chronic neurodegenerative diseases include interactions among neuroinflammation, oxidative stress, mitochondrial dysfunction, alterations in calcium homeostasis, proteasomal dysfunction, protein aggregation, decrease in blood flow, alterations in blood brain barrier, and neuronal cell cycle induction (Golde 2009; Farooqui 2010). However, placing these pathways in the proper relationship to the onset, time course, and progress of neurodegeneration and its relationship to cytoskeletal pathology are challenging issues that are not fully understood (Golde 2009).

10.2 Effects of Diet Patterns on Acute and Chronic Visceral and Brain Diseases

In nutrition studies, it is difficult to determine the effect of single nutrient on human health and in chronic visceral and brain diseases. Nutrition research has therefore focused on studying the synergistic effects of complex whole foods and dietary patterns, acknowledging the reality of the human diet and its usefulness in translating information into dietary recommendations (Jacobs and Tapsell 2013; Oude Griep et al. 2013). The dietary pattern approach provides us with a comprehensive understanding of studying the effect of whole diet on complex chronic visceral and age-related brain diseases. Studies on analysis of dietary patterns have indicated several dietary patterns among various human populations. These dietary patterns include vegetarian diet pattern, Mediterranean diet pattern, Okinawan diet pattern, and high calorie diet pattern (western diet pattern). Mediterranean diet pattern, and Okinawan diet pattern are characterized by the consumption of high intakes of vegetables, legumes, fruit, whole grains, fish, and poultry whereas a high calorie diet pattern (Western pattern) is characterized by red meat, processed meat, refined grains, starch, high-fat diary, sweets, desserts, high-sugar drinks, and eggs (Farooqui 2013). The Mediterranean and Okinawan diet patterns reduce the risk of developing chronic visceral and brain diseases but the high calorie dietary pattern strongly increases risk of onset of heart disease, stroke, and chronic neurodegenerative diseases (Farooqui 2013).

10.2.1   Beneficial Effects of Vegetarian Diet on Visceral and Brain Tissues

The consumption of diet enriched in vegetable, fruits, chick peas, pulses, nuts, and soy products, including soy milk, soy yogurt, and tofu produce healthy effects in visceral and brain tissues.

Questions have been raised about the adequacy of protein in vegetarian diets. However, vegetarian diets usually exceed protein requirements (Craig and Mangels 2009; Millward 1999). Vegetables and fruits are enriched in magnesium, potassium, and nitrate. These components produce beneficial effects in circulatory system, visceral and brain tissues. Thus, dietary magnesium regulates many enzymes including kinases, ATPases, and enzymes catalyzing many steps of the glycolytic pathway. Magnesium activates tyrosine kinase, a component of the beta subunit of the insulin receptor and promotes insulin sensitivity. Activation of tyrosine kinase produces a signaling cascade that ultimately translocates GLUT4 (the major insulin-regulated glucose transporter expressed in muscle and other insulin-responsive tissues) to the membrane, and allows the cell to take up glucose. In addition, magnesium may be acting as an inhibitor of the inositol 1,4,5-trisphosphate ($InsP_3$)-gated calcium channel—thus magnesium may be acting as a calcium antagonist (Chaudhary et al. 2010). Magnesium not only influences vascular tone and responsiveness, but it also modulates acetylcholine-induced endothelium-dependent relaxation. Magnesium also stabilizes cell membranes and functions as a signal transducer. Collective evidence suggests that magnesium is a cofactor for more than 300 enzymes including but not limited to, protein synthesis, cellular energy production and storage, reproduction, DNA and RNA synthesis, and stabilizing mitochondrial membranes. Approximately 50 % of magnesium is in the bone, 50 % is in the tissues and organs, and 1 % is in the blood. Magnesium plays a critical role in visceral and brain tissue (Fig. 10.3) (Volpe 2013). At least 70–80 % Americans are deficient in magnesium and its deficiency is a risk factor for a number of chronic diseases including migraine headaches, AD, stroke, hypertension, cardiovascular disease, and type II diabetes mellitus (Fig. 10.4) (Volpe 2013).

Dietary nitrate produces a range of beneficial vascular effects, including reducing blood pressure, inhibiting platelet aggregation, preserving or improving endothelial dysfunction, enhancing exercise performance in healthy individuals and patients with peripheral arterial disease (Fig. 10.4) (Hruby et al. 2013). Pre-clinical studies have indicated that nitrate or nitrite also show the potential to protect against ischemia-reperfusion injury and reduce arterial stiffness, inflammation and intimal thickness (Hruby et al. 2013). The consumption of vegetable, fruits, chick peas, and

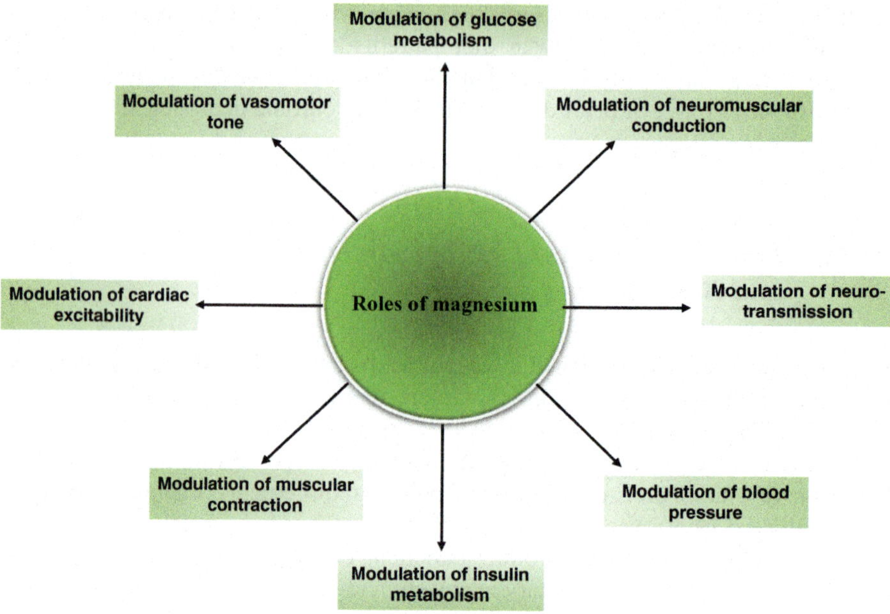

Fig. 10.3 Effects of magnesium on metabolic processes in visceral and brain tissues

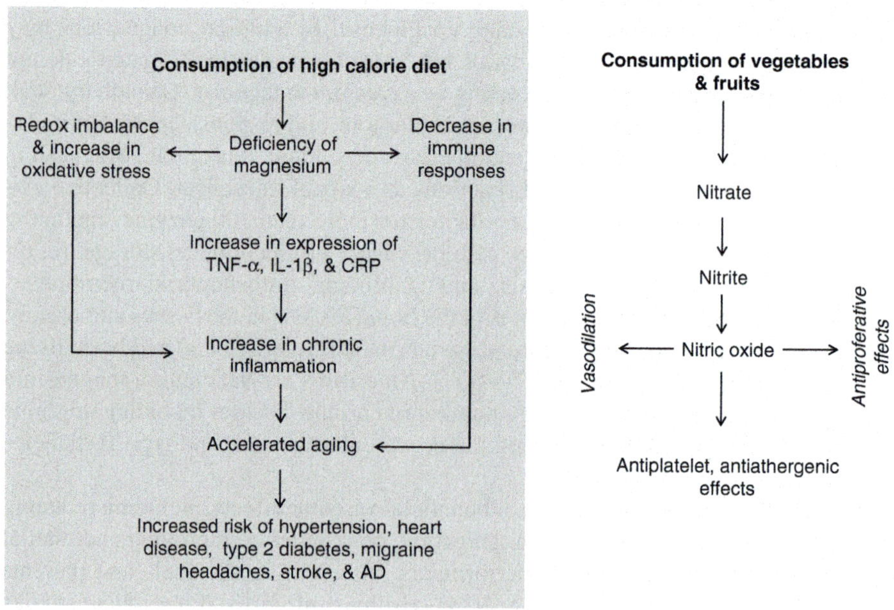

Fig. 10.4 Effects of high calorie and vegetarian diets on cardiovascular and cerebrovascular diseases

pulses (proteins of plant origin) produces multifaceted antihyperglycemic activities such as inhibition of pancreatic α-amylase and intestinal α-glucosidase, inhibition of protein-tyrosine phosphatase 1β in liver and skeletal muscles, and insulin mimetic and secretagogue activities (Tiwari 2014; Tiwari et al. 2014). In addition, components found in vegetables and fruits also modulate polyol pathway in favor of reducing the development and intensity of oxidative stress, and consequently the development of diabetic complications. Furthermore, Polyphenols (flavonoids, phenolic acids, proanthocyanidins and tannins), which are present in vegetables and fruits have been reported to alter glycemia by inhibiting carbohydrate digestion, reducing carbohydrate absorption in the intestines, stimulating the release of insulin from pancreatic β-cells, and modulating hepatic glucose output (Rahimi et al. 2005; Farooqui 2013). Soluble and insoluble fibers, which are present in vegetables and fruits impact intestinal tract time, absorption of macronutrients, alter the action of digestive enzymes and secretion of gastrointestinal and pancreatic hormones (Anderson 1986). Insoluble fiber can decrease intestinal tract time, potentially reducing time for the carbohydrates to be absorbed in the jejunum (Montonen et al. 2003). Soluble fiber delays gastric emptying slowing the absorption and digestion of carbohydrates potentially delaying the insulin response (Montonen et al. 2003). In addition, non digestive fiber is fermented by the microflora of the colon producing short chain fatty acids. The production of short chain fatty acids also impacts on carbohydrate metabolism (Thorburn et al. 1993). Among vegetables, cruciferous vegetables such as broccoli, Brussels sprouts, cabbage, kale, and cauliflower contain isothiocyanates and indole-3-carbinol, which have been reported to produce anticancer effects. In addition, indole-3-carbinol has been reported to be an effective treatment for cervical dysplasia. The consumption of onions may inhibit platelet aggregation, increase fibrinolytic activity, and lower blood pressure. Collective evidence suggests that vegetarian diet provides a low intake of saturated fat and cholesterol and a high intake of dietary fiber and many health-promoting phytochemicals-derived from increased consumption of fruits, vegetables, whole-grains, legumes, nuts, and various soy products. As a result, vegetarians typically have lower body mass index, low serum total and low-density lipoprotein cholesterol levels, and low blood pressure; reduced rates of death from ischemic heart disease; improved glycemic and lipid control, and decreased incidence of high blood pressure, stroke, type II diabetes, gallbladder disease, kidney stones, and certain cancers than do non-vegetarians. However, vegetarian diets may be low in a number of micronutrients, including vitamin B_{12}, iron, vitamin D, zinc, iodine, riboflavin, calcium, and selenium (Gaby 2013).

10.2.2   Beneficial Effects of Mediterranean Diet on Visceral and Brain Tissues

The Mediterranean diet is characterized by a high intake of vegetables, legumes, fruits, and whole grains (vitamins and minerals, as well as fiber, essential fatty acids, and accessory food factors); a moderate to high intake of fish; a low intake of

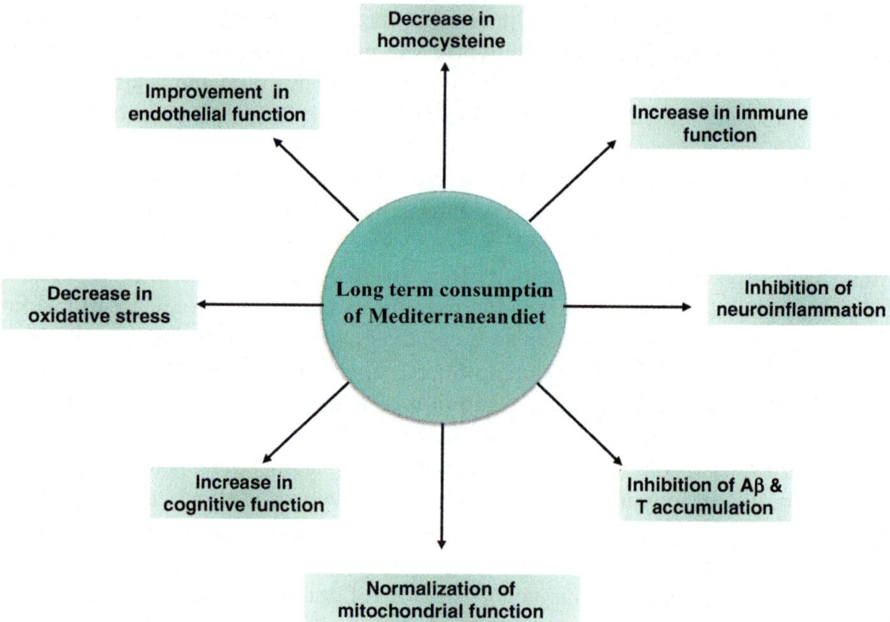

Fig. 10.5 Effects of long term consumption of Mediterranean diet on oxidative stress, inflammation and cognitive function

saturated lipids but high intake of olive oil; a low to moderate intake of dairy products, mostly cheese and yogurt; a low intake of meat; and a modest intake of ethanol, mostly as red wine (Willett et al. 1995). As stated above, fruits and vegetables are rich sources of vitamins, potassium and other minerals, carotenoids, flavonoids, fiber, and phytochemicals. The consumption of abundant amounts of fruits and vegetables may be useful for retarding cardiovascular disease, stroke, some cancers, hypertension, osteoporosis, and other diseases. Mediterranean diet not only increases longevity by lowering cardiovascular disease, inhibiting cancer growth, but also by protecting the body from age-dependent cognitive decline (Fig. 10.5) (Lopez-Miranda et al. 2007; López-Miranda et al. 2012). Health benefits of Mediterranean diet are related to the nutrition pattern of whole diet rather than few components of the diet, supporting the view that interactions between the multiple components may produce beneficial effects (Farooqui 2012). However, lots of information has been published on various components individually. Thus, olive oil not only provides the higher percent of energy but a lot of bioactive compounds (tyrosol, hydroxytyrosol, oleocanthal, and oleuropein) that promote beneficial effects on neurovascular and cardiovascular systems. In addition, Oleic acid is converted into nitrated oleic acid (nitro-oleic acid) in the presence nitric oxide (NO•) (Freeman et al. 2008). Nitrated oleic is found in the plasma and several tissues where it not only inhibits inflammation (Fig. 10.6), but also promotes blood vessel relaxation through modulation of macrophage activation and prevention of leukocyte and platelet activation

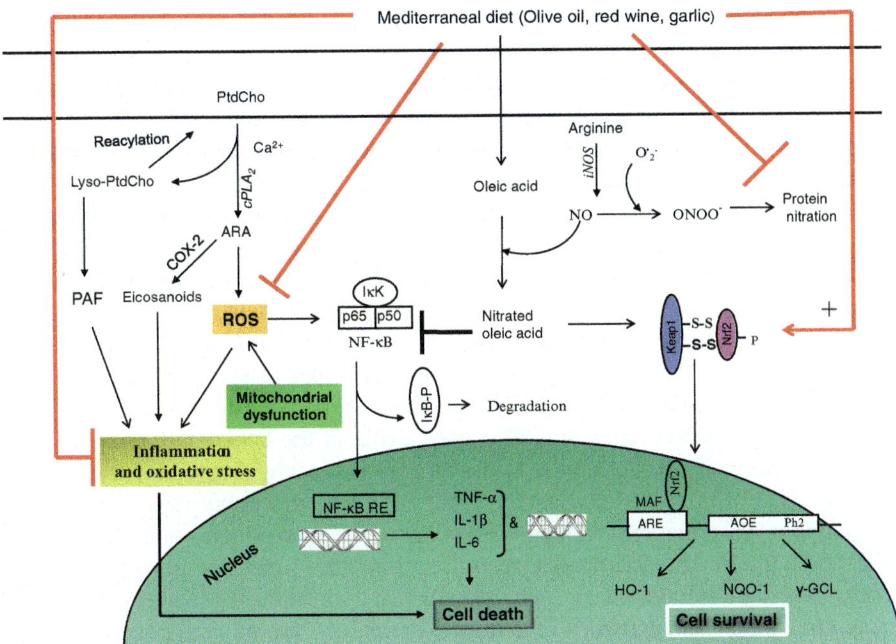

Fig. 10.6 Beneficial effects of Mediterranean diet components on oxidative stress and neuroinflammation. *PtdCho* Phosphatidylcholine, *lyso-PtdCho* lyso-phosphatidylcholine, *ARA* arachidonic acid, *PAF* platelet-activating factor, *cPLA₂* cytosolic phospholipase A₂, *COX-2* cyclooxygenase-2, *ROS* reactive oxygen species, *NF-κB* nuclear factor kappaB, *NF-κB-RE* nuclear factor κB-response element, *IκB* inhibitory subunit of NF-κB, *IκB-P* phosphorylated IκB, *TNF-α* tumor necrosis factor-α, *IL-1β* interleukin-1β, *IL-6* interleukin-6, *iNOS* inducible nitric oxide synthase, O_2^- superoxide, *Nrf2* NFE2-related factor 2, *HO-1* heme oxygenase 1, NADH quinine oxidoreductase, *γ-GCL* γ-glutamylcystein ligase

(Trostchansky and Rubbo 2008). Olive oil also decreases blood pressure and protects from type II diabetes. Olive oil not only provides the higher percent of energy, but the presence of bioactive polyphenolic compounds promotes human health through their antioxidant activities, upregulation of nitric oxide synthase, gluthathione peroxidase; heme oxygenase 1 (HO-1), NADH quinine oxidoreductase, γ-glutamylcystein ligase (γ-GCL), and activation of NFE2-related factor 2 (Nrf2) along with increase in insulin secretion (Fig. 10.6) (Farooqui 2012).

Resveratrol (3,4′,5-trihydroxystilbene) is a natural compound found in red grapes and wine. Red wine is an important component of Mediterranean diet. It produces several beneficial effects in humans. Resveratrol promotes both anti-aging, anticarcinogenic, cardioprotective, and cerebroprotective activities, which are attributed to its antioxidant, anti-inflammatory, and gene modulating properties. Resveratrol modulates genes important for mitochondrial function, such as peroxisome proliferator-activated receptor-γ coactivator (PGC)-1α, a master regulator of mitochondrial biogenesis, leading to an increase in mitochondrial content and function as well as an increase in physical endurance (Lagouge et al. 2006).

It not only acts as an analgesic, but also a calorie restriction mimetic, which may produce beneficial effects against numerous diseases such as type II diabetes, cardiovascular diseases, and cancer in tissue culture and animal models (Farooqui 2013). Resveratrol indirectly activates AMPK (Dasgupta and Milbrandt 2007), and this activation of AMPK has been shown to reduce fat accumulation and increase glucose tolerance, insulin sensitivity, mitochondrial biogenesis, and physical endurance (Hardie et al. 2012). Intracellular levels of cAMP are modulated by the activities of adenyl cyclases (ACs) and cyclic nucleotide phosphodiesterase (PDEs) (Conti and Beavo 2007), which hydrolyze cAMP or cGMP to AMP or GMP, respectively. Resveratrol inhibits the activity of PDEs in a dose-dependent manner (Park et al. 2012). Due to its structural similarity with diethylstilbestrol (a synthetic estrogen), resveratrol also produces oestrogenic effects by binding to estrogen receptors and evoking neurochemical effects parallel to those exerted by endogenous estrogen. These oestrogenic properties may also play a role in the beneficial cardiovascular effects. Additionally, resveratrol not only inhibits platelet aggregation and lipid peroxidation, but also blocks eicosanoid synthesis, modulates lipoprotein metabolism, and exhibits vasorelaxing and anti-inflammatory activities (Das and Das 2007) (Fig. 10.7). Collective evidence suggests that resveratrol functions as an antioxidant, vasorelaxing anti-inflammatory agent by hampering free radical generation and scavenging free radicals, but also by inhibiting the release of proinflammatory cytokines.

Garlic is an important component of Mediterranean diet. It contains several organosulfur compounds, such as, allicin, alliin, diallyl sulfide, diallyl disulfide,

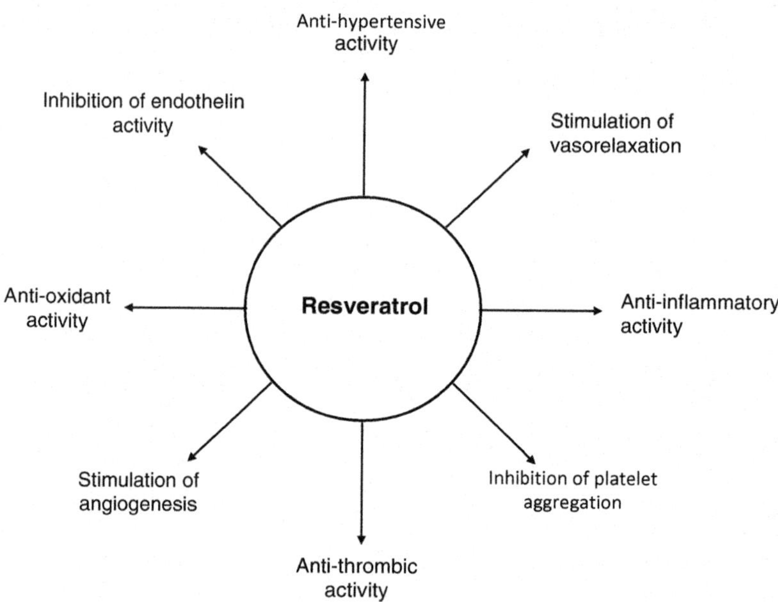

Fig. 10.7 Effects of resveratrol on various parameters of blood

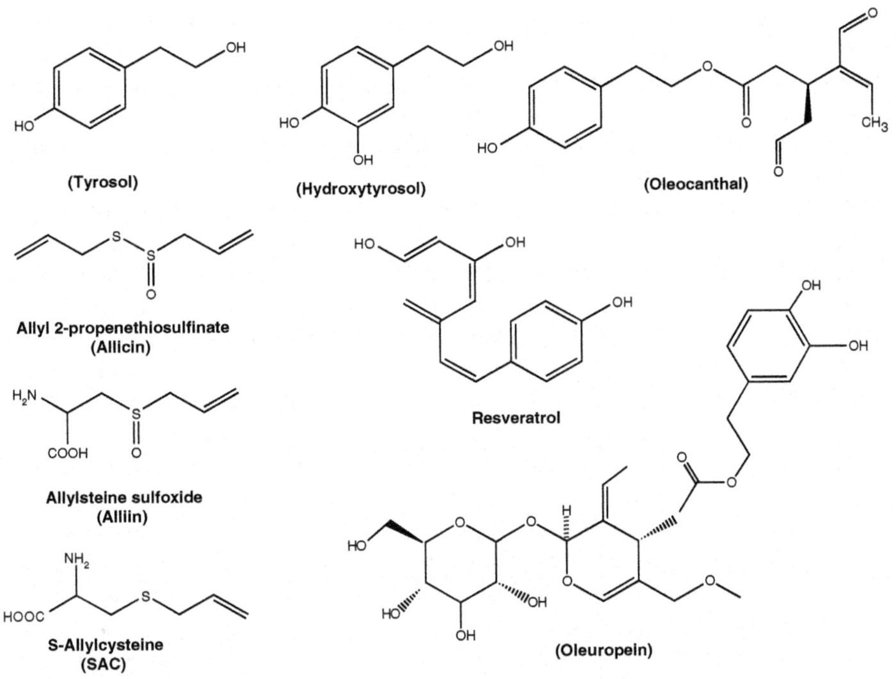

Fig. 10.8 Chemical structures of components found in Mediterranean diet

diallyl trisulfide, S-allylcysteine (SAC), dithiins, ajoene, methyl allyl disulfide, methyl allyl trisulfide, 2-vinyl-1,3-dithiin, 3-vinyl-1,2-dithiin (Fig. 10.8) (Fenwick and Hanley 1985; Rybak et al. 2004).

These are lipophilic thioesters derived from oxidized allicin, which is produced when garlic cloves are crushed. Diallyl sulfide has been reported to be protective against numerous chemically induced cancers and is potent inducers of phase II detoxifying enzymes (Borek 2006). Lipophilic thioesters provide garlic its characteristic odor and flavor as well as most of its biological properties. Beneficial effects of garlic are not only due to its antioxidant, anti-inflammatory, bacteriostatic, anti-apoptotic effects, but also produced by immune system enhancing properties (increase in interleukin-1 levels in T lymphocytes and macrophages) and facilitating healthy blood circulation (Le Bon and Siess 2000; Borek 2006; Moriarty et al. 2007). In addition, organosulfur compounds of garlic also lower cholesterol and blood pressure (Borek 2006). Oral administration of raw garlic for a period of eight weeks results in significant reduction of blood glucose and improvement of insulin sensitivity in garlic treated rats (Padiya et al. 2011). In addition, other metabolic complications of type II diabetes like increase in serum triglyceride, insulin and uric acid levels are also normalized after garlic intake in diabetic rats. Lowering serum uric acid and triglyceride after garlic administration may contribute to improvement in insulin resistance in fructose fed rats (Nakagawa et al. 2006). It is proposed that the overall Mediterranean diet pattern may exert beneficial effects by lowering risk for several forms of cancer, obesity, dyslipidemia, hypertension, abnormal glucose metabolism, coronary heart disease, and overall mortality (Scarmeas et al. 2006; Farooqui 2014). The interaction of various components of Mediterranean diet, such as olive oil, whole grain products, large intake of fibers by fruits and vegetables, vitamins, and polyphenolic compounds by fruits or red wine, may exert a pivotal role in modulating glucose homeostasis pathways beyond the amount of adipose tissue (Farooqui 2013). In addition, antioxidants and polyphenols components of Mediterranean diet reduce inflammatory angiogenesis in cultured endothelial cells, through MMP-9 and COX-2 inhibition. These processes have been shown to exert protective effects against cardiovascular diseases (Scoditti et al. 2012). Cardiovascular disorders are accompanied by oxidative stress. Once again polyphenols present in Mediterranean diet due to their antioxidants effects protect against cardiovascular diseases. Recent studies have also indicated that polyphenols produce cardioprotective effects by modulating signaling pathways (Rahman et al. 2006; Vauzour et al. 2010; Farooqui 2012). Based on above mentioned characteristics, it is proposed that Mediterranean diet may improve lipid profiles, glycemic control, insulin sensitivity, inhibits oxidative stress and inflammation in people with type II diabetes, metabolic syndrome, and cardiovascular disease (Martinez-Gonzalez and Sánchez-Villegas 2004; Salas-Salvadó et al. 2011; Estruch et al. 2013).

Collective evidence suggests that Mediterranean diet is known to be one of the healthiest dietary patterns in the world not only due to its relation with a low morbidity, mortality, and better quality of life, but also due to its beneficial effects on lipoprotein levels, endothelium vasodilatation, hyperglycemia, insulin resistance, dyslipidemia, and obesity (Fig. 10.1b). Mediterranean diet also decreases the prevalence of the metabolic syndrome, cardiovascular and cerebrovascular, neurodegenerative diseases and various types of cancers (Sofi et al. 2013; Estruch et al. 2013; Castro-Quezada et al. 2014). As mentioned in Chap. 1, Mediterranean dietary patterns consist of fruits, vegetables, breads, other forms of cereals, pulses, nuts seeds, olive oil, moderate amounts of dairy products (mainly cheese and yoghurt), fish and shellfish, white and red meat consumed in low frequency and amounts with meals. Such a dietary pattern assures a sufficient intake of phytonutrients, which reduce the risk of many chronic visceral diseases and age-related neurodegenerative diseases (Fig. 10.2) (Bach et al. 2006). Recent in vitro studies have indicated that Mediterranean diet protects the cells from oxidative stress and neuroinflammation not only by preventing cellular senescence, and cellular apoptosis, but also through the maintenance of telomere length (Boccardi et al. 2013). Molecular mechanisms underlying the neuroprotective effects of Mediterranean diet are not clear. However, it is becoming increasingly evident that Mediterranean diet not only produces antioxidant, anti-inflammatory, anti-apoptotic, anti-aging effects, but also promotes neurotrophic effects in animal as well as human studies. In contrast, as described in Chaps. 8 and 9, long term consumption of high calorie diet, which includes red meat, processed meats, foods rich in sugars and in fats, and fructose corn syrup results in increased risk of chronic inflammation and oxidative stress involved in metabolic syndrome, cardiovascular, cerebrovascular, neurodegenerative diseases, and various types of cancers (Farooqui 2013).

As stated in Chap. 7, human digestive tract hosts more than 100 trillion bacteria and archaea, which together make up the gut microbiota. The amount of microbiota in the human gut outnumbers human cells by a factor of 10, but some finely tuned mechanisms allow these bacteria to colonize and survive within the host in a mutual relationship. Microorganisms live in human digestive tract in a symbiotic relationship, which leads to physiological homeostasis that promotes and maintains the good health of the host. Despite its importance in maintaining the health of the host, growing evidence suggests the gut microbiota also contribute to the pathogenesis of various diseases such as hepatic steatosis (Ottman et al. 2012; Compare et al. 2012; Tremaroli and Backhed 2012), metabolic syndrome (Weiss et al. 2004), behavior abnormalities (Bercik et al. 2011), metabolic disorders (Hosseini et al. 2011), inflammatory disorders (Farrell et al. 2012), and inflammatory bowel diseases (Comito et al. 2014). The relationship between gut microbiota, health and disease have led to the use of probiotics, prebiotics and functional food to prevent or treat some diseases, such as inflammatory bowel diseases (Sokol et al. 2009), inflammatory bowel syndrome (Moayyedi et al. 2010), allergy (Okada et al. 2010), metabolic syndromes (Qin et al. 2012), hepatic steatosis (Malaguarnera et al. 2012) and colorectal cancer (Uccello et al. 2012).

The long term consumption of Mediterranean diet leads to a reduction in mortality and the incidence of major chronic diseases (Mitrou et al. 2007; Del Chierico et al. 2014), such as cancer (Couto et al. 2011), metabolic and cardiovascular syndrome (Georgousopoulou et al. 2014), neurodegenerative diseases (Sofi et al. 2013), type II diabetes (Salas-Salvado et al. 2014), fatty liver diseases (Salamone et al. 2012) and allergy (Garcia-Marcos et al. 2013). Moreover, Mediterranean diet is highly associated with an improved quality of life, which is translated into better psycho/physiological and metabolic profiles (Martinez-Gonzalez et al. 2009; Sofi et al. 2008).

10.2.3   Beneficial Effects of Ketogenic Diet on Visceral and Brain Tissues

The ketogenic diet is composed of 80–90 % fat, with carbohydrate and protein constituting the remainder of the intake. The diet provides sufficient protein for growth, but insufficient amounts of carbohydrates for the body's metabolic needs. Ketogenic diet provides energy mainly from the oxidation of fatty acids in mitochondria as opposed to glucose, which is the main energy source following consumption of normal diet (Fig. 10.9). Following prolonged consumption of ketogenic diet, fatty acids are oxidized at a high rate, which results in an overproduction of acetyl-CoA. The overproduction of acetyl-CoA then leads to the synthesis of ketone bodies (β-hydroxybutyrate, acetoacetate, and acetone) primarily in the liver (Freeman et al. 2006). The accumulation of ketone bodies in the bloodstream causes ketosis, a survival mechanism, which is activated during prolonged fasting, starvation or lack of carbohydrate ingestion. Ketone bodies readily cross the blood brain barrier either by simple diffusion (acetone) or with the aid of monocarboxylic transporters

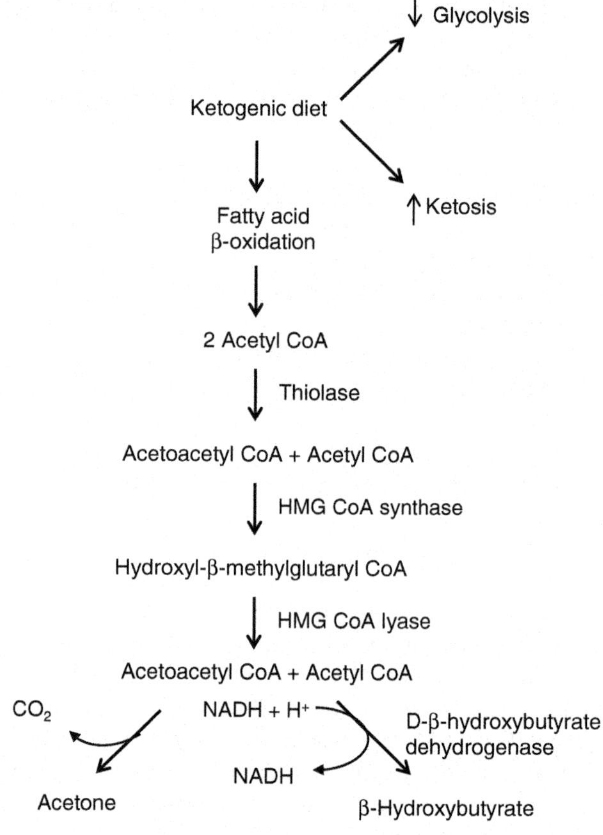

Fig. 10.9 Biochemical mechanism associated with the synthesis of ketone bodies in visceral and brain tissues

(β-hydroxybutyrate, acetoacetate), whose expression is related to the level of ketosis (Pierre and Pellerin 2005). During prolonged ketosis, the brain is capable of utilizing ketone bodies as an alternate fuel, thus reducing its requirement for glucose (LaManna et al. 2009). It is proposed that ketogenic diet acts in a way similar to caloric restriction (CR) on AMPK, PGC1α, and SIRT-1 (Newman and Verdin 2014). Thus, PGC1α is activated in the phosphorylated state. Once phosphorylated, PGC1α translocates from the cytosol to the nucleus, where it promotes the transcription of genes involved in fatty acid transport, fat oxidation, and oxidative phosphorylation (Jäer et al. 2007). PGC1α is phosphorylated through several different pathways, such as AMPK, calcium-calmodulin-dependent protein-kinase, and p38 mitogen-activated protein kinase pathways (Benton et al. 2008). PGC1α is also activated by a SIRT1-mediated deacetylation (Yu and Auwerx 2010). AMPK exerts its effect via phosphorylation of PGC1α. AMPK activation promotes enhanced expression of skeletal muscle oxidative-related enzymes, proteins, and metabolism, which are consistent with the findings that obese skeletal muscles are less oxidative and have lower AMPK activation (during fasting conditions). At the same time, AMPK activation also inhibits mTOR signaling. However, it seems counterintuitive to inhibit an important growth-mediated pathway (i.e., mTOR), regulating muscle mass, so that skeletal muscles can grow (Williamson 2011). The relative deficiency in carbohydrates in ketogenic diet activates AMPK and SIRT-1, increases phosphorylation of AMPK and promotes deacetylation of PGC1α in skeletal muscle without affecting the total amount of AMPK, PGC1α, or SIRT 1 (Draznin et al. 2012). These mechanisms appear to be activated just after few hours (5 h) of starvation in mice (Yoon et al. 2001) whilst no data are available on KD in humans. Once activated, SIRT1 and AMPK produce beneficial effects on glucose homeostasis and insulin action (Ruderman et al. 2010). In the brain, at the molecular level the consumption of ketogenic diet not only results in the modulation of ATP-sensitive potassium (K_{ATP}) channels and enhancement in purinergic (i.e., adenosine) and GABAergic neurotransmission, but also increases BDNF expression and attenuates neuroinflammation (Politi et al. 2011). The consumption of ketogenic diet not only produces reduction in blood glucose, decreases in ROS, and increases in glutathione peroxidase activity in the hippocampus, but also increases uncoupling in proteins expression and alterations in the expression of genes associated with oxidative stress (Sullivan et al. 2004; Ziegler et al. 2003; Bough et al. 2006). In addition, ketogenic diet also facilitates the stabilization of the neuronal membrane potential through improved mitochondrial function.

Based on above information, it is proposed that ketogenic diet mediates its neuroprotective effects by raising ATP levels and reducing ROS production through enhanced NADH oxidation and inhibition of mitochondrial permeability transition (mPT) (Kim et al. 2011). Ketogenic diet also activates the Nrf2 pathway via redox signaling leading to chronic cellular adaptation, induction of protective proteins, and improvement of the mitochondrial redox state (Milder et al. 2010). These processes exert neuroprotective and antiepileptogenic properties, heightening the clinical potential of the ketogenic diet as a disease-modifying intervention (Politi et al. 2011). Collective evidence suggests that ketogenic diet can enhance neuronal survival under hypoxia, anoxia, or ischemia (Suzuki et al. 2001; Gibson et al. 2012). In addition, ketogenic diet can also provide symptomatic and disease-modifying activity in a broad range of neurodegenerative disorders including animal models of AD and Parkinson disease (PD), multiple sclerosis, amyotrophic lateral sclerosis, and traumatic brain injury (Fig. 10.2) (Gasior et al. 2006; Van der Auwera et al. 2005; Zhao et al. 2006; Maafouf et al. 2009).

10.2.4   *Beneficial Effects of Okinawan Diet on Visceral and Brain Tissues*

Okinawan diet is characterized by the consumption of high amounts of Satsamu sweet potato, green and yellow vegetable, and fruit and low amounts of meat (fish and pork), a wide array of other plant foods including seaweed (especially konbu) and soy, and by green tea and kohencha tea (Willcox et al. 2009). Many characteristics of the Okinawan diet are similar to the Mediterranean diet, DASH diet, and

Portfolio diet. All these dietary patterns are associated with reduced risk for cardiovascular disease, type II diabetes, and other chronic visceral and neurodegenerative diseases and promotion of healthy aging and longevity.

The mechanisms by which these dietary nutrients exert protective properties against neurological disorders are still under investigation, but several lines of evidence have shown beneficial effects of n-3 fatty acid containing diet has beneficial on the cardiovascular system (Lee et al. 2008; Lavie et al. 2009; Farooqui 2009) and on neuronal membrane properties (Horrocks and Farooqui 2004; Farooqui 2009). These beneficial effects on the cardiovascular system are due to the ability of n-3 fatty acids to decrease blood pressure (Geleijnse et al. 2002), lower plasma triglyceride (Sacks and Katan 2002; Farooqui 2009), prevent arrhythmias (Leaf et al. 2003), improve vascular reactivity (Harris 1997; Farooqui 2009), decrease atherosclerosis (Okuda et al. 2005), and suppress inflammatory processes (Farooqui et al. 2007; Farooqui 2009). Furthermore, high levels of n-3 fatty acids replace n-6 fatty acids and cholesterol from cell membranes, leading to increased membrane fluidity, increased number of receptors, enhanced receptor binding and affinity, better ion channel functionality, and modulation of gene expression of many enzyme proteins involved in signal transduction processes (Farooqui 2009) leading to improvement in neurotransmission and signaling, which is important for optimal cognitive functioning and learning and memory (Farooqui 2009). Other dietary components, like vitamins B, C, and D have also protect visceral tissues and the brain from oxidative and inflammatory damage, and synaptic and neuronal loss (Farooqui 2014). However, when tested in a clinical setting supplementation with single nutrients is marginally effective in improving disease status (Chiu et al. 2008; Malouf et al. 2003; Farooqui 2014). It is proposed that approaches with multiple nutritional components may be more effective, since individual nutrients produce very little beneficial effects in neurological disorders (Farooqui 2013; Farooqui 2014).

10.3   Conclusion

Diet exerts potent effects on metabolism through a variety of regulatory mechanisms, resulting in local and systemic changes in metabolite levels. Nutrients modulate metabolism through regulatory mechanisms at the transcriptional, posttranscriptional, translational, or post-translational levels, resulting in changes at the tissue and/or systemic metabolite levels. Diet patterns have focused on mechanisms by which nutrients and disease states regulate metabolism at the gene or protein levels using genomic and proteomic approaches to identify metabolic diseases, which are influenced by dietary factors. Diet patterns have been reported to modulate the onset and pathogenesis of metabolic and age-related chronic neurological disorders. Thus, long term consumption of high calorie diet with high saturated fat, cholesterol, n-6 fatty acids, refined carbohydrates, and salt produces chronic oxidative stress and inflammation which are characterized by over production of ROS, AGE, eicosanoids, and proinflammatory cytokines along with declines in antioxidant capacity in visceral tissues and the brain. These processes not only contribute to cognitive decline, but also to neurodegeneration. In contrast, Mediterranean diet with low saturated fats and low refined sugars, but rich in olive oil, n-3 fatty acids, antioxidants vitamins C, E, and polyphenolic compounds produces neuroprotective effects supporting the view that specific dietary constituents of Mediterranean diet are able to influence the onset and development of above mentioned metabolic and neurological disorders. Saturated fats and n-6 fatty acids promote inflammation and oxidative stress whereas fruits, vegetables, n-3 fatty acids, and olive oil retard inflammation and oxidative stress.

It is proposed that in future human diet can be individualized for each individual. However, certain basic principles apply to most individuals. A healthful diet should include a wide variety of whole, unprocessed foods that are free of additives and, if possible, grown without the use of pesticides, herbicides, and other potentially toxic agricultural chemicals (Gaby 2013). For people who do not have specific food intolerances, such a diet generally includes liberal amounts of fresh fruits and vegetables, whole grains, nuts, seeds, and legumes. Humans should consume foods of animal origin (eggs, chicken, beef, and dairy products) in moderation. Supersizing of high calorie diet should be avoided. It is not necessary to consume foods of animal origin to maintain good health. In fact, compared with meat consuming individuals, vegetarians have a lower risk of developing a number of chronic diseases. However, vegetarians must carefully plan their diet and consume proteins of plant origins so that they do not develop nutritional deficiencies (Gaby 2013).

References

Anderson JW (1986) Fiber and health: an overview. Am J Gastroenterol 81:892–897

Aronson D (2008) Hyperglycemia and the pathobiology of diabetic complications. Adv Cardiol 45:1–16

Averill MM, Bornfeldt KE (2009) Lipids versus glucose in inflammation and the pathogenesis of macrovascular disease in diabetes. Curr Diab Rep 9:18–25

Bach A, Serra-Majem L, Carrasco JL, Roman B, Ngo J, Bertomeu I, Obrador B (2006) The use of indexes evaluating the adherence to the Mediterranean diet in epidemiological studies: a review. Public Health Nutr 9:132–146

Benton CR, Wright DC, Bonen A (2008) PGC-1alpha-mediated regulation of gene expression and metabolism: implications for nutrition and exercise prescriptions. Appl Physiol Nutr Metab 33:843–862

Bercik P, Denou E, Collins J, Jackson W, Lu J, Jury J, Deng Y, Blennerhassett P, Macri J, McCoy KD, Verdu EF, Collins SM (2011) The intestinal microbiota affect central levels of brain-derived neurotropic factor and behavior in mice. Gastroenterology 141:599–609

Boccardi V, Esposito A, Rizzo MR, Marfella R, Barbieri M, Paolisso G (2013) Mediterranean diet, telomere maintenance and health status among elderly. PLoS One 8:e62781

Borek C (2006) Garlic reduces dementia and heart-disease risk. J Nutr 136:810S–812S

Bough KJ, Wetherington J, Hassel B, Rare JF, Gawryluk JW, Greene JG, Shaw R, Smith Y, Geiger JD, Dingledine RJ (2006) Mitochondrial biogenesis in the anticonvulsant mechanism of the ketogenic diet. Ann Neurol 60:223–235

Castro-Quezada I, Román-Viñas B, Serra-Majem L (2014) The Mediterranean diet and nutritional adequacy: a review. Nutrients 6:231–248

Chandalia M, Abate N (2007) Metabolic complications of obesity: inflated or inflamed? J Diabetes Complications 21:128–136

Chaudhary DP, Sharma R, Bansal DD (2010) Implications of magnesium deficiency in type 2 diabetes: a review. Biol Trace Elem Res 134:119–129

Chiu CC, Su KP, Cheng TC, Liu HC, Chang CJ et al (2008) The effects of omega-3 fatty acids monotherapy in Alzheimer's disease and mild cognitive impairment: a preliminary randomized double-blind placebo-controlled study. Prog Neuropsychopharmacol Biol Psychiatry 32: 1538–1544

Comito D, Cascio A, Romano C (2014) Microbiota biodiversity in inflammatory bowel disease. Ital J Pediatr 40:32

Compare D, Coccoli P, Rocco A, Nardone OM, de Maria S, Cartenì M, Nardone G (2012) Gut—liver axis: the impact of gut microbiota on non alcoholic fatty liver disease. Nutr Metab Cardiovasc Dis 22:471–476

Conti M, Beavo J (2007) Biochemistry and physiology of cyclic nucleotide phosphodiesterases: essential components in cyclic nucleotide signaling. Annu Rev Biochem 76:481–511

Couto E, Boffetta P, Lagiou P, Ferrari P, Buckland G, Overvad K, Dahm CC, Tjonneland A, Olsen A, Clavel-Chapelon F, Boutron-Ruault MC, Cottet V, Trichopoulos D, Naska A, Benetou V, Kaaks R, Rohrmann S, Boeing H, von Ruesten A, Panico S, Pala V, Vineis P, Palli D, Tumino R, May A, Peeters PH, Bueno-de-Mesquita HB, Büchner FL, Lund E, Skeie G, Engeset D, Gonzalez CA, Navarro C, Rodríguez L, Sánchez MJ, Amiano P, Barricarte A, Hallmans G, Johansson I, Manjer J, Wirfärt E, Allen NE, Crowe F, Khaw KT, Wareham N, Moskal A, Slimani N, Jenab M, Romaguera D, Mouw T, Norat T, Riboli E, Trichopoulou A (2011) Mediterranean dietary pattern and cancer risk in the EPIC cohort. Br J Cancer 104:1493–1499

Craig WJ, Mangels AR (2009) Position of the American Dietetic Association: vegetarian diets. J Am Diet Assoc 109:1266–1282

Das S, Das DK (2007) Anti-inflammatory responses of resveratrol. Inflamm Allergy Drug Targets 6:168–173

Dasgupta B, Milbrandt J (2007) Resveratrol stimulates AMP kinase activity in neurons. Proc Natl Acad Sci U S A 104:7217–7222

Del Chierico F, Vernocchi P, Dallapiccola B, Putignani L (2014) Mediterranean diet and health: food effects on gut microbiota and disease control. Int J Mol Sci 15:11678–11699

Draznin B, Wang C, Adochio R, Leitner JW, Cornier M-A (2012) Effect of dietary macronutrient composition on AMPK and SIRT1 expression and activity in human skeletal muscle. Horm Metab Res 44:650–655

Estruch R, Ros E, Salas-Salvadó J, Covas MI, Corella D, Arós F, Gómez-Gracia E, Ruiz-Gutiérrez V, Fiol M, Lapetra J, Lamuela-Raventos RM, Serra-Majem L, Pintó X, Basora J, Muñoz MA, Sorlí JV, Martínez JA, Martínez-González MA, PREDIMED Study Investigators (2013) Primary prevention of cardiovascular disease with a Mediterranean diet. N Engl J Med 368:1279–1290

Farooqui AA (2009) Beneficial effects of fish oil human brain. Springer, New York

Farooqui AA (2010) Neurochemical aspects of neurotraumatic and neurodegenerative diseases. Springer, New York

Farooqui AA (2012) Phytochemical, signal transduction, and neurological disorders. Springer, New York

Farooqui AA (2013) Metabolic syndrome: an important risk factor for stroke, Alzheimer disease, and depression. Springer, New York

Farooqui AA (2014) Inflammation and Oxidative Stress in Neurological Disorders. Springer International Publishing, Switzerland

Farooqui AA, Horrocks LA, Farooqui T (2007) Modulation of inflammation in brain: a matter of fat. J Neurochem 101:577–599

Farrell GC, van Rooyen D, Gan L, Chitturi S (2012) NASH is an inflammatory disorder: pathogenic, prognostic and therapeutic implications. Gut Liver 6:149–171

Fenwick GR, Hanley AB (1985) The genus Allium. Part 2. Crit Rev Food Sci Nutr 22:273–377

Freeman J, Veggiotti P, Lanzi G, Tagliabue A, Perucca E, Institute of Neurology IRCCS C, Mondino Foundation (2006) The ketogenic diet: from molecular mechanisms to clinical effects. Epilepsy Res 68:145–180

Freeman BA, Baker PR, Schopfer FJ, Woodcock SR, Napolitano A, d'Ischia M (2008) Nitro-fatty acid formation and signaling. J Biol Chem 283:15515–15519

Gaby A (2013) A review of the fundamentals of diet. Glob Adv Health Med 2:58–63

Garcia-Marcos L, Castro-Rodriguez JA, Weinmayr G, Panagiotakos DB, Priftis KN, Nagel G (2013) Influence of Mediterranean diet on asthma in children: a systematic review and meta-analysis. Pediatr Allergy Immunol 24:330–338

Gasior M, Rogawski MA, Hartman AL (2006) Neuroprotective and disease-modifying effects of the ketogenic diet. Behav Pharmacol 17:431–439

Geleijnse JM, Giltay EJ, Grobbee DE, Donders AR, Kok FJ (2002) Blood pressure response to fish oil supplementation: metaregression analysis of randomized trials. J Hypertens 20:1493–1499

Georgousopoulou EN, Kastorini CM, Milionis HJ, Ntziou E, Kostapanos MS, Nikolaou V, Vemmos KN, Goudevenos JA, Panagiotakos DB (2014) Association between Mediterranean diet and non-fatal cardiovascular events, in the context of anxiety and depression disorders: a case/case-control study. Hell J Cardiol 55:24–31

Gibson CL, Murphy AN, Murphy SP (2012) Stroke outcome in the ketogenic state–a systematic review of the animal data. J Neurochem 123(Suppl 2):52–57

Golde TE (2009) The therapeutic importance of understanding mechanisms of neuronal cell death in neurodegenerative disease. Mol Neurodegener 4:8

Hardie DG, Ross FA, Hawley SA (2012) AMPK: a nutrient and energy sensor that maintains energy homeostasis. Nat Rev Mol Cell Biol 13:251–262

Harris WS (1997) n-3 fatty acids and serum lipoproteins: human studies. Am J Clin Nutr 65:1645S–1654S

Horrocks LA, Farooqui AA (2004) Docosahexaenoic acid in the diet: its importance in maintenance and restoration of neural membrane function. Prostaglandins Leukot Essent Fatty Acids 70:361–372

Hosseini E, Grootaert C, Verstraete W, van de Wiele T (2011) Propionate as a health-promoting microbial metabolite in the human gut. Nutr Rev 69:245–258

Hotamisligil GS (2006) Inflammation and metabolic disorders. Nature 444:860–867

Hruby A, McKeown NM, Song Y, Djoussé L (2013) Dietary magnesium and genetic interactions in diabetes and related risk factors: a brief overview of current knowledge. Nutrients 5:4990–5011

Iadecola C (2004) Neurovascular regulation in the normal brain and in Alzheimer's disease. Nat Rev Neurosci 5:347–360

Jacobs DR, Tapsell LC (2013) Food synergy: the key to a healthy diet. Proc Nutr Soc 2013:1–7

Jäer S, Handschin C, St-Pierre J, Spiegelman BM (2007) AMP-activated protein kinase (AMPK) action in skeletal muscle via direct phosphorylation of PGC-1α. Proc Natl Acad Sci U S A 104:12017–12022

Kennedy A, Martinez K, Chuang CC, LaPoint K, McIntosh M (2009) Saturated fatty acid-mediated inflammation and insulin resistance in adipose tissue: mechanisms of action and implications. J Nutr 139:1–4

Kim J, Hakim F, Kheirandish-Gozal L, Gozal D (2011) Inflammatory pathways in children with insufficient or disordered sleep. Respir Physiol Neurobiol 178:465–474

Lagouge M, Argmann C, Gerhart-Hines Z, Meziane H, Lerin C, Daussin F, Messadeq N, Milne J, Lambert P, Elliott P, Geny B, Laakso M, Puigserver P, Auwerx J (2006) Resveratrol improves mitochondrial function and protects against metabolic disease by activating SIRT1 and PGC-1alpha. Cell 127:1109–1122

LaManna JC, Salem N, Puchowicz M, Erokwu B, Koppaka S, Flask C, Lee Z (2009) Ketones suppress brain glucose consumption. Adv Exp Med Biol 645:301–306

Lavie CJ, Milani RV, Mehra MR, Ventura HO (2009) Omega-3 polyunsaturated fatty acids and cardiovascular diseases. J Am Coll Cardiol 54:585–594

Le Bon AM, Siess MH (2000) Organosulfur compounds from Allium and the chemoprevention of cancer. Drug Metabol Drug Interact 17:51–79

Leaf A, Kang JX, Xiao YF, Billman GE (2003) Clinical prevention of sudden cardiac death by n-3 polyunsaturated fatty acids and mechanism of prevention of arrhythmias by n-3 fish oils. Circulation 107:2646–2652

Lee JH, O'Keefe JH, Lavie CJ, Marchioli R, Harris WS (2008) Omega-3 fatty acids for cardioprotection. Mayo Clin Proc 83:324–332

Lopez-Miranda J, Delgado-Lista J, Perez-Martinez P, Jimenez-Gomez Y, Fuentes F, Ruano J, Marin C (2007) Olive oil and the haemostatic system. Mol Nutr Food Res 51:1249–1259

López-Miranda V, Soto-Montenegro ML, Vera G, Herradón E, Desco M, Abalo R (2012) Resveratrol: a neuroprotective polyphenol in the Mediterranean diet. Rev Neurol 54:349–356

Maafouf MA, Rho JM, Mattson MP (2009) The neuroprotective properties of calories restriction, the ketogenic diet, and ketone bodies. Brain Res Rev 59:293–315

Malaguarnera M, Vacante M, Antic T, Giordano M, Chisari G, Acquaviva R, Mastrojeni S, Malaguarnera G, Mistretta A, Li Volti G, Galvano F (2012) Bifidobacterium longum with fructo-oligosaccharides in patients with non alcoholic steatohepatitis. Dig Dis Sci 57:545–553

Malouf M, Grimley EJ, Areosa SA (2003) Folic acid with or without vitamin B12 for cognition and dementia. Cochrane Database Syst Rev CD004514.

Martinez-Gonzalez MA, Sánchez-Villegas A (2004) The emerging role of Mediterranean diets in cardiovascular epidemiology: MUFA, olive oil, red wine or the whole pattern? Eur J Epidemiol 19:9–13

Martinez-Gonzalez MA, Bes-Rastrollo M, Serra-Majem L, Lairon D, Estruch R, Trichopoulou A (2009) Mediterranean food pattern and the primary prevention of chronic disease: Recent developments. Nutr Rev 67:S111–S116

Milder JB, Liang LP, Patel M (2010) Acute oxidative stress and systemic Nrf2 activation by the ketogenic diet. Neurobiol Dis 40:238–244

Millward DJ (1999) The nutritional value of plant-based diets in relation to human amino acid and protein requirements. Proc Nutr Soc 58:249–260

Mitrou PN, Kipnis V, Thiebaut ACM, Reedy J, Subar AF, Wirfalt E, Flood A, Mouw T, Hollenbeck AR, Leitzmann MF, Schatzkin A (2007) Mediterranean dietary pattern and prediction of all-cause mortality in a US population: Results from the NIH-AARP diet and health study. Arch Intern Med 167:2461–2468

Moayyedi P, Ford AC, Talley NJ, Cremonini F, Foxx-Orenstein AE, Brandt LJ, Quigley EMM (2010) The efficacy of probiotics in the treatment of irritable bowel syndrome: A systematic review. Gut 59:325–332

Montonen J, Knekt P, Järvinen R, Aromaa A, Reunanen A (2003) Whole-grain and fiber intake and the incidence of type 2 diabetes. Am J Clin Nutr 77:622–629

Moriarty RM, Nathani R, Surve B (2007) Organosulfur compounds in cancer chemoprevention. Mini Rev Med Chem 7:827–838

Nakagawa T, Hu H, Zharikov S, Tuttle KR, Short RA, Glushakova O, Ouyang X, Feig DI, Block ER, Acosta JH, Patel JM, Johnson RJ (2006) A causal role for uric acid in fructose-induced metabolic syndrome. Am J Physiol 290:625–631

Newman JC, Verdin E (2014) Ketone bodies as signaling metabolites. Trends Endocrinol Metab 25:42–52

Okada H, Kuhn C, Feillet H, Bach JF (2010) The "hygiene hypothesis" for autoimmune and allergic diseases: an update. Clin Exp Immunol 160:1–9

Okuda N, Ueshima H, Okayama A, Saitoh S, Nakagawa H et al (2005) Relation of long chain n-3 polyunsaturated fatty acid intake to serum high density lipoprotein cholesterol among Japanese men in Japan and Japanese-American men in Hawaii: the INTERLIPID study. Atherosclerosis 178:371–379

Ottman N, Smidt H, de Vos WM, Belzer C (2012) The function of our microbiota: who is out there and what do they do? Front Cell Infect Microbiol 2:104

Oude Griep LM, Wang H, Chan Q (2013) Empirically-derived dietary patterns, diet quality scores, and markers of inflammation and endothelial dysfunction. Curr Nutr Rep 2:97–104

Padiya R, Khatua TN, BAGUL P, Kuncha M, Banerjee SK (2011) Garlic improves insulin sensitivity and associated metabolic syndromes in fructose fed rats. Nutr Metab 8:53

Park SJ, Ahmad F, Philp A, Baar K, Williams T, Luo H, Ke H, Rehmann H, Taussig R, Brown AL, Kim MK, Beaven MA, Burgin AB, Manganiello V, Chung JH (2012) Resveratrol ameliorates aging-related metabolic phenotypes by inhibiting cAMP phosphodiesterases. Cell 148:421–433

Pickup JC, Crook MA (1998) Is type II diabetes mellitus a disease of the innate immune system? Diabetologia 41:1241–1248

Pierre K, Pellerin L (2005) Monocarboxylate transporters in the central nervous system: distribution, regulation and function. J Neurochem 94:1–14

Politi K, Shemer-Meiri L, Shuper A, Aharoni S (2011) The ketogenic diet 2011: how it works. Epilepsy Res Treat 2011:963637

Qin J, Li Y, Cai Z, Li S, Zhu J, Zhang F, Liang S, Zhang W, Guan Y, Shen D, Peng Y, Zhang D, Jie Z, Wu W, Qin Y, Xue W, Li J, Han L, Lu D, Wu P, Dai Y, Sun X, Li Z, Tang A, Zhong S, Li X, Chen W, Xu R, Wang M, Feng Q, Gong M, Yu J, Zhang Y, Zhang M, Hansen T, Sanchez G, Raes J, Falony G, Okuda S, Almeida M, LeChatelier E, Renault P, Pons N, Batto JM, Zhang Z, Chen H, Yang R, Zheng W, Li S, Yang H, Wang J, Ehrlich SD, Nielsen R, Pedersen O, Kristiansen K, Wang J (2012) A metagenome-wide association study of gut microbiota in type 2 diabetes. Nature 490:55–60

Rader DJ (2000) Inflammatory markers of coronary risk. N Engl J Med 343:1179–1182

Rahimi R, Nikfar S, Larijiani B, Abdollahi M (2005) A review on the role of antioxidants in the management of diabetes and its complications. Biomed Pharmacother 59:365–373

Rahman I, Biswas SK, Kirkham PA (2006) Regulation of inflammation and redox signaling by dietary polyphenols. Biochem Pharmacol 72:1439–1452

Ruderman NB, Xu XJ, Nelson L et al (2010) AMPK and SIRT1: a long-standing partnership? Am J Physiol Endocrinol Metab 298:E751–E760

Rybak ME, Calvey EM, Harnly JM (2004) Quantitative determination of allicin in garlic: Supercritical fluid extraction and standard addition of alliin. J Agric Food Chem 52:682–687

Sacks FM, Katan M (2002) Randomized clinical trials on the effects of dietary fat and carbohydrate on plasma lipoproteins and cardiovascular disease. Am J Med 113(Suppl 9):B13S–B24S

Salamone F, Li Volti G, Titta L, Puzzo L, Barbagallo I, la Delia F, Zelber-Sagi S, Malaguarnera M, Pelicci PG, Giorgio M, Galvano F (2012) Moro orange juice prevents fatty liver in mice. World J Gastroenterol 18:3862–3868

Salas-Salvadó J, Bulló M, Babio N, Martínez-González MÁ, Ibarrola-Jurado N, Basora J, Estruch R, Covas MI, Corella D, Arós F, Ruiz-Gutiérrez V, Ros E, PREDIMED Study Investigators (2011) Reduction in the incidence of type 2 diabetes with the Mediterranean diet: results of the PREDIMED-Reus nutrition intervention randomized trial. Diabetes Care 34:14–19

Salas-Salvado J, Bullo M, Estruch R, Ros E, Covas M-I, Ibarrola-Jurado N, Corella D, Aros F, Gomez-Gracia E, Ruiz-Gutierrez V, Romaguera D, Lapetra J, Lamuela-Raventós RM, Serra-Majem L, Pintó X, Basora J, Muñoz MA, Sorlí JV, Martínez-González MA (2014) Prevention of diabetes with Mediterranean diets: a subgroup analysis of a randomized trial. Ann Intern Med 160:1–10

Scarmeas N, Stern Y, Mayeux R, Luchsinger JA (2006) Mediterranean diet, Alzheimer disease, and vascular mediation. Arch Neurol 63:1709–1717

Scoditti E, Calabriso N, Massaro M, Pellegrino M, Storelli C, Martines G, De Caterina R, Carluccio MA (2012) Mediterranean diet polyphenols reduce inflammatory angiogenesis through MMP-9 and COX-2 inhibition in human vascular endothelial cells: a potentially protective mechanism in atherosclerotic vascular disease and cancer. Arch Biochem Biophys 527:81–89

Shoelson SE, Herrero L, Naaz A (2007) Obesity, inflammation, and insulin resistance. Gastroenterology 132:2169–2180

Sizonenko SV, Babiloni C, Sijben JW, Walhovd KB (2013) Brain imaging and human nutrition: which measures to use in intervention studies? Adv Nutr 4:554–556

Skoog I, Gustafson D (2006) Update on hypertension and Alzheimer's disease. Neurol Res 28:605–611

Sofi F, Cesari F, Abbate R, Gensini GF, Casini A (2008) Adherence to Mediterranean diet and health status: meta-analysis. Br Med J 337:673–675

Sofi F, Macchi C, Abbate R, Gensini GF, Casini A (2013) Mediterranean diet and health. Biofactors 39:335–342

Sokol H, Seksik P, Furet JP, Firmesse O, Nion-Larmurier I, Beaugerie L, Cosnes J, Corthier G, Marteau DJ (2009) Low counts of *Faecalibacterium prausnitzii* in colitis microbiota. Inflamm Bowel Dis 15:1183–1189

Solomon A, Kivipelto M, Wolozin B, Zhou J, Whitmer RA (2009) Midlife serum cholesterol and increased risk of Alzheimer's and vascular dementia three decades later. Dement Geriatr Cogn Disord 28:75–80

Sullivan PG, Rippy NA, Dorenbos K, Concepcion RC, Agarwal AK, Rho JM (2004) The ketogenic diet increases mitochondrial uncoupling protein levels and activity. Ann Neurol 55:576–580

Suzuki M, Suzuki M, Sato K, Dohi S, Sato T, Matsuura A, Hiraide A (2001) Effect of beta-hydroxybutyrate, a cerebral function improving agent, on cerebral hypoxia, anoxia and ischemia in mice and rats. Jpn J Pharmacol 87:143–150

Thorburn A, Muir J, Proietto J (1993) Carbohydrate fermentation decreases hepatic glucose output in healthy subjects. Metabolism 42:780–785

Tiwari AK (2014) Revisiting "Vegetables" to combat modern epidemic of imbalanced glucose homeostasis. Pharmacogn Mag 10(Suppl 2):S207–S213

Tiwari AK, Kumar DA, Sweeya PS, Chauhan HA, Lavanya V, Sireesha K, Pavithra K, Zehra A (2014) Vegetables' juice influences polyol pathway by multiple mechanisms in favour of reducing development of oxidative stress and resultant diabetic complications. Pharmacogn Mag 10(Suppl 2):S383–S391

Tremaroli V, Backhed F (2012) Functional interactions between the gut microbiota and host metabolism. Nature 489:242–249

Trostchansky A, Rubbo H (2008) Nitrated fatty acids: mechanisms of formation, chemical characterization, and biological properties. Free Radic Biol Med 44:1887–1896

Uccello M, Malaguarnera G, Basile F, D'agata V, Malaguarnera M, Bertino G, Vacante M, Drago F, Biondi A (2012) Potential role of probiotics on colorectal cancer prevention. BMC Surg 2012:S35

Van der Auwera I, Wera S, Van Leuven F, Henderson ST (2005) A ketogenic diet reduces amyloid beta 40 and 42 in a mouse model of Alzheimer's disease. Nutr Metab (Lond) 2:28

Vauzour D, Rodriguez-Mateos A, Corona G, Oruna-Concha MJ, Spencer JPE (2010) Polyphenols and human health: prevention of disease and mechanisms of action. Nutrients 2:1106–1131

Volpe SL (2013) Magnesium in disease prevention and overall health. Adv Nutr 4:378S–383S

Weiss R, Dziura J, Burgert TS, Tamborlane WV, Taksali SE, Yeckel CW, Allen K, Lopes M, Savoye M, Morrison J, Sherwin RS, Caprio S (2004) Obesity and the metabolic syndrome in children and adolescents. N Engl J Med 350:2362–2374

Whitmer RA, Gustafson DR, Barrett-Connor E, Haan MN, Gunderson EP et al (2008) Central obesity and increased risk of dementia more than three decades later. Neurology 71:1057–1064

Willcox DC, Willcox BJ, Todoriki H, Suzuki M (2009) The Okinawan diet: health implications of a low-calorie, nutrient-dense, antioxidant-rich dietary pattern low in glycemic load. J Am Coll Nutr 28(Suppl):500S–516S

Willett WC, Sacks F, Trichopoulou A, Drescher G, Ferro-Luzzi A, Helsing E, Trichopoulos D (1995) Mediterranean diet pyramid: a cultural model for healthy eating. Am J Clin Nutr 61(Suppl):1402S–1406S

Williamson DL (2011) Normalizing a hyperactive mTOR initiates muscle growth during obesity. Aging 3:83–84

Yoon JC, Puigserver P, Chen G, Donovan J, Wu Z, Rhee J, Adelmant G, Stafford J, Kahn CR, Granner DK, Newgard CB, Spiegelman BM (2001) Control of hepatic gluconeogenesis through the transcriptional coactivator PGC-1. Nature 413:131–138

Yu J, Auwerx J (2010) Protein deacetylation by SIRT1: an emerging key post-translational modification in metabolic regulation. Pharmacol Res 62:35–41

Zhao Z, Lange DJ, Voustianiouk A, MacGrogan D, Ho L et al (2006) A ketogenic diet as a potential novel therapeutic intervention in amyotrophic lateral sclerosis. BMC Neurosci 7:29

Ziegler DR, Ribeiro LC, Hagenn M, Siqueira IR, Araujo E, Torres IL, Gottfried C, Netto CA, Goncalves CA (2003) Ketogenic diet increases glutathione peroxidase activity in rat hippocampus. Neurochem Res 28:1793–1797

Index

© Springer International Publishing Switzerland 2015

A.A. Farooqui, *High Calorie Diet and the Human Brain*,

DOI 10.1007/978-3-319-15254-7

CPSIA information can be obtained
at www.ICGtesting.com
Printed in the USA
LVOW02*2136171215

467050LV00001B/40/P